EMBRYONIC DEV

Day 5 LATE BLASTOCYST

Trophoblast

Inner cell mass

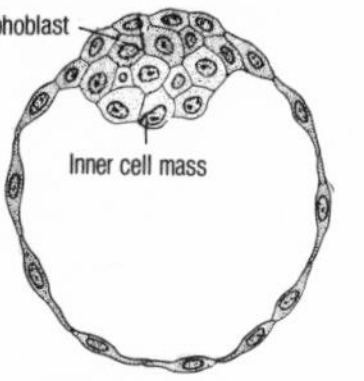

Days 6-7 EVENTS DURING FIRST WEEK

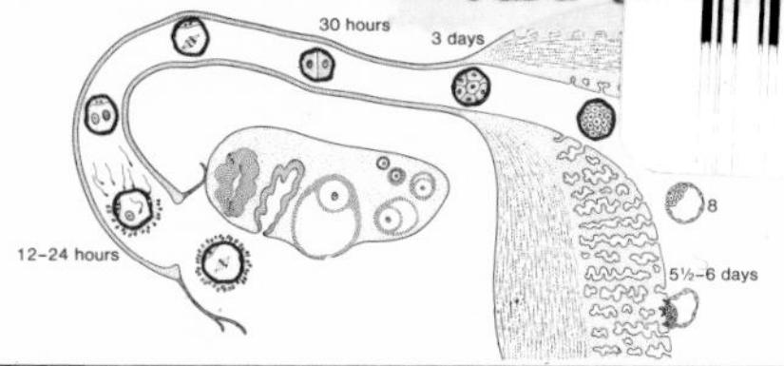

30 hours

3 days

12–24 hours

8

5½–6 days

1

Day 12

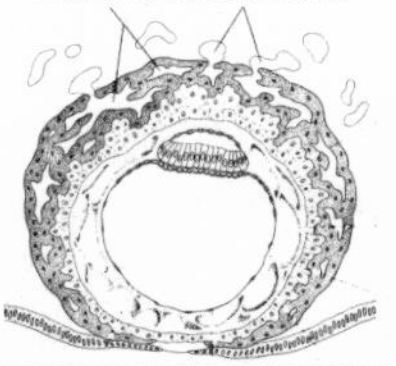

Maternal and Trophoblast Vessels

Day 13 UTEROPLACENTAL CIRCULATION BEGINS

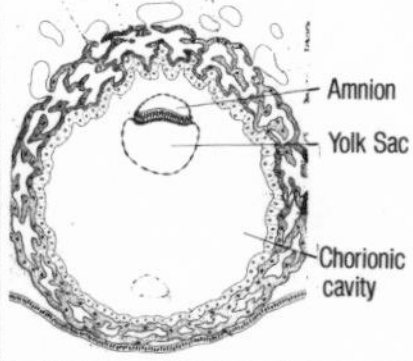

Amnion

Yolk Sac

Chorionic cavity

Day 14 EMBRYONIC DISC SEEN FROM DORSAL

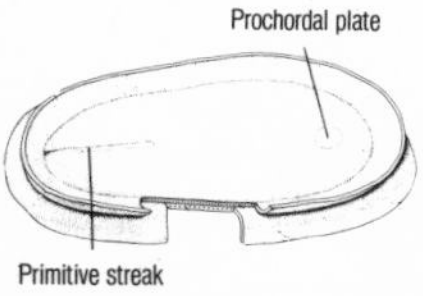

Prochordal plate

Primitive streak

DEVELOPMENT WEEK 2

Day 19 FORMATION CNS

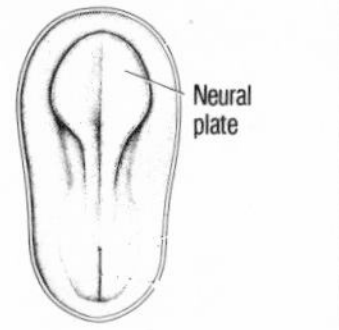

Neural plate

Day 20 APPEARANCE OF SOMITES

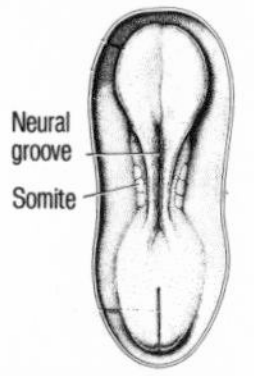

Neural groove

Somite

Day 21 TRANSVERSE SECTION THROUGH SOMITE REGION

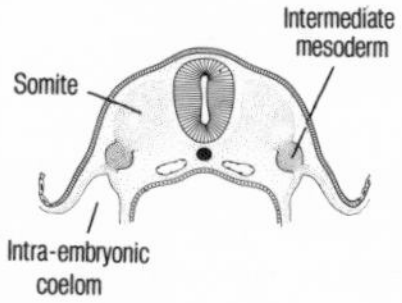

Intermediate mesoderm

Somite

Intra-embryonic coelom

DEVELOPMENT WEEK 3

Day 26 BRANCHIAL ARCHES

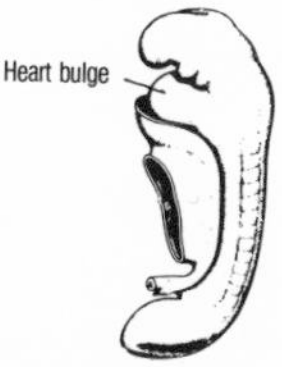

Heart bulge

Day 27

Approx. Age	No. of Somites
days	
20	1–4
21	4–7
22	7–10
23	10–13
24	13–17
25	17–20
26	20–23
27	23–26
28	26–29
30	34–35

Day 28

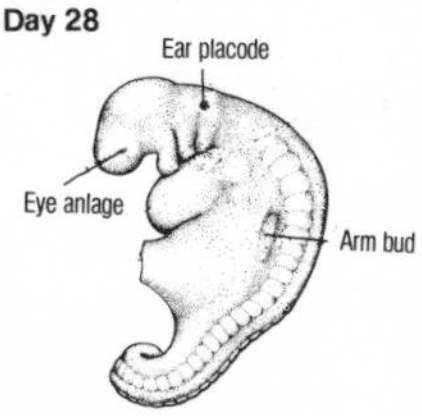

Ear placode

Eye anlage

Arm bud

DEVELOPMENT WEEK 4

Day 33

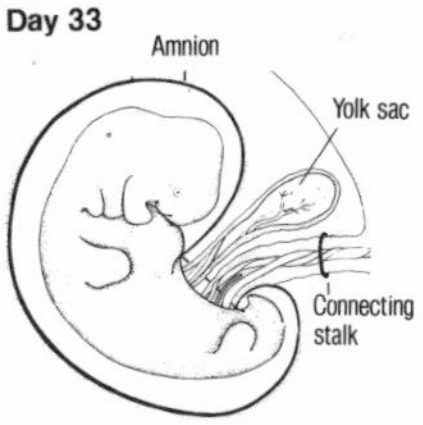

Amnion

Yolk sac

Connecting stalk

Day 34 DEVELOPING LIMB BUDS

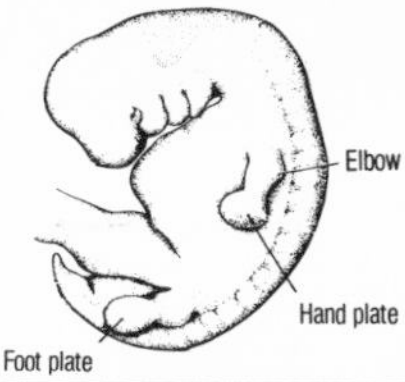

Elbow

Hand plate

Foot plate

Day 35 BRANCHIAL ARCHES AND CLEFTS

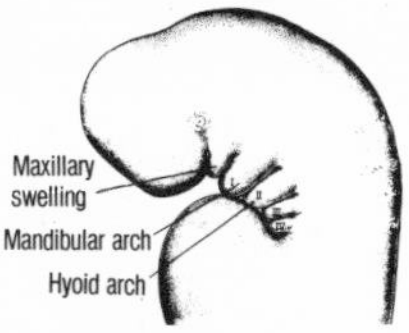

Maxillary swelling

Mandibular arch

Hyoid arch

DEVELOPMENT WEEK 5

Day 40 EMBRYO IN UTERO

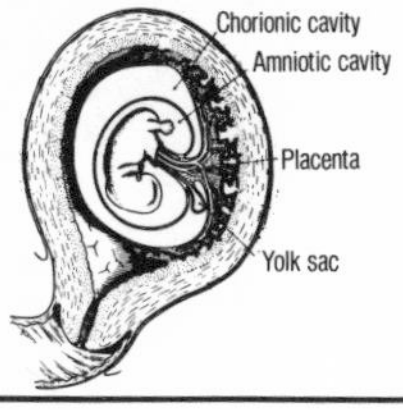

Chorionic cavity

Amniotic cavity

Placenta

Yolk sac

Day 41

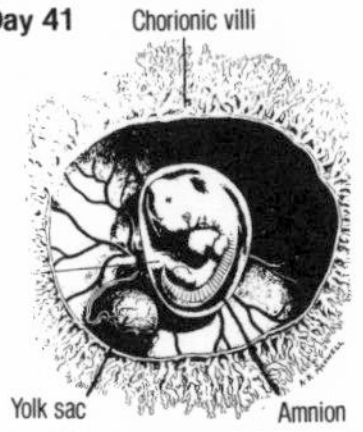

Chorionic villi

Yolk sac

Amnion

Day 42

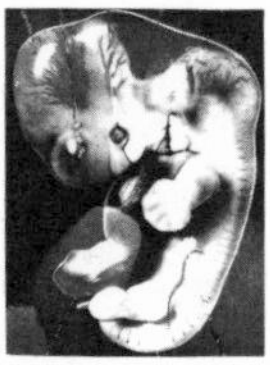

DEVELOPMENT WEEK 6

Day 47

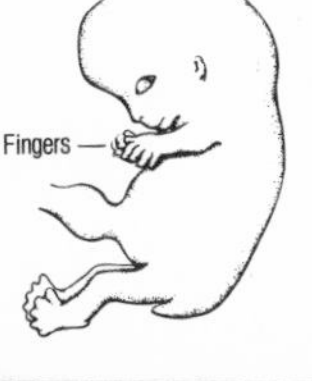

Fingers

Day 48

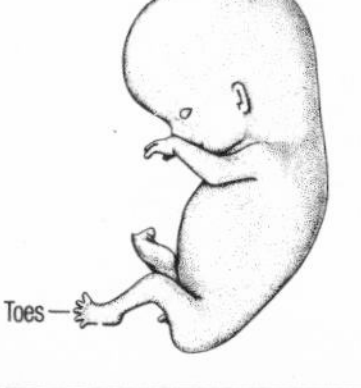

Toes

Day 49 FETAL MEMBRANES IN THIRD MONTH

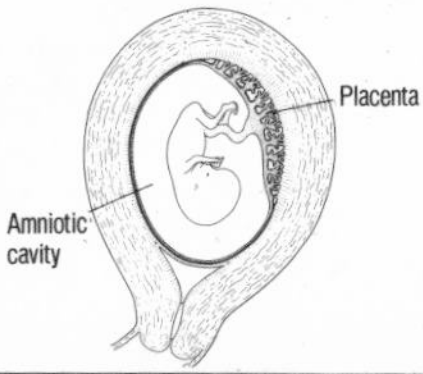

Placenta

Amniotic cavity

DEVELOPMENT WEEK 7

MEDICAL EMBRYOLOGY

Fourth Edition

MEDICAL EMBRYOLOGY

Fourth Edition

JAN LANGMAN, M.D., PH.D.
Professor and Chairman,
Department of Anatomy
University of Virginia
Charlottesville, Virginia

Original Illustrations by Jill Leland

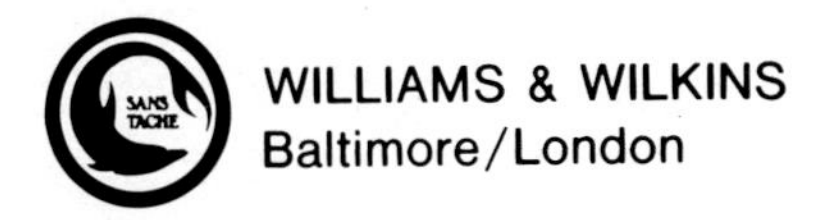

WILLIAMS & WILKINS
Baltimore/London

Made in the United States of America

First edition, 1963
Reprinted 1964, 1965, 1966, 1967

Second edition, 1969
Reprinted 1970, 1971, 1972, 1973

Third edition, 1975
Reprinted 1975, 1976, 1977, 1979, 1980

Fourth edition, 1981
Reprinted 1982, 1983, 1984

English language co-editions
Asian, 1966, 1969, 1975
Indian, 1966, 1969, 1976
Korean student, 1966
International student, 1973

Translations
Dutch, 1966, 1970, 1975, 1981
French, 1965, 1971, 1976
German, 1970, 1975
Indonesian, 1981
Italian, 1967, 1970, 1978
Japanese, 1967, 1971, 1976
Persian, 1971
Portuguese, 1966, 1970, 1977
Spanish, 1963, 1969, 1976, 1981

Library of Congress Cataloging in Publication Data

Langman, Jan.
Medical embryology.

Includes bibliographical references and index.
1. Embryology, Human. 2. Abnormalities, Human—Genetic aspects. I. Title.
QM601.L35 1980 612'.64 80-25847
ISBN 0-683-04858-9

Composed and printed at the
Waverly Press, Inc.
Mt. Royal and Guilford Aves.
Baltimore, MD 21202, U.S.A.

To my wife
INA
in appreciation of her great support during
the writing of this book

Preface to Fourth Edition

The worldwide acceptance of the previous editions of *Medical Embryology*, as reflected in its many foreign translations, indicates that the book continues to fill a great need among students preparing themselves for a medical career. As a result of recommendations from students and colleagues, this edition has been changed considerably. In the first place, the medically important congenital abnormalities are now interwoven with the normal development, and not discussed in separate subchapters. This provides a better coordination and integration between normal and abnormal development. Secondly, a summary has been added at the end of each chapter so that the student quickly can review the important points covered in the foregoing pages. Thirdly, and maybe most importantly, the material has been rearranged to correlate better with gross anatomy. Therefore, some chapters have been shortened and a few new ones have been added. This change makes it easier to integrate the gross structures of the thorax, head and neck, abdomen and pelvis with its normal and abnormal development. A fourth important change is the addition of color to almost all drawings. We have tried to give a particular color to each of the germ layers and to maintain this color for most of its derivatives throughout the book, thus making it easier to recognize the origin of certain organs and structures. The fifth change concerns the addition of photographic charts at the beginning and end of the book. The first chart gives a day-to-day recapitulation of the important facts of the first seven weeks of development. The other charts make it possible to correlate normal and abnormal development and to determine on what day or week an abnormality has developed.

Hence, the changes in this edition, which even concern its outside appearance, will make this a better illustrated, more integrated, more concise and, hopefully, a better book to study normal development and its clinical applications.

As in the previous editions almost all the illustrations were made by Miss Jill Leland; the new drawings and photographs were made by Betsy Cochran and Barbara Haynes. In particular, I wish to express my thanks to Rudolf Johannes Müller from Berlin who did the color work and some new drawings in an outstanding manner. I also wish to express my great gratitude to Sara Finnegan, Vice President and Editor-in-Chief of the Williams & Wilkins Co., who for so many years has spared no effort to make this book a success in its field. Thanks also to Nan Tyler, George Stamathis and Patrick Hudson for their excellent cooperation and outstanding qualities in the production of this edition. My special thanks to Dr. Thomas W. Sadler, who will become my co-author with the fifth edition of *Medical Embryology*.

Charlottesville, Va.
November 1980

Preface to First Edition

Recent advances in embryology, radioautography, and electron microscopy have been so overwhelming that the medical student often has difficulty in grasping the basic facts of development from the highly complicated picture presented to him. The aim of this book, therefore, is to give the future doctor a concise, well illustrated presentation of the essential facts of human development, clarifying the gross anatomical features without omitting the recent advances or changing concepts in the basic sciences. Furthermore, since embryology has become of great practical value because of the enormous progress made in surgery and teratology, each chapter on the development of the organ systems has been complemented by a description of those malformations important to the student in his further training. As a further reflection of the increased clinical importance of embryology an entire chapter has been devoted to the etiology of congenital defects.

Of the many colleagues who have been of help in the writing of this book, I particularly wish to thank Dr. C. P. Leblond for his continuous interest and encouragement; Dr. F. Clarke Fraser, for his help in discussing the various aspects of the congenital malformations; and my friends, Dr. Harry Maisel, Dr. Robert van Mierop, and Dr. Yves Clermont, who have spared no effort in assisting with the design of the drawings and the checking of the text.

I wish to express my sincere thanks to Miss Jill Leland, who prepared all the illustrations in this book, and to Mrs. E. Dawson, who has been of such excellent support to me in setting up the manuscript.

Contents

PART 1
General Embryology

PART 2
Special Embryology

PART 1

General Embryology

chapter 1

Gametogenesis

The development of a human being begins with fertilization, a process by which the **spermatozoon** from the male and the **oocyte** from the female unite to give rise to a new organism, the **zygote**. In preparation for fertilization, both male and female germ cells undergo a number of changes involving the chromosomes as well as the cytoplasm. The purpose of these changes is twofold:

(1) To reduce the number of chromosomes to half that in the normal somatic cell, *i.e.*, from 46 to 23. This is accomplished by the **meiotic** or **maturation** divisions and is necessary, since otherwise fusion of a male and a female germ cell would result in an individual with twice the number of chromosomes of the parent cells.

(2) To alter the shape of the germ cells in preparation for fertilization. The male germ cell, initially large and round, loses practically all of its cytoplasm and develops a head, neck, and tail. The female germ cell, on the contrary, gradually becomes larger as the result of an increase in the amount of cytoplasm. At maturity the oocyte has a diameter of about 120 μ.

The human somatic cell contains 23 pairs or a diploid (diploos-double) number of chromosomes. One chromosome of each pair is originally derived from the mother and the other from the father. The members of a chromosome pair are generally not in close proximity to each other either in the resting cell or during the mitotic division. The only time that they come in intimate contact with each other is during the meiotic or maturation divisions of the germ cells.

To make the events occurring during the meiotic divisions easier to understand, the most important features of these divisions are compared with those of a mitotic division. Similarly, although reduction in the number of chromosomes and the cytoplasmic changes are both integral parts of germ cell maturation, each process is described separately.

Chromosomes During Mitotic Division

Before a normal somatic cell enters mitosis, each chromosome replicates its DNA and in fact becomes doubled. During the DNA-duplication phase the chromosomes are extremely long, diffusely spread through the cytoplasm, and cannot be recognized with the light microscope. With the onset of mitosis, the chromosomes begin to coil, contract, and condense, but the two paired subunits (chromatids) can still not be recognized as individual units (fig. 1-1*A*). Only in the prometaphase when the chromosomes become compact rods are the chromatids distinguishable (fig. 1-1*B*). During the metaphase the chromosomes line up in the equatorial plane and their doubled structure is clearly visible (fig. 1-1*C*). Soon afterwards each chromosome undergoes a longitudinal division of the centromere and separates into two daughter chromosomes which migrate to opposite poles of the cell (fig. 1-1*D,E*). Each daughter cell receives one half of all the doubled chromosome material and thus maintains the same number of chromosomes as the mother cell.

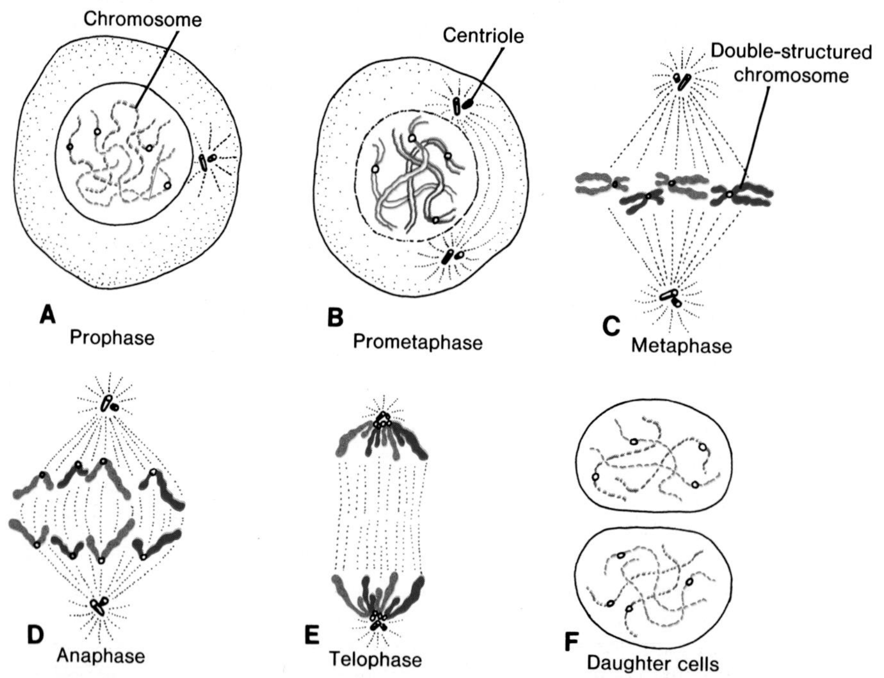

Figure 1-1. Schematic drawing of the various stages of the mitotic division. In the prophase the chromosomes are visible as slender threads. The doubled chromatids become clearly visible as individual units during the prometaphase. At no time during the division do the members of a chromosome pair unite. Blue—paternal chromosomes; red—maternal chromosomes.

Chromosomes During Meiotic Divisions

FIRST MEIOTIC DIVISION

Similarly as in a mitotic division the female as well as the male primitive germ cells (primary oocyte and primary spermatocyte) replicate their DNA just before the first meiotic division begins. Hence, at the beginning of the maturation divisions the germ cells contain double the normal amount of DNA (4n) and each of the 46 chromosomes is a double structure (fig. 1-2).

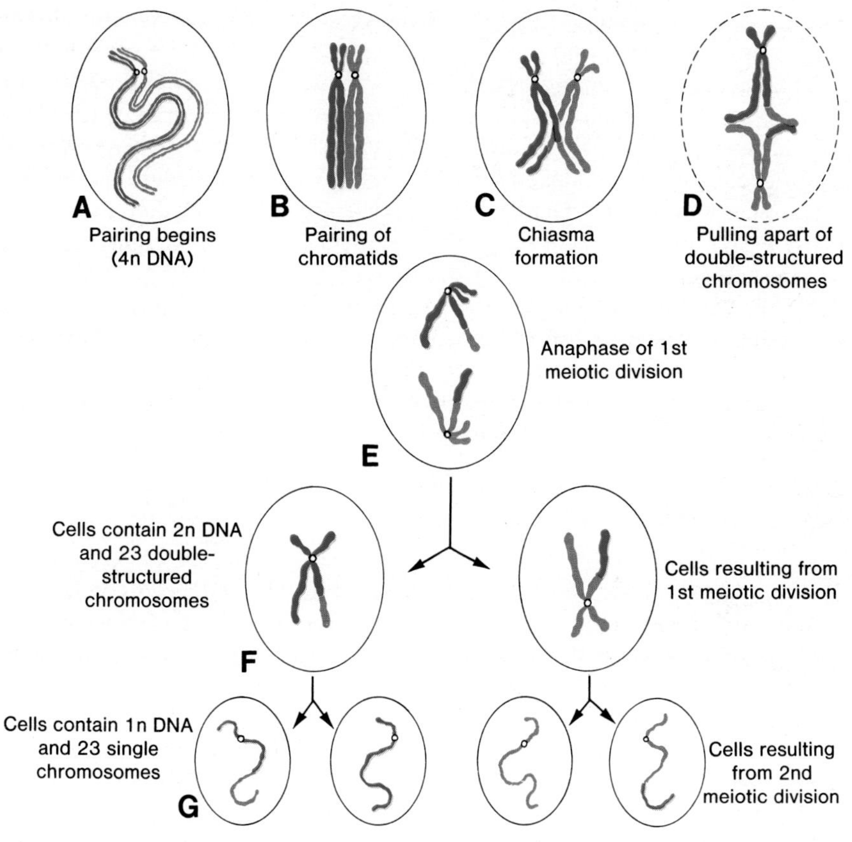

Figure 1-2. Schematic representation of the first and second meiotic divisions. *A*, The homologous chromosomes approach each other. *B*, The homologous chromosomes pair and each member of the pair consists of two chromatids. *C*, The intimately paired homologous chromosomes interchange chromatid fragments. Note the chiasma. *D*, The double-structured chromosomes pull apart. *E*, Anaphase of the first meiotic division. *F* and *G*, During the second meiotic division the double-structured chromosomes split at the centromere. At completion of the division the chromosomes in each of the four daughter cells are different from each other.

The *first characteristic feature* of this meiotic division is the **pairing of the homologous chromosomes** (fig. 1-2*A*). The pairing is exact and point for point, except for the X-Y combination. The centromere regions of the homologous chromosomes do not pair. Since each individual chromosome is double-structured and contains two chromatids, the homologous pair consists of four chromatids (fig. 1-2*B*). In a mitotic division the homologous chromosomes never pair.

The *second characteristic feature* of the first meiotic division is the **interchange of chromatid segments** between the two paired homologous chromosomes (fig. 1-2*C*). When subsequently each (double-structured) member of the homologous pair splits longitudinally, one or more transverse breaks occur in the chromatids and an interchange of chromatid segments between two homologous chromosomes occurs (fig. 1-2*C*). During the separation of the homologous chromosomes, the points of interchange temporarily remain united and the chromosomal structure has then an X appearance, known as a **chiasma** (fig. 1-2*C*). During the chiasma stage blocks of genes are exchanged between homologous chromosomes. In the meantime the separation continues and the two members of each pair become oriented on the spindle (fig. 1-2*D*). In subsequent stages the members migrate to the opposite poles of the cell (fig. 1-2*E*).

After the first meiotic division has been completed, each daughter cell contains one member of each chromosome pair and thus has 23 double-structured chromosomes (fig. 1-2*F*). Since each chromosome is still double-structured except at the centromere, the amount of DNA in each daughter cell equals that of a normal somatic cell (2n).

SECOND MEIOTIC DIVISION

Immediately after the first meiotic division, the cell begins its second maturation division. In contrast to the first meiotic division **no DNA synthesis occurs in advance of this division**. The 23 double-structured chromosomes divide at the centromere and each of the newly formed daughter cells receives 23 chromatids (fig. 1-2*G*). The amount of DNA in the newly formed cells is now half that of the normal somatic cell. Hence, the purpose of the two meiotic or maturation divisions is twofold: (1) to enable the members of the homologous chromosome pair to exchange blocks of genetic material (first meiotic division); and (2) to provide each germ cell with both a haploid number of chromosomes and half the amount of DNA of a normal somatic cell (second meiotic division).

As a result of the meiotic divisions, one primary oocyte eventually gives rise to four daughter cells, each with 22 + 1 X-chromosomes (fig. 1-3*A*). Only one of these develops into a mature gamete, the oocyte; the other three, the **polar bodies**, receive hardly any cytoplasm and degenerate during subsequent development.

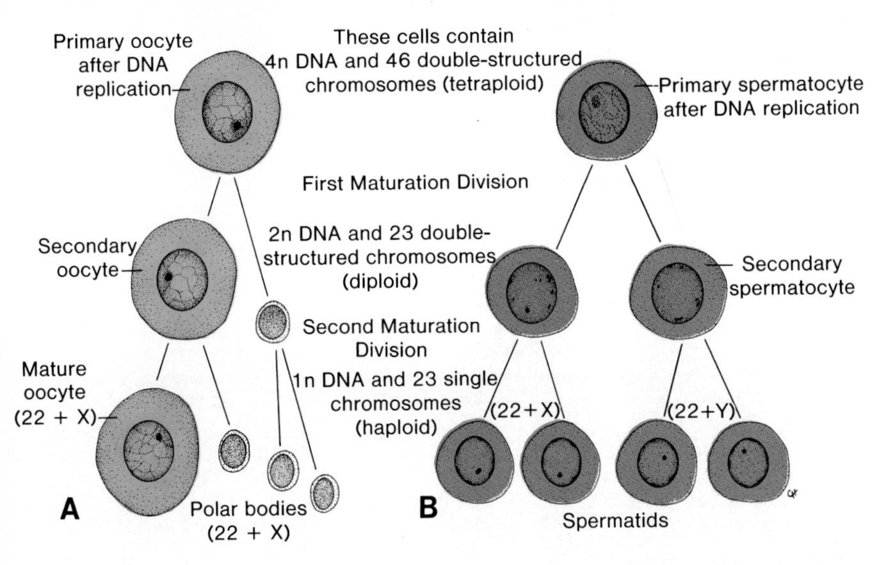

Figure 1-3. Schematic drawing showing the events occurring during the first and second maturation divisions. *A*, The primitive female germ cell (primary oocyte) produces only one mature gamete, the mature oocyte. *B*, The primitive male germ cell (primary spermatocyte) produces four spermatids, all of which develop into spermatozoa.

The primary spermatocyte gives rise to four daughter cells; two with 22 + 1 X-chromosomes and two with 22 + 1 Y-chromosomes (fig. 1-3*B*). All four develop into mature gametes.

ABNORMAL MEIOTIC DIVISIONS

The events occurring during the meiotic divisions apparently are not without hazards. No sooner was the normal chromosome pattern in man established when it became evident that some people possessed an abnormal number of chromosomes.

Chromosomal abnormalities originate during the meiotic divisions. Normally the two members of a homologous chromosome pair separate during the first meiotic division so that each daughter cell receives one component of each pair (fig. 1-4*A*). Sometimes, however, separation does not occur (**nondisjunction**), and both members of a pair then move into one cell (fig. 1-4*B*). As a result of the nondisjunction of the chromosomes, one cell receives 24 chromosomes and the other 22, instead of the normal 23 chromosomes. When, at fertilization, a gamete having 23 chromosomes fuses with a gamete having 24 or 22 chromosomes, the result will be an individual with either 47 chromosomes (trisomy), or 45 chromosomes (monosomy). Nondisjunction is thought to occur during the first or second meiotic division of the female germ cells rather than during the divisions of the male germ cells (fig. 1-4*C*). For further details see Chapter 8.

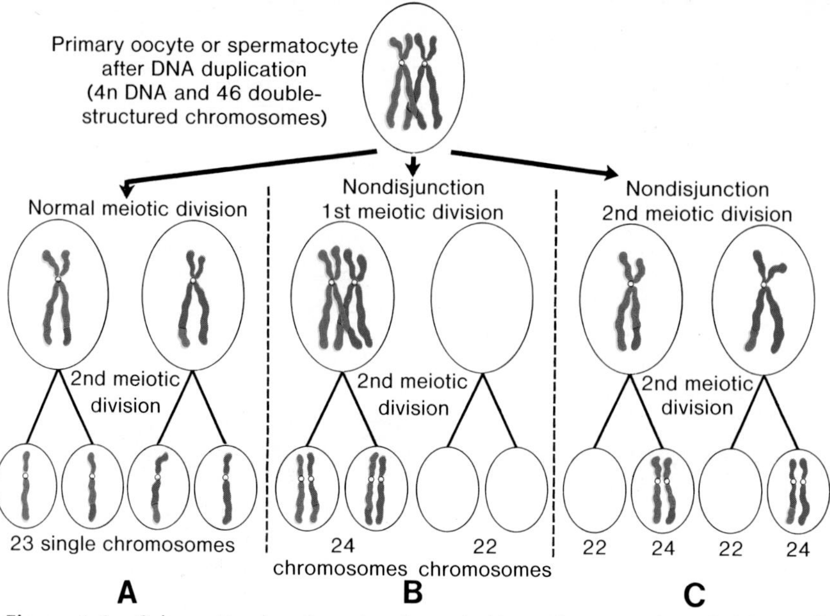

Figure 1-4. Schematic drawing showing: *A*, Normal maturation divisions. *B*, Nondisjunction in first meiotic division. *C*, Nondisjunction in second meiotic division.

Morphological Changes during Maturation

PRIMORDIAL GERM CELLS

The mature male and female germ cells are direct descendants of the primordial germ cells, which in human embryos appear in the wall of the yolk sac at the end of the third week of development (fig. 1-5).[1] These cells migrate by ameboid movement from the yolk sac toward the developing gonads (primitive sex glands), where they arrive at the end of the fourth or the beginning of the fifth week.[2–4] (See Chapter 15, fig. 15-15.)

OOGENESIS

Prenatal Maturation. Once the primordial germ cells have arrived in the gonad of a genetic female, they differentiate into **oogonia** (fig. 1-6*A*, *B*). These cells undergo a number of mitotic divisions, and by the end of the third month they become arranged in clusters which are surrounded by a layer of flat epithelial cells (fig. 1-7*A*). While all the oogonia in one cluster are probably derived from a single primordial germ cell, the flat epithelial cells are believed to originate from the surface epithelium covering the ovary.

The majority of the oogonia continue to divide by mitosis, but some of them differentiate into the much larger **primary oocytes**. Immediately after their formation, they replicate their DNA and enter the prophase of the first meiotic division (figs. 1-6*C* and 1-7*A*).[5–8] During the next few months the

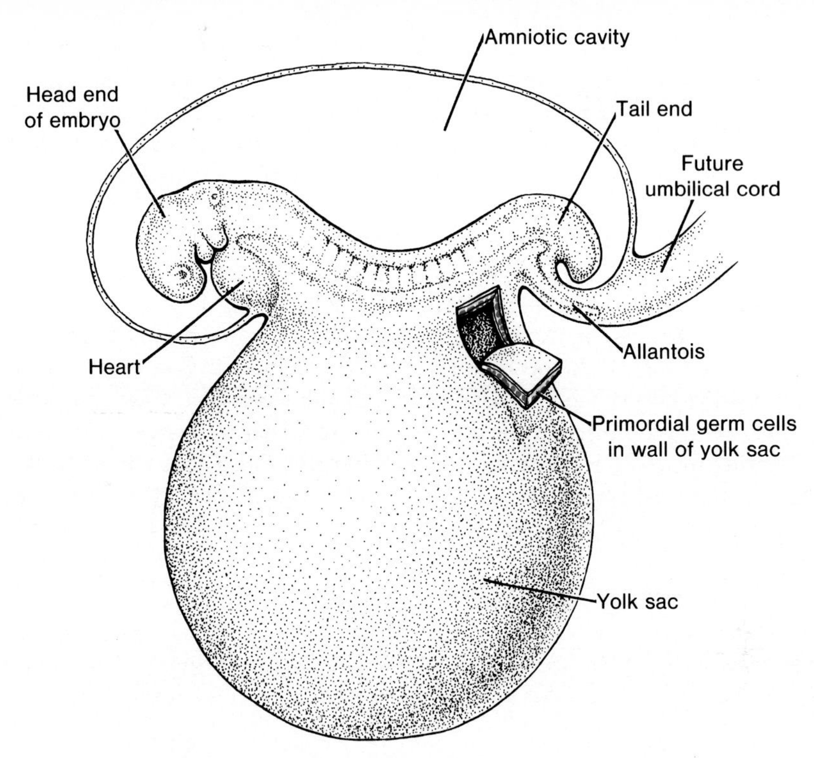

Figure 1-5. Drawing of a three-week-old embryo, showing the position of the primordial germ cells in the wall of the yolk sac, close to the attachment of the future umbilical cord.

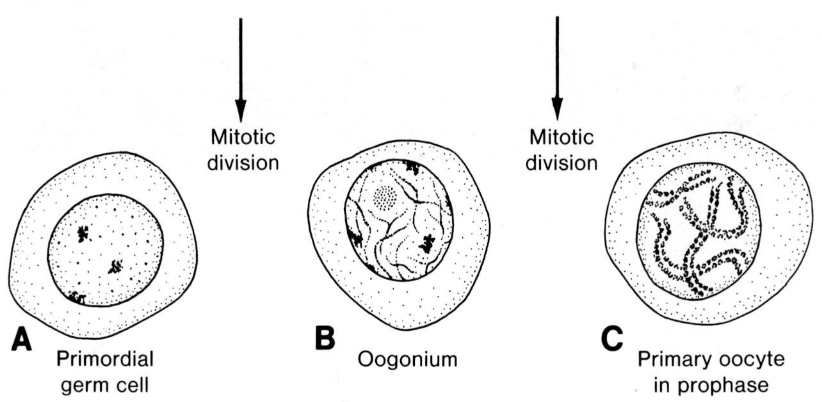

Figure 1-6. Differentiation of primordial germ cells into oogonia begins shortly after their arrival in the ovary. By the third month of development some oogonia give rise to primary oocytes which almost immediately begin with the prophase of the first meiotic division. This prophase may last 40 or more years and will finish only when the cell begins its final maturation. During this period it carries 46 double-structured chromosomes and therefore has 4n DNA.

oogonia increase rapidly in number, and by the fifth month of development the total number of germ cells in the ovary reaches its maximum, estimated at 7,000,000. At this time cell degeneration begins and many oogonia as well as primary oocytes become atretic. By the seventh month, the majority of the oogonia have degenerated, with the exception of a few near the surface. All surviving primary oocytes, however, have entered the first meiotic division and most of them are now individually surrounded by a layer of flat epithelial cells (fig. 1-7*B*). A primary oocyte, together with its surrounding flat epithelial cells, is known as a **primordial follicle** (fig. 1-8*A*).

Postnatal Maturation. Approximately at the time of birth all primary oocytes have finished the prophase of the first meiotic division, but instead of proceeding into the metaphase, they enter the **dictyotene stage**, a resting stage between prophase and metaphase characterized by a lacy network of chromatin (fig. 1-7*C*).[6-8] **Primary oocytes do not finish their first meiotic division before puberty is reached**. The total number of primary oocytes at

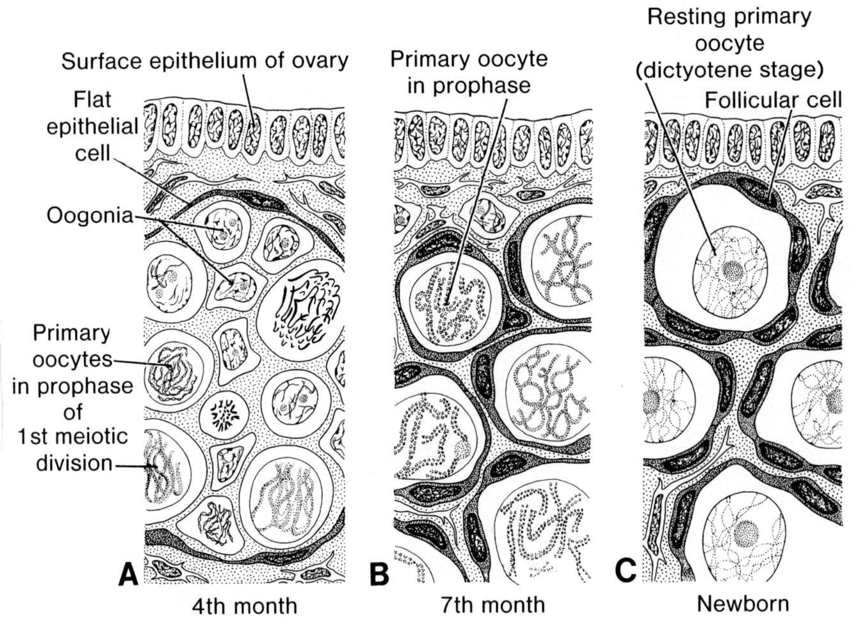

Figure 1-7. Schematic representation of a segment of the ovary at different stages of development. *A*, At four months. The oogonia are grouped in clusters in the cortical part of the ovary. Some show mitosis; others have already differentiated into primary oocytes and have entered the prophase of the first meiotic division. *B*, At seven months. Almost all the oogonia are transformed into primary oocytes in the prophase of the first meiotic division. *C*, At birth. Oogonia are absent. Each primary oocyte is surrounded by a single layer of follicular cells, thus forming the primordial follicle. The oocytes have entered the dictyotene stage in which they remain until just before ovulation. Only then do they enter the metaphase of the first meiotic division. (Modified after Ohno et al.[5])

birth is estimated to vary from 700,000 to 2,000,000. Since during the following years of childhood the majority of the oocytes become atretic, only approximately 40,000 are present by the beginning of puberty.[3] Only then do the primordial follicles develop into mature follicles and do the primary oocytes complete their first meiotic division.

It is important to realize that some of the oocytes, reaching maturity late in life, have been dormant in the dictyotene stage of the first meiotic division for 40 years or more. Whether or not the dictyotene stage is the most suitable phase to protect the oocyte against environmental influences acting upon the ovary during life is presently unknown. Considering that the incidence of children with chromosomal abnormalities increases with maternal age, one may wonder whether or not the extended meiotic division makes the primary oocyte vulnerable to damage.

With the onset of puberty, a number of primordial follicles begin to mature with each ovarian cycle. The primary oocyte (still in the dictyotene stage) begins to increase in size, while the surrounding epithelial cells, the **follicular cells**, change from flat to cuboidal. The follicle is now known as the **primary follicle** (fig. 1-8*B*). Initially the follicular cells are in intimate contact with the oocyte, but soon a layer of acellular material consisting of mucopolysaccharides is deposited on the surface of the oocyte.[9, 10] This material, which is produced by the follicular cells as well as the oocyte, gradually increases in thickness, thus forming the **zona pellucida** (fig. 1-8*C*). Small finger-like processes of the follicular cells extend across the zona pellucida and interdigitate with the microvilli of the plasma membrane of the oocyte. These processes are thought to be important for the transport of materials from the follicular cells to the oocyte.[11]

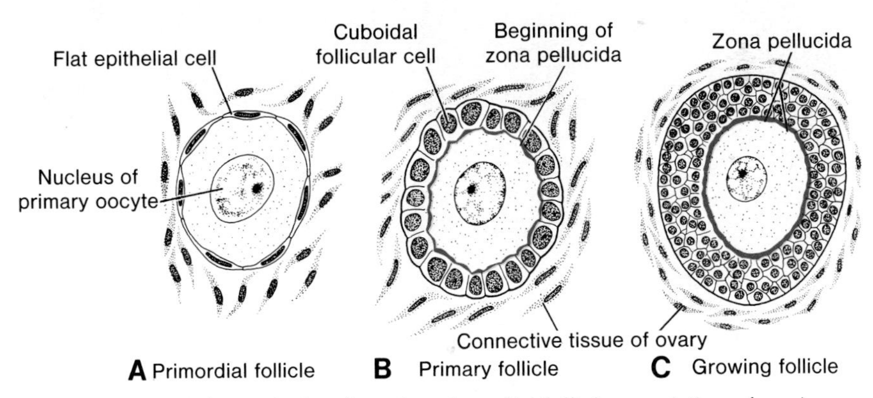

Figure 1-8. *A*, Schematic drawing of a primordial follicle, consisting of a primary oocyte surrounded by a layer of flattened epithelial cells. *B*, As maturation of the follicle proceeds, the follicular cells become cuboidal. They now begin to secrete the zona pellucida, which is visible in irregular patches on the surface of the oocyte. *C*, With further maturation the follicular cells form an increasingly thick layer around the oocyte. The zona pellucida is well-defined.

As development continues, the follicular cells begin to proliferate, thereby forming a thick cellular layer around the oocyte (fig. 1-8*C*). Subsequently, under influence of gonadotropins from the anterior pituitary, fluid-filled spaces appear between the follicular cells, and when these spaces coalesce the **follicular antrum** is formed. Initially the antrum is crescent-shaped, but with time it greatly enlarges (fig. 1-9*A*, *B*). The follicular cells surrounding the oocyte remain intact and form the **cumulus oophorus**. At maturity, the follicle is known as the **tertiary** or **vesicular follicle**. It is surrounded by two layers of connective tissue: an inner cellular layer, the **theca interna**, which is rich in blood vessels; and an outer fibrous layer, the **theca externa**, which gradually merges with the ovarian stroma (fig. 1-9). The follicle then has a diameter varying from 6 to 12 mm. The theca interna or **thecal gland** is considered a main source of estrogen, the female sex hormone that regulates the function of the reproductive organs.[12]

With each ovarian cycle a number of follicles begins to develop, but usually only one reaches full maturity. The others degenerate and become atretic. As soon as the follicle is mature, the primary oocyte resumes its first meiotic division, leading to the formation of two daughter cells of unequal size, but each with 23 (double-structured) chromosomes and 2n DNA (fig.

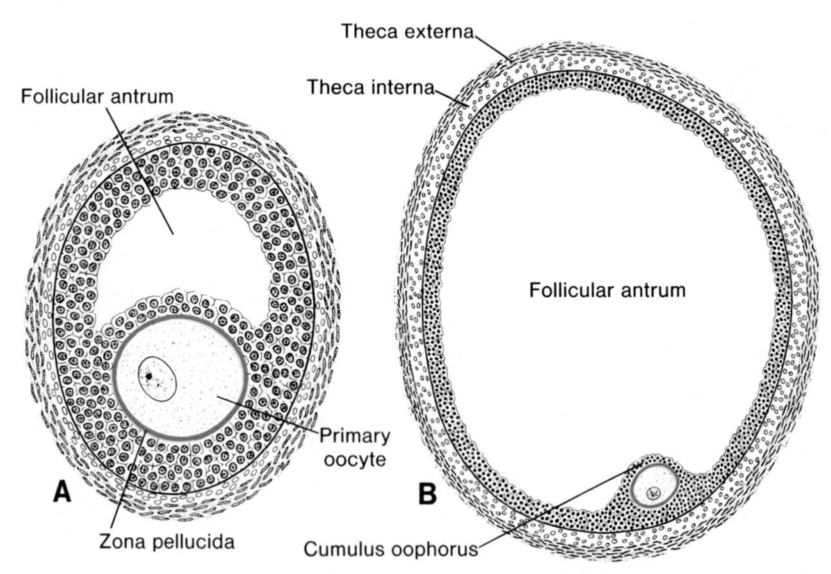

Figure 1-9. Schematic representation of a maturing follicle. *A*, The oocyte surrounded by the zona pellucida is eccentrically located; the follicular antrum has developed by coalescence of intercellular spaces. Note the arrangement of the cells of the theca interna and the theca externa. *B*, Mature vesicular or Graafian follicle. The antrum has enlarged considerably, is filled with follicular fluid, and is surrounded by a stratified layer of follicular cells. The oocyte is embedded in a mound of follicular cells, known as the cumulus oophorus.

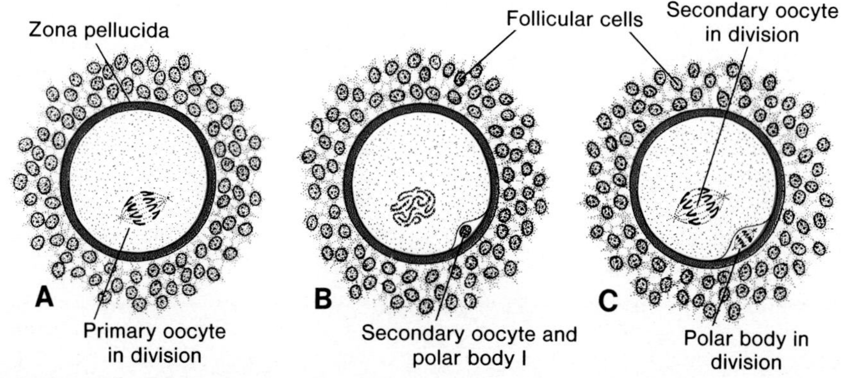

Figure 1-10. Maturation of the oocyte. *A*, Primary oocyte showing the spindle of the first meiotic division. It is characterized by 4n DNA. *B*, Secondary oocyte and polar body I. Note that the nuclear membrane is absent. *C*, Secondary oocyte, showing the spindle of the second meiotic division. Polar body I, likewise, is in division (adapted from several sources).

1-10*A*, *B*). One cell, the **secondary oocyte**, receives all of the cytoplasm; the other, the **first polar body**, receives practically none. The latter is located between the zona pellucida and the cell membrane of the secondary oocyte (fig. 1-10*B*). The first meiotic division occurs shortly before ovulation.

At completion of the first maturation division and before the nucleus of the seconday oocyte has returned to its resting stage, the cell enters the **second maturation division without DNA replication**. The moment the secondary oocyte shows the spindle formation, ovulation occurs and the oocyte is shed from the ovary (fig. 1-10*C*).[13] The second maturation division is completed only if the oocyte is fertilized; otherwise the cell degenerates approximately 24 hours after ovulation. Whether or not the first polar body undergoes a second division is uncertain, but fertilized ova accompanied by three polar bodies have been observed.

Human primary oocytes have been recovered from their follicles and cultured in an artificial medium.[14–17] In these conditions the oocyte moved from the dictyotene stage into metaphase 25 to 28 hours after the beginning of the culture period. Extrusion of the first polar body and the metaphase of the second meiotic division were observed 36 to 43 hours after the beginning of the experiment. In some cases the spindle of the second maturation division was seen within 30 minutes after the telophase of the first meiotic division.

SPERMATOGENESIS

Differentiation of the primordial germ cells in the female begins in the third month of development, but in the male at puberty. At the time of birth they can be recognized in the sex cords of the testis as large, pale cells,

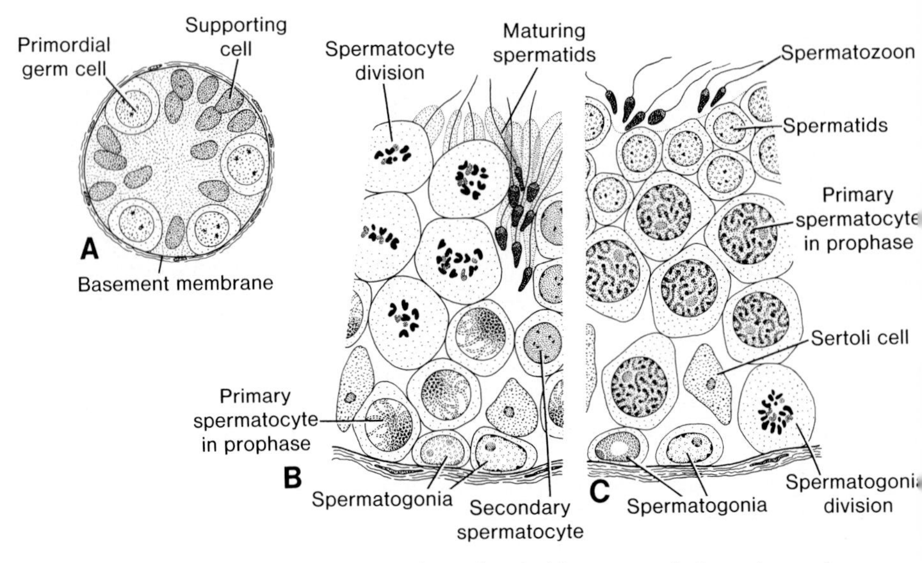

Figure 1-11. *A*, Transverse section through primitive sex cord of a male newborn, showing the primordial germ cells and supporting cells. *B* and *C*, Two segments of the seminiferous tubule shown in transverse section. Note the different stages in spermatogenesis.

surrounded by the supporting cells (fig. 1-11*A*). The latter, probably derived from the surface epithelium of the gland in the same manner as the follicular cells, become the **sustentacular** or **Sertoli cells**.

Shortly before puberty the sex cords acquire a lumen and become the **seminiferous tubules**. At about the same time the primordial germ cells give rise to the **spermatogonia**, which in turn differentiate into the **primary spermatocytes** (fig. 1-11*B*, *C*).[18] After replicating their DNA, these cells then start with the prophase of their first meiotic or maturation division. Upon completion of the prophase, which lasts about 16 days, the cell passes rapidly through the remaining phases and then gives rise to two **secondary spermatocytes** (figs. 1-3*B*, and 1-12). These cells begin immediately with their **second maturation** or **meiotic division**, which results in the production of two spermatids (fig. 1-3*B*). As a result of the two maturation divisions the spermatid contains 23 chromosomes and **n** amount of DNA (figs. 1-3*B* and 1-12).

SPERMIOGENESIS

The spermatids undergo a series of changes resulting in the production of the spermatozoa. These changes are: (1) formation of the **acrosome**, which extends over half the nuclear surface (fig. 1-13*B*, *C*); (2) condensation of the nucleus; (3) formation of neck, middle piece, and tail (fig. 1-13*C*); and (4)

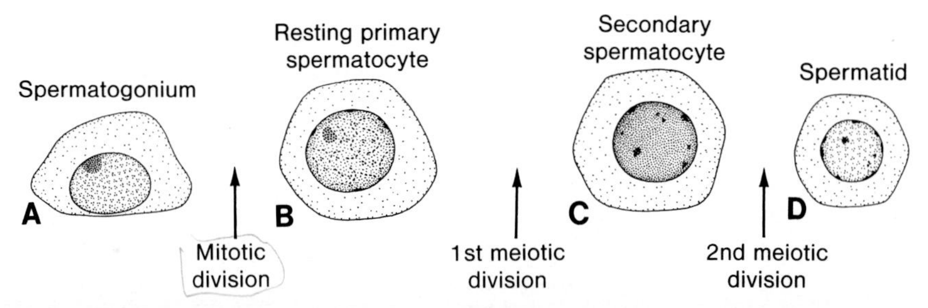

Figure 1-12. Schematic representation of spermatogenesis in man.

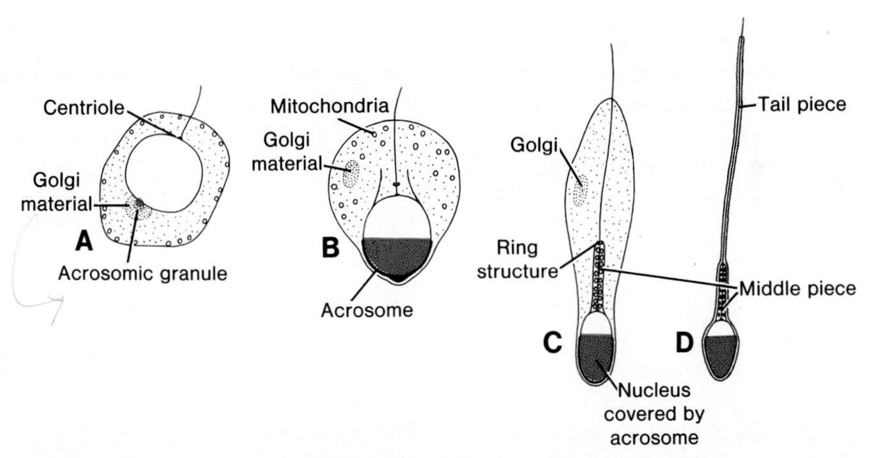

Figure 1-13. Schematic drawings showing the important stages in the transformation of the human spermatid into the spermatozoon. (Derived from Clermont and Leblond.)

shedding of most of the cytoplasm (fig. 1-13*D*). In man the time required for a spermatogonium to develop into a mature spermatozoon is 61 days.[19]

When fully formed, the spermatozoa enter the lumen of the seminiferous tubules. From here, they are pushed toward the epididymis by the contractile elements in the wall of the seminiferous tubules. Although initially only slightly motile, the spermatozoa obtain full motility in the epididymis.

ABNORMAL GAMETES

In the human as well as in most mammals, one ovarian follicle occasionally contains two or three clearly distinguishable primary oocytes (fig. 1-14*A*). Although these oocytes may give rise to twins or triplets, they usually degenerate before reaching maturity. In rare cases, one primary oocyte contains two or even three nuclei (fig. 1-14*B*). Such bi- or trinucleated oocytes, however, die before reaching maturity.

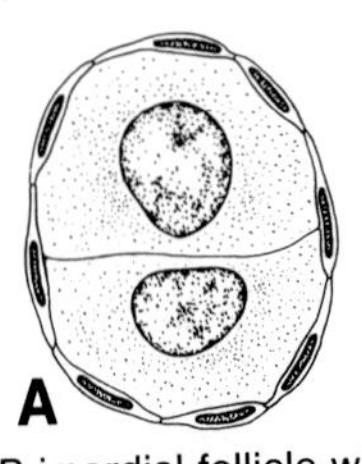

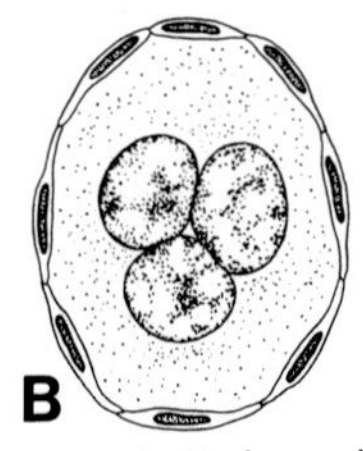

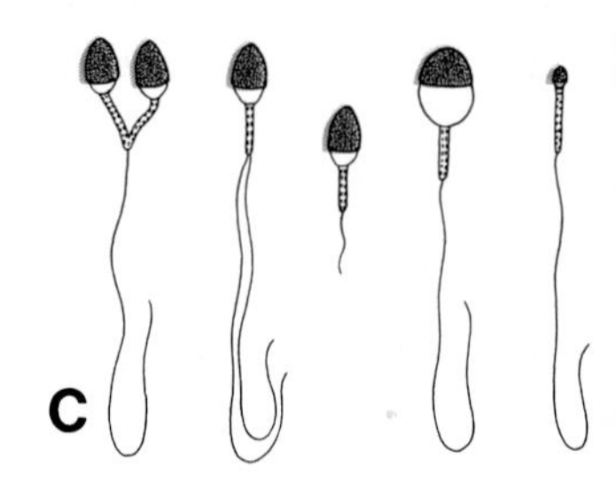

Figure 1-14. Drawings of abnormal germ cells in the female and male. *A*, Primordial follicle with two oocytes. *B*, Trinucleated oocyte. *C*, Various types of abnormal spermatozoa.

Contrary to atypical oocytes, abnormal spermatozoa are seen frequently. The head as well as the tail may be abnormal; they may be giants or dwarfs; sometimes, they are joined (fig. 1-14*C*). There is evidence suggesting that 10 per cent of the spermatozoa may be abnormal without any loss of fertility. When, however, a quarter or more of them are abnormal, fertility is usually impaired.

SUMMARY

In preparation for fertilization both male and female germ cells undergo a number of chromosomal and morphological changes, a process known as **gametogenesis**. The chromosomal changes occur during the **meiotic divisions**. During the first meiotic division the **homologous chromosomes pair** and **exchange genetic material**; during the second meiotic division the cells fail to replicate DNA, and each cell is thus provided with a haploid number of chromosomes and half the amount of DNA of a normal somatic cell (fig. 1-2). Hence, mature male and female gametes have 22 + X or 22 + Y chromosomes.

The first human germ cells, known as the **primordial germ cells** appear in the wall of the yolk sac (fig. 1-5) at the end of the third week and migrate to the indifferent gonad where they arrive in the fifth week. In the male the maturation process from primitive germ cell to mature gamete is known as **spermatogenesis**, in the female as **oogenesis**. In the female the primordial germ cells differentiate into **oogonia**. After repeated divisions some of these differentiate into **primary oocytes**, which immediately after their formation enter the first meiotic division. By the seventh month all primary oocytes have entered the first meiotic division and most of them are individually surrounded by a layer of flat follicular cells (fig. 1-7). Together they form the **primordial follicle. The primary oocytes do not finish their first meiotic division**

but remain in the dictyotene stage till puberty. At birth their total number varies from 700,000 to 2,000,000.

With the onset of puberty a number of primordial follicles begins to mature with each ovarian cycle, but only one reaches full maturity. During this maturation process one primary oocyte gives rise to one secondary oocyte plus one polar body. The secondary oocyte in turn gives rise to the mature oocyte plus another polar body. Hence, a primary oocyte develops into one mature oocyte and three polar bodies (fig. 1-3).

In the male the primordial cells remain dormant till puberty and only then do they differentiate into spermatogonia. These stem cells give rise to primary spermatocytes, which, through two successive meiotic divisions, produce four **spermatids** (fig. 1-11). The spermatids subsequently go through a series of changes (**spermiogenesis**) (fig. 1-13): (1) formation of the acrosome; (2) condensation of the nucleus; (3) formation of neck, middle piece, and tail; and (4) shedding of most cytoplasm. The time required for a spermatogonium to become a mature spermatocyte is 61 days.

REFERENCES

1. Witschi, E. Migration of the germ cells of the human embryos from the yolk sac to the primitive gonadal folds. *Contrib. Embryol., 32:* 67, 1948.
2. Blandau, R. H., White, B. J., and Rumery, R. E. Observations on the movements of the living primordial germ cells in the mouse. *Fertil. Steril., 14:* 482, 1963.
3. Pinkerton, H. M., McKay, D. G., Adams, E. C., and Hertig, A. T. Development of the human ovary—a study using histochemical techniques. *Obstet. Gynecol., 18:* 152, 1961.
4. Franchi, L. L., Mandl, A. M., and Zuckerman, S. The development of the ovary and the process of oogenesis. In *The Ovary*, edited by S. Zuckerman, A. M. Mandl, and P. Eckstein. Academic Press, New York, 1962.
5. Ohno, S., Makino, S., Kaplan, W. D., and Kinosita, R. Female germ cells in man. *Exp. Cell Res., 24:* 106, 1961.
6. Ohno, S., Klinger, H. P., and Atkin, N. B. Human oogenesis. *Cytogenetics, 1:* 42, 1962.
7. Ohno, S., and Smith, J. B. Role of fetal follicular cells in meiosis of mammalian oocytes. *Cytogenetics, 3:* 324, 1964.
8. Manotaya, T., and Potter, E. L. Oocytes in prophase of meiosis from squash preparations of human fetal ovaries. *Fertil. Steril., 14:* 378, 1963.
9. Weakley, B. S. Electron microscopy of the oocyte and granulosa cells in the developing ovarian follicles of the golden hamster (*Mesocricetus auratus*). *J. Anat., 100:* 503, 1966.
10. Zamboni, L. Fine morphology of the follicle wall and follicle cell-oocyte association. *Biol. Reprod., 10:* 125, 1974.
11. Anderson, E., and Albertini, D. F. Gap junctions between the oocyte and companion follicle cells in the mammalian ovary. *J. Cell Biol., 71:* 680, 1976.
12. Albrecht, E. D., Koos, R. D., and Wehrenberg, W. B. Ovarian Δ^5-3-β-hydroxysteroid dehydrogenase and cholesterol in the aged mouse during pregnancy. *Biol. Reprod., 13:* 158, 1975.
13. Hafez, E. S. E. *Human Ovulation. Mechanisms, Prediction, Detection and Induction.* North Holland Publishing Co., Amsterdam, 1979.
14. Steptoe, P. C., and Edwards, R. G. Laparoscopic recovery of preovulatory human oocytes after priming of ovaries with gonadotropins. *Lancet, 1:* 683, 1970.

15. Edwards, R. G. Fertilization of human eggs in vitro: morals, ethics and the law. *Q. Rev. Biol.*, *49:* 3, 1974.
16. Edwards, R. G. Physiological aspects of human ovulation, fertilization and cleavage. *J. Reprod. Fertil.* (Suppl.), *18:* 87, 1973.
17. Lopata, A., Brown, J. B., Leeton, J. F., Talbot, J. M. and Wood, C. In vitro fertilization of preovulatory oocytes and embryo transfer in infertile patients treated with clomiphene and human chorionic gonadotropin. *Fertil. Steril.*, *30:* 27, 1978.
18. Clermont, Y., and Trott, M. Kinetics of spermatogenesis in mammals: seminiferous epithelium cycle and spermatogonial renewal. *Physiol. Rev.*, *52:* 198, 1972.
19. Heller, C. G., and Clermont, Y. Kinetics of the germinal epithelium in man. *Recent Prog. Horm. Res.*, *20:* 545, 1964.

chapter 2

Ovulation to Implantation
(First Week of Development)

Ovarian Cycle

At the time of puberty the female begins to undergo regular monthly cycles. These cycles known as **sexual cycles**, are controlled by the hypothalamus.[1,2] Releasing factors produced by the hypothalamus act on the cells of the anterior pituitary gland, which in turn secrete the **gonadotropins**. These hormones, the **follicle-stimulating hormone** (FSH) and **luteinizing hormone** (LH), stimulate and control the cyclic changes in the ovary (see fig. 2-13).

At the beginning of each ovarian cycle a number of primordial follicles, varying from 5 to 12, begins to grow under influence of the follicle-stimulating hormone (fig. 2-1). Under normal conditions only one of these follicles reaches full maturity and only one oocyte is discharged; the others degenerate and become atretic. In the next cycle another group of follicles begins to

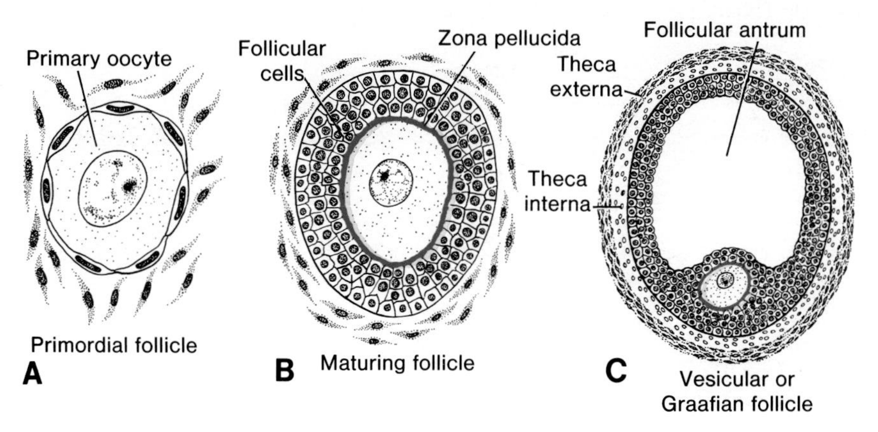

Figure 2-1. Drawing of the changes occurring in the primordial follicle during the first half of the ovarian cycle. Under influence of FSH the primordial follicle *(A)* matures into the Graafian follicle *(C)*. The oocyte remains a primary oocyte in the dictyotene stage until shortly before ovulation. During the last few days of the growing period the estrogens produced by the follicular and theca cells stimulate the formation of LH in the pituitary (see figure 2-13).

grow and again only one reaches maturity. Consequently, the majority of follicles degenerate without ever reaching full maturity. When a follicle becomes atretic, the oocyte and surrounding follicular cells degenerate and are replaced by connective tissue, thus forming a **corpus atreticum**. During the growth of the follicle large numbers of follicle and theca cells are formed. These cells produce estrogens, which stimulate the pituitary gland to secrete **luteinizing hormone**. This hormone is needed for the final stages of follicle maturation and to induce the shedding of the oocyte, known as ovulation (see fig. 2-13).

Ovulation

In the days immediately preceding ovulation, the Graafian follicle increases rapidly in size under influence of follicle-stimulating and luteinizing hormones and expands to a diameter of 15 mm. Coincident with the final development of the Graafian follicle, the primary oocyte, which till this time has remained in its dictyotene stage, resumes and finishes its first meiotic division. In the meantime the surface of the ovary begins to bulge locally and at the apex an avascular spot appears, the so-called **stigma**. As a result of local weakening and degeneration of the ovarian surface, follicular fluid oozes through the stigma, which gradually opens. Subsequently, when more fluid escapes, the tension in the follicle is released and the oocyte together with the surrounding cumulus oophorus cells breaks free and floats out of the ovary (figs. 2-2 and 2-3).[3–5] At the moment that the oocyte with its cumulus oophorus cells is discharged from the ovary—**ovulation**— the first meiotic division is completed and the secondary oocyte has started its second

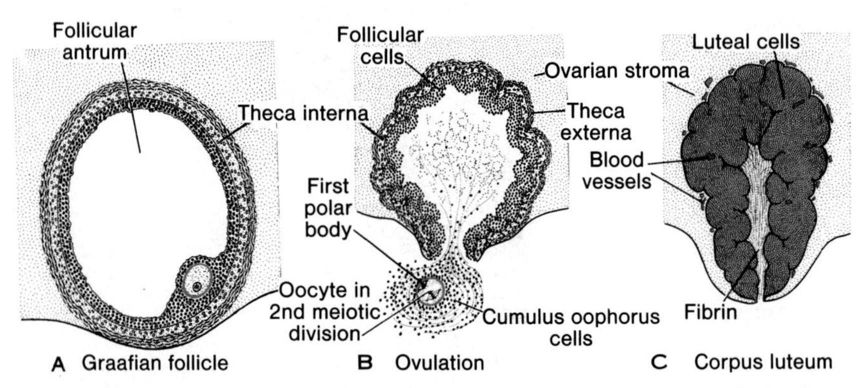

Figure 2-2. *A*, The Graafian follicle just before rupture. *B*, Ovulation. The oocyte, beginning its second meiotic division, is discharged from the ovary, together with a large number of cumulus oophorus cells. The follicular cells remaining inside the collapsed follicle divide rapidly and differentiate into luteal cells. *C*, Corpus luteum. Note the large size of the corpus luteum caused by the massive proliferation of luteal cells. The remaining cavity of the follicle is filled with fibrin.

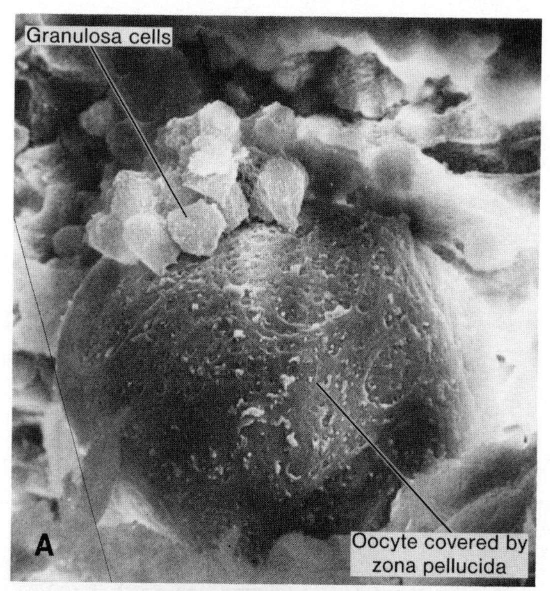

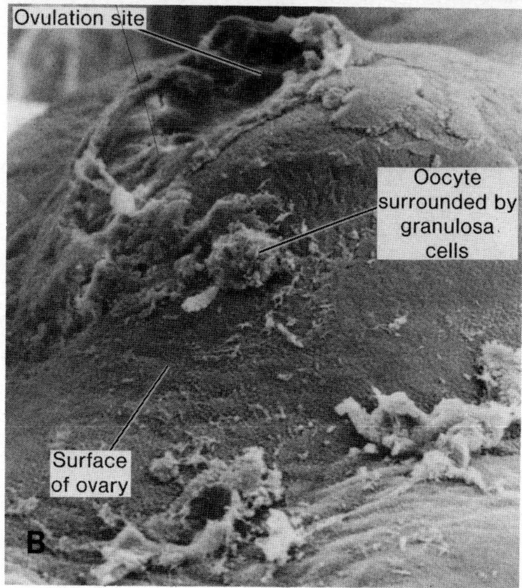

Figure 2-3. *A*, Scanning electron micrograph of the ovulation process in the mouse. The surface of the oocyte is covered by the zona pellucida. Note the cumulus oophorus or granulosa cells. B, Scanning electron micrograph of a rabbit oocyte 1½ hours after ovulation. The oocyte is surrounded by granulosa cells and lies on the surface of the ovary. Note the site of ovulation. (From Van Blerkom, J., and Motta, P. *The Cellular Basis of Mammalian Reproduction*. Urban and Schwarzenberg, Baltimore, 1979.)

meiotic division (fig. 2-2*B*). In some women ovulation is accompanied by some pain, known as **middle pain**.

Corpus Luteum

Following ovulation the follicular cells remaining in the wall of the ruptured follicle are vascularized by surrounding vessels and become polyhedral. Under influence of the luteinizing hormone these cells develop a yellowish pigment and change into the **luteal cells**, which form the **corpus luteum** and secrete **progesterone** (fig. 2-2*C*). This hormone, together with the estrogenic hormones produced by the theca cells and the surrounding ovarian tissue, cause the uterine mucosa to enter the **progestational** or **secretory stage** (see fig. 2-12), in preparation for implantation of the embryo.

Oocyte Transport

Shortly before ovulation, the fimbriae of the oviduct begin to cover the surface of the ovary and the tube itself begins to contract rhythmically. It is believed that the oocyte surrounded by some cumulus or granulosa cells (figs. 2-3 and 2-4) is carried into the tube by the sweeping movements of the fimbriae and by the motion of the cilia on the epithelial lining.[6–9] Once in the tube, the cumulus cells lose contact with the oocyte by withdrawing their cytoplasmic processes from the zona pellucida.[10]

Once the oocyte is in the uterine tube, it is pushed toward the uterine lumen by contractions of the muscular wall.[11] The rate of transport is somewhat affected by the endocrine status during and after ovulation, but in man the fertilized oocyte reaches the uterine lumen in approximately three to four days (see fig. 2-11).[12, 13]

Corpus Albicans

If fertilization fails to occur, the corpus luteum reaches maximum development about nine days after ovulation. It can easily be recognized as a yellowish projection on the surface of the ovary. Subsequently, the corpus luteum decreases in size through degeneration of the luteal cells and forms a mass of fibrotic scar tissue, known as the **corpus albicans**. Simultaneously the progesterone production decreases, thus precipitating the menstrual bleeding (see fig. 2-13).[14–15]

If the oocyte is fertilized, degeneration of the corpus luteum is prevented by a **gonadotropic** hormone secreted by the trophoblast of the developing embryo. The corpus luteum continues to grow, and forms the **corpus luteum of pregnancy (graviditatis)**. By the end of the third month, this structure may be one-third to one-half of the total size of the ovary. The yellowish luteal cells continue to secrete progesterone until the end of the fourth month; thereafter they regress slowly as secretion of progesterone by the trophoblastic component of the placenta becomes adequate for maintenance of

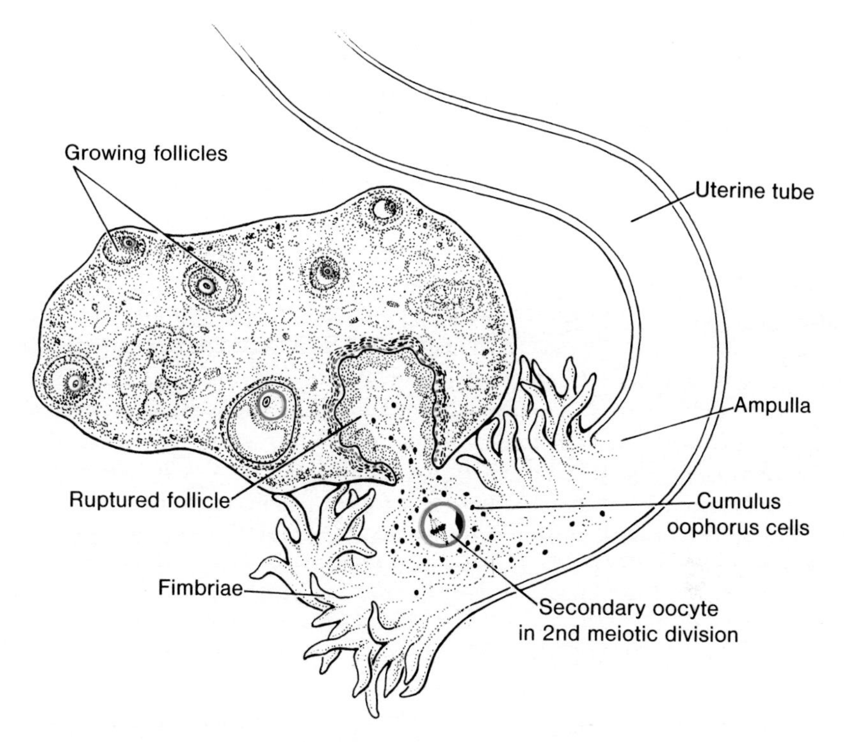

Figure 2-4. Relationship of fimbriae and ovary. During ovulation the fimbriae are thought to sweep over the rupturing follicle.

pregnancy. Removal of the corpus luteum of pregnancy before the fourth month usually leads to abortion.

It is now well known that ovulation can be inhibited by treatment with progestational compounds.[16] When taken orally from day 5 to day 25 of the menstrual cycle the progestins act as contraceptives and inhibit ovulation in almost 100 per cent of the cases. Most of the presently used anti-ovulatory preparations contain a small amount of estrogen in addition to one of the progestin compounds.[17] The estrogenic component also inhibits ovulation and the progestin component ensures a normal menstrual bleeding upon withdrawal of the preparation on the 25th day of the cycle. The feedback action of both hormones is thought to act at the hypothalamic level. Preliminary observations indicate that the maturation of the follicles is completely suppressed.

Fertilization

Fertilization, the process by which the male and female gametes fuse, occurs in the **ampullary region of the uterine tube**. This is the widest part of the tube and is located close to the ovary (fig. 2-4). While spermatozoa can stay alive in the female reproductive tract for about 24 hours, the secondary oocyte is thought to die 12 to 24 hours after ovulation, if not fertilized.[18]

The spermatozoa pass rapidly from the vagina into the uterus and subsequently into the uterine tubes. This ascent is probably caused by contractions of the musculature of the uterus and the tube. It must be kept in mind that spermatozoa, upon arrival in the female genital tract, are not capable of fertilizing the oocyte. They must undergo (1) **capacitation** and (2) the **acrosome reaction**.

Capacitation is a period of conditioning in the female reproductive tract which, in the human, lasts approximately seven hours. During this time a glycoprotein coat and seminal plasma proteins are removed from the plasma membrane that overlies the acrosomal region of the spermatozoa. Completion of capacitation permits the acrosome reaction to occur.[19, 20]

The **acrosome reaction** occurs in the immediate vicinity of the oocyte under influence of substances emanating from the corona radiata cells and the oocyte.[21] Morphologically multiple point fusions between the plasma membrane and the outer acrosomal membrane take place, permitting the

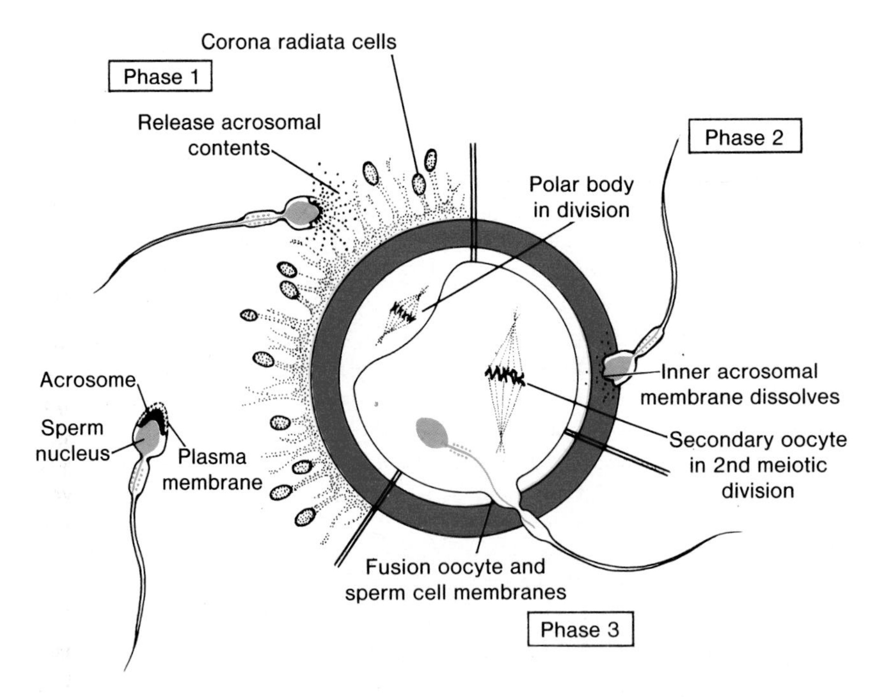

Figure 2-5. Schematic representation of the three phases of oocyte penetration. In phase 1 the spermatozoa break through the corona radiata barrier; in phase 2 one or more spermatozoa penetrate in the zona pellucida; in phase 3 one spermatozoon penetrates through the oocyte membrane, thereby losing its own plasma membrane. Inset shows normal spermatocyte with acrosomic head cap.

release of the acrosomal contents (fig. 2–5) needed to penetrate the corona radiata and zona pellucida. During the acrosome reaction the following substances are released: (1) hyaluronidase needed to penetrate the corona radiata barrier;[22] (2) trypsin-like substances needed for digestion of the zona pellucida;[23] and (3) zona lysin, attached to the inner surface of the acrosomal membrane, also needed to help the spermatozoon cross the zona pellucida.[24]

PHASE 1: PENETRATION OF THE CORONA RADIATA

Of the 200 to 300 million spermatozoa deposited in the female genital tract only 300 to 500 reach the site of fertilization. Only one of those is needed for fertilization and it is thought that the others aid the fertilizing sperm in penetrating the first barrier protecting the female gamete, the **corona radiata** (fig. 2-5). Initially the enzyme hyaluronidase was assumed to be the important enzyme in the dispersal of the corona cells. Presently it is thought that the corona cells are dispersed by the combined action of sperm and tubal mucosa enzymes.[18]

PHASE 2: PENETRATION OF ZONA PELLUCIDA

This second barrier protecting the female gamete is penetrated by the sperm with the aid of enzymes released from the inner acrosomal membrane (fig. 2-5).[15] Once the spermatozoon touches the zona pellucida, it becomes firmly attached and penetrates rapidly. The permeability of the zona pellucida changes when the head of the sperm comes in contact with the oocyte surface. This results in the release of substances that cause an alteration in the properties of the zona pellucida, the **zona reaction**, and inactivate species-specific receptor sites of spermatozoa.[26] Indeed, other spermatozoa have been found embedded in the zona pellucida, but only one seems to be able to penetrate into the oocyte proper (fig. 2-6*B*)[27] On rare occasions two sperms penetrate the female gamete simultaneously, and embryos with 69 chromosomes have been described.[28]

PHASE 3: FUSION OF OOCYTE-SPERM CELL MEMBRANES

As soon as the spermatozoon comes in touch with the oocyte cell membrane, the two plasma membranes fuse (fig. 2-5). Since the plasma membrane covering the acrosomal head cap has disappeared during the acrosome reaction, actual fusion is accomplished between the oocyte membrane and the membrane that covers the posterior region of the head (fig. 2-5). In the human both the head and the tail of the spermatozoon enter the cytoplasm of the oocyte, but the plasma membrane is left behind on the oocyte surface.[27]

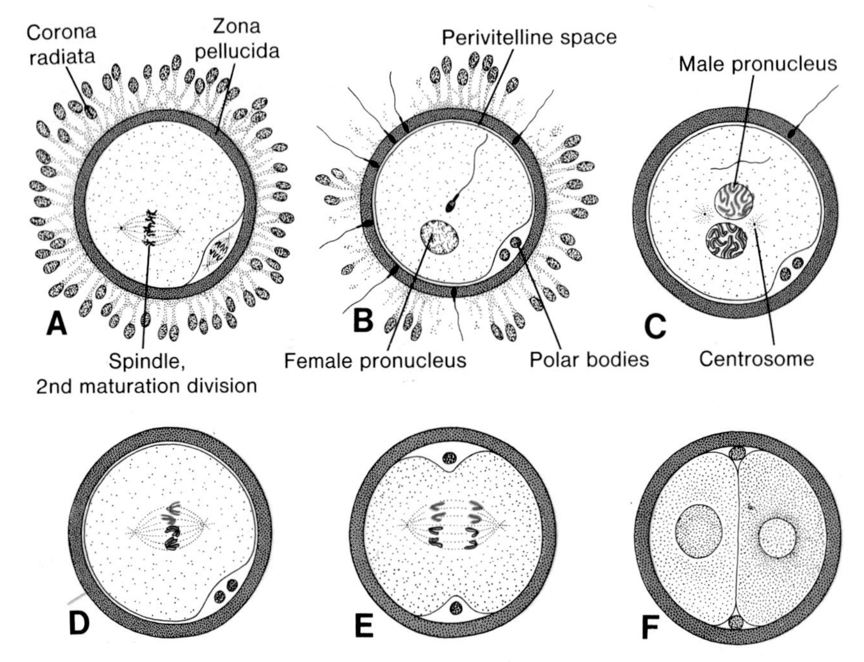

Figure 2-6. *A*, Oocyte immediately after ovulation, showing the spindle of the second meiotic division. *B*, A spermatozoon has penetrated the oocyte, which has finished its second meiotic division. The chromosomes of the oocyte are arranged in a vesicular nucleus, the female pronucleus. The heads of several are stuck in the zona pellucida. *C*, Stage of male and female pronuclei. *D*, *E*, The chromosomes become arranged on the spindle, split longitudinally and move to opposite poles. *F*, The two-cell stage.

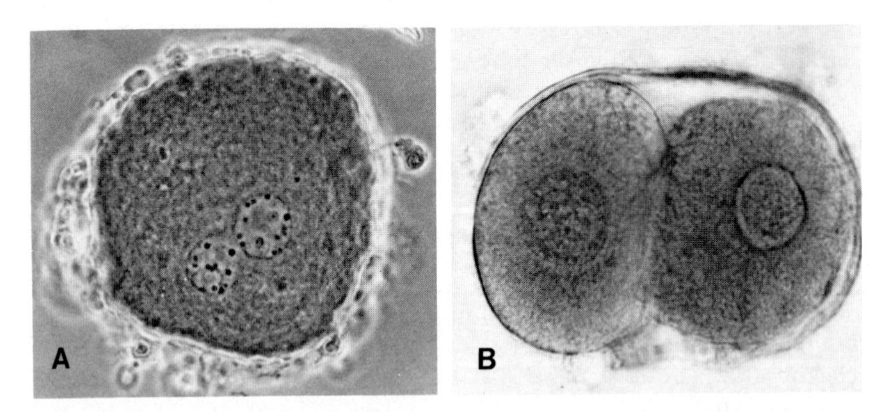

Figure 2-7. *A*, Phase contrast view of the pronuclear stage of a fertilized human oocyte. Note the male and female pronuclei.[32] (Courtesy Dr. L. Dickmann and Dr. R. Noyes, Vanderbilt University.) *B*, Two-cell stage of human zygote. (From A. T. Hertig and J. Rock. Courtesy Carnegie Institution of Washington.)

As soon as the spermatozoon has entered the oocyte the egg responds in three different ways:

① **Cortical and zona reactions.** As a result of the release of cortical oocyte granules the (a) oocyte membrane becomes impenetrable to other spermatozoa and (b) the zona pellucida alters its structure and composition, possibly through removal of specific receptor sites for spermatozoa. In this manner polyspermy is prevented.

② **Resumption of the second meiotic division.** The oocyte finishes its second meiotic division immediately after entry of the spermatozoon. One of the daughter cells receives hardly any cytoplasm and is known as the **second polar body**, the other daughter cell is the **definitive oocyte**. Its chromosomes (22 + X) become arranged in a vesicular nucleus known as the **female pronucleus** (fig. 2-6*B*).

③ **Metabolic activation of the egg.** The activating factor is probably carried by the spermatozoon. The postfusion activation may be considered to encompass the initial cellular and molecular events associated with early embryogenesis.

The spermatozoon meanwhile moves forward until it lies in close proximity to the female pronucleus. Its nucleus becomes swollen and forms the **male pronucleus** (fig. 2-6*C*) The tail is detached and degenerates. Morphologically the male and female pronuclei are indistinguishable (fig. 2-7*A*).[30] During the growth of the male and female pronuclei (both haploid and containing only 1n DNA) each pronucleus must duplicate its DNA.[31] If not, each cell of the two-cell stage zygote would have cells with half the normal amount of DNA. Immediately after DNA synthesis the chromosomes become organized on the spindle in preparation for a normal mitotic division. The 23 maternal and 23 paternal (double) chromosomes split longitudinally at the centromere and the sister chromatids move to the opposite poles, thus providing each cell of the zygote with the normal number of chromosomes and the normal amount of DNA (2n) (fig. 2-6*D, E*). While the sister chromatids move to the opposite poles, a deep furrow appears on the surface of the cell, gradually dividing the cytoplasm into two parts (figs 2-6*F* and 2-7*B*).

The main results of fertilization are: (1) **restoration of the diploid number of chromosomes**, half from the father and half from the mother. Hence, the zygote contains a new combination of chromosomes, different from both parents. (2) **Determination of the sex** of the new individual. An X-carrying sperm will produce a female (XX) embryo, and a Y-carrying sperm a male (XY) embryo. Hence, the chromosomal sex of the embryo is determined at fertilization. (3) **Initiation of cleavage**. Without fertilization the oocyte usually degenerates 24 hours after ovulation.

External Human Fertilization

Fertilization of human ova *in vitro* has been accomplished by only a few investigators.[33] The oocyte is recovered from the ovarian follicle with an aspirator just prior to ovulation when the oocyte is in the late stages of the first meiotic metaphase. It is then placed in a simple culture medium and the sperm added immediately. In this manner it has been possible to grow human embryos to the early blastocyst stage. Reimplantation of 8 or 16 cell blastocysts into the mother has been difficult. Of the 79 patients treated, 32 have had fertilized eggs successfully reimplanted in the uterus. Four pregnancies have resulted, two of which ended in spontaneous abortion. The remaining two led to the successful birth of normal children.[34, 35]

Alternatives to Normal Fertilization

SUPERFECUNDATION

This condition may follow **polyovulation**, in which one or more oocytes, released in a given ovarian cycle, are fertilized by spermatozoa from one male and another oocyte is fertilized by a different male. The condition occurs in various mammals, but has not been described for man.

PARTHENOGENESIS

A female gamete cannot produce an embryo without participation of a male gamete. Occasionally, however, the oocyte is activated without sperm penetration and development may start. This form of reproduction is called **parthenogenesis**. Early parthenogenetic development of mammalian oocytes has been introduced by chilling, by the local application of heat, by hyperthermia, and by other means.[36–38] There is, however, no record of the birth of viable young originating in any of these ways. Sometimes cleaving oocytes are found in the ovary, where they may develop into an **ovarian teratoma**.[39]

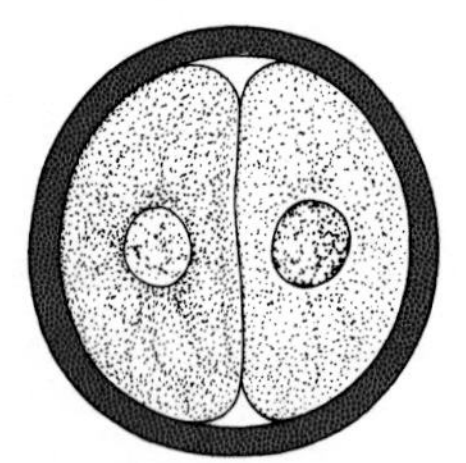

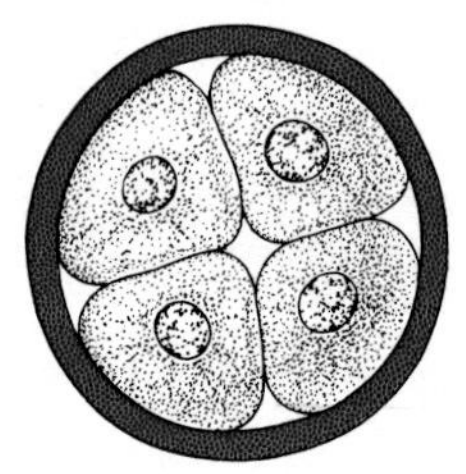

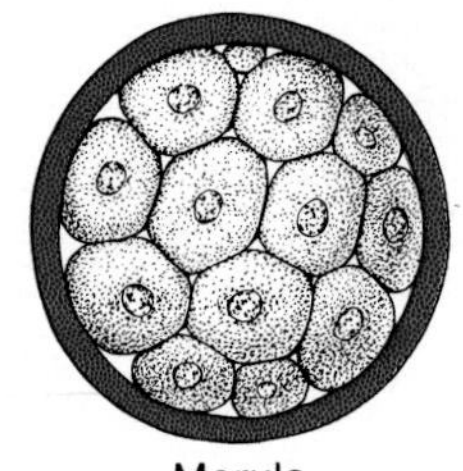

Figure 2-8. Schematic representation of the development of the zygote from the two-cell stage to the late morula stage. The two-cell stage is reached approximately 30 hours after fertilization; the four-cell stage at approximately 40 hours; the 12- and 16-cell stage at approximately three days; and the late morula stage at approximately four days. During this period the blastomeres are surrounded by the zona pellucida, which disappears at the end of the fourth day.

Cleavage

Once the zygote has reached the two-cell stage, it undergoes a series of mitotic divisions, resulting in a rapid increase in the number of cells. These cells, which become smaller with each cleavage division, are known as blastomeres (fig. 2-8). After three to four divisions the zygote, similar in appearance to a mulberry, is known as the **morula**. This stage is reached about three days after fertilization and the embryo is about to enter the uterus (see fig. 2-11). At this time (12 to 16-cell stage) the morula consists of a group of centrally located cells, the **inner cell mass**, and a surrounding layer, the **outer cell mass**. The inner cell mass will give rise to the tissues of the **embryo proper**, while the outer cell mass forms the **trophoblast**, which later contributes to the **placenta**.

Blastocyst Formation

At about the time that the morula enters the uterine cavity, fluid begins to penetrate through the zona pellucida into the intercellular spaces of the inner cell mass. Gradually the intercellular spaces become confluent and finally a single cavity, the **blastocele**, is formed (figs. 2-9*B* and 2-10*A*). At this time

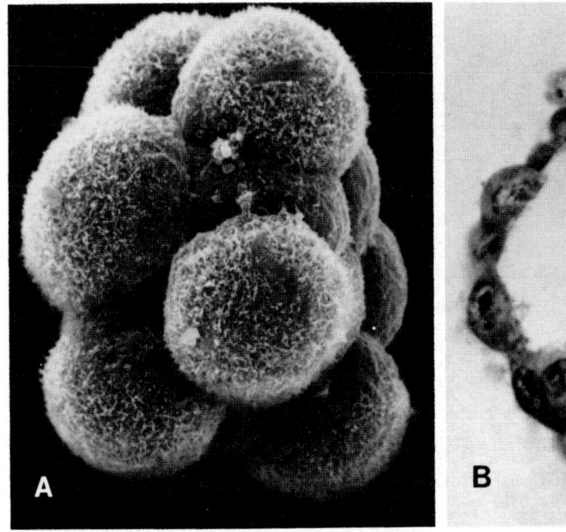

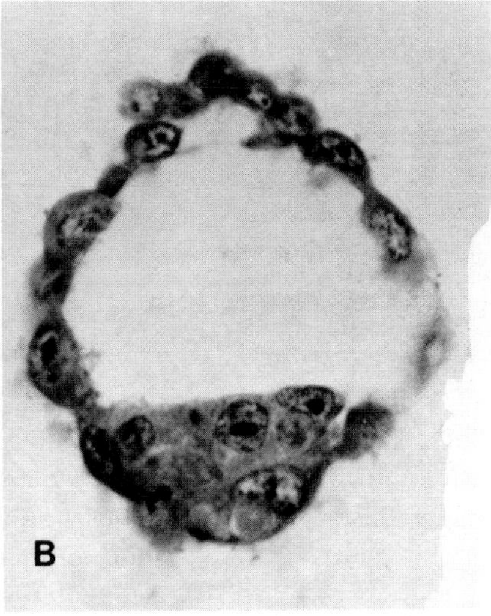

Figure 2-9. *A*, Scanning electron micrograph of a mouse morula after removal of the zona pellucida. The outlines of each blastomere are quite distinct. Note the numerous microvilli covering the surface (From Van Blerkom, J., and Motta, P. *The Cellular Basis of Mammalian Reproduction*. Urban and Schwarzenberg, Baltimore, 1979). *B*, Section of a 107-cell human blastocyst. Note the inner cell mass and the trophoblast cells. (From Hertig, A. I., Rock, J., and Adams, E. C. Courtesy Carnegie Institution of Washington.[41])

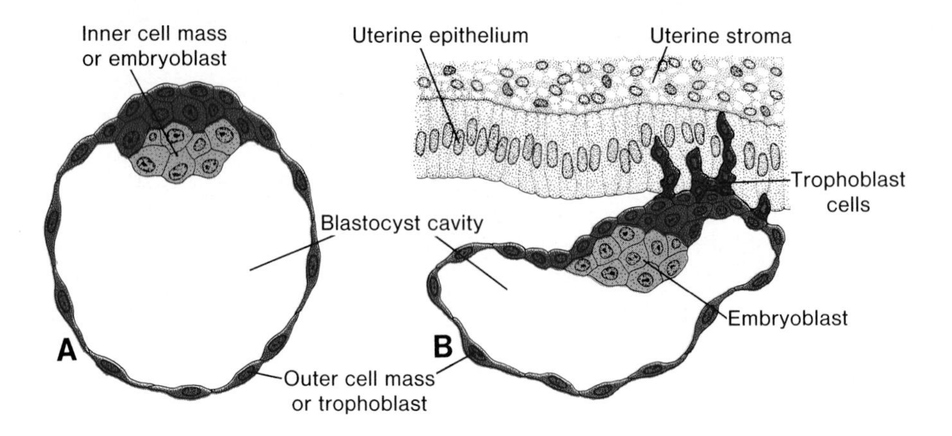

Figure 2-10. *A*, Schematic representation of a section through a human blastocyst recovered from the uterine cavity at approximately 4½ days. The blue cells represent the inner cell mass or embryoblast, and the red cells the trophoblast. (Modified after Hertig, A. T. and Rock J.) *B*, Schematic drawing of a section of a Macaque monkey blastocyst at the ninth day of development. The trophoblast cells, located at the embryonic pole of the blastocyst, begin to penetrate the uterine mucosa.[42] The human blastocyst begins to penetrate the uterine mucosa probably by the fifth or sixth day of development.

the embryo is known as the **blastocyst**. The cells of the inner cell mass, now referred to as the **embryoblast**, are located at one pole, while those of the outer cell mass, or **trophoblast**, flatten and form the epithelial wall of the blastocyst (fig. 2-10*A*). The zona pellucida has now disappeared, allowing implantation to begin.

Two human blastocysts, consisting of 58 and 107 cells with an estimated age of 4 and 4½ days, respectively, have been recovered from the uterine cavity (fig. 2-9*B*).[40, 41]

The 107-cell blastocyst contained an embryoblast with eight large vacuolated cells; the remaining 99 belonged to the trophoblast (fig. 2-9). Of the latter, 69 formed the wall of the blastocele, and 30 were grouped over the embryoblast. In the human the trophoblastic cells over the embryoblast pole begin to penetrate between the epithelial cells of the uterine mucosa at about the sixth day (fig.2-10*B*). It is probable that the penetration and subsequent erosion of the epithelial cells of the mucosa result from proteolytic enzymes produced by the trophoblast.[43] The uterine mucosa, however, promotes the proteolytic action of the blastocyst, so that implantation is the result of mutual trophoblastic and endometrial action. Hence, by the end of the first week of development, the human zygote has passed through the morula and blastocyst stages and has begun its implantation in the uterine mucosa (fig. 2-10).

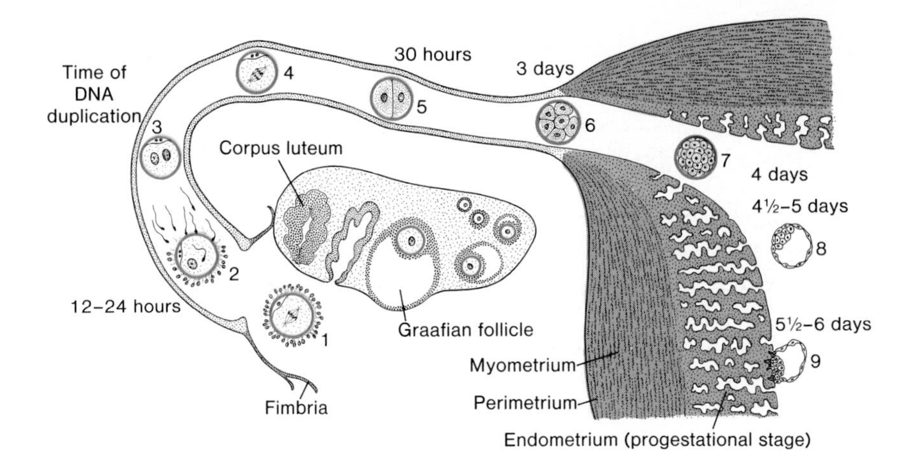

Figure 2-11. Schematic representation of the events taking place during the first week of human development. (1) Oocyte immediately after ovulation. (2) Fertilization approximately 12 to 24 hours after ovulation. (3) Stage of the male and female pronuclei. (4) Spindle of the first mitotic division. (5) Two-cell stage (approximately 30 hours of age). (6) Morula containing 12 to 16 blastomeres (approximately three days of age). (7) Advanced morula stage reaching the uterine lumen (approximately four days of age). (8) Early blastocyst stage (approximately 4½ days of age). The zona pellucida has now disappeared. (9) Early phase of implantation (blastocyst approximately six days of age). The ovary shows the stages of the transformation between a primary follicle and a Graafian follicle as well as a corpus luteum. The uterine endometrium is depicted in the progestational stage.

Abnormal Zygotes

Abnormal human and other mammalian zygotes have been frequently described. Of a total of eight zygotes in the preimplantation stage recovered from the uterine tube by Hertig and coworkers, four appeared to be normal, whereas the other four were abnormal.[40] The abnormal zygotes, which varied from three to five days of age, showed multinucleated blastomeres and variable degrees of cellular degeneration. Although it is doubtful that any of these zygotes would have been able to implant, all four were recovered from patients of normal fertility. According to Hertig[44] 16 per cent of all oocytes coming in contact with sperm are not cleaving, either because they are not properly penetrated by sperm or the mitotic mechanism is not functioning. Another 15 per cent are lost during the first week at cleavage and blastula stages. Since many abnormal zygotes are lost during the early stages of development, this process is often considered as a "self-cleaning" process, whereby abnormal embryos are eliminated without the mother being aware of it.

Uterus at Time of Implantation

The wall of the uterus consists of three layers: (1) the **endometrium** or mucosa lining the inside wall; (2) the **myometrium**, a thick layer of smooth muscle; and (3) the **perimetrium**, the peritoneal covering lining the outside wall (fig. 2-11).

At the time of implantation the mucosa of the uterus is in the **secretory** or **progestational** phase (figs. 2-11 and 2-12). This phase is caused by progesterone. The first signs of its action can be recognized two or three days after ovulation. At that time the uterine glands and arteries become coiled and the tissue becomes succulent. As a result three distinct layers can be recognized in the endometrium: a superficial **compact layer**, an intermediate **spongy layer**, and a thin **basal layer** (fig. 2-12).

If the oocyte is fertilized, the glands in the endometrium show increasing secretory activity and the arteries become tortuous and form a dense capillary bed just beneath the surface. As a result the endometrium becomes highly edematous. Normally the human blastocyst implants in the endometrium

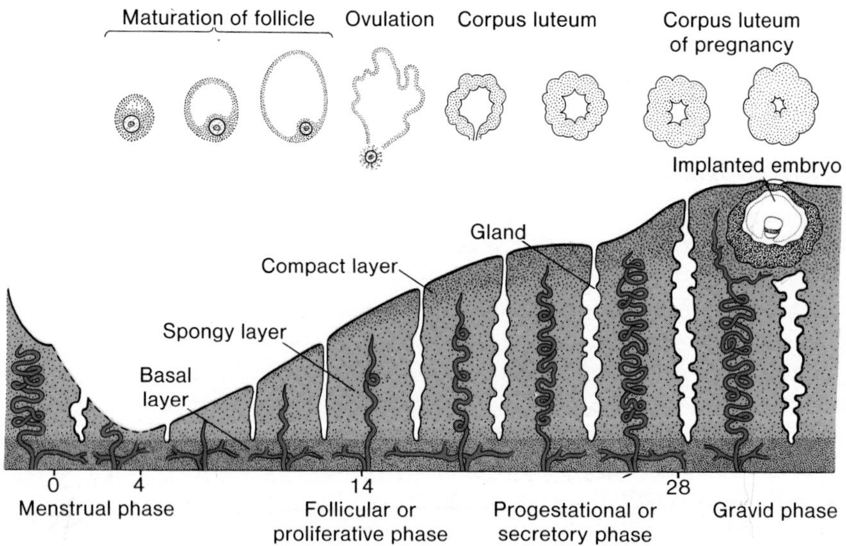

Figure 2-12. Schematic representation of the changes taking place in the uterine mucosa correlated with those in the ovary. Note that the implantation of the blastocyst has caused the development of a large corpus luteum of pregnancy. The secretory activity of the endometrium increases gradually as a result of the large amounts of progesterone produced by the corpus luteum of pregnancy.

along the posterior or anterior wall of the body of the uterus, where it becomes embedded between the openings of the glands (fig. 2-12).

If the oocyte is not fertilized, the venules and sinusoidal spaces become gradually packed with blood cells and an extensive diapedesis of blood into the tissue is seen. When the **menstrual phase** begins blood escapes from the superficial arteries and small pieces of stroma and glands break away. During the following three or four days the compact and spongy layers are expelled from the uterus and the basal layer is the only part of the endometrium which is retained (fig. 2-13). This layer is supplied by its own arteries, the **basal arteries**, and functions as the regenerative layer in the rebuilding of glands and arteries in the **proliferative phase** (fig. 2-13).

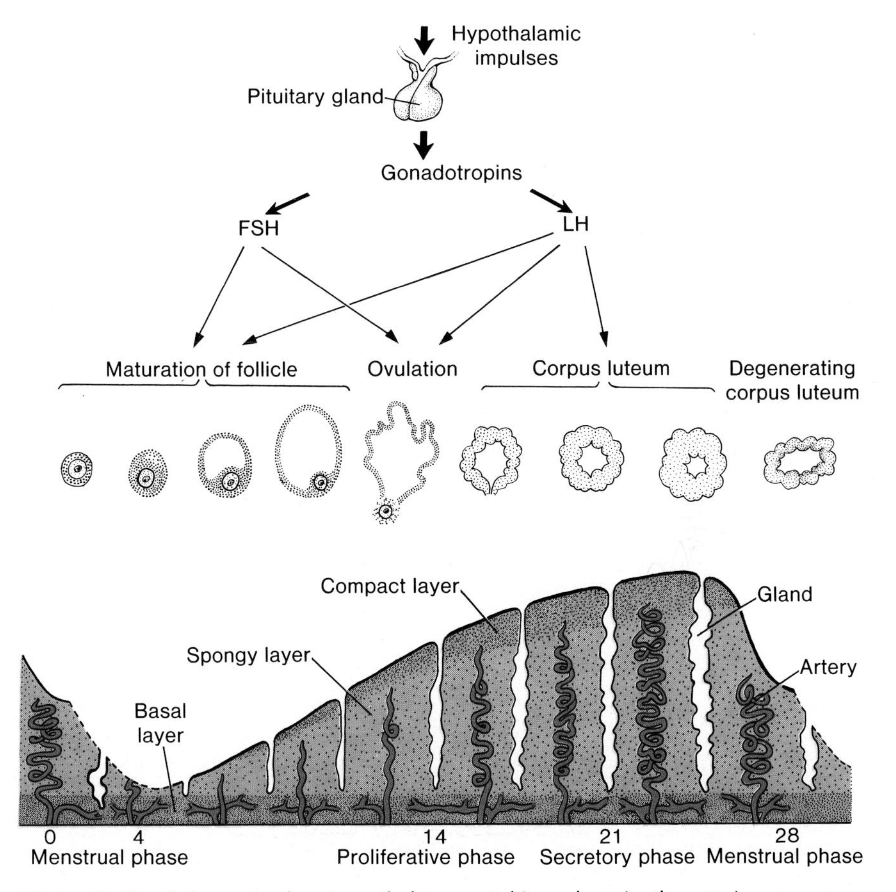

Figure 2-13. Schematic drawing of changes taking place in the uterine mucosa (endometrium) during a regular menstrual cycle in which fertilization fails to occur. Note the corresponding changes in the ovary.

Abnormal Implantation Sites

The human blastocyst usually implants along the posterior or anterior wall of the body of the uterus. Occasionally the blastocyst implants close to the internal os (fig. 2-14). At later stages of development the placenta then overbridges the os **(placenta previa)**, and causes severe bleeding in the second part of pregnancy and during delivery.

Not infrequently implantation sites are found outside the uterus, resulting in **extra-uterine** or **ectopic pregnancy**. This may occur at any place in the abdominal cavity, ovary, or uterine tube (fig. 2-14). Ectopic pregnancy usually leads to death of the embryo and severe hemorrhaging by the mother during the second month of pregnancy. In the abdominal cavity the blastocyst most frequently attaches itself to the peritoneal lining of the recto-uterine cavity **(Douglas' pouch)** (fig. 2-15). The blastocyst also may attach itself to the peritoneal covering of the intestinal tract or to the omentum. Rarely does an extra-uterine embryo come to full term.

Sometimes the blastocyst develops in the ovary proper, causing a **primary**

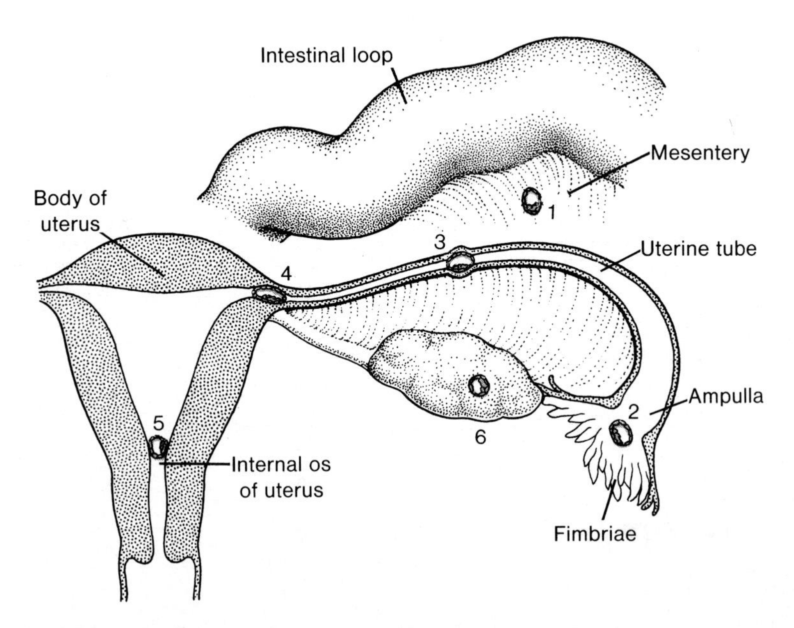

Figure 2-14. Drawing to show the abnormal implantation sites of the blastocyst. (1) Implantation site in the abdominal cavity. The ovum most frequently implants in the recto-uterine cavity (Douglas' pouch), but may implant at any place covered by peritoneum. (2) Implantation in the ampullary region of the tube. (3) Tubal implantation. (4) Interstitial implantation—that is, in the narrow portion of the uterine tube. (5) Implantation in the region of the internal os, frequently resulting in placenta previa. (6) Ovarian implantation. (Modified after Hamilton, Boyd, and Mossman.)

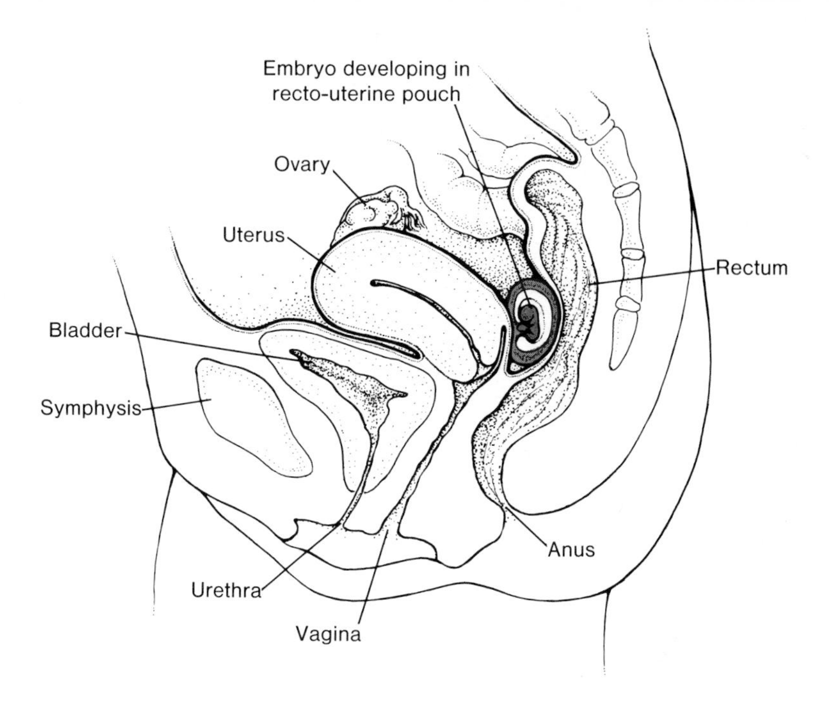

Figure 2-15. Midline section of bladder, uterus, and rectum to show an abdominal pregnancy in the recto-uterine pouch.

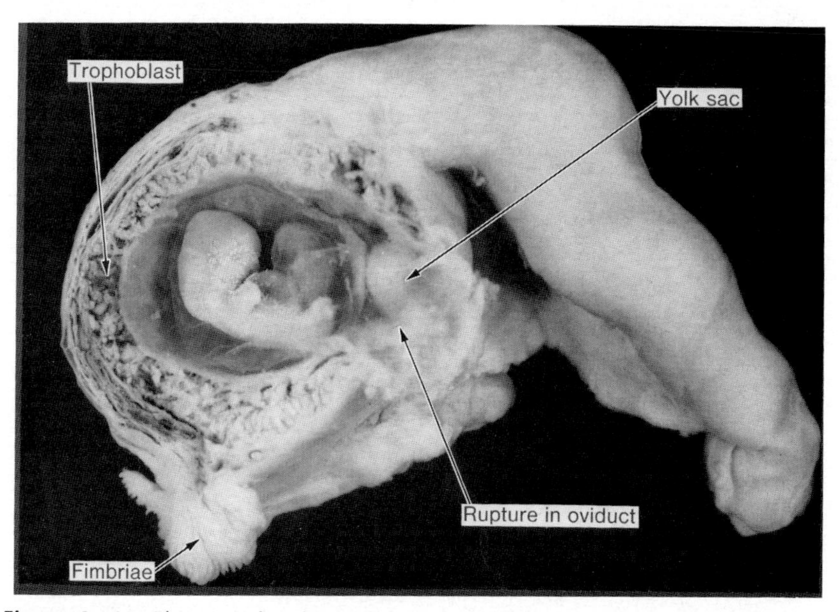

Figure 2-16. Photograph of tubal pregnancy. Embryo is approximately two months old and is about to escape through a rupture in the tubal wall.

ovarian pregnancy. More commonly an ectopic pregnancy is lodged in the uterine tube **(tubal pregnancy)**. In the latter case, the tube ruptures at about the second month of pregnancy, resulting in severe internal hemorrhaging by the mother (fig. 2-16).

SUMMARY

Although at each ovarian cycle a number of follicles begins to grow, only one reaches full maturity and only one oocyte is discharged at **ovulation**. At this moment the oocyte is in its **secondary meiotic** division and is surrounded by the zona pellucida and some granulosa cells (fig. 2-4). Through the sweeping action of the tubal fimbriae the oocyte is carried into the uterine tube.

Before spermatozoa can fertilize the oocyte they must undergo (1) a **capacitation process**, during which a glycoprotein coat and seminal plasma proteins are removed from the spermatozoon head, and (2) the **acrosome reaction** during which hyaluronidase and trypsin-like substances are released to penetrate the oocyte barriers. During fertilization the spermatozoon must penetrate (1) **the corona radiata**, (2) **the zona pellucida** and (3) **the oocyte cell membrane** (fig. 2-5). As soon as the spermatocyte has entered the oocyte (1) it finishes its second meiotic division and forms the **female pronucleus**; (2) the zona pellucida becomes impenetrable to other spermatozoa; (3) the head of the sperm separates from the tail, swells, and forms the **male pronucleus** (fig. 2-6). After both pronuclei have replicated their DNA, the paternal and maternal chromosomes intermingle, split longitudinally and go through a mitotic division, thus giving rise to the two-cell stage. The **results of fertilization** are: (1) **restoration of diploid number of chromosomes**; (2) **determination of chromosomal sex**; (3) **initiation of cleavage**.

Cleavage is a series of mitotic divisions, resulting in an increase in cells, **blastomeres**, which become smaller with each division. After three to four days the zygote has the appearance of mulberry **(morula)** and enters the uterine cavity. At this time the morula loses the zona pellucida and obtains the blastocyst cavity. The cells are then arranged in (1) an **outer cell mass**, which will form the trophoblast, and (2) an **inner cell mass** which will give rise to the **embryo proper**. The embryo is now known as blastocyst.

Implantation occurs at the end of the first week. Trophoblast cells then invade the epithelium and underlying endometrial stroma with the help of proteolytic enzymes. Implantation may also occur outside the uterus such as in the recto-uterine pouch, on the mesentery, in the uterine tube, or in the ovary **(ectopic pregnancies)**.

REFERENCES

1. Klopper, A. The reproductive hormones. *Sci. (London)*, *6:* 44, 1970.
2. Reed, M. Hypothalamic releasing factors. In *Scientific Foundations of Obstetrics and Gynecology*, edited by E. E. Philipp, J. Barnes, and M. Newton. William Heinemann, London, 1970.
3. Blandau, R. J. Anatomy of ovulation. *Clin. Obstet. Gynaecol.*, *10:* 347, 1967.
4. Blandau, R. J. Growth of the ovarian follicle and ovulation. *Prog. Gynecol.* *5:* 58, 1970.
5. Motta, P., and Van Blerkom, J. A Scanning electron microscopic study of the luteo-follicular complex. II. Events leading to ovulation. *Am. J. Anat.*, *143:* 241, 1975.
6. Odor, D. L., and Blandau, R. J. Egg transport over the fimbrial surface of the rabbit oviduct under experimental conditions. *Fertil. Steril.*, *24:* 292, 1973.
7. Talo, A. Electric and mechanical activity of the rabbit oviduct *in vitro* before and after ovulation. *Biol. Reprod.*, *11:* 335, 1974.
8. Hafez, E. S. E. Kinetics of luminal secretions in the female reproductive tract. Ultrastructural and physiological parameters. *Acta Anat. (Basel)*, *97:* 143, 1977.
9. Gaddum-Rosse, P., and Blandau, R. J. Comparative observation on ciliary currents in mammalian oviducts. *Biol. Reprod.*, *14:* 605, 1976.
10. Motta, P., and Van Blerkom, J. A scanning electron microscopic study of the luteo-follicular complex. I. Follicle and oocyte. *J. Submicrosc. Cytol*, *6:* 297, 1974.
11. Harper, M. J. K., Bennett, J. P., Boursnell, J. C., and Rowson, L. E. A. An autoradiographic method for the study of egg transport in the rabbit Fallopian tube. *J. Reprod. Fertil.*, *1:* 249, 1960.
12. Austin, C. R. Fertilization and transport of the ovum. In *Mechanisms Concerned with Conception*, edited by C. G. Hartman, p. 285. The Macmillan Co., New York, 1963.
13. Mastroianni, L. The Tube. In *Scientific Foundations of Obstetrics and Gynecology*, edited by E. E. Philipp, J. Barnes, and M. Newton. William Heinemann, London, 1970.
14. Crisp, T. M., Dessouky, D. A., and Denys, F. R. The fine structure of the human corpus luteum of early pregnancy and during the progestational phase of the menstrual cycle. *Am. J. Anat.*, *127:* 37, 1970.
15. Mossman, H. W., and Duke, K. L. Comparative morphology of the mammalian ovary. University of Wisconsin Press, Madison, 1973.
16. Pincus, G. Clinical effects of new progestational compounds. In *Clinical Endocrinology*, Vol. 1, edited by E. B. Astwood, p. 526. Grune & Stratton, New York, 1960.
17. Peel, J., and Potts, M. *Textbook in Contraceptive Practice.* Cambridge University Press, London, 1969.
18. Shettles, L. B. Fertilization and early development from the inner cell mass. In *Scientific Foundations of Obstetrics and Gynecology*, edited by E. E. Philipp, J. Barnes, and M. Newton. William Heinemann, London, 1970.
19. Austin, C. R. Sperm capacitation-biological significance in various species. *Adv. Biosci.*, *4:* 5, 1969.
20. Hartree, E. F. Spermatozoa, eggs and proteinases. *Biochem. Soc. Trans.*, *5:* 375, 1977.
21. Pavlok, A., and McLaren, A. The role of cumulus cells and the zona pellucida in fertilization of mouse eggs *in vitro*. *J. Reprod. Fertil.*, *29:* 91, 1972.
22. Gould, S. F., and Berstein, M. H. The localization of bovine sperm hyaluronidase. *Differentiation*, *3:* 123, 1975.
23. McRoric, R. A., and Williams, W. L. Biochemistry of mammalian fertilization. *Ann. Rev. Biochem*, *43:* 777, 1974.
24. Bedford, J. M. Sperm capacitation and fertilization in mammals. *Biol. Reprod.* (Suppl.), *2:* 128, 1970.
25. Garner, D. L., and Easton, M. P. Immunofluorescent localization of acrosin in mammalian spermatozoa. *J. Exp. Zool.*, *200:* 157, 1977.
26. Baranska, W., Konwinski, M., and Kujawa, M. Fine structure of the zona pellucida of unfertilized egg cells and embryos. *J. Exp. Zool.*, *192:* 193, 1975.
27. Yanagimachi, R. Specificity of sperm-egg interation. In *The Immunobiology of the Gametes*, edited by M. Edidin and M. M. Johnson. Cambridge University Press, Cambridge, England, 1977.

28. Carr, D. H. Chromosome studies on selected spontaneous abortions: Polyploidy in man. *J. Med. Genet.*, *8:* 164, 1971.
29. Thompson, R. S., Moore-Smith, D., and Zamboni, L. Fertilization of mouse ova *in vitro*: An electron microscopic study. *Fertil. Steril.*, *25:* 222, 1974.
30. Zamboni, L. *Fine Morphology of Mammalian Fertilization.* Harper and Row, New York, 1971.
31. Mintz, B. Nucleic acid and protein synthesis in the developing mouse embryo. In *Preimplantation Stages of Pregnancy*, edited by G. E. W. Wolstenholme and M. O'Connor, p. 145. Little, Brown and Co., Boston, 1965.
32. Dickmann, Z., Clewe, T. H., Bonney, W. A., and Noyes, R. W. The human egg in the pronuclear stage. *Anat. Rec.*, *152:* 293, 1965.
33. Edwards, R. G., Bavister, B. D., and Steptoe, P. C. Early stages of fertilization *in vitro* of human oocytes matured *in vitro. Nature (London)*, *221:* 632, 1969.
34. Elliott, J. Finally: some details on *in vitro* fertilization. *J.A.M.A.*, *241:* 9, 1979.
35. Grobstein, C. External human fertilization. *Sci. Am.*, *240:* 59, 1979.
36. Chang, M. C. Development of parthenogenetic rabbit blastocysts induced by low temperature storage of unfertilized ova. *J. Exp. Zool.*, *125:* 127, 1954.
37. Braden, A. W. H., and Austin, C. R. Reactions of unfertilized mouse eggs to some experimental stimuli. *Exp. Cell Res.*, *7:* 277, 1954.
38. Beatty, R. A. *Parthenogenesis and Polyploidy in Mammalian Development.* Cambridge University Press, Cambridge, 1957.
39. Simard, L. C. Polyembryonic embryos of the ovary of parthenogenic origin. *Cancer, 10:* 215, 1957.
40. Hertig, A. T., Adams, E. C., and Mulligan, W. J. On the pre-implantation stages of the human ovum: a description of four normal and four abnormal specimens ranging from the second to the fifth day of development. *Contrib. Embryol.*, *35:* 199, 1954.
41. Hertig, A. T., Rock, J., and Adams, E. C. A description of 34 human ova within the first 17 days of development. *Am. J. Anat.*, *98:* 435, 1956.
42. Heuser, C. H., and Streeter, G. L. Development of the Macaque embryo. *Contrib. Embryol.*, *29:* 15, 1941.
43. Billington, W. D. Trophoblast. In *Scientific Foundations of Obstetrics and Gynecology*, edited by E. E. Philipp, J. Barnes, and M. Newton. William Heinemann, London, 1970.
44. Hertig, A. T. The overall problem in man. In *Comparative Aspects of Reproductive Failure*, edited by K. Benirschke. Springer Verlag, New York, 1967.

chapter 3

Bilaminar Germ Disc (Second Week of Development)

In the following paragraphs a day-by-day account is given of the major events occurring in the second week of development. It must be realized, however, that embryos of the same fertilization age do not necessarily develop at the same rate. Indeed, considerable differences in the rate of growth have been found even at these early stages of development.[1, 2]

Eighth Day of Development

At the eighth day of development the blastocyst is partially embedded in the endometrial stroma.[3] In the area over the embryoblast, the trophoblast

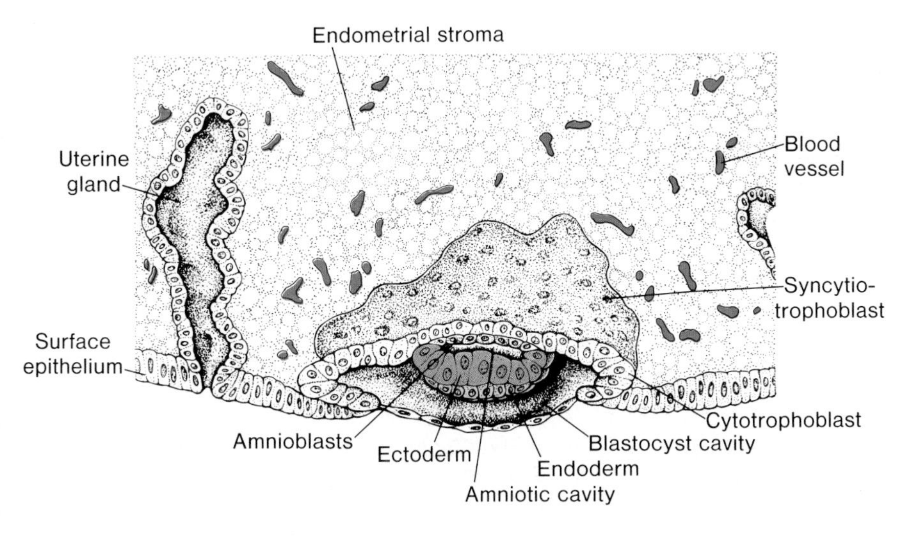

Figure 3-1. Drawing representing a 7½-day human blastocyst, partially embedded in the endometrial stroma. The trophoblast consists of an inner layer with mononuclear cells, the cytotrophoblast, and an outer layer without distinct cell boundaries, the syncytiotrophoblast. The embryoblast is formed by the ectodermal and endodermal germ layers. The amniotic cavity appears as a small cleft.

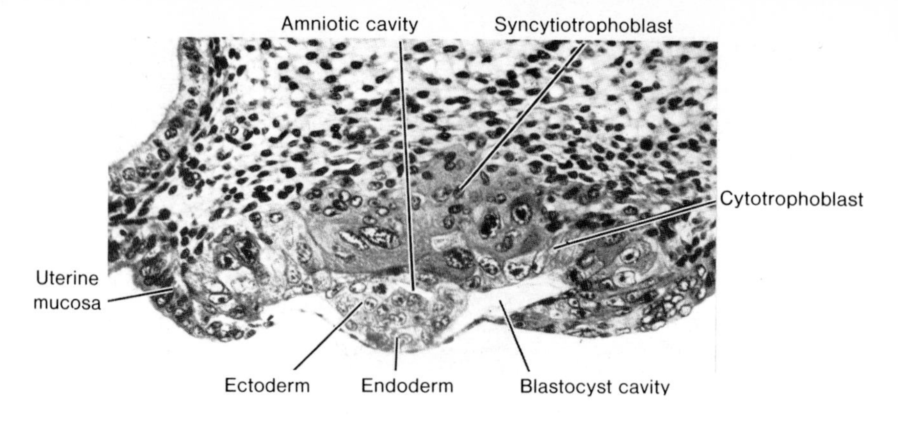

Figure 3-2. A section of a 7½-day human blastocyst (×100). Note the multinuclearity of the syncytiotrophoblast, the large cells of the cytotrophoblast, the slit-like amniotic cavity, and the endoderm cells. (From Hertig, A. T., and Rock, J. Courtesy Carnegie Institution of Washington.[3])

has differentiated into two layers: (1) an inner layer of mononucleated cells, the **cytotrophoblast**, and (2) an outer, multinucleated zone without distinct cell boundaries, the **syncytiotrophoblast** or **syncytium** (figs. 3-1 and 3-2). Mitotic figures are usually found in the cytotrophoblast but never in the syncytium, yet the thickness of the latter increases considerably. This suggests that the trophoblast cells divide in the cytotrophoblast and then migrate into the syncytiotrophoblast, where they fuse and lose their individual cell membranes.

The cells of the inner cell mass or embryoblast also differentiate into two layers: (1) a layer of small, cuboidal cells, known as the **endodermal germ layer**; and (2) a layer of high columnar cells, the **ectodermal germ layer** (figs. 3-1 and 3-2). The cells of each of the germ layers form a flat disc and together they are known as the **bilaminar germ disc.**

At the same time a small cavity appears within the ectoderm. This cavity enlarges to become the **amnionic cavity.** Those ectodermal cells adjacent to the cytotrophoblast are called **amnioblasts**, and together with the rest of the ectoderm they line the amnionic cavity (figs. 3-1 and 3-3).

The endometrial stroma adjacent to the implantation site is edematous and highly vascular and the large tortuous glands secrete abundant glycogen and mucus.

Ninth Day of Development

The blastocyst is more deeply embedded in the endometrium, and the penetration defect in the surface epithelium is closed by a fibrin coagulum (fig. 3-3).[3] The trophoblast shows considerable progress in development, particularly at the embryonic pole, where vacuoles appear in the syncytium. When these vacuoles fuse they form large lacunae, and this phase of the trophoblast development is therefore known as the **lacunar stage** (fig. 3-3).

At the abembryonic pole, meanwhile, flattened cells probably originating from the endoderm form a thin membrane, known as the exocoelomic (Heuser's) membrane, which lines the inner surface of the cytotrophoblast (fig. 3-3). This membrane, together with the endoderm, forms the lining of the **exocoelomic cavity (primitive yolk sac)**.

Eleventh to Twelfth Days of Development

By the 11th to 12th day of development the blastocyst is completely embedded in the endometrial stroma, and the surface epithelium covers

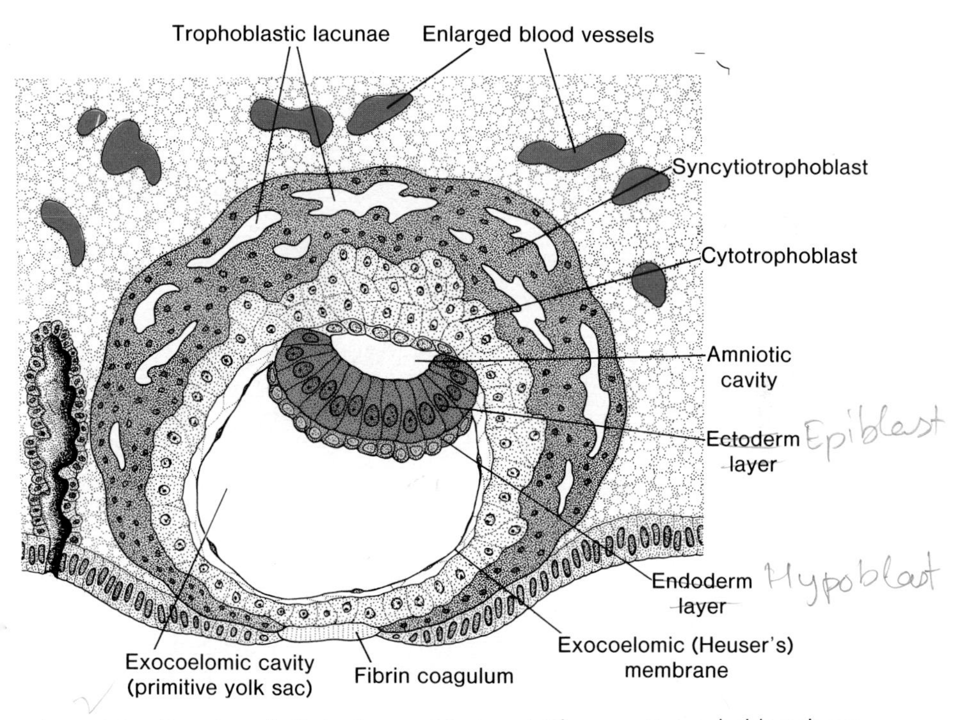

Figure 3-3. Drawing of a 9-day human blastocyst. The syncytiotrophoblast shows a large number of lacunae. Note the flat cells which form the exocoelomic membrane. The bilaminar germ disc consists of a layer of columnar ectodermal cells and a layer of cuboidal endodermal cells. The original surface defect is closed by a fibrin coagulum.

almost entirely the original defect in the uterine wall (figs. 3-4 and 3-5). The blastocyst now produces a slight protrusion into the lumen of the uterus.

The trophoblast is characterized by lacunar spaces in the syncytium which form an intercommunicating network.[5] This is particularly evident at the embryonic pole; at the abembryonic pole, however, the trophoblast still consists mainly of cytotrophoblastic cells (figs. 3-4 and 3-5).

Concurrently the syncytial cells penetrate deeper into the stroma and erode the endothelial lining of the maternal capillaries. They are congested and dilated and are known as **sinusoids.**[6] The syncytial lacunae then become continuous with the sinusoids and maternal blood enters the lacunar system (fig. 3-4). As the trophoblast continues to erode more and more sinusoids maternal blood begins to flow through the trophoblastic system, thus establishing the **uteroplacental circulation.**

In the meantime, a new population of cells appears between the inner surface of the cytotrophoblast and the outer surface of the exocoelomic

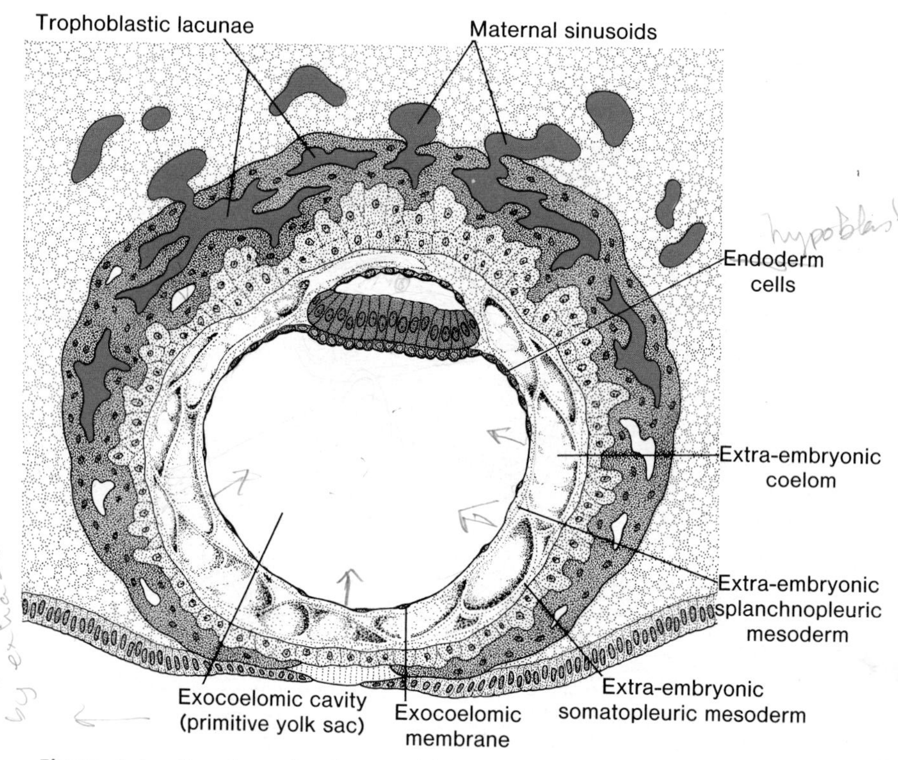

Figure 3-4. Drawing of a human blastocyst of approximately 12 days. The trophoblastic lacunae at the embryonic pole are in open connection with the maternal sinusoids in the endometrial stroma. Extra-embryonic mesoderm proliferates and fills the space between the exocoelomic membrane and the inner aspect of the trophoblast.

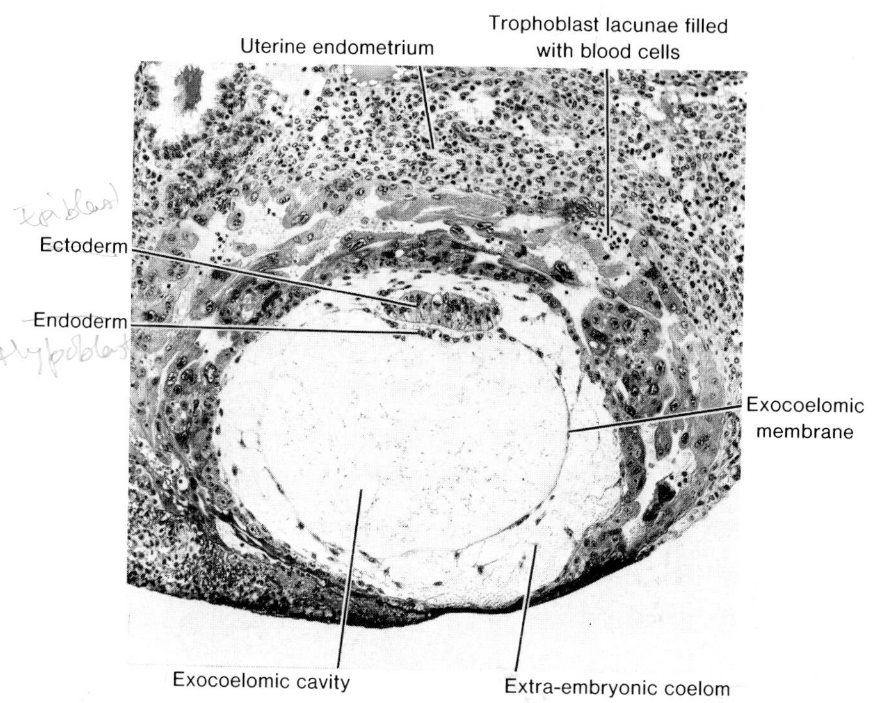

Figure 3-5. Section of a fully implanted 12-day human blastocyst (×100). Note the maternal blood cells in the lacunae, the exocoelomic membrane lining the primitive yolk sac, and the endodermal and ectodermal germ layers. (From Hertig, A. T., and Rock, J. Courtesy Carnegie Institution of Washington.[5])

cavity. These cells form a fine, loose connective tissue, the **extra-embryonic mesoderm**, which eventually fills all of the space between the trophoblast externally and the amnion and exocoelomic membrane internally (figs. 3-4 and 3-5). Soon, large cavities develop in the extra-embryonic mesoderm and when these become confluent, a new space, known as the **extra-embryonic coelom**, is formed (fig. 3-4). This space surrounds the primitive yolk sac and amniotic cavity except where the germ disc is connected to the trophoblast by the connecting stalk (fig. 3-6). The extra-embryonic mesoderm lining the cytotrophoblast and amnion is called the **extra-embryonic somatopleuric mesoderm**; that covering the yolk sac is known as the **extra-embryonic splanchnopleuric mesoderm** (fig. 3-4).[7]

The growth of the bilaminar germ disc is relatively slow compared to that of the trophoblast and is still very small (0.1 to 0.2 mm).

The cells of the endometrium, meanwhile, become polyhedral and loaded with glycogen and lipids; the intercellular spaces are filled with extravasate and the tissue is edematous. These changes, known as the **decidua reaction**, are first confined to the area immediately surrounding the implantation site, but soon occur throughout the endometrium.

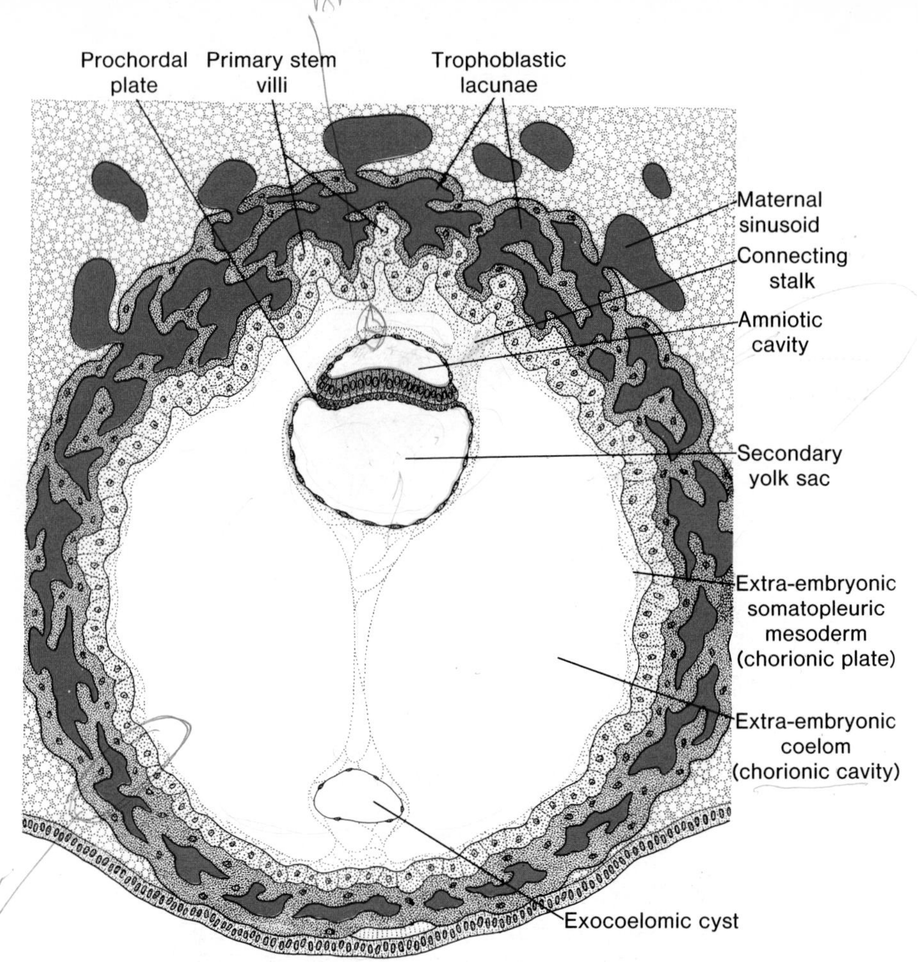

Figure 3-6. Drawing of a 13-day human blastocyst. The trophoblastic lacunae are now present at the embryonic as well as the abembryonic pole, and the uteroplacental circulation has begun. Note the primary stem villi and the extra-embryonic coelom **or chorionic cavity.** The secondary yolk sac is entirely lined with endoderm.

Thirteenth Day of Development

By the 13th day of development the surface defect in the endometrium has usually healed. Occasionally, however, a bleeding may occur at the implantation site as a result of increased blood flow into the lacunar spaces. Since this bleeding occurs at about the 28th day of the menstrual cycle, it may be confused with a normal menstrual bleeding and so cause an inaccuracy in determining the expected delivery date.

The trophoblast is characterized by the first appearance of villous structures. The cells of the cytotrophoblast proliferate locally and penetrate into the syncytium, thus forming cellular columns surrounded by syncytium. The cellular columns with the syncytial covering become known as the **primary stem villi** (figs. 3-6, 3-7, and 4-7*A*).

In the meantime the endodermal germ layer produces additional cells that migrate along the inside of the exocoelomic membrane (fig. 3-4). These cells proliferate and gradually form a new cavity within the exocoelomic cavity. This new cavity is known as the **secondary** or **definitive yolk sac** (figs. 3-6 and 3-7).

This yolk sac is much smaller than the original exocoelomic cavity or primitive yolk sac. During its formation large portions of the exocoelomic cavity are pinched off. These portions are represented by the so-called **exocoelomic cysts**, which are often found in the extra-embryonic coelom or **chorionic cavity** (figs. 3-6 and 3-7).[1]

In the meantime the extra-embryonic coelom expands and forms a large cavity known as the **chorionic cavity.** The extra-embryonic mesoderm lining the inside of the cytotrophoblast is then known as the **chorionic plate.** The only place where extra-embryonic mesoderm traverses the chorionic cavity is in the **connecting stalk** (fig. 3-6). With the development of blood vessels the stalk will become the **umbilical cord.**

By the end of the second week, the germ disc is represented by two

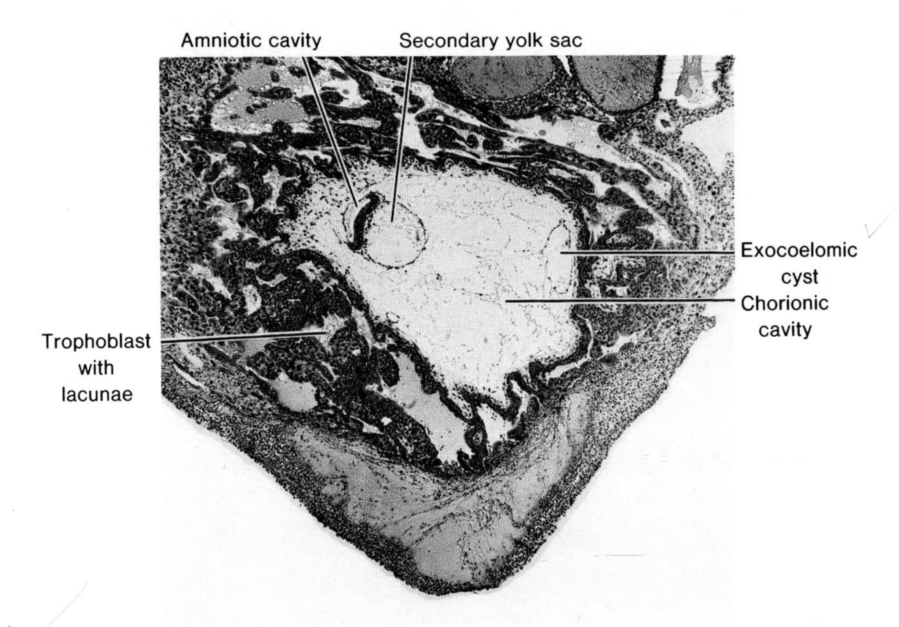

Figure 3-7. Section through the implantation site of a 13-day embryo. Note the amniotic cavity, yolk sac, and exocoelomic cyst in chorionic cavity. Most of the lacunae are filled with blood. (From Hertig, A.T., Rock, J., and Adams, E. C. Courtesy Carnegie Institution of Washington.[1])

apposed cell discs: the ectodermal germ layer, which forms the floor of the continuously expanding amniotic cavity, and the endodermal germ layer, which forms the roof of the secondary yolk sac. In its cephalic region the endodermal disc shows a slight thickening known as the **prochordal plate.** This is an area of columnar cells which are firmly attached to the overlying ectodermal disc (fig. 3-6).

Abnormal Blastocysts

Hertig, Rock, and Adams described a unique series of 26 implanted blastocysts varying in age from 7½ to 17 days.[1] All these blastocysts were recovered from patients of normal fertility. Surprisingly, nine (34.6 per cent) of the specimens were abnormal. Some consisted of syncytium only, whereas others showed variable degrees of trophoblastic hypoplasia. In two of them the embryoblast was absent and in some the germ disc showed an abnormal orientation.

It is likely that the most abnormal blastocysts would not have caused any sign of pregnancy, as their trophoblast was of such inferior quality that the corpus luteum could not have persisted. These ova probably would have been aborted with the following menstrual flow.

On the basis of many years of experience in examining hysterectomy material for embryos, Hertig[8] summarized his data on reproductive failure as follows. In selected fertile women under optimal conditions for pregnancy, 15 per cent of oocytes fail to become fertilized and 10 to 15 per cent start cleavage but fail to implant. Of the 70 to 75 per cent that implant, only 58 per cent will survive till the second week and 16 per cent of those will be abnormal. Hence, at the time when the first expected menstruation is missed, only 42 per cent of the eggs exposed to sperm are surviving.[9] Of this percentage a number of cases will be aborted during subsequent weeks and a number will be abnormal at the time of birth.

SUMMARY

At the beginning of the second week the blastocyst is partially embedded in the endometrial stroma. The **trophoblast** differentiates in (1) an inner, actively proliferating layer, the **cytotrophoblast**, and (2) an outer layer, the **syncytiotrophoblast**, which erodes the maternal tissues (fig. 3-1). By day 9 lacunae develop in the syncytiotrophoblast. When subsequently maternal sinusoids are eroded by the syncytiotrophoblast, maternal blood enters the lacunar network and by the end of the week a primitive **uteroplacental circulation** begins (fig. 3-6). The cytotrophoblast, meanwhile, forms cellular columns penetrating into and surrounded by the syncytium. These columns are the **primary stem villi.** By the end of the second week the blastocyst is completely embedded and the surface defect in the mucosa has healed (fig. 3-6).

The **inner cell mass** or **embryoblast**, meanwhile, differentiates into (1) the **endodermal germ layer** and (2) the **ectodermal germ layer**, together forming the **bilaminar germ disc** (fig. 3-1). The ectoderm cells are continuous with the amnioblasts and together they surround a new cavity, the **amnion cavity.** The endoderm cells are continuous with the **exocoelomic membrane** and together they surround the **primitive yolk sac** (fig. 3-4). By the end of the second week the extra-embryonic mesoderm is formed, which fills the space between the trophoblast and the amnion and exocoelomic membrane internally. When vacuoles develop in this tissue, the **extra-embryonic coelom** or **chorionic cavity** is formed (fig. 3-6). The extra-embryonic mesoderm lining the cytotrophoblast and amnion is the **somatopleuric meoderm** and that covering the yolk sac is the **splanchnopleuric mesoderm** (fig. 3-6).

REFERENCES

1. Hertig, A. T., Rock, J., and Adams, E. C. A description of 34 human ova within the first 17 days of development. *Am. J. Anat., 98:* 435, 1956.
2. O'Rahilly, R. *Developmental Stages in Human Embryos. Part A. Embryos of the First Three Weeks.* Carnegie Institution of Washington, Washington, D.C., 1973.
3. Hertig, A. T., and Rock, J. Two human ova of the previllous stage, having a developmental age of about seven and nine days respectively. *Contrib. Embryol., 31:* 65, 1945.
4. Luckett, W. P. Amniogenesis in the early human and rhesus monkey embryos. *Anat. Rec., 175:* 375, 1973.
5. Hertig, A. T., and Rock, J. Two human ova of the previllous stage, having an ovulation age of eleven and twelve days respectively. *Contrib. Embryol., 29:* 127, 1941.
6. Hamilton, W. J., and Boyd, J. D. Development of the human placenta. In *Scientific Foundations of Obstetrics and Gynaecology,* edited by E. E. Philipp, J. Barnes, and M. Newton, p. 185. F. A. Davis Co., Philadelphia, 1970.
7. Luckett, W. P. The origin of extraembryonic mesoderm in the early human and rhesus monkey embryos. *Anat. Rec., 169:* 369, 1971.
8. Hertig, A. T. The overall problem in man. In *Comparative Aspects of Reproductive Failure,* edited by K. Benirschke, p. 11. Springer Verlag, New York, 1967.
9. Witschi, E. Teratogenic effects from overripeness of the egg. In *Congenital Malformations,* edited by F. C. Fraser and V. A. McKusick, p. 157. Excerpta Medica, Amsterdam, 1970.

chapter 4

Trilaminar Germ Disc
(Third Week of Development)

Formation of Mesoderm Germ Layer

The most characteristic event occurring during the third week is the formation of the **primitive streak** on the surface of the ectoderm (figs. 4-1, 4-2*A*, and 4-6). Initially the streak is vaguely defined (fig. 4-1), but in a 15- to 16-day embryo it is clearly visible as a narrow groove with slightly bulging regions on either side (fig. 4-2*A*). The cephalic end of the streak, known as **primitive node**, consists of a slightly elevated area surrounding a small pit (fig. 4-2). In a transverse section through the region of the primitive groove it is seen that the cells are flask-shaped and that a new cell layer develops between the ectodermal and endodermal layers (fig. 4-2*B*). It is now generally believed that cells of the ectodermal layer migrate in the direction of the primitive streak (fig. 4-3). On arrival in the region of the streak, they become flask-shaped, detach from the ectodermal layer and slip underneath it (fig.

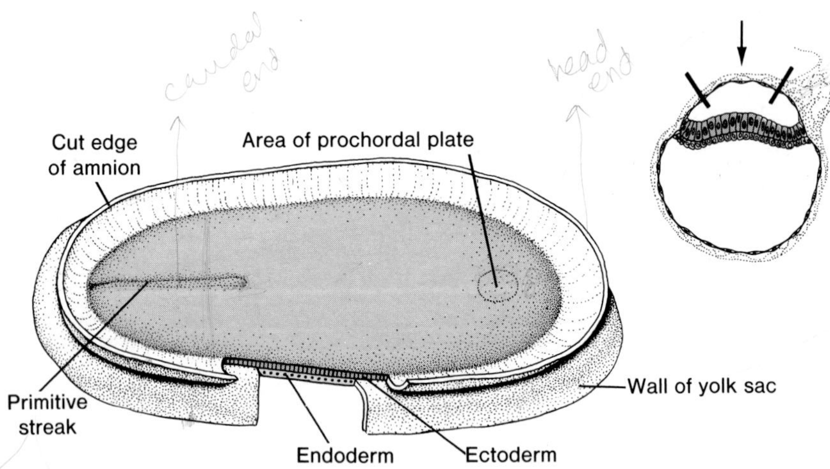

Figure 4-1. Schematic view of the germ disc at the end of the second week of development. The amniotic cavity has been opened to permit a view on the dorsal side of the ectodermal germ layer (see inset). Note that the endodermal and ectodermal layers are in contact with each other.

4-2*B*). This inward movement is known as **invagination**. Once the cells have arrived between the ectodermal and endodermal layers they form an intermediate cell layer, known as **intra-embryonic mesoderm**. This is the **mesodermal or third germ layer** (fig. 4-2*B*).[1]

As more and more cells move in between the ectodermal and endodermal layers, they begin to spread in lateral and cephalic directions (fig. 4-3). Gradually they migrate beyond the margin of the disc and establish contact

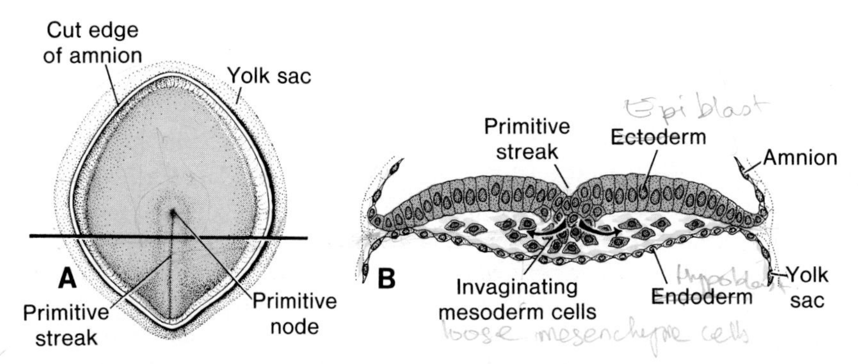

Figure 4-2. *A*, Schematic drawing of the dorsal side of a 16-day presomite embryo (modified after Streeter). The primitive streak and node are clearly visible. *B*, Transverse section through the region of the primitive streak as indicated in *A*, showing the invagination and subsequent lateral migration of the mesoderm cells.

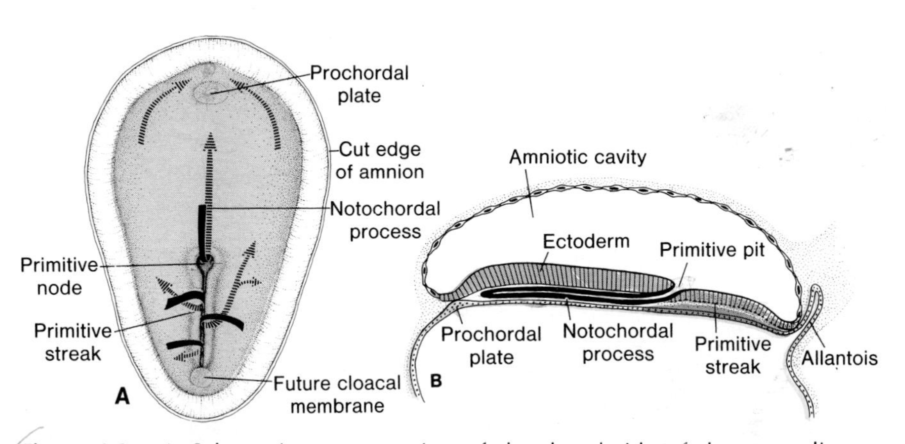

Figure 4-3. *A*, Schematic representation of the dorsal side of the germ disc, indicating the movement of surface cells (solid black lines) toward the primitive streak and node and the subsequent migration of the mesoderm cells between the endodermal and ectodermal germ layers (broken lines). *B*, Cephalo-caudal midline section through a 16-day embryo. The notochordal process occupies the midline region extending from the prochordal plate to the primitive node. Note the notochordal or central canal in the center of the notochordal process.

with the extra-embryonic mesoderm covering the yolk sac and amnion (figs. 4-4*B* and 4-5*B*). In cephalic direction they pass on each side of the prochordal plate to meet each other in front of this area, where they form the **cardiogenic** or **heart forming plate** (figs. 4-5*A* and 12-2).

Formation of Notochord

The cells invaginating in the primitive pit move straight forward in cephalic direction until they reach the **prochordal plate** (fig. 4-3). In this manner they form a tube-like process, known as the **notochordal** or **head process** (fig. 4-3*B*). The small, central canal is considered as the forward extension of the primitive pit (fig. 4-3*B*).

By the 17th day of development the mesoderm layer and the notochordal process separate the endoderm and ectoderm layers entirely with the exception of the prochordal plate in the cephalic region and the **cloacal plate** in the region caudal to the primitive streak (fig. 4-4*A*). The latter plate also consists of the tightly adhering endodermal and ectodermal layers.

By the 18th day of development the floor of the notochordal process fuses with the underlying endoderm and in the merged areas the two layers disintegrate (fig. 4-4*A*).[2] Gradually the lumen of the notochordal process disappears completely (figs. 4-4*A* and 4-5*A*). The remaining portion of the

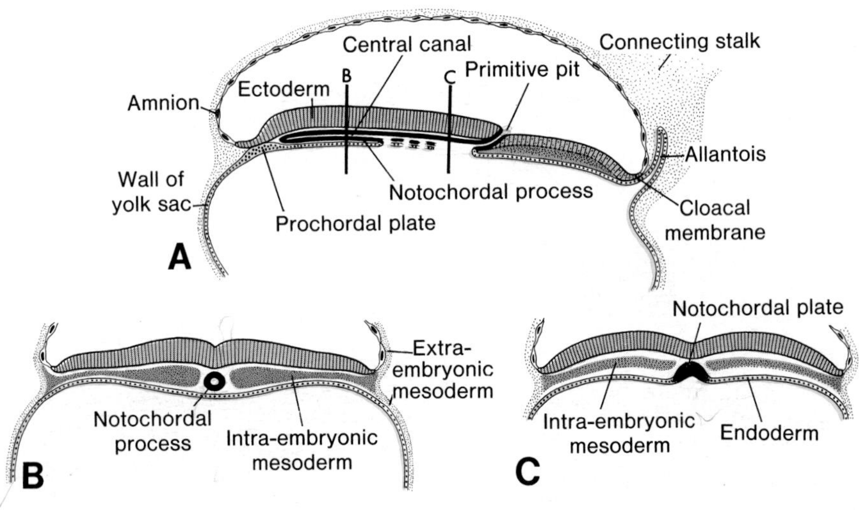

Figure 4-4. *A*, Drawing of a cephalo-caudal midline section through a 17-day embryo. The life span of the notochordal process is short and it frequently breaks down before it is completed. *B*, Transverse section through region in which the central canal still persists; the intra-embryonic mesoderm is in contact with the extra-embryonic mesoderm covering the yolk sac and the amnion. *C*, Transverse section through region where the remaining portion of the notochordal process is temporarily intercalated in the endoderm.

notochordal process forms a narrow plate of cells, intercalated in the endodermal germ layer (fig. 4-4*C*).

With further development the notochordal cells proliferate and form a solid cord, known as the **definitive notochord** (fig. 4-5).[3] This structure in turn becomes detached from the endoderm, which once again forms an uninterrupted layer in the roof of the yolk sac (fig. 4-5*B*). The notochord forms now a midline axis, which will serve as the basis of the axial skeleton.[4] It extends from the prochordal plate (the future buccopharyngeal membrane) to the primitive node. A small canal, the **neurenteric canal**, temporarily connects the yolk sac and the amniotic cavity (fig. 4-5*A*).

Concomitantly with the formation of the cloacal membrane, the posterior wall of the yolk sac forms a small diverticulum which extends into the connecting stalk. This diverticulum, the **allantoenteric diverticulum**, or **allantois**, appears at about the 16th day of development (figs. 4-4*A* and 4-5*A*).[5] Although in some lower vertebrates the allantois serves as a reservoir for the excretion products of the renal system, in man it remains rudimentary and plays no role in the development.

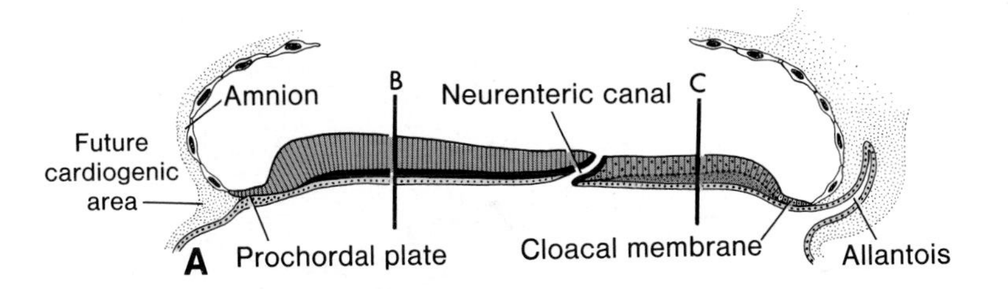

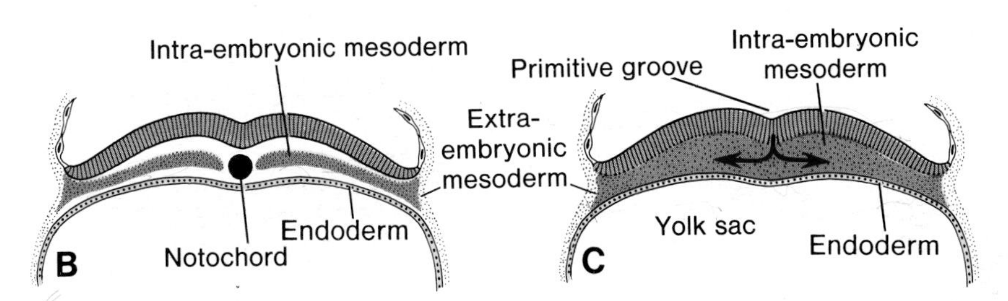

Figure 4-5. *A*, Drawing of a cephalo-caudal midline section through an 18-day embryo. The definitive notochord is established. The neurenteric canal connects the amniotic cavity with the yolk sac. *B*, Transverse section through the cephalic part of the embryo, showing the definitive notochord flanked by intra-embryonic mesoderm. *C*, Transverse section through the primitive streak region. Mesoderm formation in the caudal part of the embryo continues into the fourth week.

Growth of Germ Disc

The embryonic disc, initially flat and almost round (fig. 4-2A), gradually becomes elongated with a broad cephalic and a narrow caudal end (fig. 4-6*A*, *B*). Expansion of the embryonic disc occurs mainly in the cephalic region; the region of the primitive streak remains more or less the same size. It must be realized, however, that growth and elongation of the cephalic part of the disc are caused by a continuous migration of cells from the primitive streak region in cephalic direction. Invagination of surface cells in the primitive streak and their subsequent migration in forward and lateral directions continue until the end of the fourth week. At that stage the primitive streak shows regressive changes, rapidly diminishes in size, and soon disappears. It is, however, not uncommon that remnants of the primitive streak persist and at birth cause tumors in the sacrococcygeal region.[6] These tumors frequently contain tissues derived from all three germ layers.

The fact that the caudal end of the disc continues to supply new cells until the end of the fourth week has an important bearing on the further development of the embryo. In the cephalic part the germ layers begin their specific differentiation by the middle of the third week, whereas in the caudal part this occurs by the end of the fourth week. In this chapter

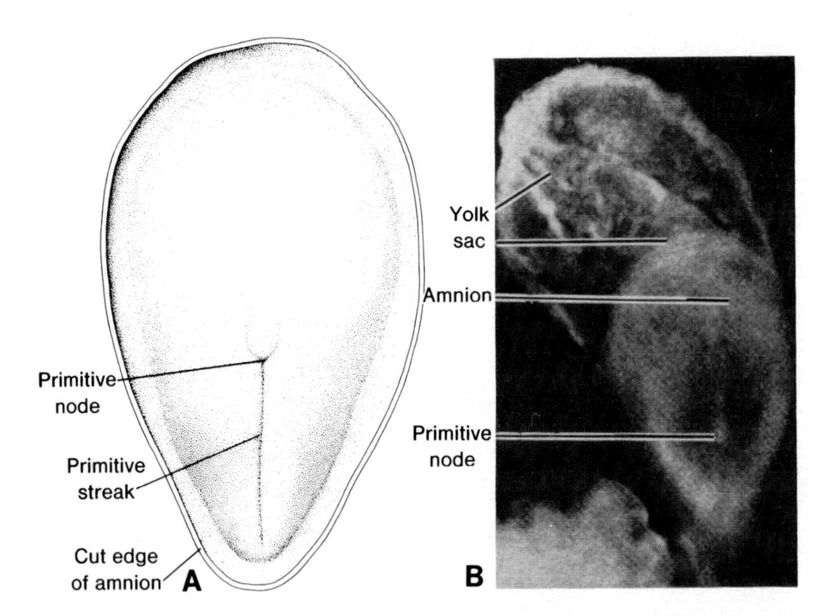

Figure 4-6. *A*, Drawing of the dorsal aspect of an 18-day embryo. The embryo has a pear-shaped appearance and shows at its caudal end the primitive streak and node. *B*, Photograph of 18-day human embryo, dorsal view (Heuser).[2] Note the primitive node and, extending forward from it, the notochord. The yolk sac has a somewhat mottled appearance. The length of the embryo is 1.25 mm and the greatest width is 0.68 mm. (Courtesy Carnegie Institution of Washington.)

attention has been focused on formation of the germ layers, but it must be realized that formation of the somites from the mesoderm and formation of the central nervous system from the ectoderm begin during the third week.

Further Development of Trophoblast

By the beginning of the third week the trophoblast is characterized by **primary stem villi** which consist of a cytotrophoblastic core covered by a syncytial layer (figs. 4-7*A* and 3-6). During further development mesodermal cells penetrate the core of the primary villi and grow in the direction of the decidua. The newly formed structure is known as the **secondary stem villus** (fig. 4-7*B*).

By the end of the third week the mesodermal cells in the core of the villus begin to differentiate into blood cells and small blood vessels, thus forming the villous capillary system (fig. 4-7*C*). The villus is now known as the **tertiary stem villus.** The capillaries in the tertiary villi make contact with capillaries developing in the mesoderm of the chorionic plate and in the connecting stalk (figs. 4-8 and 4-9). These vessels in turn establish contact with the intra-embryonic circulatory system, thus connecting the placenta and the embryo (see Chapters 5 and 12). Hence, when the heart begins to beat in the fourth week of development, the villous system is ready to supply the embryo proper with the necessary nutrients and oxygen.

Meanwhile, the cytotrophoblastic cells in the villi penetrate progressively into the overlying syncytium until they reach the maternal endometrium. Here they establish contact with similar extensions of neighboring villous stems, thus forming a thin **outer cytotrophoblast shell** (figs. 4-8 and 4-9). This shell gradually surrounds the trophoblast entirely and attaches the chorionic sac firmly to the maternal endometrial tissue (fig. 4-9).[7, 8]

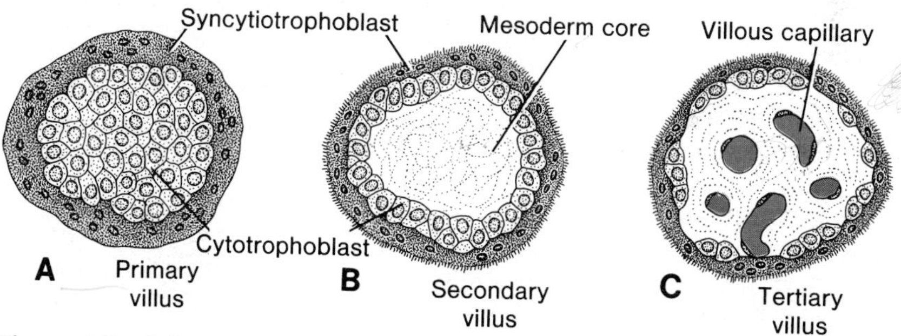

Figure 4-7. Schematic drawings to show the development of the villus. *A*, Transverse section of a primary stem villus, showing a core of cytotrophoblastic cells covered by a layer of syncytium. *B*, Transverse section of a secondary stem villus with a core of mesoderm covered by a single layer of cytotrophoblastic cells, which in turn is covered by the syncytium. *C*, The mesoderm of the villus shows a number of capillaries.

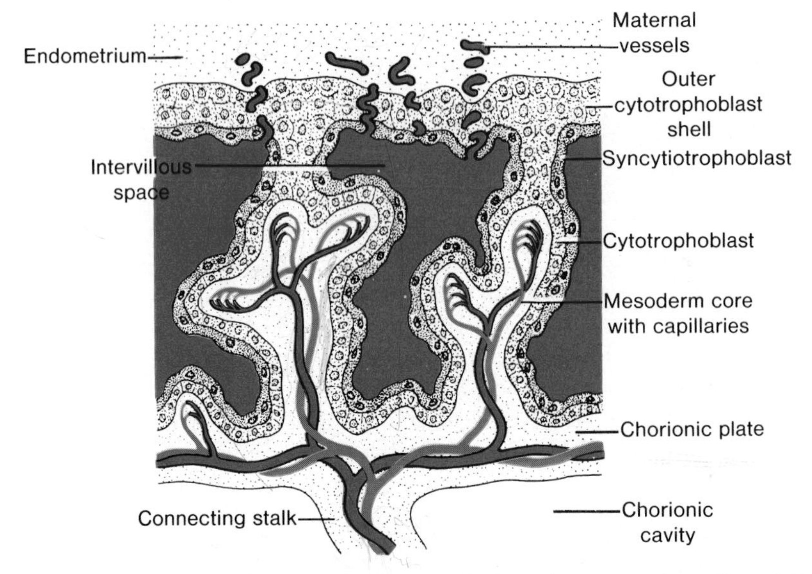

Figure 4-8. Schematic drawing of a longitudinal section through a villus at the end of the third week of development. Note that the maternal vessels penetrate the cytotrophoblastic shell to enter the intervillous spaces, which surround the villi. The capillaries in the villi are in contact with the vessels in the chorionic plate and in the connecting stalk, which in turn are connected to the intra-embryonic vessels.

The chorionic cavity meanwhile becomes larger, and by the 19th or 20th day the embryo is attached to its trophoblastic shell by a narrow **connecting stalk** only (fig. 4-9). The connecting stalk later develops into the **umbilical cord**, which forms the connection between the placenta and the embryo.

SUMMARY

The most characteristic event occurring during the third week is the appearance of the **primitive streak** with, at its cephalic end, the **primitive node**. In the region of the node and streak ectodermal cells move inward (**invaginate**) to form a new cell layer between the endoderm and ectoderm. This new layer is the third or **intra-embryonic mesodermal germ layer**. Its cells migrate between the two other germ layers until they establish contact with the extra-embryonic mesoderm covering the yolk sac and amnion (figs. 4-2 and 4-3).

Cells invaginating in the primitive pit move straight forward until they reach the prochordal plate. They form a tube-like process, the

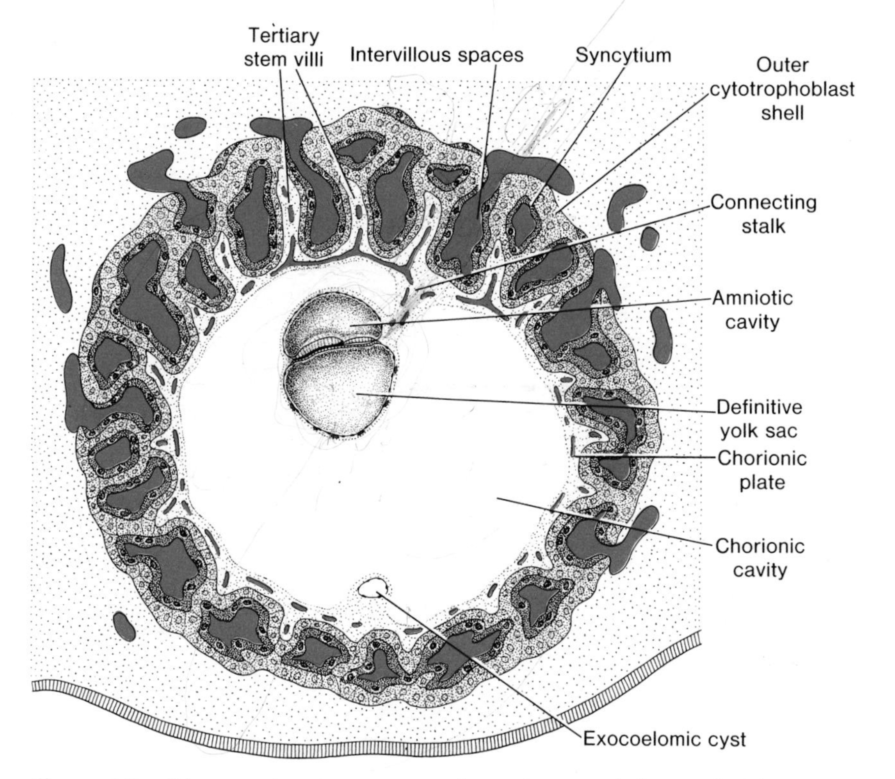

Figure 4-9. Diagram showing a presomite embryo and the trophoblast at the end of the third week. The tertiary and secondary stem villi give the trophoblast a characteristic radial appearance. The intervillous spaces are found throughout the trophoblast and are lined with syncytium. The cytotrophoblastic cells surround the trophoblast entirely and are in direct contact with the endometrium. The embryo is suspended in the chorionic cavity by means of the connecting stalk.

notochordal or head process (fig. 4-4). With further development the lumen of the process disappears and a solid cord, the **notochord**, is formed. It forms a midline axis, which will serve as the basis of the axial skeleton (fig. 4-5). Hence, by the end of the third week the three basic germ layers are laid down and further tissue and organ differentiation can begin.

The trophoblast in the meantime has rapidly progressed. The primary villus has obtained a mesenchyme core in which subsequently small capillaries arise (fig. 4-7). When these villous capillaries make contact with capillaries in the chorionic plate and connecting stalk, the villous system is ready to supply the embryo with its nutrients and oxygen (fig. 4-9).

REFERENCES

1. Wilson, K. M. A normal human ovum of 16 days development, the Rochester ovum. *Contrib. Embryol., 31:* 103, 1945.
2. Heuser, C. H. A presomite embryo with a definite chorda canal. *Contrib. Embryol., 23:* 253, 1932.
3. Jones, H. O., and Brewer, J. T. A human embryo in the primitive streak stage. *Contrib. Embryol., 29:* 159, 1941.
4. O'Rahilly, R. The manifestation of the axes of the human embryo. *Z. Anat. Entwicklungsgesch., 132:* 50, 1970.
5. O'Rahilly, R. *Development Stages in Human Embryos. Part A. Embryos of the first three weeks* (Stages 1 to 9). Carnegie Institute of Washington. Washington, D.C., 1973.
6. Willis, R. A. *The Borderland of Embryology and Pathology.* Butterworth and Co., London, 1962.
7. Hamilton, W. J., and Boyd, J. D. Development of the human placenta in the first three months of gestation. *J. Anat., 94:* 297, 1960.
8. Hamilton, W. J., and Boyd, J. D. Development of the human placenta. In *Scientific Foundations of Obstetrics and Gynecology,* edited by E. E. Philipp, J. Barnes, and M. Newton. p. 185. William Heinemann, London, 1970.

chapter 5

Embryonic Period
(Fourth to Eighth Week)

Organ syst - Develop
* - Crucial period - Teratogens

During the fourth to eighth week of development, a period known as the **embryonic period**, each of the three germ layers gives rise to a number of specific tissues and organs. By the end of the embryonic period the main organ systems have been established. As a result of the organ formation, the shape of the embryo changes greatly and the major features of the external body form are recognizable by the end of the second month.

Derivatives of Ectodermal Germ Layer

At the beginning of the third week of development, the ectodermal germ layer has the shape of a flat disc which, in the cephalic region, is somewhat broader than caudally (fig. 5-1*A*, *B*). Simultaneously with the formation of

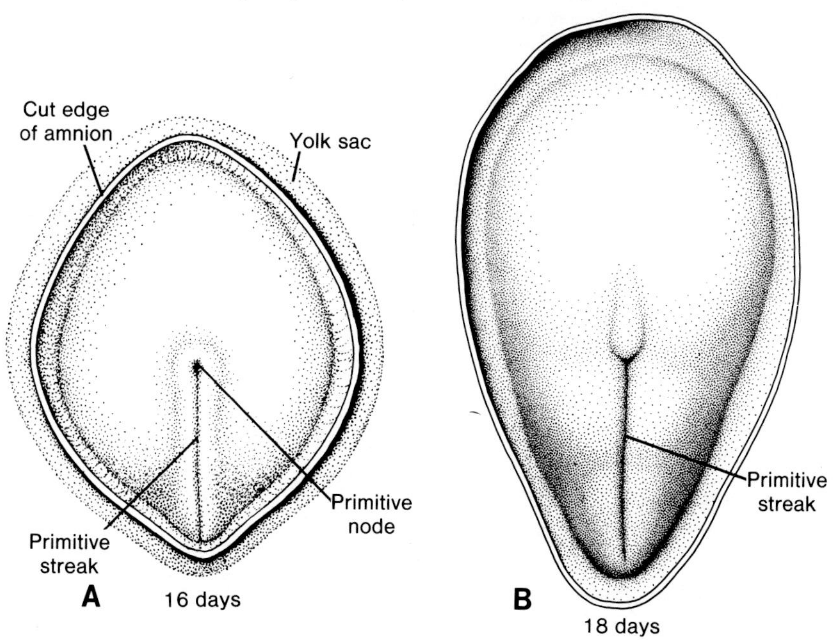

Figure 5-1. *A*, Dorsal view of a 16-day presomite embryo. The primitive streak and node are visible. *B*, Dorsal view of an 18-day presomite embryo. The embryo is pear-shaped with its cephalic region somewhat broader than its caudal end.

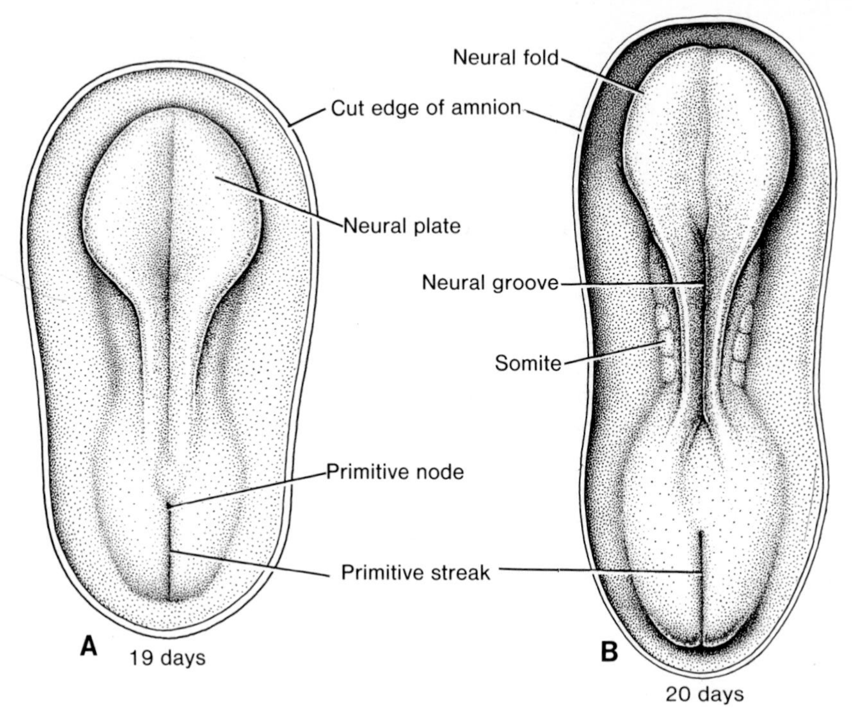

Figure 5-2. *A*, Dorsal view of a late presomite embryo (approximately 19 days). (Modified after Davis.) The amnion has been removed. The neural plate is clearly visible. *B*, Dorsal view of a human embryo at approximately 20 days. (Modified after Ingalls.) Note the appearance of the somites and the formation of the neural groove and the neural folds.

the notochord, and in all probability under its inductive influence, the ectoderm overlying the notochord gives rise to the **central nervous system** (fig. 5-2).

Initially the nervous system appears as a thickening of the ectoderm, rather narrow in the cervical region and somewhat wider in the cephalic region of the embryo (fig. 5-2*A*). This elongated, slipper-shaped plate, the **neural plate**, gradually expands towards the primitive streak (fig. 5-2*B*). By the end of the third week the lateral edges of the neural plate become more elevated to form the **neural folds**, while the depressed midregion forms a groove, the **neural groove** (figs. 5-2*B* and 5-3*A*, *B*). Gradually the neural folds approach each other in the midline, where they fuse (fig. 5-3*C*). This fusion begins in the region of the future neck (fourth somite) and proceeds in cephalic and caudal directions (fig. 5-4*A*, *B*). As a result the **neural tube** is formed. At the cephalic and caudal ends of the embryo the tube remains temporarily in open connection with the amniotic cavity by way of the **anterior** and **posterior neuropores**, respectively (figs. 5-4*B* and 5-6*A*). Closure of the anterior neuropore occurs approximately at day 25 (18- to 20-somite stage), whereas the posterior neuropore closes at day 27 (25-somite stage).

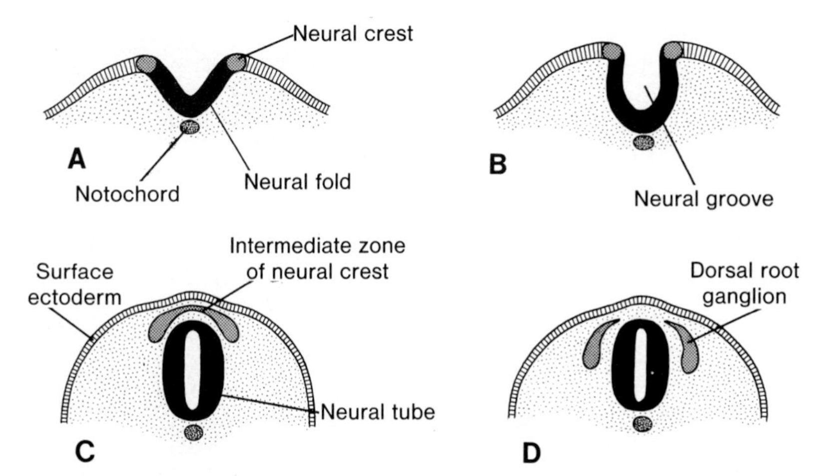

Figure 5-3. Schematic drawing of a number of transverse sections through successively older embryos, showing the formation of the neural folds, neural groove, and neural tube. The cells of the neural crest, initially forming an intermediate zone between the neural tube and surface ectoderm (*C*) develop into the dorsal root and cranial sensory ganglia (*D*).

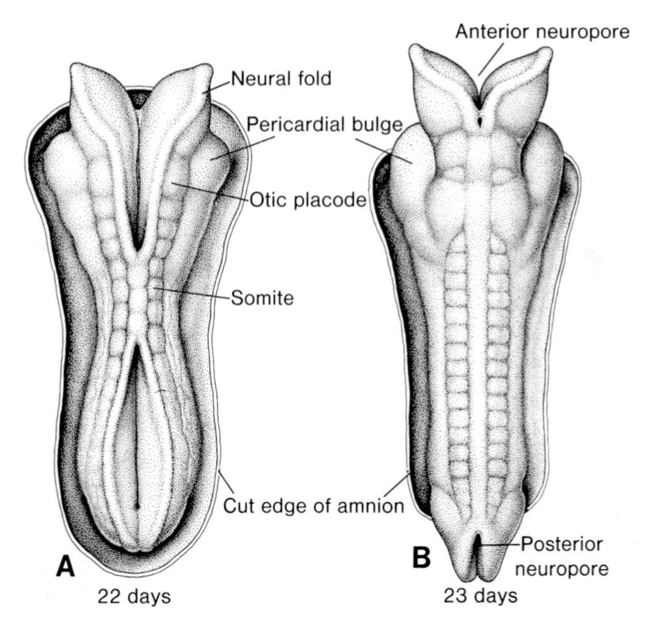

Figure 5-4. *A*, Dorsal view of a human embryo at approximately day 22. (Modified after Payne.) Seven distinct somites are visible on each side of the neural tube. *B*, Dorsal view of a human embryo at approximately day 23. (Modified after Corner.) Note the pericardial bulge on each side of the midline in the cephalic part of the embryo.

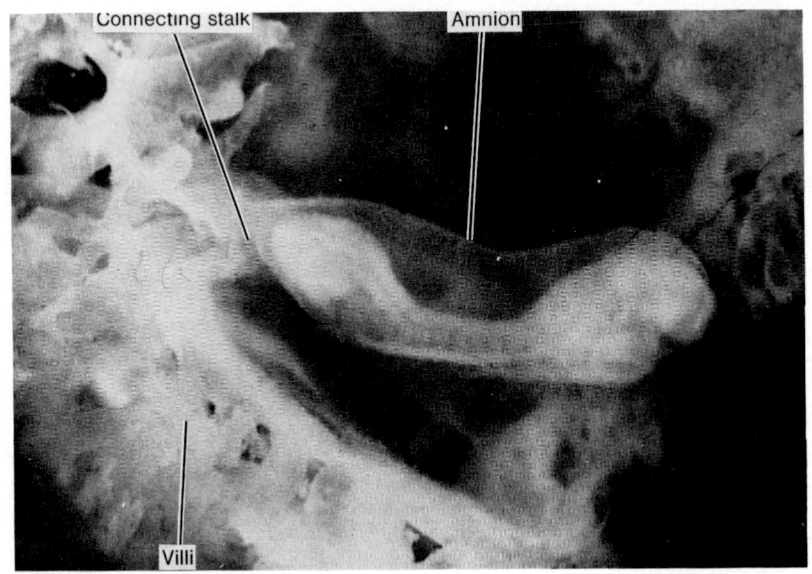

Figure 5-5. Photograph of a 12 to 13-somite embryo (approximately 23 days). The embryo within its amniotic sac is attached to the chorion by the connecting stalk. Note the well-developed chorionic villi. (Courtesy E. Blechschmidt,[6] Professor of Anatomy, University of Göttingen.)

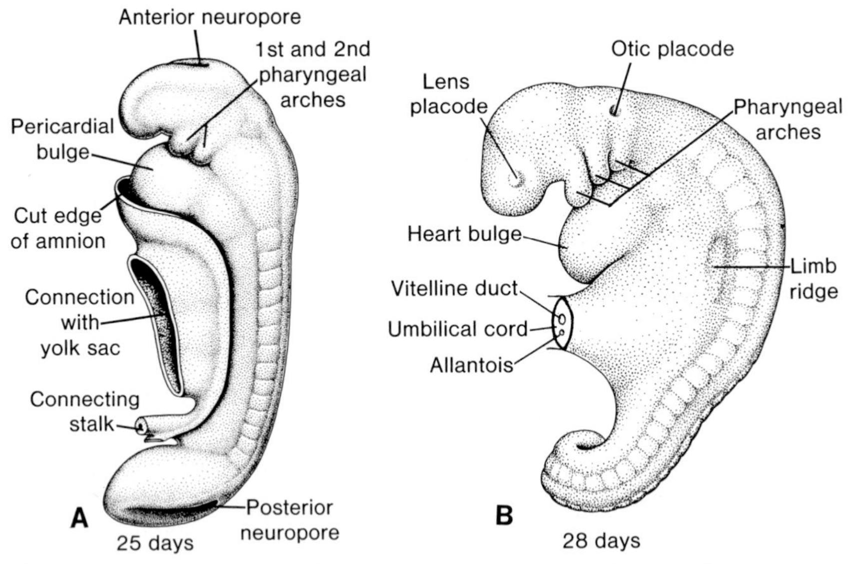

Figure 5-6. *A*, Lateral view of a 14-somite embryo (approximately 25 days). Note the bulging pericardial area and the first and second pharyngeal arches. *B*, Schematic drawing showing the left side of a 25-somite embryo approximately 28 days old. The first three pharyngeal arches and the lens and otic placodes are visible. (Modified after Streeter.[2])

The central nervous system then forms a closed tubular structure with a narrow caudal portion, the **spinal cord**, and a much broader cephalic portion characterized by a number of dilatations, the **brain vesicles** (see "Central Nervous System," Chapter 20).

By the time the neural tube is closed, two other ectodermal thickenings, the **otic placode** and the **lens placode** become visible in the cephalic region of the embryo (fig. 5-6*B*). During further development, the otic placode invaginates and forms the **otic vesicle**, which will develop into the structures needed for hearing and maintenance of the equilibrium (see "Ear," Chapter 17). At approximately the same time, the **lens placode** appears. This placode also invaginates and during the fifth week forms the **lens** (see "Eye," Chapter 18).

In general terms it may be stated that the ectodermal germ layer gives rise to those organs and structures that maintain contact with the outside world: (1) the central nervous system; (2) the peripheral nervous system; (3) the sensory epithelium of ear, nose, and eye; and (4) the epidermis, including hair and nails. In addition it gives rise to: the subcutaneous glands; the mammary gland; the pituitary gland; and the enamel of the teeth.

Derivatives of Mesodermal Germ Layer

Since the external contours of the embryo are greatly influenced by the formation of the **somites**, a series of mesodermal tissue blocks found on each

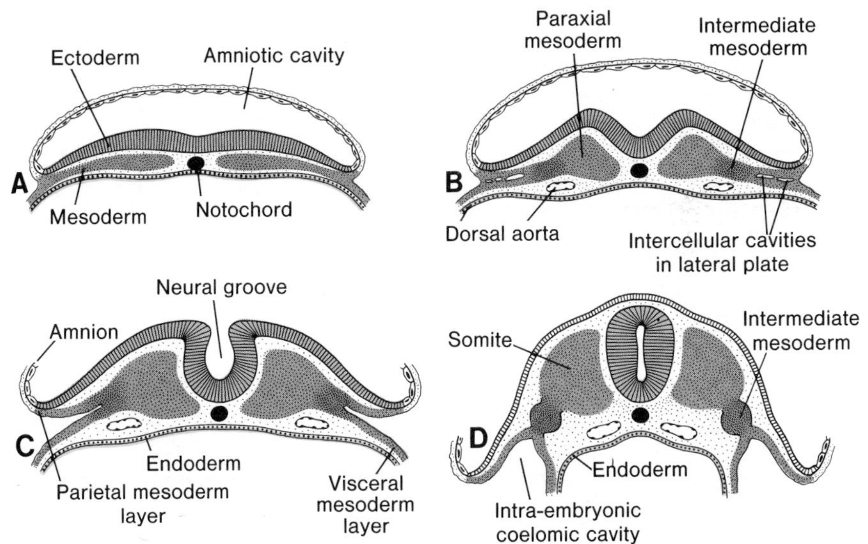

Figure 5-7. Transverse sections showing the development of the mesodermal germ layer. *A*, Day 17; *B*, day 19; *C*, day 20; *D*, day 21. The thin mesodermal sheet gives rise to the paraxial mesoderm (the future somites), the intermediate mesoderm (the future excretory units), and the lateral plate, which is split into the parietal and visceral mesoderm layers lining the intra-embryonic coelomic cavity.

side of the neural tube, the development and differentiation of these structures will be discussed briefly.

Initially the cells of the mesodermal germ layer form a thin sheet of loosely woven tissue on each side of the midline (fig. 5-7*A*). By about the 17th day, however, the cells close to the midline proliferate and form a thickened plate of tissue, known as the **paraxial mesoderm** (fig. 5-7*B*). More laterally, the mesoderm layer remains thin and is known as the **lateral plate**. With the appearance and coalescence of intercellular cavities in the lateral plate, this tissue is divided into two layers (fig. 5-7*B*, *C*): (1) a layer continuous with the mesoderm covering the amnion, known as the **somatic** or **parietal mesoderm layer**; and (2) a layer continuous with the mesoderm covering the yolk sac, known as the **splanchnic** or **visceral mesoderm layer** (fig. 5-7*C*, *D*). Together, these layers line a newly formed cavity, the **intra-embryonic coelomic cavity**, which, on each side of the embryo, is continuous with the extraembryonic coelom. The tissue connecting the paraxial mesoderm and the lateral plate is known as the **intermediate mesoderm** (fig. 5-7*B*, *D*).

By the end of the third week the paraxial mesoderm breaks up into segmented blocks of epithelioid cells, the **somites**. The first pair of somites arises in the cephalic part of the embryo at about the 20th day of development. From here new somites appear in craniocaudal sequence, approximately three per day, until at the end of the fifth week 42 to 44 pairs are present (figs. 5-4 and 5-6).[1] These are 4 occipital, 8 cervical, 12 thoracic, 5 lumbar, 5 sacral, and 8 to 10 coccygeal pairs. The first occipital and the last 5 to 7 coccygeal somites later disappear. During this period of development the age of the embryo is expressed in the number of somites, and Table 5-1 represents the approximate age of the embryo correlated to the number of somites.[2, 3]

DIFFERENTIATION OF THE SOMITE

By the beginning of the fourth week the epithelioid cells forming the ventral and medial walls of the somite lose their epithelial shape, become polymorphous, and migrate toward the notochord (fig. 5-8*B*). These cells, collectively known as the **sclerotome**, form a loosely woven tissue known as **mesenchyme** or **young connective tissue**. They will surround the spinal cord and notochord to form the vertebral column (see "Skeletal System," Chapter 9).

The remaining dorsal somite wall, now referred to as the **dermatome**, gives rise to a new layer of cells, (fig. 5-8*C*) characterized by pale nuclei and darkly stained nucleoli. These cells fail to divide once they are laid down.[4] The tissue so composed is known as the **myotome**. Each myotome provides the musculature for its own segment (see "Muscular System," Chapter 10).

After the cells of the dermatome have formed the myotome, they lose their epithelial characteristics and spread out under the overlying ectoderm (fig.

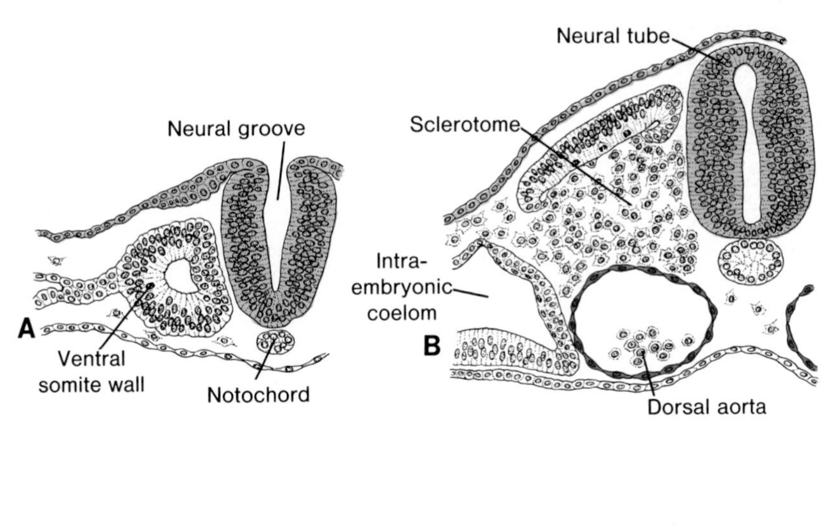

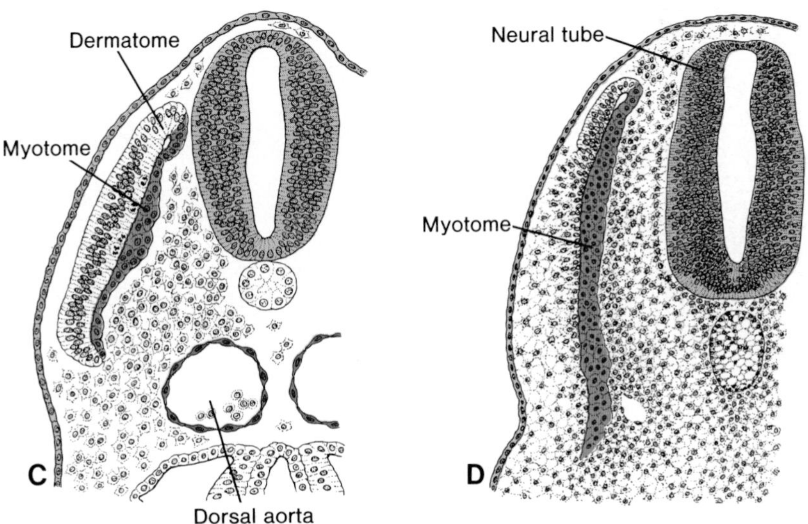

Figure 5-8. Successive stages in the development of the somite. *A*, The mesoderm cells are arranged around a small cavity. *B*, The cells of the ventral and medial walls of the somite lose their epithelial arrangement and migrate in the direction of the notochord. These cells are collectively referred to as the sclerotome. *C*, The dorsal somite wall gives rise to a new cell layer, the myotome. *D*, After extension of the myotome in ventral direction, the dermatome cells lose their epithelial configuration and spread out under the overlying ectoderm to form the dermis.

5-8*D*). Here they form the **dermis** and subcutaneous tissue of the skin (see "Integumentary System," Chapter 19). Hence, each somite forms its own **sclerotome** (the cartilage and bone component), its own **myotome** (providing the segmental muscle component), and its own **dermatome**, the segmental skin component. As will be seen later, each myotome and dermatome has its own segmental nerve component.

INTERMEDIATE MESODERM

This tissue, which temporarily connects the paraxial mesoderm with the lateral plate (fig. 5-7*D*), differentiates in a manner entirely different from that of the somites. In the cervical and upper thoracic regions it forms

Table 5-1. **Number of Somites Correlated to Approximate Age in Days**

Approx. Age	No. of Somites	Approx. Age	No. of Somites
days		*days*	
20	1–4	25	17–20
21	4–7	26	20–23
22	7–10	27	23–26
23	10–13	28	26–29
24	13–17	30	34–35

segmentally arranged cell clusters (the future **nephrotomes**), whereas more caudally it forms an unsegmented mass of tissue, known as the **nephrogenic cord**. From this partly segmented, partly unsegmented intermediate mesoderm develop the excretory units of the urinary system (see fig. 15-2).

PARIETAL AND VISCERAL MESODERM LAYERS

These two layers line the intra-embryonic coelom (figs. 5-7*C*, *D* and 5-9*A*). The parietal mesoderm together with the overlying ectoderm will form the lateral and ventral body wall. The visceral mesoderm and the embryonic endoderm will form the wall of the gut (fig. 5-9*B*). The cells facing the coelomic cavity will form thin membranes, the **mesothelial** or **serous membranes**, which will line the peritoneal, pleural, and pericardial cavities (fig. 5–9*B*) (See "Body Cavities and Serous Membranes," Chapter 11).

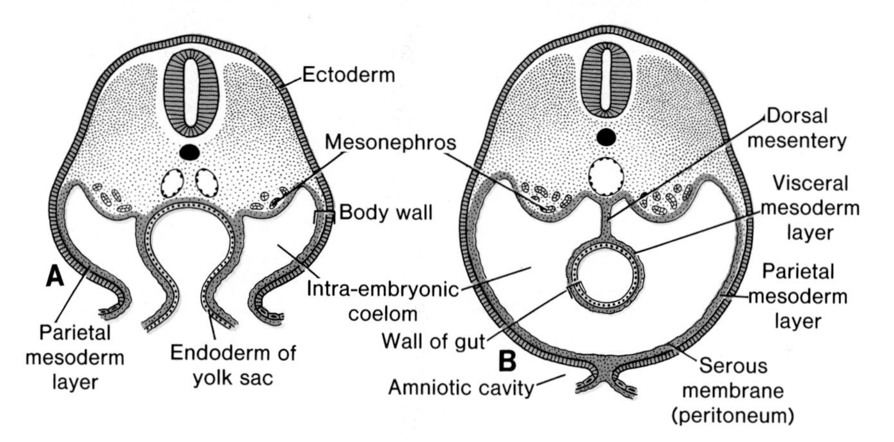

Figure 5-9. *A*, Transverse section through a 21-day embryo in the region of the mesonephros. Note the parietal and visceral mesoderm layers. The intra-embryonic coelomic cavities communicate with the extra-embryonic coelom or chorionic cavity. *B*, Section at the end of the fourth week. The parietal mesoderm and the overlying ectoderm form the ventral and lateral body wall. Note the peritoneal or serous membrane.

BLOOD AND BLOOD VESSELS

At about the beginning of the third week mesoderm cells located in the visceral mesoderm of the wall of the yolk sac differentiate into blood cells and blood vessels. These cells, known as the **angioblasts**, form isolated clusters and cords (**angiogenetic cell clusters**), which gradually become canalized by confluence of intercellular clefts (fig. 5-10). The centrally located cells then give rise to the primitive blood cells, while those on the periphery flatten and form the **endothelial cells** lining the **blood islands** (fig. 5-10*B*, *C*). The blood islands approach each other rapidly by sprouting of

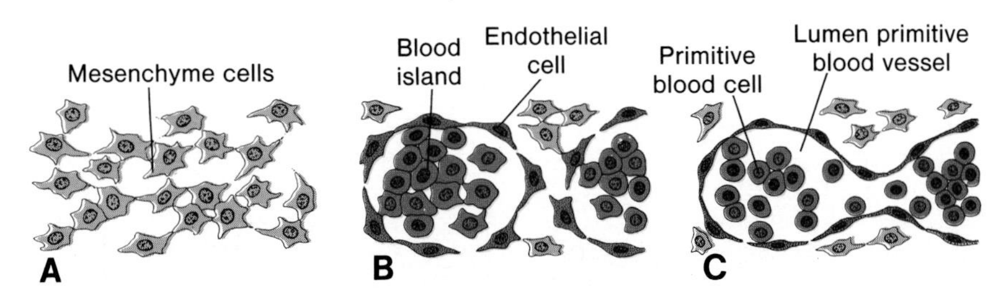

Figure 5-10. Successive stages of blood vessel formation. *A*, Undifferentiated mesenchyme cells. *B*, Blood island formation. *C*, Primitive capillary. Note the differentiation of mesenchymal cells into the primitive blood cells and the endothelial cells.

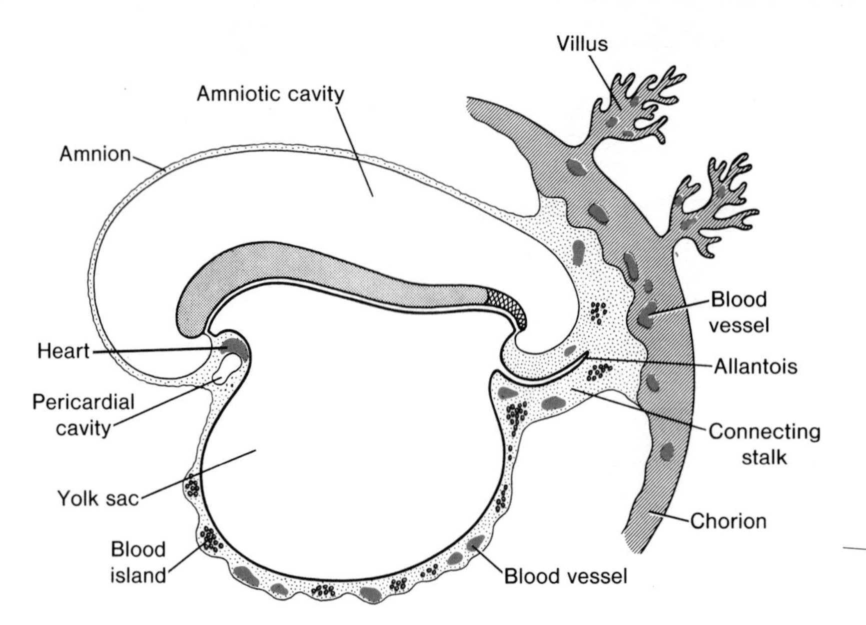

Figure 5-11. Extra-embryonic blood vessel formation in the villi, the chorion, the connecting stalk, and the wall of the yolk sac in a presomite embryo of approximately 19 days. Note the angiogenetic cell clusters in relation to the prochordal (buccopharyngeal) plate.

the endothelial cells and after fusion give rise to small vessels. At the same time blood cells and capillaries develop in the extra-embryonic mesoderm of the villous stems and the connecting stalk (fig. 5-11). By continuous budding the extra-embryonic vessels establish contact with those inside the embryo, thus connecting the embryo and the placenta.

The intra-embryonic blood cells and blood vessels, including the heart tube, are established in exactly the same manner as described for the extra-embryonic vessels (see "Cardiovascular System," Chapter 12).

In summary, the following tissues and organs are considered to be of mesodermal origin: (1) supporting tissues such as connective tissue, cartilage, and bone; (2) striated and smooth musculature; (3) blood and lymph cells and the walls of the heart, blood, and lymph vessels; (4) kidneys, gonads, and their corresponding ducts; (5) the cortical portion of the suprarenal gland; and (6) the spleen.

Derivatives of Endodermal Germ Layer

The gastrointestinal tract is the main organ system derived from the endodermal germ layer. Its formation is greatly dependent on the **cephalo-caudal** and **lateral folding** of the embryo. The **cephalo-caudal folding is caused mainly by the rapid, longitudinal growth of the central nervous system**;

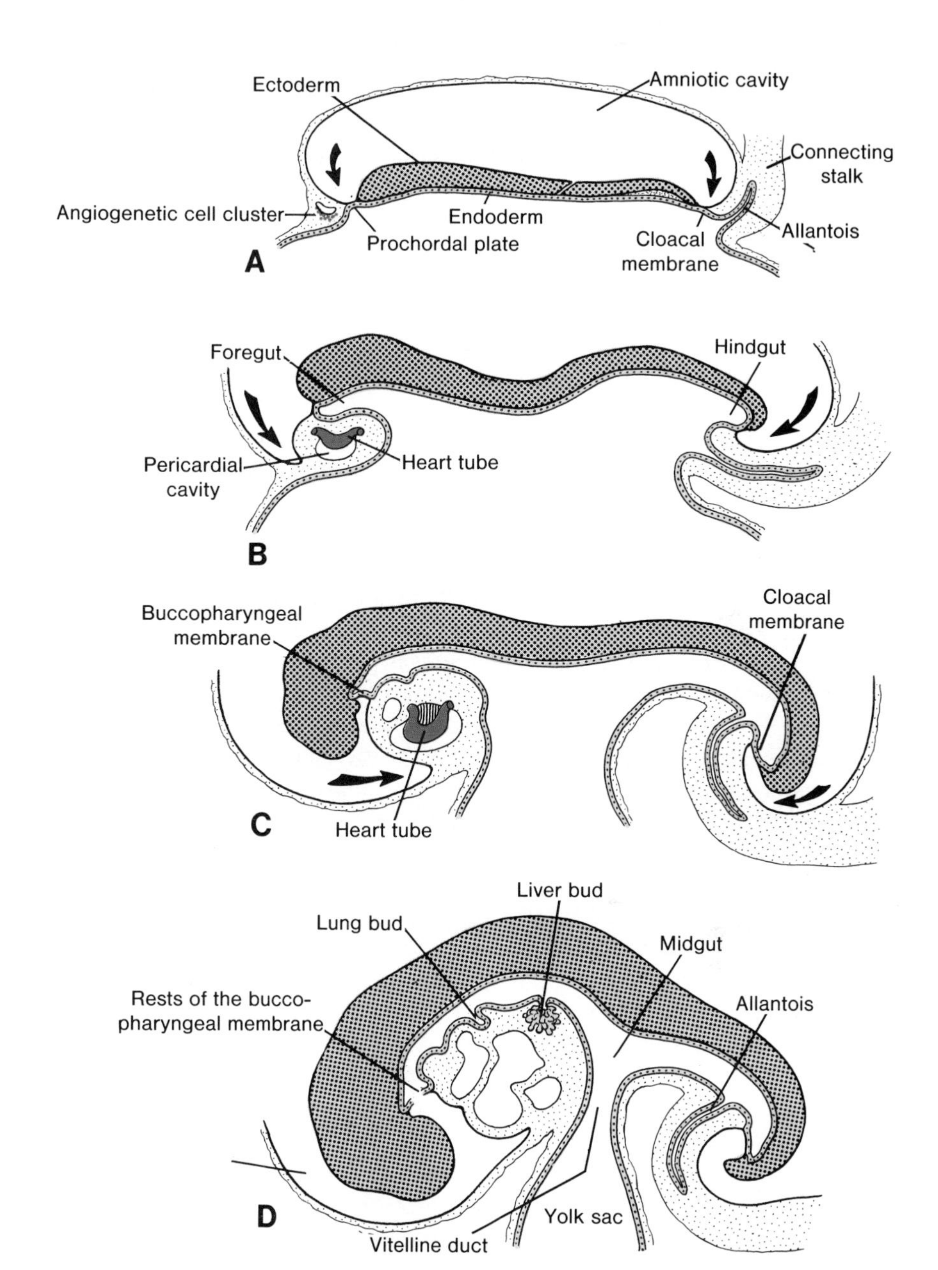

Figure 5-12. Drawings of sagittal midline sections of embryos at various stages of development to demonstrate the cephalo-caudal flexion and its effect on the position of the endoderm lined cavity. *A*, Presomite embryo; *B*, 7-somite embryo; *C*, 14-somite embryo; *D*, at the end of the first month.

the transverse or lateral folding by the formation of the rapidly growing somites. Hence, the formation of the tube-like gut is a passive event and consists of the inversion and incorporation of part of the endoderm lined yolk sac into the body cavity. As an additional result of the folding movements, the initial wide communication between the embryo and the yolk sac becomes constricted until only a narrow, long duct, the **vitelline duct**, is left.

Initially the endodermal germ layer has the shape of a flat disc, forming the roof of the yolk sac and closely apposed to the ectoderm (fig. 5-12*A*). With the development and the growth of the brain vesicles, however, the embryonic disc begins to bulge into the amniotic cavity and to fold in cephalo-caudal direction. This folding is most pronounced in the regions of the head and tail, where the so-called **head fold** and **tail fold** are formed (fig. 5-12).

As a result of the cephalo-caudal folding, a continuously larger portion of the endoderm lined cavity is incorporated into the body of the embryo proper (fig. 5-12*C*). In the anterior part the endoderm forms the **foregut**; in the tail region the **hindgut**. The part between the foregut and hindgut is known as the **midgut**. The midgut remains temporarily in open connection with the yolk sac by way of a broad stalk, the **omphalomesenteric** or **vitelline duct** (fig. 5-12*D*). This duct is initially wide, but with further growth of the embryo it becomes narrow and much longer (see fig. 5-17).

At its cephalic end the foregut is temporarily bounded by the prochordal plate, an ectodermal-endodermal membrane, which is now called the **buccopharyngeal membrane** (fig. 5-12*A*, *C*). At the end of the third week the buccopharyngeal membrane ruptures, thus establishing an open connection between the amniotic cavity and the primitive gut (fig. 5-12*D*). The hindgut also terminates temporarily at a membrane known as the **cloacal membrane** (fig. 5-12*C*).

As a result of the rapid growth of the somites the initial flat embryonic disc begins to fold in lateral direction and the embryo obtains a round appearance (fig. 5–13). Simultaneously the ventral body wall of the embryo is established, with the exception of a small part in the ventral abdominal region where the yolk sac stalk is attached.

While the foregut and hindgut are established mainly as a result of the formation of the head fold and tail fold, respectively, the midgut remains in communication with the yolk sac. Initially, this connection is wide (fig. 5-13*A*), but as a result of the lateral folding it gradually becomes long and narrow, the **vitelline duct** (figs. 5-13*B* and 5-14). Only much later, when the vitelline duct is obliterated, does the midgut lose its connection with the original endoderm lined cavity and obtain its free position in the abdominal cavity (fig. 5-13*C*).

Another important result of the cephalo-caudal and lateral folding is the partial incorporation of the allantois into the body of the embryo, where it forms the **cloaca** (fig. 5-14*A*). The distal portion of the allantois remains in

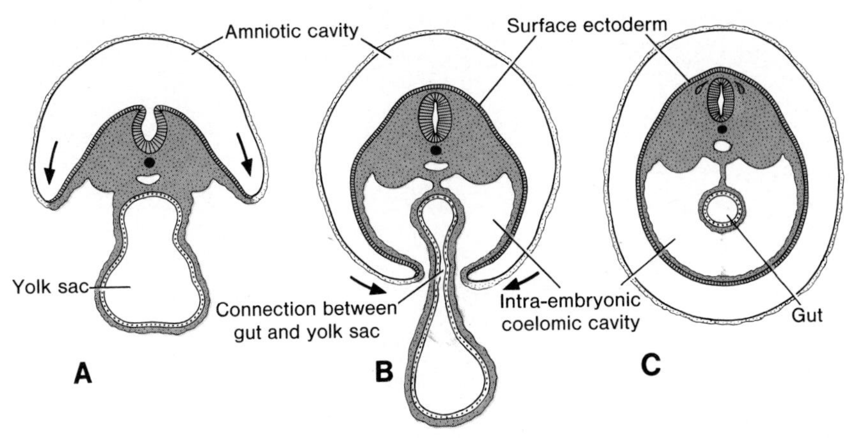

Figure 5-13. Schematic drawings of transverse sections through embryos at various stages of development to show the effect of the lateral folding on the endoderm lined cavity. *B*, Transverse section through region of the midgut to show the connection between the gut and the yolk sac. *C*, Section just below the midgut to show the closed ventral abdominal wall and the gut suspended from the dorsal abdominal wall by its mesentery.

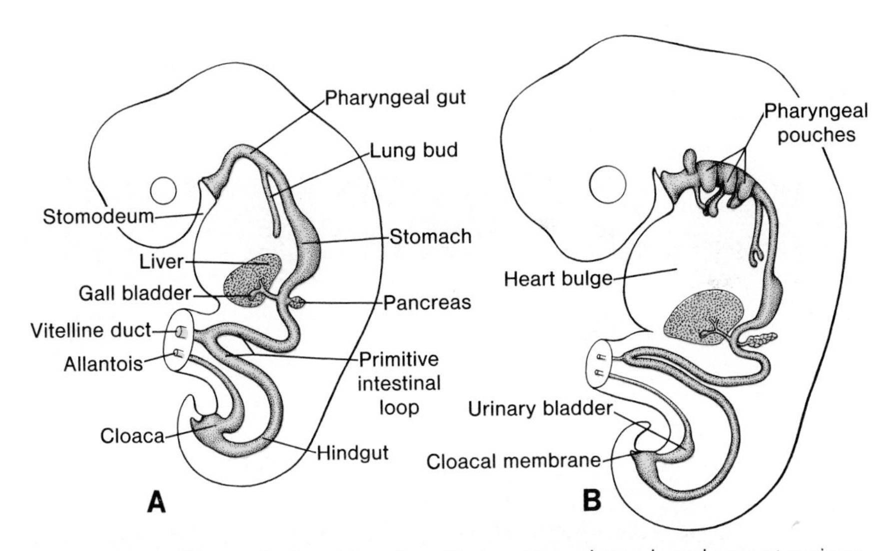

Figure 5-14. Schematic drawings of sagittal sections through embryos at various stages of development to show the derivatives of the endodermal germ layer. Note the pharyngeal pouches and the epithelial lining of the lung buds and trachea. Note the liver, gallbladder and pancreas. The urinary bladder is derived from the cloaca and at this stage of development is in open connection with the allantois.

the connecting stalk. By the end of the fourth week the yolk sac stalk and connecting stalk fuse to form together the umbilical cord (Figs. 5-14 and 7-9).

In man, the yolk sac is vestigial and in all probability has a nutritive role only in the early stages of development. Its diameter is never more than 5 mm. In the second month of development it is found in the chorionic cavity (figs. 5-18 and 7-3).

Hence, the endodermal germ layer initially forms the epithelial lining of the primitive gut and the intra-embryonic portions of the allantois and the vitelline duct (fig. 5-14*A*, *B*). During further development it gives rise to: (1) the epithelial lining of the respiratory tract; (2) the parenchyma of the tonsil, thyroid, parathyroids, thymus, liver, and pancreas (see "Head and Neck," Chapter 16 and "Digestive System," Chapter 14); (3) the epithelial lining of the urinary bladder and urethra (see "Urogenital System," Chapter 15); and (4) the epithelial lining of the tympanic cavity and Eustachian tube (see "Ear," Chapter 17).

External Appearance during Second Month

At the end of the fourth week when the embryo has approximately 28 somites, the main external features are the somites and pharyngeal arches (fig. 5-15). The age of the embryo is therefore usually expressed in somites (Table 5-1). Since counting the number of somites becomes difficult during the second month of development, the age of the embryo is then indicated as the **crown-rump (C.R.) length** and expressed in millimeters.

The C.R. length is the measurement from the vertex of the skull to the

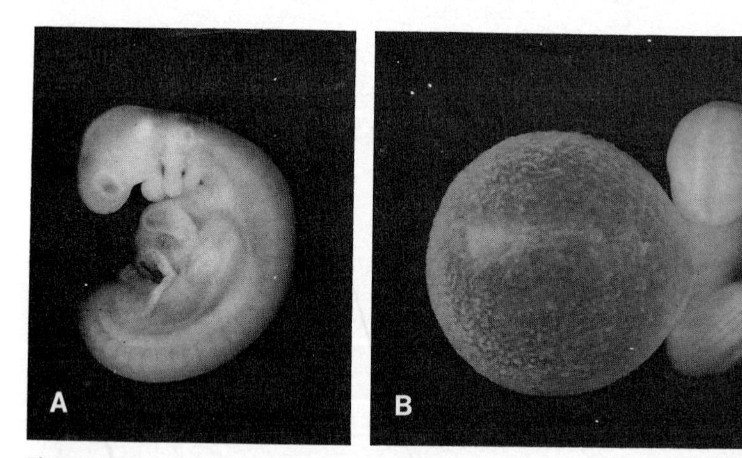

Figure 5-15. *A*, Lateral view of 28-somite human embryo. The main external features are the pharyngeal arches and the somites. Note the pericardial-liver bulge. The limb buds are not visible. *B*, Photograph of the same embryo but taken from a different angle to demonstrate the size of the yolk sac. (From Streeter. Courtesy Carnegie Institution of Washington.[5])

midpoint between the apices of the buttocks. Owing to the considerable variation in the degree of flexure from one embryo to another, it is understandable that the measurements given in Table 5-2 can be only approximate indications of the real age of the embryo.[15]

During the second month the external appearance of the embryo is greatly changed by the enormous size of the head and the formation of the limbs,

Table 5-2. **Crown-Rump (C.R.) Length Correlated to Approximate Age in Weeks**

C.R. Length	Approx. Age	C.R. Length	Approx. Age
mm	*weeks*	*mm*	*weeks*
5–8	5	17–22	7
10–14	6	28–30	8

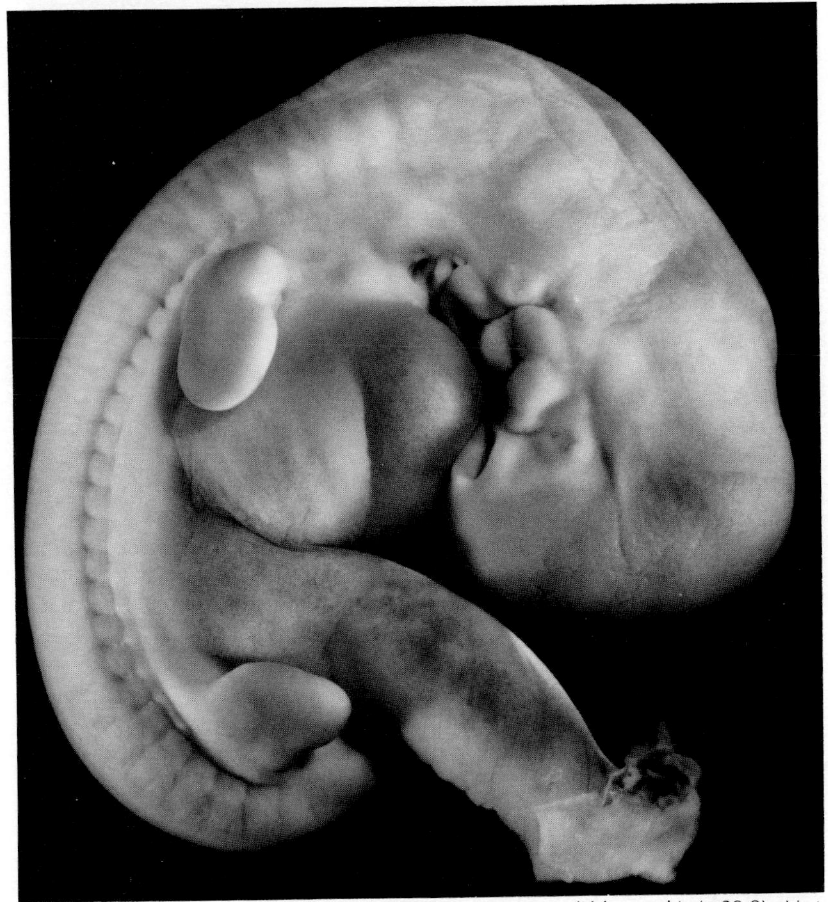

Figure 5-16. Photograph of human embryo (9.8 mm, fifth week) (×29.9). Note that the fore- and hindlimbs have a paddle-shaped appearance. (Courtesy E. Blechschmidt,[6] Professor of Anatomy, University of Göttingen.)

face, ear, nose, and eyes. By the beginning of the fifth week, the fore- and hindlimbs appear as paddle-shaped buds (fig. 5-16). The former are located dorsal to the pericardial swelling at the level of the fourth cervical to the first thoracic somites, thus explaining the innervation by the brachial plexus. The hindlimb buds appear slightly later just caudal to the attachment of the umbilical stalk at the level of the lumbar and upper sacral somites. With further growth, the terminal portion of the buds flattens and becomes separated from the proximal, more cylindrically shaped segment by a circular constriction (fig. 5-17). Soon four radial grooves separating five slightly

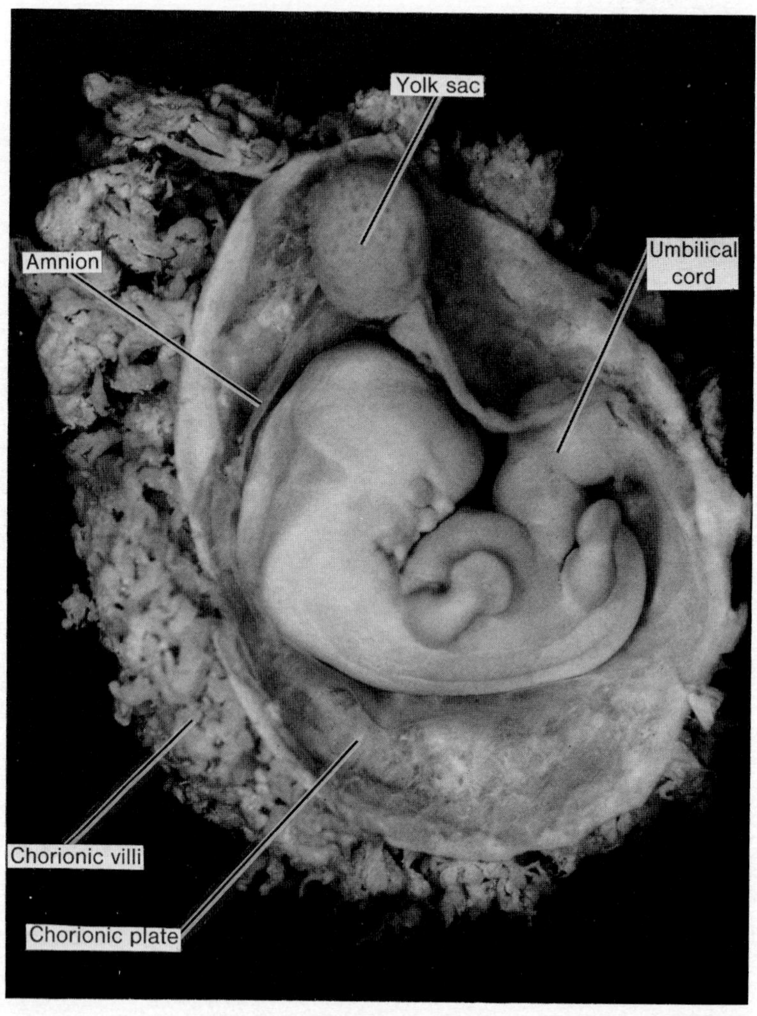

Figure 5-17. Photograph of human embryo (C.R. length 13 mm, sixth week). Note that the yolk sac is visible in the chorionic cavity. (Courtesy E. Blechschmidt,[6] Professor of Anatomy, University of Göttingen.)

thicker areas appear on the distal portion of the buds, foreshadowing the formation of the digits (fig. 5-17).

These grooves, known as **rays**, appear first in the hand region and shortly afterwards in the foot, again indicating that the arm is always slightly more advanced in development than the leg. While the fingers and toes are being formed (fig. 5-18) a second constriction divides the proximal portion of the buds into two segments, and the three parts characteristic of the adult extremities can be recognized (fig. 5-19).

It is evident from the above account that all major organs and organ systems are formed during the fourth to eighth weeks. This period is therefore also called the **period of organogenesis**. It is the time that the embryo is most susceptible to factors interfering with development and most congenital malformations seen at birth find their origin during this critical period. Being

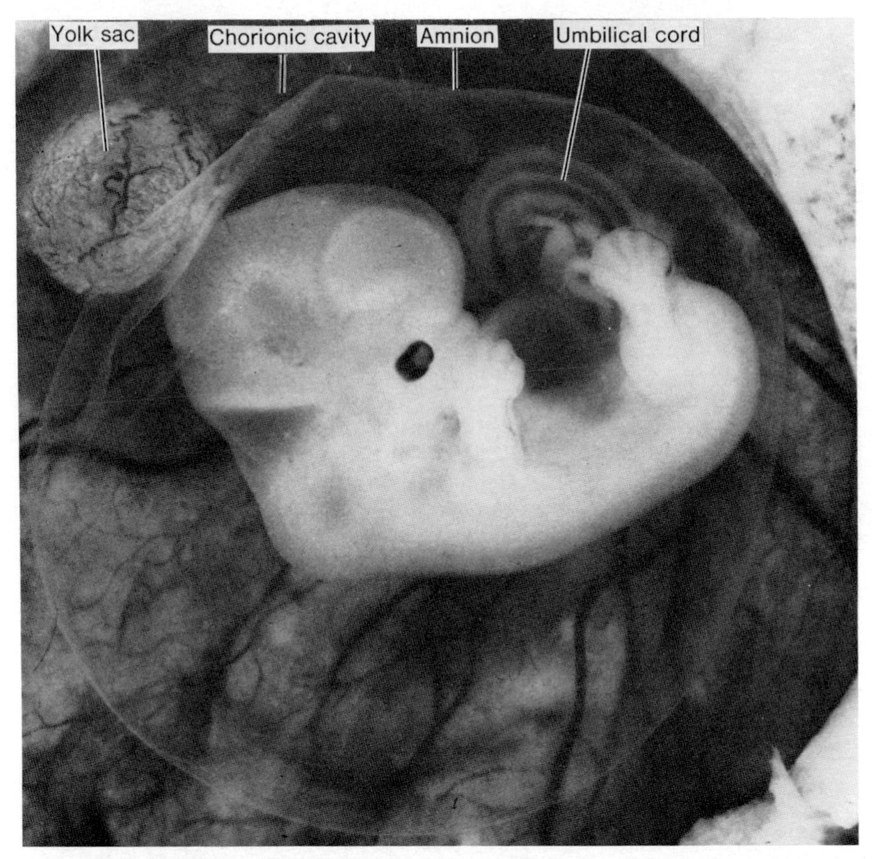

Figure 5-18. Photograph of human embryo (C.R. length 21 mm, seventh week) (×4). The chorionic sac is wide open to show the embryo in its amniotic sac. The yolk sac, the umbilical cord, and vessels in the chorionic plate of the placenta are clearly visible. Note the enormous size of the head in comparison to the remaining part of the body. (Courtesy Professors Hamilton, Boyd, and Mossman.[7])

* Teratogens

8"

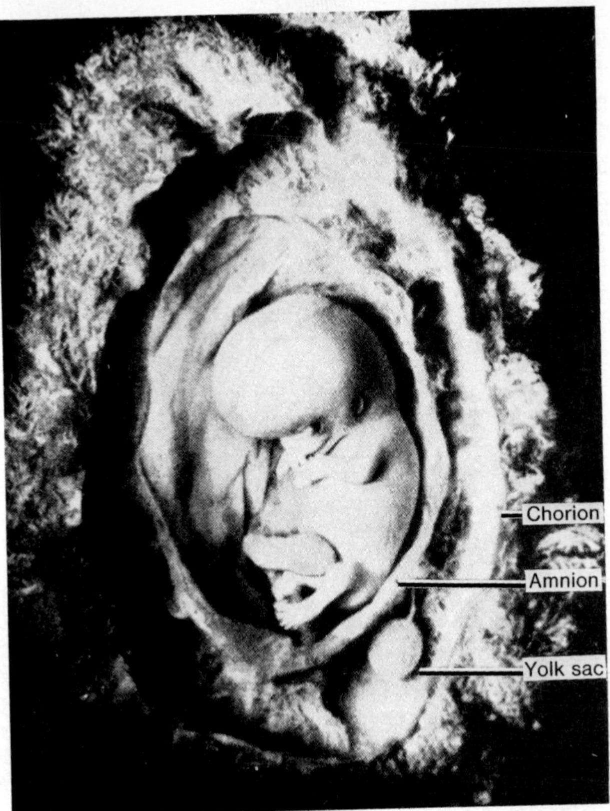

Figure 5-19. Photograph of human embryo (C.R. length 25 mm, seventh to eighth week). The chorion as well as the amnion have been opened. Note the size of the head, the eye, the auricle of the ear, the well-formed toes, the swelling in the umbilical cord caused by intestinal loops, and the yolk sac hanging in the chorionic cavity. (Courtesy Dietrich Starck, Professor of Anatomy, University of Frankfurt am Main.[8])

familiar with the main events of organogenesis will be of great help in identifying the time that a particular abnormality arose. If presented with a child having anencephaly, one can quickly calculate that this abnormality must have started on the 23rd to 25th day of development. Similarly in a child without extremities (amelia), the trauma must have affected the limb buds in the fifth week of development. (See photographic charts at the beginning and end of this book.)

SUMMARY

The **embryonic period** extends from the fourth to eighth week of development and is the period that each of the three germ layers gives rise to its own tissues and organ systems. As a result of the organ formation the major features of the body form are established.

The ***ectodermal germ layer*** gives rise to those organs and structures that maintain contact with the outside world: (1) **central nervous system**; (2) **peripheral nervous system**; (3) **sensory epithelium of ear, nose, and eye**; (4) **skin, including hair and nails**; (5) in addition, the **pituitary gland, mammary** and **sweat glands** and the **enamel of the teeth** (each system will be discussed in a separate chapter).

The ***mesodermal germ layer***. A most important component of the mesodermal germ layer is formed by the **somites**, which give rise to the **myotome** (muscle tissue), **sclerotome** (cartilage and bone), and **dermatome** (subcutaneous tissue of the skin), **all supporting tissues of the body**. The mesoderm also gives rise to the **vascular system**, that is, the heart, arteries, veins, lymph vessels, and all blood and lymph cells. Furthermore, it gives rise to the **urogenital system**—kidneys, gonads, and their ducts (but not the bladder). Finally, the **spleen** and **suprarenal glands** are mesodermal derivatives.

The ***endodermal germ layer*** provides the epithelial lining of the **gastrointestinal tract, respiratory tract**, and **urinary bladder**. It further forms the parenchyma of the **tonsil, thyroid, parathyroids, thymus, liver**, and **pancreas.** Finally, the epithelial lining of the **tympanic cavity** and **Eustachian tube** are lined by epithelium of endodermal origin.

Body form. As a result of the formation of the organ systems and the rapid growth of the central nervous system the initial flat embryonic disc begins to fold in **cephalo-caudal direction**, thus establishing the **head** and **tail folds**, and in **transverse direction**, thus establishing the **rounded body form**. Connection with the yolk sac and placenta is maintained through the vitelline duct and umbilical cord, respectively.

REFERENCES

1. Arey, L. B. The history of the first somite in human embryos. *Contrib. Embryol., 27:* 233, 1938.
2. Streeter, G. L. Developmental horizons in human embryos: age group XI, 13–20 somites, and age group XII, 21–29 somites. *Contrib. Embryol., 30:* 211, 1942.
3. Streeter, G. L. Developmental horizons in human embryos: age group XIII, embryos 4 or 5 mm. long, and age group XIV, indentation of lens vesicle. *Contrib. Embryol., 31:* 26, 1945.
4. Langman, J., and Nelson, G. R. A radioautographic study of the development of the somite in the chick embryo. *J. Embryol. Exp. Morphol., 19:* 217, 1968.
5. Streeter, G. L. Developmental horizons in human embryos: age groups XV, XVI, XVII, and XVIII (the third issue of a survey of the Carnegie Collection). *Contrib. Embryol., 32:* 133, 1948.
6. Blechschmidt, E. *The Stages of Human Development before Birth.* W.B. Saunders, Philadelphia, 1961.
7. Hamilton, W. J., and Mossman, H. W. *Human Embryology.* Williams & Wilkins, Baltimore, 1972.
8. Starck, D. *Embryologie.* Georg Thieme Verlag, Stuttgart, 1965.

chapter 6

Fetal Period
(Third Month to Birth)

Development of Fetus

The period from the beginning of the third month to the end of intrauterine life is known as the **fetal period**. It is characterized by maturation of the tissues and organs and rapid growth of the body.[1] Few, if any, malformations arise during this period, although cell death in the central nervous system caused by cytotoxic factors may result in postnatal behavioral disturbances.

The length of the fetus is usually indicated as the crown-rump (C.R.) length (sitting height) and expressed in centimeters, or as the crown-heel (C.H.) length, the measurement from the vertex of the skull to the heel (standing height). These measurements are then correlated with the age of the fetus expressed in weeks or in lunar months (Table 6-1). Growth in length is particularly striking during the third, fourth, and fifth months, while increase in weight is most striking during the last two months of gestation. In general **the length of pregnancy is considered to be 280 days or 40 weeks after the onset of the last menstruation or, more accurately, 266 days or 38 weeks after fertilization**.

Table 6-1. **Growth in Length and Weight during the Fetal Period**

Age		Crown-Rump Length	Weight
weeks	*lunar months*	*cm*	*gm*
9–12	3	5–8	10–45
13–16	4	9–14	60–200
17–20	5	15–19	250–450
21–24	6	20–23	500–820
25–28	7	24–27	900–1300
29–32	8	28–30	1400–2100
33–36	9	31–34	2200–2900
37–40	10	35–36	3000–3400

Monthly Changes

One of the most striking changes taking place during fetal life is the relative slowdown in the growth of the head compared to the rest of the body. At the beginning of the third month, the head constitutes approximately one-half of the C.R. length (fig. 6-1); by the beginning of the fifth month about one-third and at birth approximately one-fourth of the C.H. length (fig. 6-2). Hence, with time the growth of the body accelerates, but that of the head slows down.

During **the third month** the face becomes more human-looking (figs. 6-3 and 6-4). The eyes, initially directed laterally, become located on the ventral aspect of the face; the ears come to lie close to their definitive position at the side of the head (fig. 6-3); the limbs reach their relative length in comparison to the rest of the body, although the lower limbs are still a little shorter and less developed than the upper extremities; the external genitalia develop to such a degree that by the 12th week the sex of the fetus can be determined

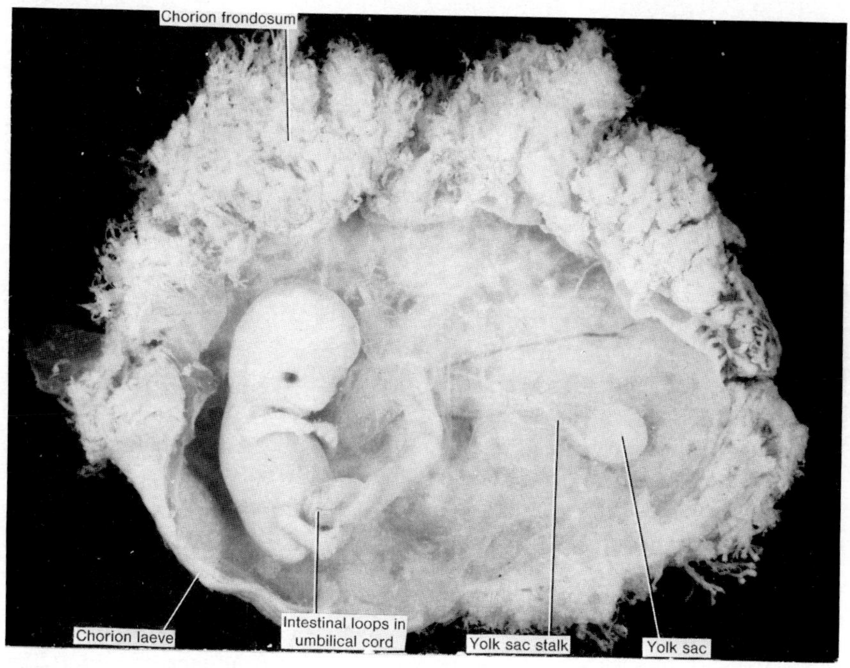

Figure 6-1. Photograph of a nine-week fetus. Note the large size of the head in comparison to that of the remaining part of the body. The yolk sac and the extremely long yolk sac stalk are visible in the chorionic cavity. Note the umbilical cord and the herniation of the intestinal loops. One side of the chorion has many villi (chorion frondosum), while the other side is almost smooth (chorion laeve).

by external examination. Initially **the intestinal loops cause a large swelling in the umbilical cord**, but by the 11th week they withdraw into the abdominal cavity. At the end of the third month reflex activity can be evoked in aborted fetuses, indicating muscular activity.[2] These movements, however, are so small that they are not noticed by the mother.

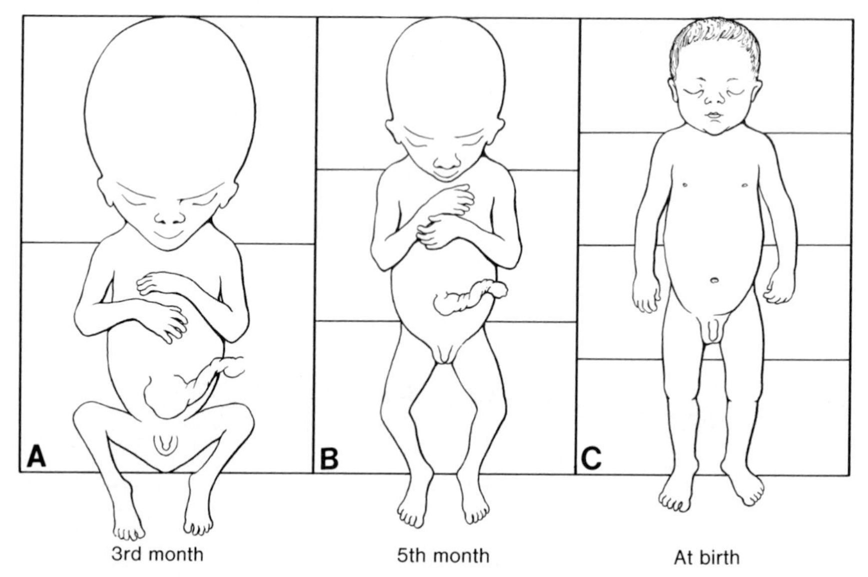

Figure 6-2. Schematic drawing showing the size of the head in relation to the rest of the body at various stages of development.

During the **fourth** and **fifth months** the fetus lengthens rapidly (fig. 6-5) and at the end of the first half of intra-uterine life its C.R. length is approximately 15 cm, that is, about half the total length of the newborn. The weight of the fetus, however, increases little during this period and by the end of the fifth month is still less than 500 gm.

The fetus is covered with fine hair, called **lanugo hair**; eyebrows and head hair are also visible. **During the fifth month the movements of the fetus are usually clearly recognized by the mother.**

During the **second half of intra-uterine life** the weight increases considerably, particularly during the last 2½ months, when 50 per cent of its full term weight (approximately 3200 gm) is added. During **the sixth month** the fetus has at first a wrinkled appearance, because of the lack of underlying connective tissue; its skin is reddish. A fetus born during the sixth month or in the first half of the seventh month has great difficulty surviving. Although several organ systems are able to function, the respiratory system and the central nervous system have not differentiated sufficiently and coordination between the two systems is not yet well established.

During the last couple of months the fetus obtains well-rounded contours as the result of the deposition of subcutaneous fat (fig. 6-6). By the end of intra-uterine life the skin is covered by a whitish, fatty substance, the **vernix caseosa**, which is composed of the secretory products of the sebaceous glands. When the fetus is 28 weeks old, it is able to survive, although with great difficulty.

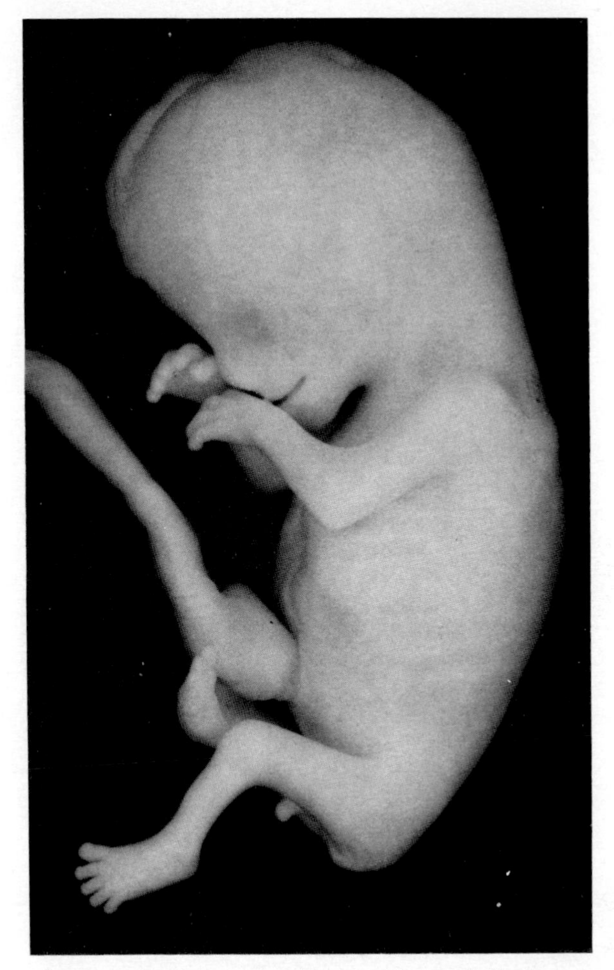

Figure 6-3. Photograph of an 11-week fetus. The umbilical cord still shows a swelling on its base, caused by the herniated intestinal loops. The toes are developed and the sex of the fetus can be recognized. The skull of this fetus lacks the normal smooth contours.

At the end of **the 10th lunar month**, the skull has the largest circumference of all parts of the body, an important fact with regard to its passage through the birth canal. At the time of birth the weight of the fetus is 3000 to 3500 gm, its C.R. length about 36 cm, and its crown-heel length about 50 cm. The sexual characteristics are pronounced and the testes should be in the scrotum.

Time of Birth

The date of birth is most accurately indicated as 266 days or 38 weeks after fertilization. The oocyte is usually fertilized within 12 hours after

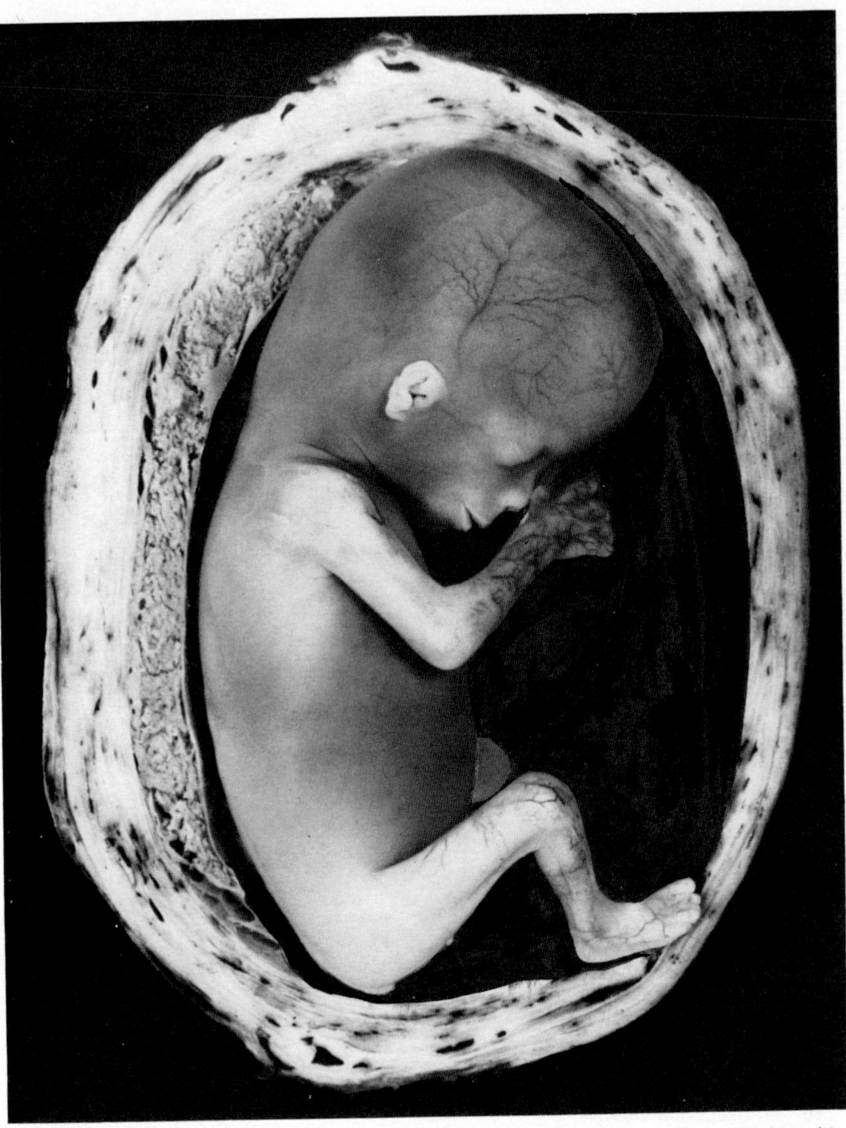

Figure 6-4. Photograph of a 12-week fetus in utero. Note the extremely thin skin and the underlying blood vessels. The face has all the human characteristics, but the ears are still primitive. Movements begin at this time but are usually not felt by the mother (Courtesy E. Blechschmidt, Professor of Anatomy, University of Göttingen.)

ovulation, and coitus must have occurred within 24 hours preceding fertilization. A pregnant woman usually will see her obstetrician when two successive menstrual bleedings have failed to occur. By that time her recollection about the coitus is usually vague and it is readily understandable that the day of fertilization is difficult to determine.

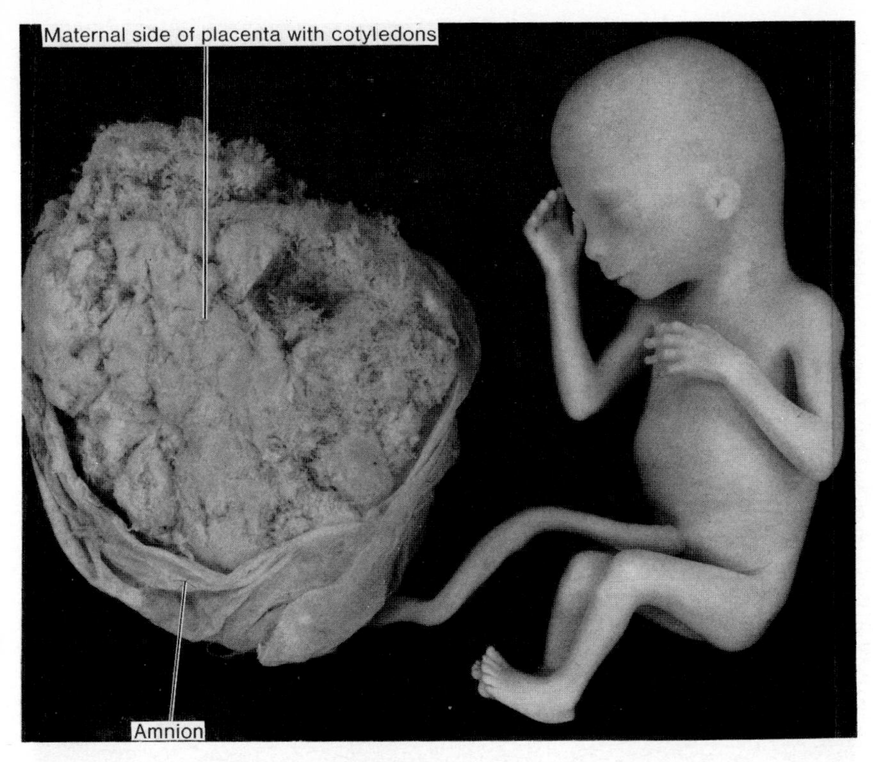

Figure 6-5. Photograph of an 18-week-old fetus connected to the placenta by its umbilical cord. The skin of the fetus is thin as a result of the absence of subcutaneous fat. Note the placenta with its cotyledons and the amnion.

The obstetrician calculates the date of birth as 280 days or 40 weeks from the first day of the last menstrual bleeding. This day is usually remembered quite accurately. In women with regular 28-day menstrual periods, the method is rather accurate, but when the cycles are irregular substantial miscalculations may be made. It must be remembered that the time between ovulation and the succeeding menstrual bleeding is constant (14 days ± 1 day), but the time between ovulation and the preceding menses is highly variable. An additional complication occurs when the woman has a short bleeding about 14 days after fertilization as a result of the erosive activity of the implanting trophoblast (see Chapter 3). Hence, it is evident that the day of delivery is not always easy to determine. In general most fetuses are born within 10 to 14 days of the calculated delivery date. If they are born much earlier they are categorized as **premature**, if later they are considered **postmature**.

Occasionally the age of an embryo or small fetus will have to be determined. By combining data on the onset of the last menstrual period with fetal length, weight, and other morphological characteristics typical for a

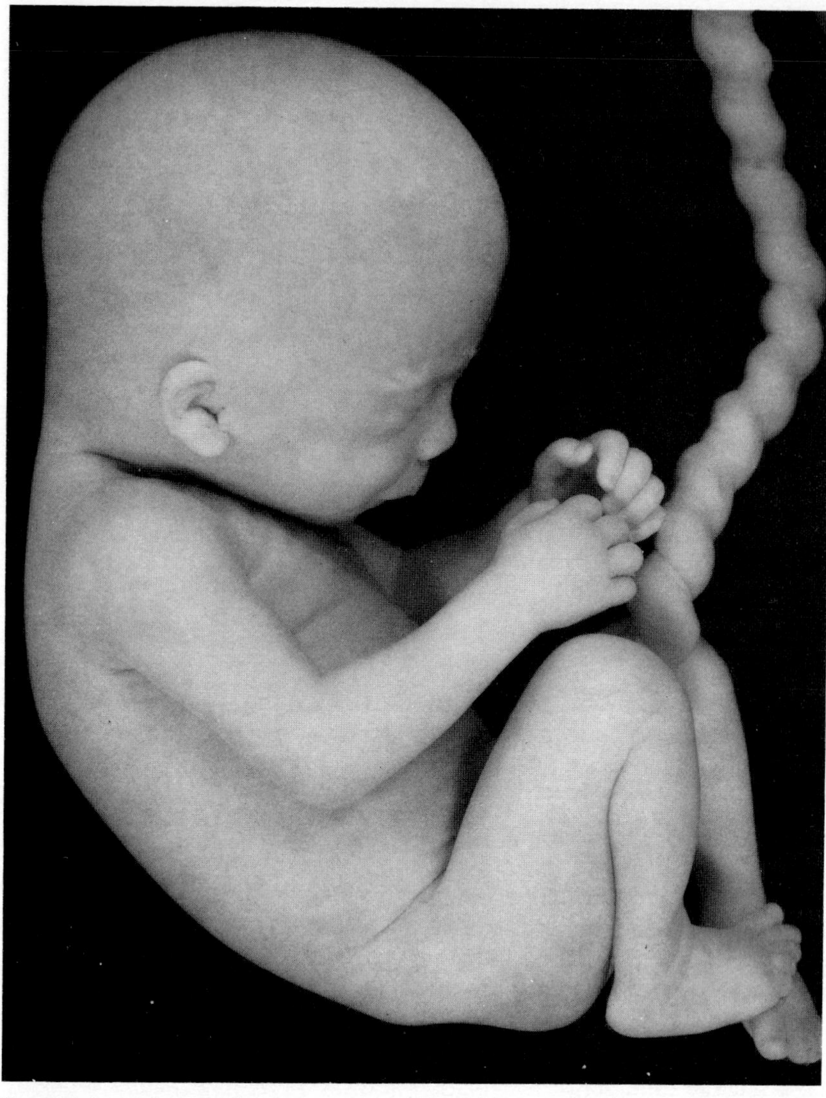

Figure 6-6. Photograph of a seven-month-old fetus. This fetus would possibly be able to survive. It has now well-rounded contours as a result of the deposition of subcutaneous fat. Note the spiral twisting of the umbilical cord.

given month of development, it is usually not too difficult to determine the approximate age of the fetus.

Abnormal Fetal Growth

Considerable variability exists in fetal length and weight and sometimes these values do not correspond with the calculated age of the fetus in months or weeks. Most factors influencing length and weight are genetically determined, but it is now known that environmental factors also play an important role. It is generally accepted that severe malnutrition[3] as well as heavy smoking[4] leads to reduced fetal growth. Similarly, placental insufficiency may cause severe growth retardation.[3] Not infrequently a full term baby is found to have a birth weight of 2500 gm, and is by mistake considered a premature baby. The **premature baby**, however, is characterized by a birth weight of 2500 gm or more and a gestation period of 28 to 38 weeks. The full term but low weight baby is known as "small for date"[5] and forms a special treatment category. They frequently have a wrinkled skin caused by the disappearance of subcutaneous fat. The cause of the small for date baby is unknown, but it has been suggested to result from placental dysfunction.

Sometimes a full term baby is larger and heavier than normal. This is particularly the case with infants from diabetic mothers.[6,7] Episodes of maternal hyperglycemia and the subsequent compensatory secretion of insulin in the fetus are thought to stimulate fetal growth. Hence, the values represented in Table 6-1 are important in estimating age of the fetus, but it must be realized that great variation exists.

SUMMARY

The **fetal period extends from the ninth week till birth** and is characterized by rapid growth of the body and maturation of the organ systems. Growth in length is particularly striking during the third, fourth, and fifth months (approximately 5 cm per month), while increase in weight is most striking during the last two months of gestation (approximately 700 gm per month).

A striking change is the relative slowdown in the growth of the head. In the third month it is about one-half the crown-rump length; by the fifth month about one-third and at birth one-fourth of the crown-heel length (fig. 6-2).

During the fifth month fetal movements are clearly recognized by the mother and the fetus is covered with fine small hairs.

A fetus born during the sixth or beginning of the seventh month has difficulties surviving, mainly because the respiratory and central nervous systems have not differentiated sufficiently.

In general the **length of pregnancy** for a full term fetus is considered to be **280 days or 40 weeks after onset of the last menstruation**, or, **more accurately, 266 days or 38 weeks after fertilization**.

REFERENCES

1. Scammon, R. E., and Calkins, M. A. *Development and Growth of the External Dimensions of the Human Body in the Foetal Period.* University of Minnesota Press, Minneapolis, 1929.

2. Hooker, D. Early fetal activity in mammals. *Yale J. Biol. Med., 8:* 579, 1936.
3. Page, E. W., Villee, C. A., and Villee, D. B. *Human Reproduction: The Core Content of Obstetrics, Gynecology and Perinatal Medicine.* W. B. Saunders, Philadelphia, 1976.
4. Gruenwald, P. Growth of the human fetus. I. Normal growth and its variation. *Am. J. Obstet. Gynecol., 94:* 1112, 1966.
5. Lucey, J. F. Conditions and diseases of the newborn. In *Principles and Management of Human Reproduction*, edited by D. E. Reid, K. J. Ryan, and K. Benirschke. W. B. Saunders, Philadelphia, 1972.
6. Delaney, J. J., and Ptacek, J. Three decades of experience with diabetic pregnancies. *Am. J. Obstet. Gynecol, 106:* 550, 1970.
7. Villee, D. B. *Human Endocrinology. A Developmental Approach.* W. B. Saunders, Philadelphia, 1975.

chapter 7

Fetal Membranes and Placenta

By the beginning of the second month, the **trophoblast** is characterized by a great number of secondary and tertiary villi which give it a radial appearance (figs. 7-1 and 4-9). The villi are anchored in the mesoderm of the **chorionic plate** and are attached peripherally to the maternal decidua by way of the outer **cytotrophoblast shell**. The surface of the villi is formed by

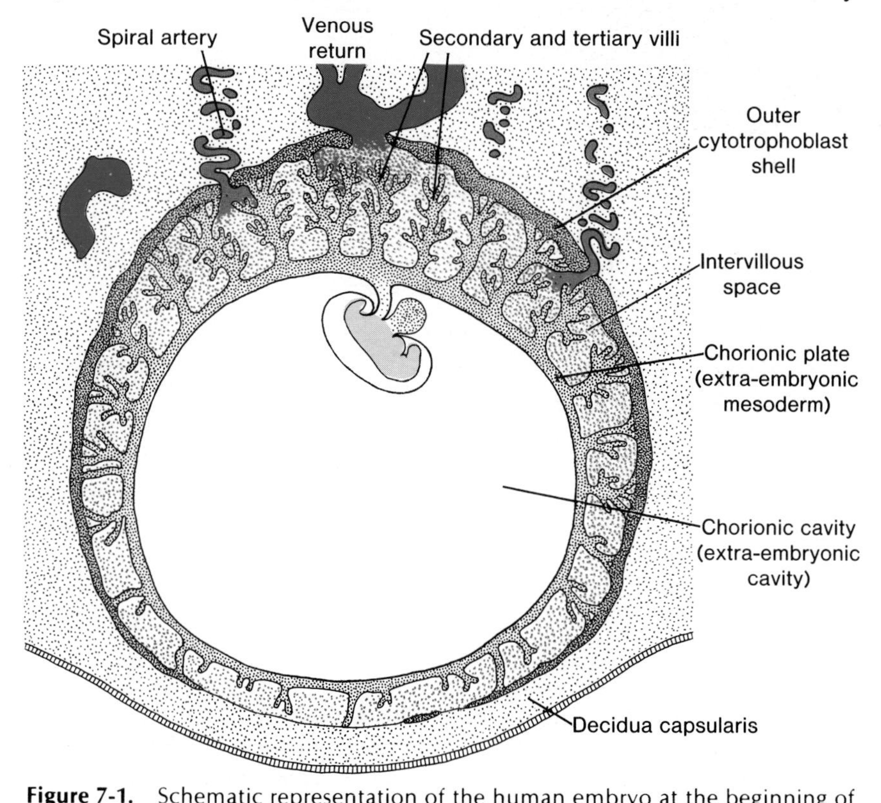

Figure 7-1. Schematic representation of the human embryo at the beginning of the second month of development. At the embryonic pole the villi are numerous and well formed; at the abembryonic pole they are few in number and poorly developed. (Modified after von Ortmann.)

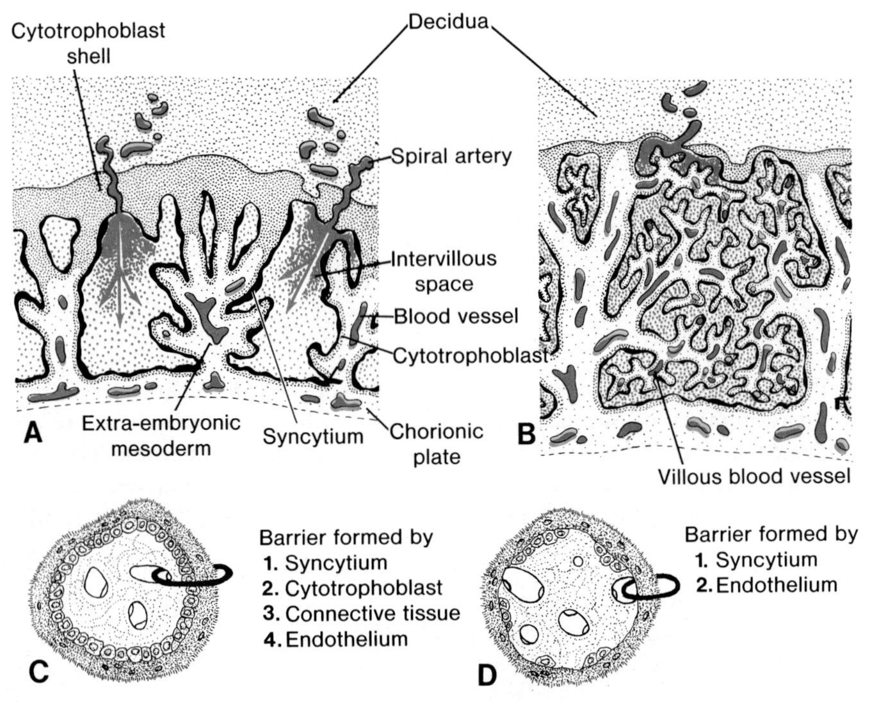

Figure 7-2. The structure of the villi at various stages of development. *A*, During the fourth week. Note how the extra-embryonic mesoderm penetrates the stem villi in the direction of the decidual plate. *B*, During the fourth month. In many small villi the wall of the capillaries is in direct contact with the syncytium. *C and D*, Enlargement of the villus as shown in *A and B*, respectively.

the syncytium, resting on a layer of cytotrophoblastic cells which in turn cover a core of vascular mesoderm (fig. 7-2*A*,*C*). The capillary system developing in the core of the villous stems soon comes in contact with the capillaries of the chorionic plate and connecting stalk, thus giving rise to the extra-embryonic vascular system (fig. 4-9).

During the following months, numerous small extensions sprout from the existing villous stems into the surrounding **lacunar** or **intervillous spaces**. These newly formed villi initially are primitive (fig. 7-2*C*) but by the beginning of the fourth month the cytotrophoblastic cells as well as some of the connective tissue cells disappear. The syncytium and the endothelial wall of the blood vessels are then the only layers which separate the maternal and fetal circulations (fig. 7-2*B*,*D*). Frequently the syncytium becomes very thin and large pieces containing several nuclei may break off and drop into the intervillous blood lakes. These pieces, known as **syncytical knots**, enter the maternal circulation and usually degenerate without causing any symptoms.[1] The disappearance of the cytotrophoblastic cells progresses from the smaller to the larger villi, although some always persist in the large villi. The latter, however, do not participate in the exchange between the two circulations.

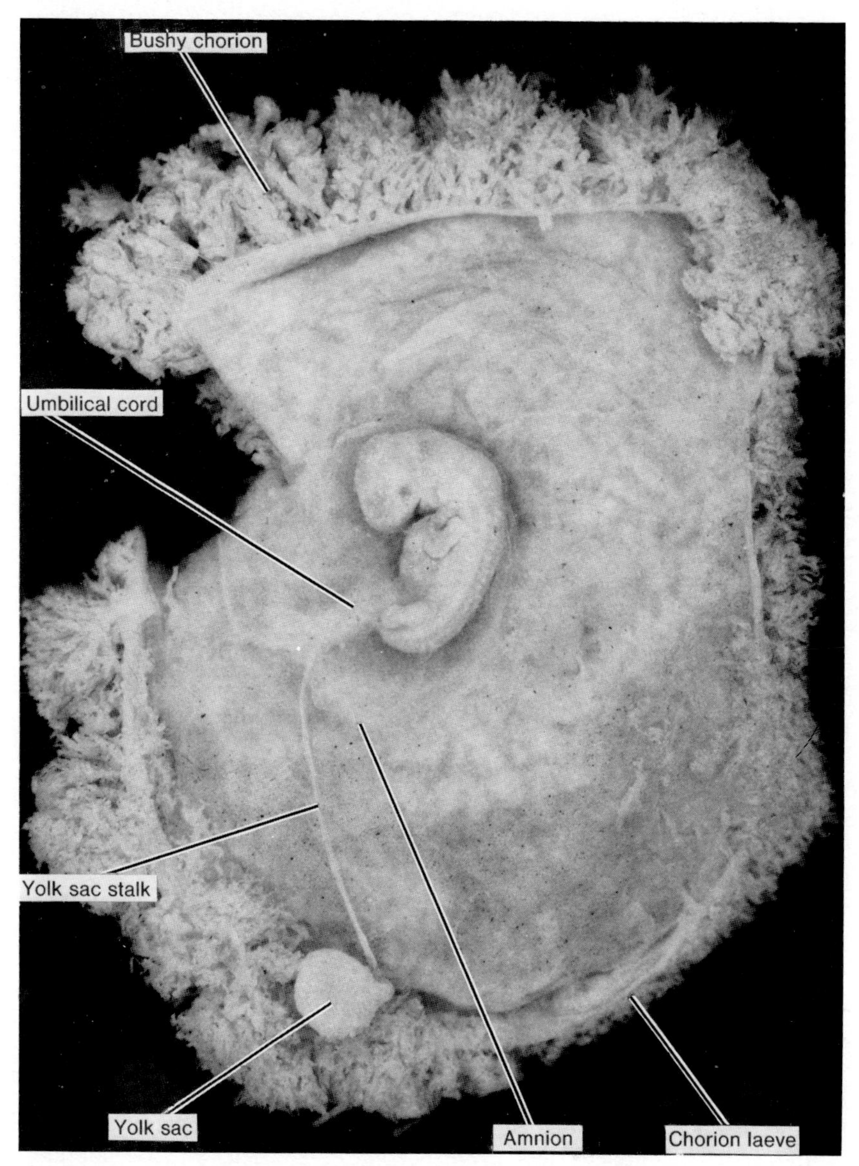

Figure 7-3. Photograph of a six-week embryo. The amniotic sac and chorionic cavity have been opened to expose the embryo. Note the bushy appearance of the trophoblast at the embryonic pole in contrast to the small villi at the abembryonic pole. Note also the connecting stalk and the yolk sac with its extremely long stalk.

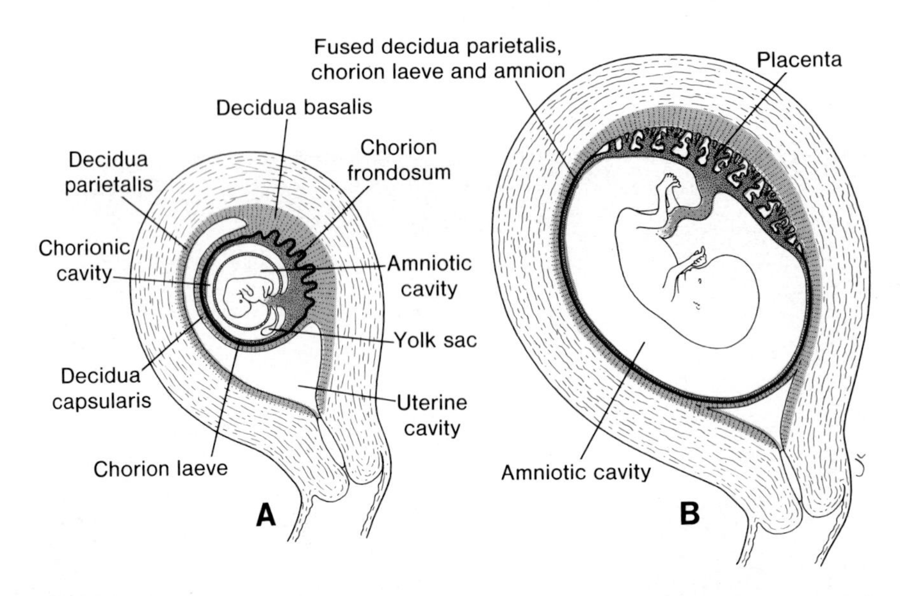

Figure 7-4. Schematic drawing showing the relation of the fetal membranes and the wall of the uterus. *A*, End of the second month. Note the yolk sac in the chorionic cavity between the amnion and chorion. At the abembryonic pole the villi have disappeared (chorion laeve). *B*, End of the third month. the amnion and chorion have fused and the uterine cavity is obliterated by fusion of the chorion laeve and the decidua parietalis.

Chorion Frondosum and Decidua Basalis

In the early weeks of development the villi cover the entire surface of the chorion (fig. 7-1). As pregnancy advances the villi on the embryonic pole continue to grow and expand, thus giving rise to the **chorion frondosum** (bushy chorion); those on the abembryonic pole degenerate, and by the third month this side of the chorion is smooth and known as the **chorion laeve** (figs. 7-3 and 7-4*A*).

The difference in the embryonic and abembryonic poles of the chorion is also reflected in the structure of the decidua. The decidua over the chorion frondosum, the **decidua basalis**, consists of a compact layer of large cells with abundant amounts of lipids and glycogen. This layer, the **decidual plate**, is tightly connected to the chorion. The decidual layer over the abembryonic pole is known as the **decidua capsularis** (fig. 7-4*A*). With increase in the size of the chorionic vesicle, this layer becomes stretched, and degenerates. Subsequently, the chorion laeve comes into contact with the **decidua parietalis** on the opposite side of the uterus and the two fuse (figs. 7-4*B*, 7-5, and 7-6). The lumen of the uterus is then obliterated. Hence, the only portion of the chorion participating in the exchange processes is the chorion frondosum and, together with the decidua basalis, the two make up the **placenta**.

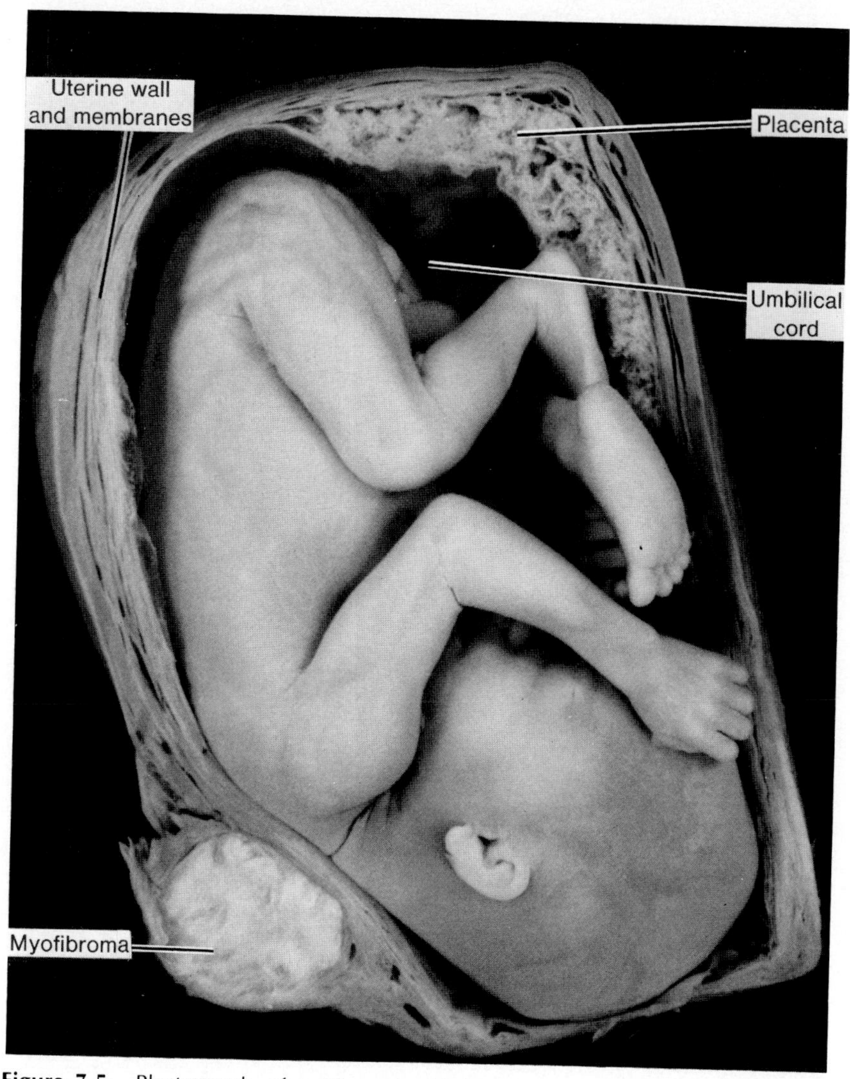

Figure 7-5. Photograph of a 19-week-old fetus in its natural position in the uterus. Note the umbilical cord and the placenta. The lumen of the uterus is obliterated. In the wall of the uterus is seen a large swelling, known as a myofibroma.

Structure of the Placenta

By the beginning of the fourth month, the placenta has two components: (1) a **fetal portion** formed by the chorion frondosum, and (2) a **maternal**

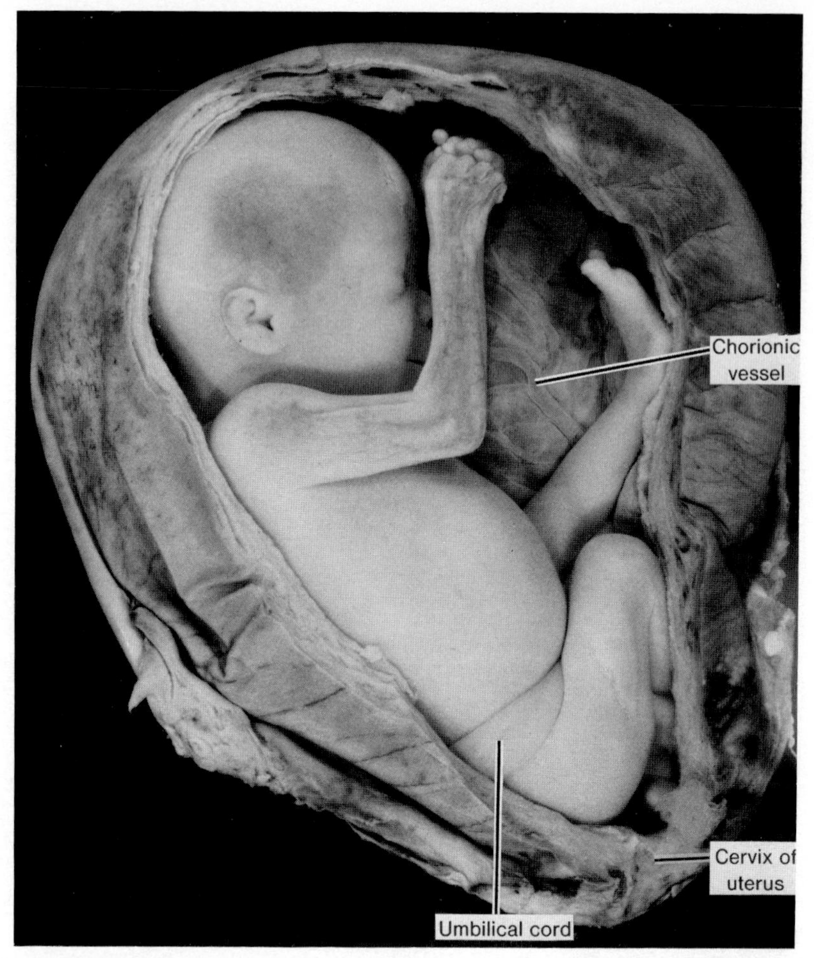

Figure 7-6. Photograph of a 23-week-old fetus in the uterus. A portion of the wall of the uterus as well as of the amnion have been removed to show the fetus. In the background can be seen the placental vessels converging toward the umbilical cord. Note that the umbilical cord is tightly wound around the abdominal cavity of the fetus, possibly causing its abnormal position in the uterus (breech position).

portion formed by the decidua basalis (fig. 7-4*B*). On the fetal side the placenta is bordered by the **chorionic plate** (fig. 7-7); on its maternal side by the decidua basalis, of which the **decidual plate** is most intimately incorporated into the placenta. In the so-called **junctional zone** the trophoblast and decidua cells intermingle. It is characterized by decidual and syncytial giant cells and is rich in amorphous mucopolysaccharide material. Most of the cytotrophoblast cells have degenerated. Between the chorionic and decidual plates are the intervillous spaces which are filled with maternal blood. They are derived from the lacunae in the syncytiotrophoblast and lined with syncytium of fetal origin. The villous trees grow into the intervillous blood lakes (figs. 7-1 and 7-7).

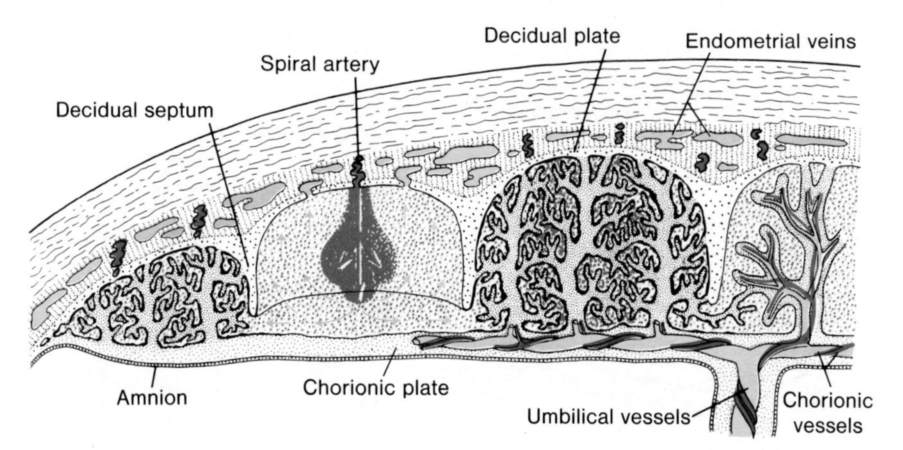

Figure 7-7. Composite drawing of the placenta in the second half of pregnancy. The cotyledons are partially separated from each other by the decidual (maternal) septa. Note that most of the intervillous blood returns to the maternal circulation by way of the endometrial veins. A small portion enters neighboring cotyledons. The intervillous spaces are lined by syncytium. (Modified after Ramsey[3] and Hamilton and Boyd.[1])

During the fourth and fifth months the decidua forms a number of septa, the **decidual septa**, which project into the intervillous spaces but do not reach the chorionic plate (fig. 7-7). These septa have a core of maternal tissue, but their surface is covered by a layer of syncytial cells so that at all times a syncytial layer separates the maternal blood in the intervillous lakes from the fetal tissue of the villi.[2, 3] As a result of this septum formation, the placenta is divided into a number of compartments or **cotyledons** (fig. 7-8). Since the decidual septa do not reach the chorionic plate, contact between the intervillous spaces in the various cotelydons is maintained.

As a result of the continuous growth of the fetus and expansion of the uterus, the placenta also enlarges. Its increase in surface area roughly parallels that of the expanding uterus and throughout pregnancy it covers approximately 25 to 30 per cent of the internal surface of the uterus. The increase in thickness of the placenta results from the arborization of existing villi and is not caused by further penetration in the maternal tissues.

Full Term Placenta

At full term the placenta has a discoid shape, a diameter of 15 to 25 cm, is approximately 3 cm thick, and has a weight of about 500 to 600 gm. At birth it is torn from the uterine wall and, approximately 30 minutes after birth of the child, expelled from the uterine cavity. When, after birth, the placenta is viewed from the **maternal side**, 15 to 20 slightly bulging areas, the **cotyledons**, covered by a thin layer of decidua basalis are clearly recognizable (fig. 7-8*B*). The grooves between the cotyledons are formed by

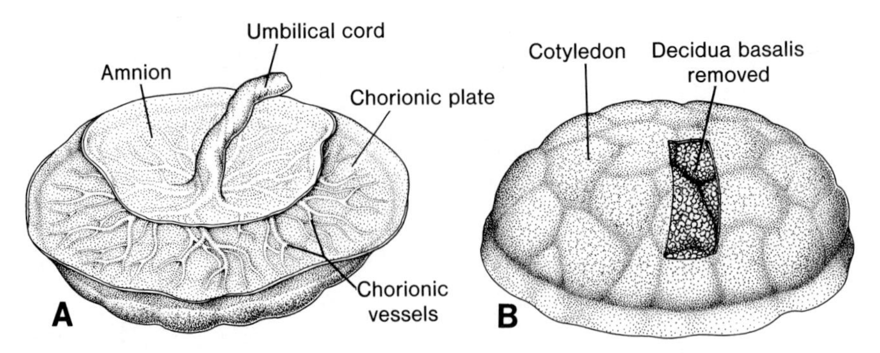

Figure 7-8. Drawing of a full term placenta. *A*, As seen from the fetal side. Note that chorionic plate and umbilical cord are covered by the amnion. *B*, As seen from the maternal side. Note the cotyledons. In one area the decidua has been removed. The maternal side of the placenta is always carefully inspected and frequently it is seen that one or more cotyledons have a whitish appearance. This is caused by excessive fibrinoid formation resulting in infarction of a group of intervillous lakes.

the decidual septa. Much of the decidua remains temporarily in the uterus and is expelled with subsequent uterine bleedings.

The **fetal surface** of the placenta is covered entirely by the chorionic plate. A number of large arteries and veins, the **chorionic vessels**, are seen to converge toward the umbilical cord (fig. 7-8*A*). The chorion in turn is covered by the amnion. The attachment of the umbilical cord is usually eccentric and occasionally even marginal. Rarely, however, does it insert into the chorionic membranes outside the placenta (**velamentous insertion**).

Circulation of the Placenta

The cotyledons receive their blood through 80 to 100 spiral arteries which pierce the decidual plate and enter the intervillous spaces at more or less regular intervals (fig. 7-7).[3, 4] The lumen of the spiral artery is narrow, resulting in an increased blood pressure when entering the intervillous space. This pressure forces the blood deep into the intervillous spaces and bathes the numerous small villi of the villous tree in oxygenated blood.[3] As the pressure decreases the blood flows back from the chorionic plate towards the decidua, where it enters the endometrial veins (fig. 7-7). Hence, the blood from the intervillous lakes drains back into the maternal circulation through the endometrial veins.

The intervillous spaces of the full-grown placenta contain approximately 150 cc of blood, which is replenished about three or four times per minute.[5–7] This blood moves along the chorionic villi which have a surface area varying from 4 to 14 square meters. It must be remembered, however, that placental exchange does not take place in all villi, but only in those in which the fetal vessels are in intimate contact with the covering syncytial membrane. In these villi the syncytium often has a brush border consisting of numerous microvilli, thus greatly increasing the exchange rate between the maternal and fetal circulations (fig. 7–2*D*).[2]

Function of the Placenta

The main functions of the placenta are: (1) **the exchange of metabolic and gaseous products** between the maternal and fetal blood streams, and (2) **the production of hormones**. The dividing membrane between the two circulations, **the placental barrier**, is initially composed of four layers: (1) the endothelial lining of the fetal vessels; (2) the connective tissue in the core of the villus; (3) the cytotrophoblastic layer; and (4) the syncytium (fig. 7-2*C*). From the fourth month on, however, the placental barrier becomes much thinner, since the endothelial lining of the vessels comes in intimate contact with the syncytial membrane, thus greatly increasing the rate of exchange (fig. 7-2*D*). Since the maternal blood in the intervillous spaces is separated from the fetal blood by a chorionic derivative, the human placenta is considered to be of the **hemochorial** type.

EXCHANGE OF GASES

Exchange of gases such as oxygen, carbon dioxide, and carbon monoxide is accomplished by simple diffusion. At term the fetus extracts 20 to 30 ml of oxygen per minute from the maternal circulation and it is understandable that even a short term interruption of the oxygen supply will be fatal for the fetus.[8, 9]

EXCHANGE OF NUTRIENTS AND ELECTROLYTES

Exchange of nutrients and electrolytes such as amino acids, free fatty acids, carbohydrates, and vitamins is rapid and increases as pregnancy advances.[9]

TRANSMISSION OF MATERNAL ANTIBODIES

Maternal antibodies are taken up by pinocytosis of the syncytiotrophoblast and subsequently transported to the fetal capillaries.[10, 11] In this manner the fetus acquires maternal antibodies of the IgG (7S) class gamma globulins against various infectious diseases and obtains passive immunity against diphtheria, smallpox, measles, and others, but not against chickenpox and whooping cough.

Of great importance is the so-called **Rh-incompatibility**. If the fetus is Rh-positive and the mother Rh-negative, fetal red blood cells invading the maternal bloodstream may elicit an antibody response in the mother. The maternal antibodies against the fetal antigens then return to the fetus and cause a breakdown of the red blood cells of the fetus. Small bleedings at the surface of the villi are probably responsible for this antigen-antibody interaction between fetus and mother.[12] The breakdown of fetal red blood cells, known as **erythroblastosis fetalis** or **hemolytic disease** of the fetus, may lead to intrauterine death. Examination of the amniotic fluid may provide an indication about the degree of the disease, and intra-uterine blood transfusions into the fetus may prevent death.

HORMONE PRODUCTION

By the end of the fourth month, the placenta produces **progesterone** in sufficient amounts to maintain pregnancy in case the corpus luteum is removed or fails to function properly.[13] In all probability the steroid hormone is synthesized in the syncytial cytoplasm. In addition to progesterone, the placenta produces increasing amounts of **estrogenic hormones** (**estradiol** and **estrogen**) until just before the end of pregnancy, when a maximum level is reached. The sudden drop in estrogenic hormone production just before birth has long been considered one of the factors responsible for the beginning of parturition. More recent work has shown that the estrogens play only a minor role in parturition.[14, 15]

The syncytiotrophoblast also produces **gonadotropins** (**human chorionic gonadotropin—HCG**), which have an effect similar to that of the luteinizing hormones of the anterior lobe of the pituitary. These hormones are excreted by the mother in the urine and, in the early stages of gestation, their presence is used as an indicator or pregnancy. Another protein hormone produced by the placenta is **human placental lactogen**. It is a growth hormone-like substance, which gives the fetus priority on maternal blood glucose and makes the mother somewhat diabetogenic.

Most maternal hormones do not cross the placenta, or, if they do, such as thyroxine, only at a slow rate.[9] Of great danger are some synthetic progestins which cross the placenta at a rapid rate and may cause masculinization in female fetuses.[16] Even more dangerous has been the use of the synthetic estrogen, **diethylstilbestrol**, which easily crosses the placenta. This compound has recently produced carcinoma of the vagina in women who were exposed to it during their intra-uterine life.[7] (See also Chapter 8.)

INFECTIOUS AGENTS TRANSFER

Although the placental barrier is frequently considered to act as a protective mechanism against damaging factors, many viruses such as rubella, Coxsackie, variola, varicella, measles, and poliomyelitis virus pass the placenta without any difficulty.[18, 19] Once in the fetus some of the viruses cause infections, which in turn may result in cell degeneration and congenital malformations.

DRUG TRANSFER

Unfortunately most drugs and drug metabolites pass the placenta without difficulty and many cause serious damage to the embryo.[20-22] (For details see Chapter 8.)

Amnion and Umbilical Cord

The line of reflexion between the amnion and the ectoderm, the **amnio-ectodermal junction**, is oval-shaped, and known as the **primitive umbilical ring**. At the fifth week of development the following structures pass through the ring (fig. 7-9*A*,*C*): (1) the **connecting stalk** containing the allantois and the umbilical vessels consisting of two arteries and one vein; (2) the **yolk sac stalk (vitelline duct)** accompanied by the vitelline vessels; and (3) the **canal, connecting the intra- and extra-embryonic coelomic cavities** (fig. 7-9*C*). The yolk sac proper occupies a space in the **chorionic cavity**, that is the space between the amnion and chorionic plate (fig. 7–9*B*).

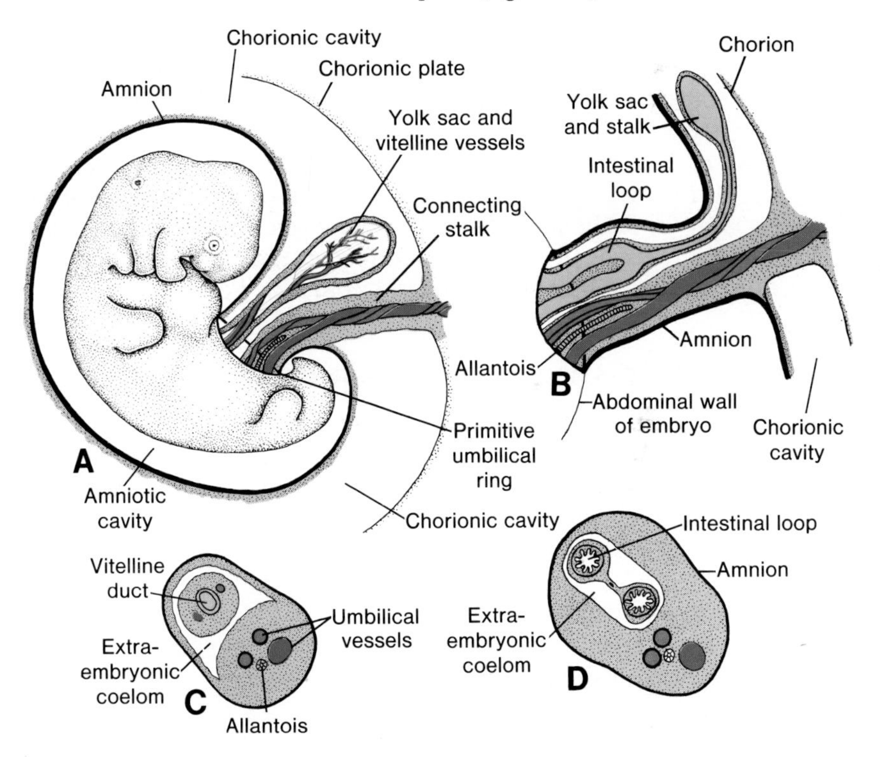

Figure 7-9. *A*, Schematic drawing of a five-week embryo to show the structures passing through the primitive umbilical ring. *B*, Schematic drawing of the primitive umbilical cord of a ten-week embryo. *C*, Transverse section through the structures at the level of the umbilical ring. *D*, Transverse section through the primitive umbilical cord, showing the intestinal loops protruding in the cord.

During further development the amniotic cavity enlarges rapidly at the expense of the chorionic cavity, and the amnion begins to envelop the connecting and yolk sac stalks thereby crowding them together and causing the formation of the **primitive umbilical cord** (fig. 7-9*B*). Distally the cord contains then the yolk sac stalk and the umbilical vessels. More proximally it contains some intestinal loops and the remnant of the allantois (fig. 7-9*B*,*D*). The yolk sac is found in the chorionic cavity and is connected to the umbilical cord by its stalk. At the end of the third month the amnion has expanded to such an extent that it comes in contact with the chorion, thereby obliterating the chorionic cavity (fig. 7-4*B*). The yolk sac then usually shrinks and is gradually obliterated.

The abdominal cavity is temporarily too small for the rapidly developing intestinal loops and some of them are pushed into the extra-embryonic coelomic space in the umbilical cord. These extruding intestinal loops from the so-called **physiological umbilical hernia** (see "Digestive System," Chapter 14). At about the end of the third month, the loops are withdrawn into the body of the embryo and the coelomic cavity in the cord is obliterated. When, in addition, the allantois, the vitelline duct, and vessels are obliterated, all that remains in the cord are the umbilical vessels surrounded by the **jelly of Wharton**. This tissue is rich in mucopolysaccharides and functions as a protective layer for the blood vessels. The walls of the arteries are muscular and contain many elastic fibers, which contribute to a rapid constriction and contraction of the umbilical vessels after the cord is tied off.

PLACENTAL CHANGES AT END OF PREGNANCY

At the end of pregnancy a number of changes occur in the placenta which may be an indication of a reduced exchange between the two circulations. These changes include: (1) an increase of the fibrous tissue in the core of the villus; (2) an increase in the thickness of the basement membrane of the fetal capillaries; (3) obliterative changes in the small capillaries of the villi; and (4) the deposition of fibrinoid on the surface of the villi in the junctional zone and in the chorionic plate. Excessive fibrinoid formation frequently causes infarction of an intervillous lake or sometimes of an entire cotyledon. The cotyledon then obtains a whitish appearance.

At birth the umbilical cord is approximately 2 cm in diameter and 50 to 60 cm long. It is tortuous, causing the so-called **false knots**. An extremely long cord may encircle the neck of the fetus, whereas a short one may cause difficulties during delivery by pulling the placenta from its attachment in the uterus.[23]

AMNIOTIC FLUID

The amniotic cavity is filled with a clear, watery fluid produced by the amniotic cells and derived from maternal blood.[8, 24] During the early months of pregnancy, the embryo is suspended by its umbilical cord in this fluid, which serves as a protective cushion. The fluid (1) absorbs jolts, (2) prevents the adherence of the embryo to the amnion, and (3) allows for fetal movements. The water in the amniotic fluid changes every three hours, indicating the enormous exchange between the amniotic cavity and the maternal circulation.[24] Probably from the beginning of the fifth month, the fetus swallows its own amniotic fluid, and it is estimated that it drinks about 400 cc per day, which is about half of the total amount.[9] Fetuses unable to swallow, either because of **esophageal atresia** or through lack of nervous control of the swallowing mechanism, as in **anencephaly**, are usually surrounded by large amounts of amniotic fluid (**hydramnios**). Under normal conditions the amniotic fluid is absorbed through the gut of the fetus into the blood stream and passes into the maternal blood by way of the placenta. At the end of pregnancy, urine is daily added to the amniotic fluid. This urine is mostly water, since the placenta is functioning as the kidney. During childbirth, the amnion and chorion combined form a hydrostatic wedge which helps to dilate the cervical canal.

Fetal Membranes in Twins

The arrangement of the fetal membranes in twins varies considerably and is dependent on the type of twins as well as on the time of separation in the case of **monozygotic twins**.

DIZYGOTIC TWINS

The most common type of twins is formed by the **dizygotic** or **fraternal twins**. They result from the simultaneous shedding of two oocytes and the fertilization by two different spermatozoa. Since both zygotes have a totally different genetic constitution, the individual members have no more resemblance than brothers or sisters of different ages. They may or may not have a different sex. Both zygotes implant individually in the uterus and each develops its own placenta, its own amnion, and its own chorionic sac (fig. 7-10*A*). Sometimes, however, the two placentas are located so close together that fusion occurs. Similarly the walls of the chorionic sacs may also come into close apposition and fuse (fig. 7-10*B*). Occasionally the members of

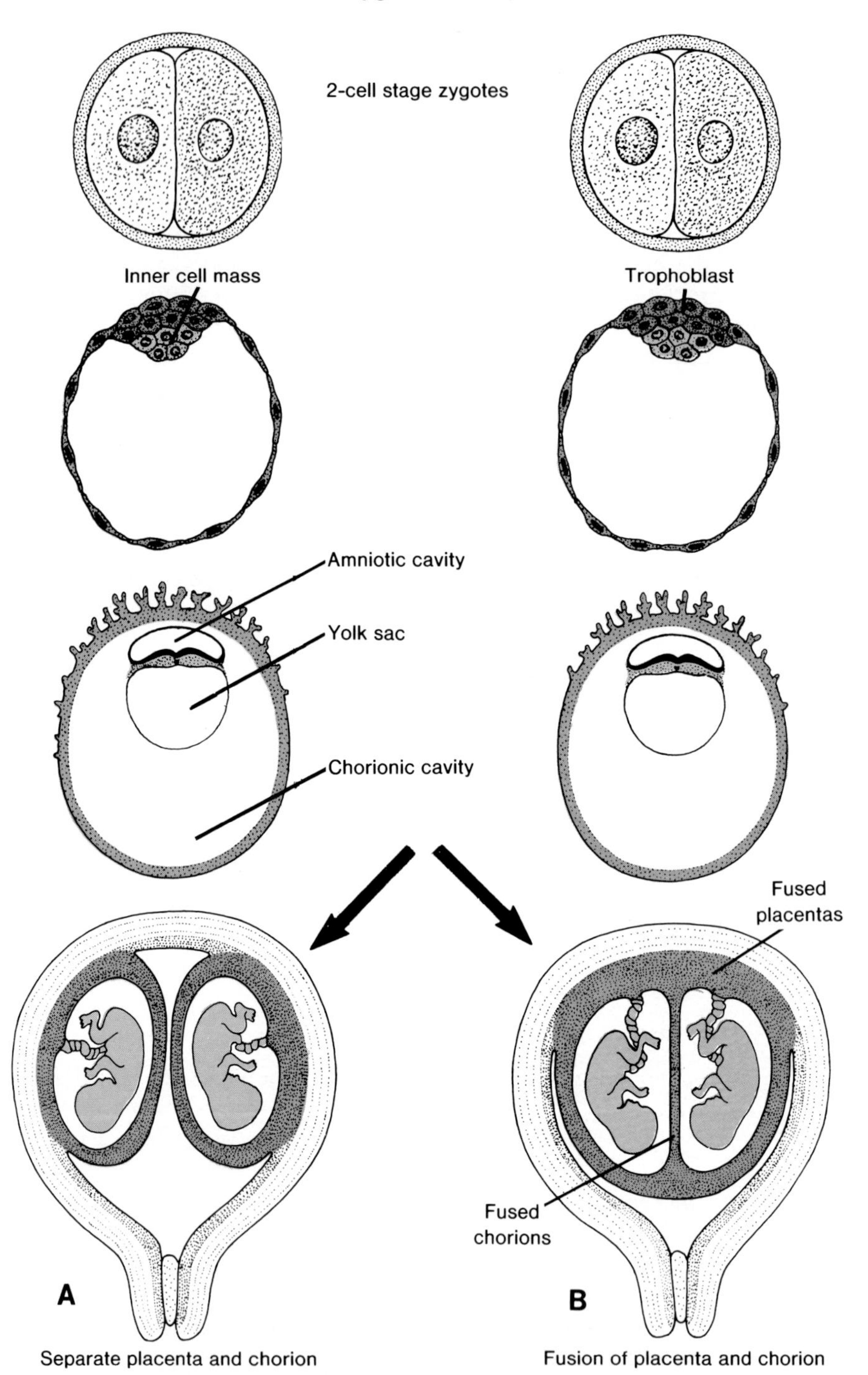

Figure 7-10. Schematic drawing showing the development of dizygotic twins. Although each embryo normally has its own amnion, chorion and placenta (*A*), sometimes the placentas may be fused (*B*) Each embryo usually receives the appropriate amount of blood, but on occasion more blood is shunted to one of the partners through large anastomoses.

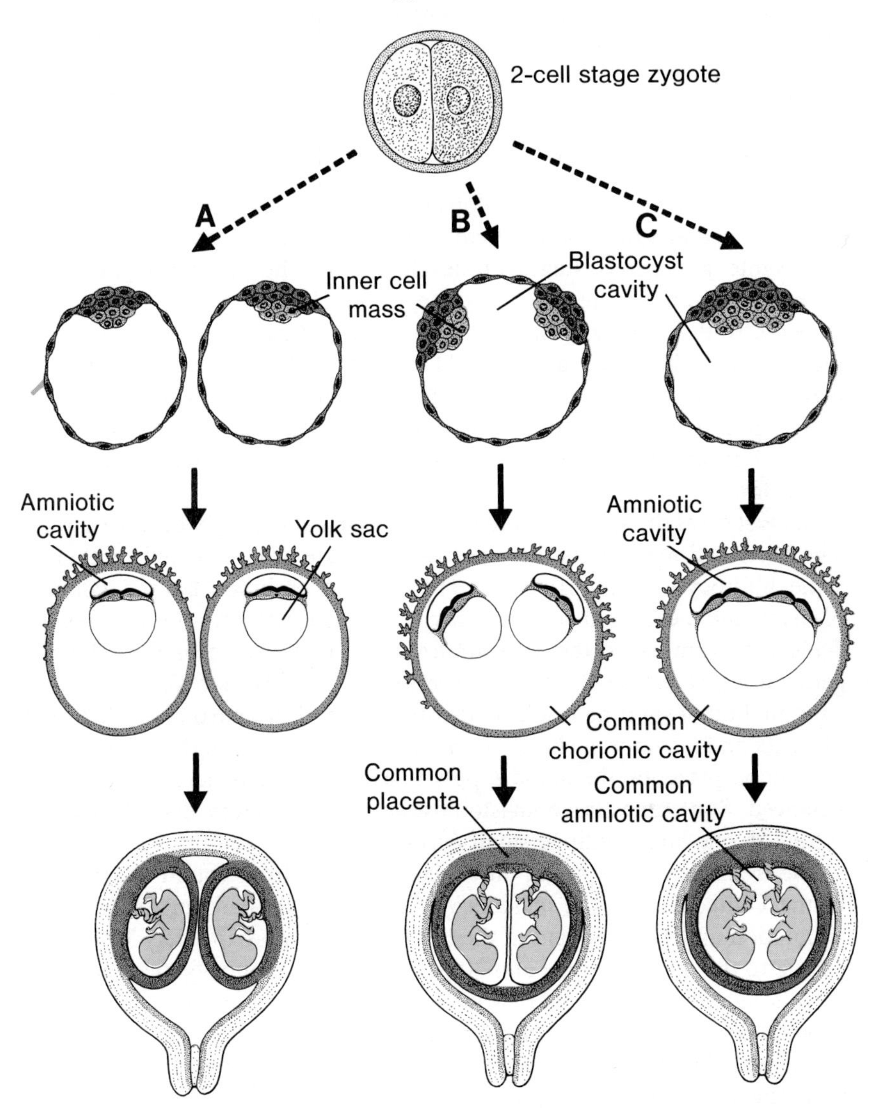

Figure 7-11. Schematic diagrams showing the possible relations of the fetal membranes in monozygotic twins. *A*, Splitting occurs at the two-cell stage and each embryo has its own placenta, amniotic cavity, and chorionic cavity. *B*, Splitting of the inner cell mass in two completely separated groups. The two embryos have a common placenta, a common chorionic sac, but separate amniotic cavities. *C*, Splitting of the inner cell mass at a late stage of development. The embryos have a common placenta, a common amniotic cavity, and a common chorionic cavity.

dizygotic twins possess red blood cells of two different types (**erythrocyte mosaicism**), indicating that the fusion of the two placentas was so intimate that red cells were exchanged. Though in cattle fusion of the chorionic circulations of two partners with different sex may disturb the sexual development of the female (intersexual female calves or **freemartins**),[2, 5] such abnormalities are unknown in man.

MONOZYGOTIC TWINS

The second type of twins develops from one single fertilized ovum and is known as **monozygotic** or **identical twins**. They result from the splitting of the zygote at various stages of development. The earliest separation is believed to occur at the two-cell stage, in which case two separate zygotes develop. Both blastocysts implant separately and each embryo has its own placenta and chorionic sac (fig. 7-11*A*). Although the arrangement of the membranes of these twins resembles that of the dizygotic twins, the two can be recognized as partners of a monozygotic pair by their strong resemblance in blood groups, fingerprints, sex, and external appearance such as eye and hair color.

In most cases the splitting of the zygote occurs at the early blastocyst stage. The inner cell mass splits then into two separate groups of cells within the same blastocyst cavity (fig. 7-11*B*). The two embryos have a common placenta and a common chorionic cavity, but separate amniotic cavities (fig. 7-11*B*). In rare cases, the separation occurs at the stage of the bilaminar germ disc just before the appearance of the primitive streak (fig. 7-11*C*). This method of splitting results in the formation of two partners with a single placenta and a common chorionic and amnion sac. Although the twins have a common placenta, the blood supply to each of the partners is usually well balanced. Sometimes large anastomoses cause circulatory disturbances, resulting in one large and one small partner.[25]

The frequency of twinning differs considerably from country to country. In the United States, the incidence of twins is 1.08 per cent in the white population and 1.36 per cent in the black group. Approximately 70 per cent belong to the dizygotic type and 30 per cent to the monozygotic type.[27]

Although the occurrence of triplets is not uncommon (1 in about 7600 pregnancies), the birth of quadruplets, quintuplets, and so forth is rare. In recent years multiple births such as sextuplets have occurred more frequently in mothers given gonadotropins for ovulatory failure.[28]

CONJOINED TWINS

The splitting of the zygote during later stages of development may result in an abnormal or incomplete splitting of the axial area of the germ disc. Such incompletely separated discs lead to the formation of **conjoined twins** or **double monsters**. According to the nature and degree of the union, they are classified as **thoracopagus** (pagus—fastened), **pygopagus**, and **craniopa-**

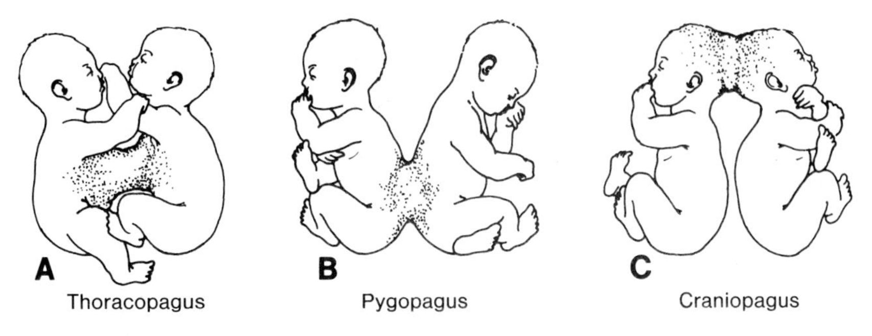

Figure 7-12. Schematic drawings of a thoracopagus, pygopagus, and craniopagus. Conjoined twins can be separated only if they have no vital parts in common.

gus (fig. 7-12). Occasionally the partners of monozygotic twins are connected to each other only by a common skin bridge or by a common liver bridge (**Siamese twins**). Several of such conjoined twins have successfully been separated by surgical procedures.

SUMMARY

The **placenta** consists of two components: (1) a fetal portion derived from the **chorion frondosum** or **villous chorion**, and (2) a maternal portion derived from the **decidua basalis**. The space between the chorionic and decidual plates is filled with **intervillous lakes** filled with maternal blood. The **villous trees** (fetal tissue) grow into the maternal blood lakes and are bathed in them. The fetal circulation is at all times separated from the maternal circulation by (1) a syncytial membrane—a chorion derivative, and (2) the endothelial cell of the fetal capillary. Hence, the human placenta is of the **hemochorial** type.

The intervillous lakes of the full grown placenta contain approximately 150 cc of maternal blood, which is renewed three or four times per minute. The villous area varies from 4 to 14 m^2, thus facilitating exchange between mother and child.

Main functions of the placenta are: (1) exchange of gases; (2) exchange of nutrients and electrolytes; (3) transmission of maternal antibodies providing the fetus passive immunity; (4) production of hormones such as progesterone, estradiol, and estrogen; in addition, it produces chorionic gonadotropin (HCG) and placental lactogen; (5) detoxification of some drugs.

The **amnion** provides a large sac with amniotic fluid in which the fetus is suspended by its umbilical cord. The fluid (1) absorbs jolts; (2) allows for fetal movements; (3) prevents adherence of the embryo to surrounding tissues. The fetus swallows amniotic fluid which is absorbed through its gut and cleared by the placenta. The fetus adds urine to the amniotic fluid, but this is mostly water.

The **umbilical cord** of the adult fetus is surrounded by the amnion and contains (1) two umbilical arteries; (2) one umbilical vein; and (3) Wharton's jelly, which serves as a protective cushion for the vessels.

Fetal membranes in twins vary according to their origin and time of formation. Seventy per cent of the twins belong to the **dizygotic** or **fraternal twins** and they have two amnions, two chorions, and two placentas which sometimes are fused. The **monozygotic twins** usually have two amnions, one chorion, and one placenta. In cases of **conjoined twins** in which the fetuses are not entirely split from each other there is one amnion, one chorion, and one placenta.

REFERENCES

1. Hamilton, W. J., and Boyd, J. D. Trophoblast in human utero-placental arteries. *Nature, 212:* 906, 1966.
2. Boyd, J. D., and Hamilton, W. J. *The Human Placenta.* Heffer, Cambridge, England, 1970.
3. Ramsey, E. M. The placenta and fetal membranes. In *Obstetrics,* edited by J. P. Greenhill. W. B. Saunders, Philadelphia, 1965.
4. Hamilton, W. J., and Boyd, J. D. Development of the human placenta in the first three months of gestation. *J. Anat., 94:* 297, 1960.
5. Martin, C. B. Uterine blood flow and placental circulation. *Anesthesiology, 26:* 447, 1965.
6. Assali, N. S., Rauramo, L., and Peltonen, T. Measurement of uterine blood flow and uterine metabolism. VIII. Uterine and foetal blood flow and oxygen consumption in early human pregnancy. *Am. J. Obstet. Gynecol., 79:* 86, 1960.
7. Martin, C. B. The anatomy and circulation of the placenta. In *Intra-uterine Development,* edited by A. C. Barnes. Lea and Febiger, Philadelphia, 1968.
8. Seeds, A. E. Placental transfer. In *Intra-uterine Development,* edited by A. C. Barnes. Lea and Febiger, Philadelphia, 1968.
9. Page, E. W., Villee, C. A., and Villee, D. B. In *Human Reproduction: The Core Content of Obstetrics, Gynecology and Perinatal Medicine.* W. B. Saunders, Philadelphia, 1976.
10. Hagerman, D. C., and Villee, C. A. Transport functions of the placenta. *Physiol. Rev., 40:* 313, 1960.
11. Burnett, L. S. Development of immune processes. In *Intra-uterine Development,* edited by A. C. Barnes. Lea and Febiger, Philadelphia, 1968.
12. Allen, F. H., and Umansky, I. Erythroblastosis fetalis. In *Principles and Management of Human Reproduction,* edited by D. E. Reid, K. J. Ryan, and K. Benirschke. W. B. Saunders, Philadelphia, 1972.
13. Ryan, K. J. Hormones of the placenta. *Am. J. Obstet. Gynecol., 84:* 1695, 1962.
14. Csapo, A. I., and Wood, C. The endocrine control of the initiation of labour in the human. In *Recent Advances in Endocrinology,* edited by J. James. Churchill, London, 1968.
15. Villee, D. B. In *Human Endocrinology.* A Developmental Approach. W. B. Saunders, Philadelphia, 1975.
16. Burns, R. K. Role of hormones in the differentiation of sex. In *Sex and Internal Secretions,* edited by W. C. Young. Williams & Wilkins, Baltimore, 1961.
17. Herbst, A. L., Ulfelder, H., and Poskanzer, D. C. Adeno-carcinoma of the vagina. *N. Engl. J. Med., 284:* 878, 1971.
18. Benirschke, K., and Driscoll, S. G. *The Pathology of the Human Placenta.* Springer, Berlin, 1967.
19. Sever, J., and White, L. R. Intrauterine viral infections. *Annu. Rev. Med., 19:* 471, 1968.

20. Apgar, V. The drug problem in pregnancy. *Clin. Obstet. Gynecol., 9:* 623, 1968.
21. Tuchmann-Duplessis, H. Reactions of the foetus to drugs taken by the mother. In *Foetal Autonomy*, edited by G. E. W. Wolstenholme and M. O'Connor. Ciba Foundation Symposium. Churchill, London, 1969.
22. Neubert, D., Merker, H. J., Nau, H., Langman, J. *Role of Pharmacokinetics in Prenatal and Perinatal Toxicology.* Georg Thieme Verlag, Stuttgart, 1978.
23. Greenhill, J. P., and Friedman, E. A. In *Biological Principles and Modern Practice of Obstetrics.* W. B. Saunders, Philadelphia, 1974.
24. Gadd, R. L. The liquor amnii. In *Scientific Foundations of Obstetrics and Gynecology*, edited by E. E. Philipp, J. Barnes, and M. Newton. William Heinemann, London, 1970.
25. Moore, K. L. The sex chromatin of freemartins and other animal intersexes. In *The Sex Chromatin*, edited by K. L. Moore. W. B. Saunders, Philadelphia, 1966.
26. Benirschke, K. Implantation, placental development and uteroplacental blood flow. In *Principles and Management of Human Reproduction*, edited by D. E. Reid, K. J. Ryan, and K. Benirschke. W. B. Saunders, Philadelphia, 1972.
27. Bulmer, M. J. *The Biology of Twinning in Man.* Clarendon Press, Oxford, 1970.
28. Turksoy, R. N., Toy, B. L., Rogers, J., and Papageorge, W. Birth of septuplets following human gonadotropin administration in Chiarai-Frommel syndrome. *Obstet. Gynecol., 30:* 692, 1967.

chapter 8

Congenital Malformations

Incidence

Congenital malformations are defined as "gross structural defects" present at birth.[1]

The figures on the incidence of congenital malformations vary greatly. In studies of official records and birth certificates, the percentage of children with abnormalities varied from 0.75 to 1.98 per cent.[2–6] These estimates were found to be rather low when compared with data from hospital and clinic birth records in which a variation of 1.43 to 3.3 per cent was noted.[7–9] Although the latter figures are probably the more accurate, they differ considerably among themselves. This may be due to actual differences in frequency in different countries or to the types of malformations considered. The racial make-up of the sample may also influence the incidence figures, since the frequency and types of malformations vary from race to race.[9]

In a worldwide survey on the incidence of congenital malformations comprising approximately 20 million births, it was found that based on birth certificates the percentage of congenital malformations was 0.83; according to hospital and clinic records it was 1.26; based on more intensive examinations by groups of pediatricians the percentage was found to be 4.50.[10] In the latter group the incidence for the United States was the highest (8.76 per cent) and that for Germany the lowest (2.20 per cent). When the figure is based on examination of infants at 6 and 12 months of age, the percentage often doubles, and an incidence of 7.5 per cent has been reported.[11]

Summarizing, it is probable that 2 to 3 per cent of all liveborn infants show one or more significant congenital malformations at birth, and that at the end of one year this figure is doubled by discovery of malformations indiscernible at birth.

Environmental Factors

Until the early 1940's it was assumed that congenital defects were caused mainly by hereditary factors. With the discovery by Gregg[12] that German measles affecting a mother during early pregnancy caused abnormalities in the embryo, it suddenly became evident that congenital malformations in humans could also be caused by environmental factors. The pioneering work of Warkany and Kalter[1, 13] who showed that a specific maternal dietary deficiency during pregnancy was teratogenic in rats, has since stimulated a great many investigations and led to the discovery of a large number of environmental factors teratogenic for the developing mammalian embryo. (For significant contributions in this field, see Warkany and Kalter,[1, 13] Wilson,[14] Fraser,[15, 16] and Wilson and Fraser.[17])

Despite the rapid development of the field of teratology, our knowledge of congenital malformations in humans has increased relatively little. At present it is estimated that approximately 10 per cent of all known human malformations are caused by environmental factors and another 10 per cent by genetic and chromosomal factors; the remaining 80 per cent are presumably caused by the intricate interplay of several genetic and environmental factors.

INFECTIOUS AGENTS

Rubella or German Measles. Gregg[12] was the first to suggest that German measles affecting pregnant women in the early stages of gestation could lead to congenital malformations in the offspring. At present it is well established that rubella virus can cause malformations of the eye (cataract and microphthalmia); internal ear (congenital deafness due to destruction of the organ of Corti); heart (persistence of the ductus arteriosus as well as atrial and ventricular septal defects); and occasionally of the teeth (enamel layer).[18–20] The virus may also be responsible for some cases of brain abnormalities and mental retardation.[21, 22] More recently it has become evident that the virus also causes intra-uterine growth retardation, myocardial damage, and vascular abnormalities.[23, 24]

The type of malformation is determined by the stage of embryonic development at which infection occurs. For example, cataracts result from infection during the 6th week of pregnancy and deafness from infection during the 9th week. Cardiac defects follow infection in the 5th to 10th weeks, and dental deformities between the 6th and 9th weeks.[25–28] Abnormalities of the central nervous system follow infection in the second trimester.[29]

It is extremely difficult to determine the exact incidence of malformations in the offspring of infected mothers, since German measles may be mild and thus escape detection, or may be accompanied by unusual clinical features and remain unrecognized. Furthermore, as pointed out above, some birth defects are not recognized until the child is two to four years of age. On the other hand, rashes caused by other viruses may be incorrectly attributed to rubella.

In a prospective study, the risk of malformations in infants examined immediately after birth was estimated at 47 per cent when the infection occurred during the first four weeks of pregnancy; 22 per cent following infection in the 5th to 8th weeks; 7 per cent in the 9th to 12th weeks; and 6 per cent in the 13th to 16th weeks.[30] Prematurity and fetal death may also follow infection in the first 8 weeks.[31]

If abnormalities such as mental retardation and dental defects, which do not become evident until later in life, were to be considered it is likely that the above percentages would be higher (65 per cent of congenital deafness due to rubella is not discovered until the fourth year).[32]

In the last decade two important advances have been made. Laboratory tests are now available which permit the detection of the virus in specimens from patients and the determination of antibody levels in the serum of patients. An important application of this test is to determine if a patient is immune and therefore need not fear the occurrence of rubella during pregnancy. An epidemiological study of 600 women has shown that 85 per cent were immune. A second important advance was the discovery that the virus penetrates into the fetus by way of the placenta and that the infection of the child may persist in the child after birth for a number of months or years. These children, usually showing no sign of infection, can transmit the virus to hospital personnel, such as nurses, doctors, and other hospital attendants.[33, 34] Safe and effective vaccines for rubella have recently been developed and administered to more than 35,000,000 women in the United States.

Cytomegalovirus. Only three virus diseases, rubella, cytomegalovirus and herpes simplex virus have been positively identified as causing malformations and chronic fetal infection which persists after birth. The congenital cytomegalic inclusion disease is in all probability the result of a human cytomegalovirus infection acquired *in utero* from an asymptomatically infected mother.[35, 36] The principal findings of the infection are microcephaly, cerebral calcifications, blindness and chorioretinitis, and hepato-splenomegaly. Some infants have kernicterus and multiple petechiae of the skin. Initially the disease was recognized only at autopsy and was based on the presence of enlarged cells, with large nuclei containing giant inclusion bodies. The inclusion bodies are most common in cells lining the renal tubules and may be present in the urine. The disease is often fatal when affecting the embryo or fetus, but in case of survival meningoencephalitis may cause mental retardation.[37] Since the disease is usually unrecognized in pregnant women, it is not known what the difference is between an early or late infection during development. It seems not unlikely that when the embryo is affected at an early stage of development the damage is so severe that it is unable to survive. Those cases which come to our attention are probably fetuses which have been infected only late in pregnancy.

Herpes Simplex Virus. There are a few reports in the literature which show that intra-uterine infection of the fetus with herpes simplex occasionally occurs.[38, 39] Usually the infection is transmitted close to the time of delivery, and the abnormalities reported are microcephaly, microphthalmus, retinal dysplasia, hepato-splenomegaly, and mental retardation. Usually, however, the child acquires the infection from the mother at birth as a venereal disease and the symptoms of the disease develop then during the first three weeks of age. The characteristics of the virus disease are then characterized by inflammatory reactions.

Other Viral Infections and Hyperthermia. Malformations following maternal infections with measles, mumps, hepatitis, poliomyelitis, chickenpox, ECHO, Coxsackie and influenza virus have been described.[40] Prospective studies indicate, however, that the malformation rate following exposure to these agents is low if not nonexistent.[41, 42]

A complicating factor introduced by these and other infectious agents is that most are pyrogenic and increased body temperature (hyperthermia) has recently been implicated as a teratogen. In a recent report, 7 of 63 (11%) anencephalic infants were born to mothers with a history of hyperthermia at the time that, in the embryo, the neural folds are closing. Interestingly, in two of these cases the episode of hyperthermia appeared to be related to sauna bathing and not to infection.[43]

Toxoplasmosis. Maternal infection with the protozoon parasite *Toxoplasma gondii* has been shown to produce congenital malformations. The affected child may have cerebral calcification, hydrocephalus, or mental retardation; chorioretinitis, microphthalmus, and other ocular defects have been reported as well.[44, 45] It is impossible to give precise figures on the incidence of malformations caused by toxoplasmosis since, as in the case of cytomegalovirus, the disease is usually unrecognized in pregnant women.

Syphilis. Formerly it was thought that syphilis was a major cause of malformations. This has since proved to be ill-founded. However, there is no doubt that syphilis may lead to congenital deafness and mental retardation in the offspring. In addition, many other organs such as the lungs and liver are characterized by diffuse fibrosis.[46]

RADIATION

The teratogenic effect of x-irradiation has been known for many years, and it is well recognized that microcephaly, skull defects, spina bifida, blindness, cleft palate, and defects of the extremities may result from treating pregnant women with large doses of roentgen rays or radium.[47] Although the maximum safe dose for humans is not known, in mice the fetus probably can be damaged with a dose as small as 5R.[48] It must be realized that the nature of the malformation depends on the dose of radiation and the stage of development at which the radiation is given.[49, 50]

Studies of the offspring of Japanese women pregnant at the time of the atomic bomb explosions over Hiroshima and Nagasaki revealed that, among the survivors, 28 per cent aborted, 25 per cent gave birth to children who died in their first year of life, and 25 per cent of the surviving children had abnormalities of the central nervous system, such as microcephaly and mental retardation.[51-53]

In addition to the effect of direct radiation on the embryo, indirect effects on the germ cells must be considered. Indeed, relatively small doses of radiation in mice have been shown to cause mutations which subsequently led to the occurrence of congenital malformations in succeeding generations.[54]

CHEMICAL AGENTS

The role of chemical agents, in particular, of drugs in the production of abnormalities in man is difficult to assess, primarily because most studies are retrospective and because of the large number of drugs employed by pregnant women. A recent NIH study discovered that among pregnant women 900 different drugs were taken with an average of 4 per individual. Only 20 per cent of pregnant women used no drugs during their pregnancy.[55, 56] Even with this widespread use of chemical agents, relatively few of the many drugs used during pregnancy have been positively identified as being teratogenic for the child. The best example is ***thalidomide***, an anti-nauseant and sleeping pill. In 1962 it was noted in West Germany that the frequency of **amelia** and **meromelia** (total or partial absence of the extremities), a rare hereditary abnormality, had suddenly increased (see chapter 9, fig. 9–10). This led to the examination of the prenatal histories of the affected children, resulting in the discovery that many of the mothers had taken thalidomide early in pregnancy. The causal relationship between thalidomide and meromelia was discovered only because the drug produced such an unusual type of abnormality. If the defect had been of a more common type, such as harelip or heart malformations, the association with the drug might easily have been overlooked.

The defects produced by thalidomide are: absence or gross deformities of the long bones, intestinal atresia, and cardiac abnormalities.[57–59] As a result of the discovery that thalidomide was directly related to meromelia, the drug was immediately removed from the market. The incidence of meromelia has since been reduced dramatically.

Another dangerous drug is ***aminopterin***. This compound belongs to the antimetabolites and is an **antagonist of folic acid**. Since in doses somewhat higher than the teratogenic level the drug will terminate pregnancy, it has been used during early pregnancy to induce therapeutic abortion in women suffering from tuberculosis.[60, 61] In four cases in which abortion did not occur, malformed offspring were found. The defects noted were anencephaly, meningocele, hydrocephalus, and cleft lip and palate. Since the drug has been used in a number of cases during pregnancy without producing teratogenic effects, its teratogenicity in man cannot yet be considered to have been demonstrated satisfactorily.

Other drugs also have teratogenic potential, including the anti-convulsants ***diphenylhydantoin*** **(Phenytoin)** and ***trimethadione***, which are **used by epileptic**

women.[62] In a retrospective study on 427 pregnancies in 186 epileptic women the frequency of major malformations such as heart abnormalities, facial clefts, and microcephaly was twice as high as expected. Specifically, diphenylhydantoin produces a broad spectrum of abnormalities, including craniofacial defects, nail and digital hypoplasia, growth abnormalities, and mental deficiency. These defects constitute a distinct pattern of dysmorphogenesis known as the "**fetal hydantoin syndrome.**"[63, 64]

Trimethadione, used in treatment of petit mal seizures, also appears to be teratogenic. The drug produces a characteristic pattern of abnormalities, including malformed ears, cleft palate, cardiac defects, and urogenital and skeletal anomalies which are collectively referred to as the "**trimethadione syndrome.**" As with diphenylhydantoin, delayed physical and mental development are also components of the syndrome.[65–67]

Anti-psychotic and ***anti-anxiety agents*** (major and minor tranquilizers, respectively) are suspected producers of congenital malformations. The anti-psychotics, **phenothiazine** and **lithium**, have been implicated as teratogens, and, although evidence for the teratogenicity of phenothiazines is conflicting, that concerning lithium has been better documented.[56] In any case, it has been strongly suggested that use of these agents during pregnancy carries a high risk.

Similar observations have been made for the anti-anxiety agents **meprobamate**, **chlordiazepoxide** and **diazepam (Valium)**. In a prospective study it was observed that severe anomalies occurred in 12 per cent of infants from mothers exposed to meprobamate and 11 per cent in those exposed to chlordiazepoxide compared to 2.6 per cent of controls.[68] Likewise, retrospective studies with **diazepam** have demonstrated up to a fourfold increase in cleft lip with or without cleft palate in offspring from mothers taking the drug during pregnancy.

In addition to the drugs discussed in some detail, caution has been expressed regarding a number of other compounds which might be damaging to the embryo or fetus.[55, 56, 69] The most prominent among these are propylthiouracil and potassium iodide (goiter and mental retardation); streptomycin (deafness); sulfonamides (kernicterus); the anti-depressant imipramine (limb deformities); tetracyclines (bone and tooth anomalies); amphetamines (oral clefts and cardiovascular abnormalities); the anti-coagulant warfarin (chondrodysplasia and microcephaly); and quinine (deafness). Finally, there is increasing evidence that **aspirin** (salicylates), the most commonly ingested drug during pregnancy, is potentially harmful to the developing offspring when used in large doses.[70]

One of the increasing problems in today's society is the effect of so-called social drugs such as LSD (lysergic acid diethylamide), PCP (phencyclidine, angel dust), marijuana, alcohol, and cigarette smoking. In the case of LSD, limb abnormalities and malformations of the central nervous system have been reported.[71, 72] However, a comprehensive review of more than 100 publications led to the conclusion that pure LSD used in moderate doses is not teratogenic and does not cause genetic damage.[73]

A similar lack of conclusive evidence for teratogenicity has been described for marijuana, PCP, and cigarette smoking. It is well established, however, that heavy smoking during pregnancy may cause small babies. Furthermore, a recent case report of malformations and behavioral abnormalities in an infant whose mother used PCP throughout her pregnancy suggests a possible relationship between the drug and the infant's defects.[74]

A well-documented association exists between maternal ***alcohol*** ingestion and congenital abnormalities. The defects include craniofacial abnormalities (short palpebral fissures and hypoplasia of the maxilla), limb deformities (altered joint mobility and position), and cardiovascular defects such as ventricular septal abnormalities. These malformations together with mental retardation and growth deficiency, make up the **"fetal alcohol syndrome"**[75] (Fig. 8-1). Interestingly, even moderate alcohol consumption during pregnancy may be detrimental to embryonic development.[76]

HORMONES

Progestins. Synthetic progestins were frequently used during pregnancy to avert abortion. The progestins ethisterone and norethisterone have considerable androgenic activity, and many cases of masculinization of the genitalia

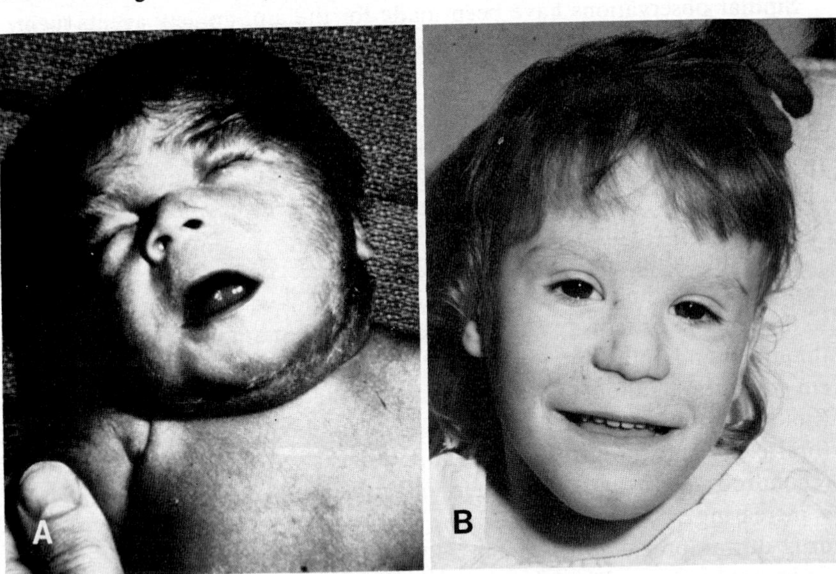

Figure 8-1. Photographs of children with "fetal alcohol syndrome." A, Severe case. B, Slightly affected child. Note in both children the short palpebral fissures and hypoplasia of the maxilla. Usually the defect includes other craniofacial abnormalities. Cardiovascular defects and limb deformities are also common symptoms of the fetal alcohol syndrome. (Courtesy Dr. David Smith, Department of Pediatrics, University of Washington.)

in female embryos have been reported.[77] The abnormalities consisted of an enlarged clitoris associated with varying degrees of fusion of the labioscrotal folds. Comparable results have been obtained experimentally by treating pregnant rats with progestins.[78]

Diethylstilbestrol, a synthetic estrogen, was commonly used in the 1940's and 1950's to treat mothers for threatening abortion. This practice was contraindicated in 1971 when it was determined that young women aged 16 to 22 years had an increased incidence of carcinomas of the vagina and cervix following exposure to the drug in utero.[79–81] Furthermore, a high percentage of these exposed women suffer from reproductive dysfunction which appears to be due in part to congenital malformations of the uterus, uterine tubes, and upper vagina.[82–84] Male embryos exposed in utero can also be affected, as evidenced by an increase in malformations of the testes and abnormal sperm analysis among these individuals. However, in contrast to females, males do not demonstrate an increased risk of developing carcinomas of the genital system.[85]

Cortisone. Experimental work has repeatedly shown that cortisone injected into mice and rabbits at certain stages of pregnancy may cause a high percentage of cleft palates in the offspring.[86, 87] However, it has been impossible to implicate cortisone as an environmental factor causing cleft palate in man.

Maternal Diabetes. Disturbances in carbohydrate metabolism during pregnancy in diabetic mothers cause a high incidence of stillbirths, neonatal deaths, abnormally large infants and congenital malformations. The risk of congenital anomalies in children of diabetic mothers is three to four times that in offspring of nondiabetic mothers and has been reported to be as high as 80 per cent in offspring from diabetics with long-standing disease.[88–91] A variety of malformations has been observed, including cardiac, skeletal, and central nervous system anomalies.[90, 91] Skeletal defects are particularly common and primarily consist of partial or complete agenesis of sacral vertebrae in conjunction with hindlimb hypoplasia (caudal regression syndrome).[92–95]

Factors responsible for these deformities have not been delineated, although evidence suggests that altered glucose and/or insulin levels play a role.[96–99] In this respect, a significant correlation between the severity and duration of the mother's disease and the incidence of malformations seems to exist. Furthermore, when women with previous histories of congenital defects and an indication of a disturbed carbohydrate metabolism were treated with insulin or thyroid, or both, subsequent pregnancies resulted in fewer miscarriages, stillbirths, and infants with congenital malformations. Unfortunately, however, there is no proof that these women would not have produced normal children without treatment.

NUTRITIONAL DEFICIENCIES

Although many nutritional deficiencies, particularly vitamin deficiencies, have been proved to be teratogenic in experimental work, there is no definite evidence that they are teratogenic in humans. With the exception of endemic cretinism, which is related to maternal iodine deficiency, no analogies to the animal experiments have been found in humans. (For an extensive review on nutritional deficiencies and their effect on the production of congenital malformations in experimental animals, see Kalter and Warkany,[13] and Hurley.[100])

HYPOXIA

Hypoxia induces congenital malformations in a great variety of experimental animals.[101] Whether the same is valid for humans remains to be seen. Although children born at relatively high altitudes are usually lighter in weight and smaller than those born near or at sea level, an increase in the incidence of congenital malformations has not been noted.[102] In addition, women with cyanotic cardiovascular disease often give birth to small infants, but usually without gross congenital malformations.

ENVIRONMENTAL CHEMICALS

A few years ago, it was noted in Japan that a number of mothers with diets consisting mainly of fish had given birth to children with multiple neurological symptoms resembling cerebral palsy. Further examination revealed that the fish contained an abnormally high level of ***organic mercury***, which was spewed into Minamata Bay and other coastal waters of Japan by large industries.[103, 104] Many of the mothers did not show any symptoms themselves, indicating that the fetus was more sensitive to the mercury than the mother. In the United States similar observations were made when seed corn sprayed with a mercury-containing fungicide was fed to hogs and the meat was subsequently eaten by a pregnant woman.[105] Similarly in Iraq probably several thousand babies were affected after the mothers ate grain treated with mercury-containing fungicides.[106]

Among the pesticides, the defoliant 2,4,5-T has most often been linked with teratogenesis. In this regard claims of embryotoxicity came from Vietnam, Swedish Lapland, and Arizona. When subsequently an exhaustive review of the effects of 2,4,5-T on mammalian reproduction was made, no evidence could be found that the herbicide was teratogenic in man.[107] None of the presently available studies suggests that 2,4,5-T as presently marketed poses any risk to human reproduction.

Chromosomal and Genetic Factors

The normal human somatic cell contains 46 chromosomes. Deviations from this number, as are found in some patients, are known as **numerical abnormalities.**[108] Some of these abnormalities involve the autosomes, usually an extra chromosome; others, the sex chromosomes, usually the X-chromosome. If an additional chromosome is present, making three chromosomes instead of the usual pair, the individual is said to be trisomic for this chromosome and the condition is known as **trisomy.** Four such conditions are now well established: (1) trisomy 21; (2) trisomy 17-18; (3) trisomy 13-15; and (4) trisomy X. If one of the chromosomes is absent the condition is known as **monosomy.** This abnormality, however, is rare.[109]

AUTOSOME ABNORMALITIES

Trisomy 21. This condition is found in the somatic cells of patients with Down's syndrome (mongoloid defectives)[110, 111] and originates from nondisjunction of the chromosomes during meiosis (see Chapter 1). Since the frequency of Down's syndrome increases with advancing maternal age, it is thought that nondisjunction occurs during oogenesis rather than during spermatogenesis. In mothers under 25 Down's syndrome is present once in about 2000 births, but in mothers over 40 once in about 100 births.[112–114] Children with Down's syndrome exhibit characteristic facial features, simian creases in the hands and are frequently mentally retarded and suffer from congenital heart malformations (fig. 8-2). Occasionally, the syndrome is produced by translocation of chromosome 21.[115–118]

Trisomy 17–18. Patients with this chromosomal arrangement show the following features suggesting a distinct clinical entity: mental retardation, congenital heart defects, low-set ears, and flexion of fingers and hands (fig.

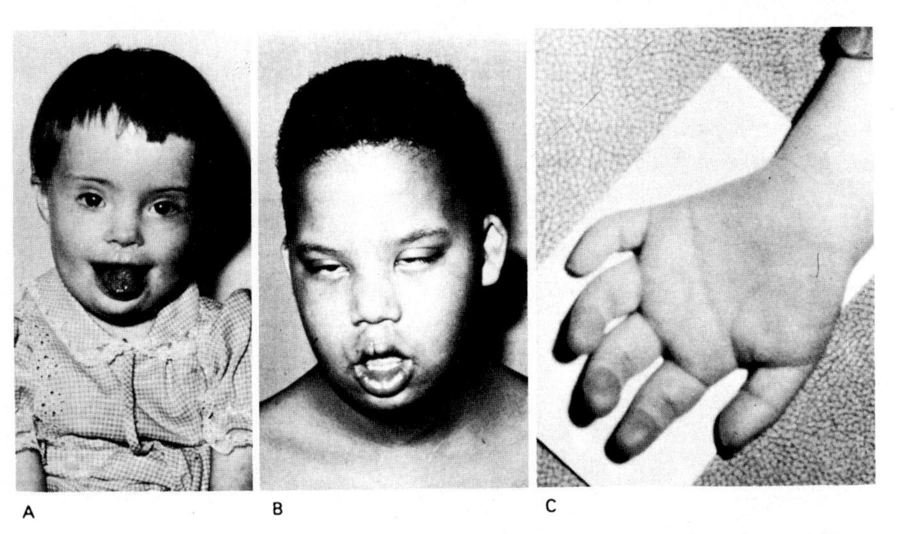

Figure 8-2. *A and B,* Photographs of children with Down's syndrome. The syndrome is characterized by the following features: a flat, broad face, oblique palpebral fissures, epicanthus, furrowed lower lip, broad hand with single transverse or simian crease (*C*). The children with Down's syndrome are frequently mentally retarded and have congenital heart abnormalities. (Courtesy Dr. J. Miller, Department of Neurology, University of Virginia.)

8-3).[119, 120] In addition, the patients frequently show micrognathia, renal anomalies, syndactyly, and malformations of the skeletal system. The incidence of this condition is about 0.3 per 1000 births. The infants usually die by the age of 2 months.

Trisomy 13-15. The main abnormalities of this syndrome are mental retardation, congenital heart defects, deafness, cleft lip and palate, and eye defects such as microphthalmia, anophthalmia, and coloboma[121] (fig. 8-4). The incidence of this abnormality is about 0.2 per 1000 newborns. Most of the infants die by the age of 3 months.

ABNORMAL CHROMOSOMES IN ABORTIONS

In recent years a number of cytogenic studies have been preformed in spontaneous abortions to determine whether a relationship exists between

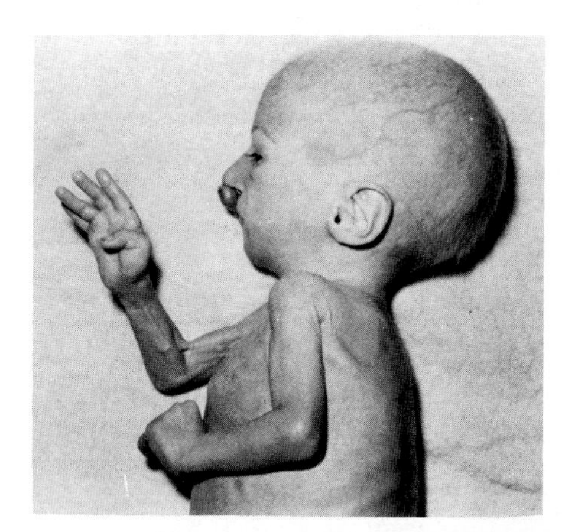

Figure 8-3. Photograph of child with trisomy 18. Note the prominent occiput, cleft lip, micrognathia, low set ears, and one or more flexed fingers. (Courtesy Dr. J. Miller, Department of Neurology, University of Virginia.)

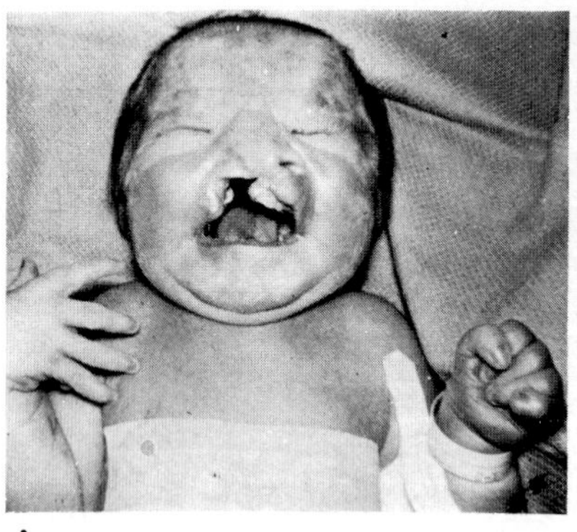

A

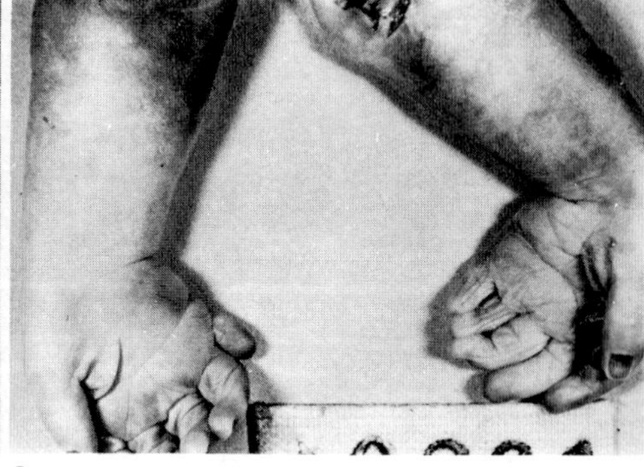

B

Figure 8-4. *A,* Photograph of child with trisomy 13-15. Note the cleft lip and palate, the sloping forehead and microphthalmia. *B,* Frequently the syndrome is accompanied by polydactyly. (Courtesy Dr. J. Miller, Department of Neurology, University of Virginia.)

chromosomal abnormalities and abortions.[122] The incidence of chromosomal abnormalities in abortions is reported to range from 10 to 64 per cent. Whatever the precise incidence may be, the percentage of chromosomal abnormalities in abortions is certainly 50 times that at full term. The chromosomal abnormalities involve the autosomes as well as the sex chromosomes. Monosomy for the X-chromosome seems to be one of the most frequently seen abnormalities.

SEX CHROMOSOME ABNORMALITIES

Chromosome analysis has shown that some cases of infertility are characterized by an abnormal sex chromosome complement. As in the case of autosomal abnormalities, it is likely that these are also caused by nondisjunction of the chromosomes (see Chapter 1).

Klinefelter's Syndrome. The clinical features of this syndrome found only in males are sterility, testicular atrophy, hyalinization of the seminiferous tubules, and usually gynecomastia.[123] The cells have 47 chromosomes with a sex chromosomal complement of the XXY type, and a sex chromatin body is found in 80 per cent of cases.[124] The incidence is about 1 in 500 males in the normal population.[125–127] Among mentally defective subjects the incidence is as high as 1 in 100 males.[128] On the basis of statistical evidence it is believed that nondisjunction of the XX homologs is the most common causative event. Occasionally, however, patients with Klinefelter's syndrome have 48 chromosomes, that is, 44 autosomes and 4 sex chromosomes (XXXY)[129] (fig. 1-4), or a variety of abnormal chromosome configurations, a condition known as mosaicism.[130]

Turner's Syndrome. This condition, found in women with an unmistakably female appearance, is characterized by the absence of the ovaries (**gonadal dysgenesis**). Other abnormalities frequently found are webbed neck, lymphedema of the extremities, skeletal deformities, and mental retardation (fig. 8-5). Despite the female appearance of these patients, almost all of their cells are sex chromatin-negative.[131] In addition the cells have only 45 chromosomes with an XO chromosomal complement.[132] Genetic analysis has shown that this syndrome is usually caused by nondisjunction in the male gamete during meiosis. As in patients with Klinefelter's syndrome, patients with Turner's syndrome occasionally show mosaicism. The incidence of XO females is presently estimated at about 2 in 3000 in the normal population. The incidence of chromatin-negative females in mental institutions is not significantly different.

Triple-X Syndrome. Patients with triple-X syndrome are infantile, with scanty menses and some degree of mental retardation.[133] They have two sex chromatin bodies in their cells and are therefore sometimes called "superfemale." The triple-X syndrome results from fertilization of an XX oocyte and an X-containing sperm. Some of the patients are of proven fertility and, surprisingly, the offspring has been uniformly normal. On theoretical

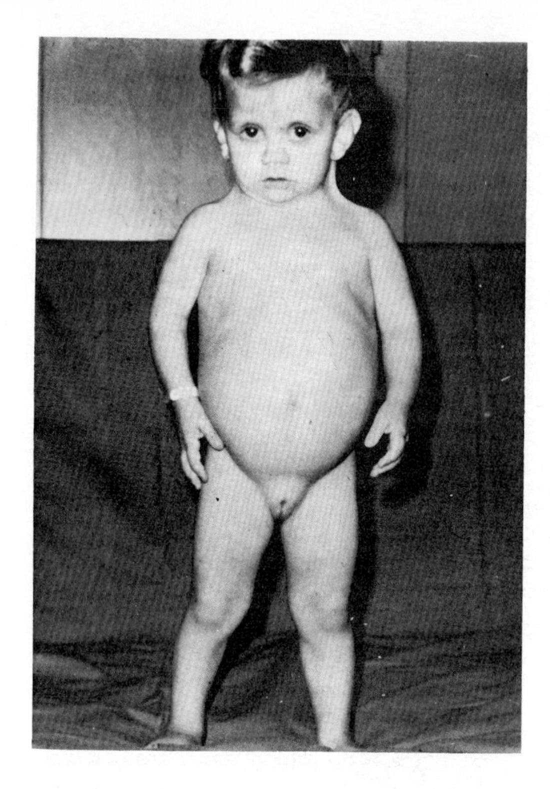

Figure 8-5. Photograph of patient with Turner's syndrome. The main characteristics are webbed neck, short stature, broad chest, and absence of sexual maturation. (Courtesy Dr. J. Miller, Department of Neurology, University of Virginia.)

grounds the triple-X patients should produce equal numbers of oocytes containing one or two X-chromosomes. Fertilization of the abnormal XX oocytes should give rise to XXX and XXY zygotes.

STRUCTURAL ABNORMALITIES

In addition to the numerical autosome and sex chromosome abnormalities, a different category is formed by the structural chromosome abnormalities which result from chromosome breaks. These breaks are probably caused by environmental factors such as viruses, radiation, or drugs. The result of the abnormality depends on what happens to the broken pieces.

In some cases the broken piece of a chromosome is lost and infants with partial **deletion** of a chromosome are abnormal. A well-known syndrome, caused by partial deletion of one of the 4-5 chromosomes, is the **cri du chat syndrome.** The children have a cat-like cry, microcephaly, mental retardation, and congenital heart disease. Many other relatively rare syndromes are known to result from a partial chromosome loss.[134]

ABNORMALITIES OF THE GENES

It has long been known that many congenital malformations in man are inherited and that some show a clear Mendelian pattern of inheritance. In many cases the abnormality is directly attributable to a change in a single gene, hence the name **single gene mutation.** It is estimated that this type of defect makes up approximately 10 per cent of all human malformations.[135-137]

In addition to causing congenital malformations, defined as structural defects present at birth, a large number of inborn errors of metabolism attributable to defective gene action have recently been described. These diseases, among which phenylketonuria,[138] homocystinuria, and galactosemia are the best known, are frequently accompanied by or cause various degrees of mental retardation. A detailed discussion of the affected metabolic pathways and enzymatic disorders, however, would lead beyond the scope of this book.

Actions of Teratogens

From data available on the action of teratogenic factors in mammals, a few basic principles have emerged.[139] Although it is too early to list these as "laws," they must be kept in mind when considering the probability of children being affected by specific teratogenic factors.

(1) **The stage of embryonic development determines the susceptibility to teratogenic factors.** Mammalian development starts with a rapid multiplication of cells which show little, if any, differentiation. This period, which lasts from fertilization to the formation of the germ layers, is referred to as the **pregerm layer stage** or **predifferentiation stage.** The next stage is known as the **embryonic period,** during which the cells begin to show distinct morphological differences resulting from changes at the chemical level. The final stage, or **fetal period,** is characterized by growth of the organ systems.

It is generally accepted that when a teratogen acts during the **predifferentiation stage,** it either damages all or a majority of the cells of the embryo, resulting in its death; or it injures only a few cells, in which case the regulative potentialities of the embryo will compensate for the loss and no abnormalities will be apparent. Several teratogenic factors, such as hypervitaminosis A and radiation, which in later stages of development are known to be highly teratogenic, were found to have no effect on the embryo in the first phase of development.[140-142] At present only a few scattered examples are known in which teratogens given during the first phase of development have caused malformations.[143] It may well be, however, that these teratogens remain in the maternal tissue and become active only when the susceptibility of the embryo increases during the second stage of development.

During the **embryonic period,** that is, the stage of intensive differentiation, most teratogenic agents are highly effective and produce numerous malformations. The type of malformation produced, however, depends on which organ is most susceptible at the time of the teratogenic action. Each organ seems to go through its most susceptible stage early in its differentiation, and the various body organs become susceptible one after the other. This was clearly demonstrated by giving rats a pteroylglutamic acid-deficient diet.[144, 145] It was thus found that abnormalities of the central nervous system and heart can be produced from day 7 to day 9; skeletal, urinary, and other cardiovascular abnormalities from day 9 to day 11; and skeletal defects from day 11 to day 14. The same seems to be true for the action of the rubella virus in the human embryo. Depending on the day of development the virus will affect one organ after the other, each at its own critical stage.[20]

During the third or **fetal period** of development, which is characterized by growth of the organs, susceptibility to teratogenic agents rapidly decreases. A small number of organs such as the cerebellum, cerebral cortex, and some urogenital structures, however, continue with their differentiation. Therefore, some of these structures remain susceptible to the action of teratogenic factors until late in pregnancy. Indeed, when mice were treated with various chemical substances during the later stages of pregnancy, the cerebral cortex was seriously affected.[146, 147] It is not unlikely that also in the human environmental factors may damage the developing brain in the second half of pregnancy and even postnatally and thus cause mental retardation and other cerebral abnormalities.

② **The effect of a teratogenic factor depends on the genotype.** A number of experiments seem to indicate that a teratogenic agent accentuates the incidence of those defects which occur sporadically without treatment and that the malformations appear as they do because of underlying genetic instabilities. For example, when an appropriate dose of cortisone was injected into pregnant mice of strains A and C57, which have spontaneous rates of oral clefting of 10 and <1 per cent, respectively, it caused cleft palate in all the offspring of strain A and in 19 per cent of strain C57. When a strain C57 male was bred with a strain A female, 43 per cent of the young had cleft palate. When a strain C57 mother was crossed with a strain A father, however, the incidence dropped to 4 per cent; this showed that the genes of the mother as well as those of the embryo may influence the susceptibility to a teratogen.[148–150]

③ **A teratogenic agent acts in a specific way on a particular aspect of cell metabolism.** Teratogens may act on a variety of cell functions or products. Thus, agents may inhibit nucleic acid or protein synthesis, alter the extracellular matrix, or adversely affect the cytoarchitecture of embryonic cells. Agents are not restricted to affecting only one cell process, however, and it is, therefore, difficult in many instances to determine the underlying mechanism of action of a drug.

SUMMARY

Many factors may interact with the differentiating and growing embryo. The result, however, is not necessarily a **gross malformation.** In some instances the teratogenic agent may be so toxic or may affect a vital organ system of the embryo or fetus so severely that

death results. In other cases the environmental influence may be so mild that the embryo or fetus is able to survive, but some of its organ systems are affected. This may result in partial or total **growth retardation** or a **functional impairment** such as mental retardation.

A variety of agents are known to produce congenital malformations in approximately 2 to 3 per cent of all liveborn infants. These agents include viruses, such as rubella and cytomegalovirus; radiation; drugs, such as thalidomide, aminopterin, anti-convulsants, anti-psychotics, and anti-anxiety compounds; social drugs, such as PCP, cigarettes, and alcohol; hormones, such as diethylstilbesterol; maternal diabetes; and chromosomal abnormalities, such as trisomy 21 (Down's syndrome). In the case of radiation and chemical factors, the malformations produced depend on the stage of gestation and organ differentiation during which the agent was present. In this respect most major malformations are produced during the embryonic period of development. Although a large number of birth defects have been described and attributed to specific factors, little is known about how an agent actually produces a defect or how a defect may be prevented or reversed. Therefore, the medical approach to this problem is postnatal repair or early detection—via amniocentesis, alpha-fetoprotein, or ultrasound techniques—and subsequent termination of those embryos found to be severely malformed.

REFERENCES

1. Warkany, J., and Kalter, H. Congenital malformations. *N. Engl. J. Med., 265:* 993, 1961.
2. Gentry, J. T., Parkhurst, E., and Bulin, G. V. Epidemiological study on congenital malformations in New York State. *Am. J. Public Health, 49:* 497, 1959.
3. Ivy, R. H. Congenital anomalies, as recorded on birth certificates in Division of Vital Statistics of Pennsylvania Department of Health for period 1951–1955, inclusive, *Plast. Reconstr. Surg., 20:* 400, 1957.
4. Wallace, H. M., and Baumgartner, L. Congenital malformations and birth injuries in New York City. *Pediatrics, 12:* 525, 1953.
5. Neel, J. V. Study on major congenital defects in Japanese infants. *Am. J. Hum. Genet., 10:* 398, 1958.
6. McKeown, T., and Record, R. G. Malformations in population observed for five years after birth. In *Ciba Foundation Symposium on Congenital Malformations,* edited by G. E. W. Wolstenholme and E. M. O'Connor, p. 2. Little, Brown and Co., Boston, 1960.
7. Schenk, H. Über die Missbildungen in den Jahren 1938–41 an der Universitäts, Fräuenklinik, Berlin. *Zbl. Gynaek., 46:* 2078, 1942.
8. Böök, J. A., and Fraccaro, M. Research on congenital malformations. *Étud. néonatales, 5:* 39, 1956.
9. Stevenson, S. S., Worcester, J., and Rice, R. G. Six hundred and seventy-seven congenitally malformed infants and associated gestational characteristics. I. General considerations. *Pediatrics, 6:* 37, 1950.
10. Kennedy, W. P. Epidemiologic aspects of the problem of congenital malformations. In *Birth Defects Original Article Series,* edited by D. Bergsma, p. 1. Alan R. Liss, Inc., New York, 1967.
11. McIntosh, R., et al. Incidence of congenital malformations; a study of 5964 pregnancies.

Pediatrics, 14: 505, 1954.
12. Gregg, N. M. Congenital cataract following German measles in mothers. *Trans. Ophthal. Soc. Aust., 3:* 35, 1941.
13. Kalter, H., and Warkany, J. Experimental production of congenital malformations in mammals by metabolic procedure. *Physiol. Rev., 39:* 69, 1959.
14. Wilson, J. G. *Environment and Birth Defects.* Academic Press, New York, 1973.
15. Fraser, F. C. The use of teratogens in the analysis of abnormal developmental mechanisms. In *Proceedings of the First International Conference on Congenital Malformations,* p. 179. J. B. Lippincott Co., Philadelphia, 1961.
16. Fraser, F. C. Methodology of experimental mammalian teratology. In *Methodology in Mammalian Genetics,* edited by W. J. Bundette, p. 233. Hold and Day, San Francisco, 1962.
17. Wilson, J. G., and Fraser F. C. *Handbook of Teratology,* Vols. 1, 2 and 3. Plenum Press, New York, 1977.
18. Logan, W. P. D. Effects of virus infections in pregnancy. *Medicine (Illus.), 8:* 502, 1954.
19. Rhodes, A. J. Virus infections and congenital malformations. In *Proceedings of the First International Conference on Congenital Malformations.* J. B. Lippincott Co., Philadelphia, 1961.
20. Töndury, G. Zur Kenntnis der Embryopathica rubeolica, nebst Bemerkungen über die Wirkung anderer Viren auf den Keimling. *Geburtshilfe Frauenheilkd, 12:* 865, 1952.
21. Ariens Kappers, J. Developmental disturbance of the brain induced by German measles in an embryo of the 7th week. *Acta Anat., 31:* 1, 1957.
22. Lacomme, M. Le point de vue de l'obstétricien sur les malformations congénitales. *Maternité, 6:* 231, 1954.
23. Töndury, G., and Smith, D. W. Fetal rubella pathology. *J. Pediatr., 68:* 867, 1966.
24. Dudgeon, J. A. Maternal rubella and its effect on the foetus. *Arch. Dis. Child., 42:* 110, 1967.
25. Bass, M. H. Diseases of pregnant women affecting the offspring. *Adv. Intern. Med., 5:* 15, 1952.
26. Tedeschi, C. G., Helfern, M. M., and Ingalls, T. H. Pathological manifestations in an infant after maternal rubella in the sixteenth week of gestation. *N. Engl. J. Med., 249:* 439, 1953.
27. Jackson, H. D. M., and Fish, L. Deafness following maternal rubella; results of a prospective investigation. *Lancet, 2:* 1241, 1958.
28. Keith, J. K., Rowe, R. D., and Vlad, P. *Heart Disease in Infancy and Childhood.* The Macmillan Co., New York, 1958.
29. Gumpel, S. M., Hayes, K., and Dudgeon, J. A. Congenital perceptive deafness: role of intrauterine rubella. *Br. Med. J., 2:* 300, 1971.
30. Michaels, R. H., and Mellin, G. W. Prospective experience with maternal rubella and associated congenital malformations. *Pediatrics, 26:* 200, 1960.
31. Siegel, M., and Grünberg, M. Fetal death, malformation and prematurity after maternal rubella; results of a prospective study, 1949–1958. *N. Engl. J. Med., 262:* 389, 1960.
32. Jackson, A. D. M., and Fish, L. Deafness following maternal rubella; results of a prospective study. *Lancet, 2:* 1241, 1958.
33. Korones, S. B., Ainger, L. E., Monif, G. R., Roane, J., Sever, J. L., and Fuste, F. Congenital rubella syndrome: new clinical aspects with recovery of virus from affected infants. *J. Pediatr., 67:* 166, 1965.
34. Sever, J. L., Nelson, K. B., and Gilkeson, M. R. Rubella epidemic 1964. Effect on 6000 pregnancies. *Am. J. Dis. Child., 110:* 395, 1965.
35. Medearis, D. N., Jr. Cytomegalic inclusion disease: an analysis of the clinical features based on the literature and six additional cases. *Pediatrics, 19:* 467, 1957.
36. Weller, T. H., and Hanshaw, J. B. Virologic and clinical observations on cytomegalic inclusion disease. *N. Engl. J. Med., 266:* 1233, 1962.
37. Medearis, D. N. Observations concerning human cytomegalovirus infection and disease. *Bull. Johns Hopkins Hosp., 114:* 181, 1964.
38. Fuccillo, D. A., and Sever, L. J. Viral teratology. *Bacteriol. Rev., 37:* 19, 1973.
39. Sever, L. J. Virus infections and malformations. *Fed. Proc., 30:* 114, 1971.
40. Wesselhoeft, C. Acute infectious diseases in pregnancy. *Ann. Intern. Med., 42:* 555, 1955.

41. Mannon, M. M., Logan, W. P. D., and Loy, R. M. *Rubella and Other Virus Infections during Pregnancy* (Great Britain Ministry of Health Reports on Public Health and Medical Subjects, Publication No. 101). Her Majesty's Stationery Office, London, 1960.
42. Doll, R., Hill, A. B., and Sakula, J. Asian influenza in pregnancy and congenital defects. *Br. J. Prev. Soc. Med., 14:* 167, 1960.
43. Miller, P., Smith, D. W., and Shepard, T. H. Maternal hyperthermia as a possible cause of anencephaly. *Lancet, 1:* 519, 1978.
44. Lechner, G., and Leinzinger, E. The relationship between maternal toxoplasmosis and embryopathy. *Arch. Gynaekol., 202:* 99, 1965.
45. Lelong, M. Rapport sur la prophylaxie de la toxoplasmose du nouveau-né et de la femme enceinte. *Rev. Hyg. Med. Soc., 7:* 71, 1959.
46. Rasmussen, D. M. Syphillis and the fetus. In *Intrauterine Development,* edited by A. C. Barnes, p. 419. Lea and Febiger, Philadelphia, 1968.
47. Cusher, I. M. Irradiation of the fetus. In *Intrauterine Development,* edited by A. C. Barnes, p. 378. Lea and Febiger, Philadelphia, 1968.
48. Rugh, R., and Grupp, E. Congenital defects following low level X-irradiation. *Anat. Rec., 138:* 380, 1960.
49. Wilson, J. G. Differentiation and the reaction of rat embryos to radiation. *J. Cell. Comp. Physiol., 43:* 11, 1954.
50. Hicks, S. P. The effects of ionizing radiation, certain hormones and radiometric drugs on the developing nervous system. *J. Cell. Comp. Physiol., 43:* 151, 1954.
51. Plummer, G. Anomalies occurring in children exposed in utero to atomic bomb in Hiroshima. *Pediatrics, 10:* 687, 1952.
52. Yamasaki, J. N., Wright, S. W., and Wright, P. M. Outcome of pregnancy in women exposed to atomic bomb in Nagasaki. *Am. J. Dis. Child., 87:* 448, 1954.
53. Wood, J. W., Johnson, K. G., and Omori, Y. In utero exposure to the Hiroshima atomic bomb. An evaluation of head size and mental retardation twenty years later. *Pediatrics, 39:* 385, 1967.
54. Carter, T. C., Lyon, M. F., and Phillips, R. J. S. Genetic hazard of ionizing radiations. *Nature (Lond.), 182:* 409, 1958.
55. Heinonen, O. P., Slone, D., and Shapiro, S. *Birth Defects and Drugs in Pregnancy.* Publishing Sciences Group, Inc., Littleton, Mass., 1977.
56. Golbus, M. S. Teratology for the obstetrician: Current status. *Obstet. Gynecol., 55:* 269, 1980.
57. Lenz, W. Thalidomide and congenital abnormalities. *Lancet, 1:* 1219, 1962.
58. Somers, G. F. Thalidomide and congenital abnormalities. *Lancet, 1:* 912, 1962.
59. Weicker, H., and Hungerland, H. Thalidomid-embryopathie. I. Vorkommen inner und ausserhalb Deutschlands. *Dtsch. Med. Wochenschr., 87:* 992, 1962.
60. Thiersch, J. B. The effects of antimetabolites on the fetus and litter of the rat in utero. In *Proceedings of the Sixth International Conference on Planned Parenthood,* p. 156. International Planned Parenthood Federation, New Delhi, India, 1959.
61. Warkany, J., Beaudry, P. H., and Hornstein, S. Attempted abortion with aminopterin: malformations of the child. *Am. J. Dis. Child., 97:* 274, 1959.
62. Montouris, G. D., Fenichel, G. M., and McLain, L. W., Jr. The pregnant epileptic. *Obstet. Gynecol. Surv., 35:* 282, 1980.
63. Hanson, J. W., and Smith, D. W. The fetal hydantoin syndrome. *J. Pediatr., 87:* 285, 1975.
64. Loughnan, P. M., Gold, H., and Vance, J. C. Phenytoin teratogenicity in man. *Lancet, 1:* 70, 1973.
65. German, J., Kowal, A., and Ehlers, K. L. Trimethadione and human teratogenesis. *Teratology, 3:* 349, 1970.
66. Zackai, E. H., Mellman, W. J., Neiderer, B., and Hanson, J. W. The fetal trimethadione syndrome. *J. Pediatr., 87:* 280, 1975.
67. Feldman, G. L., Weaver, D. D., and Lovrien, E. W. The fetal trimethadione syndrome. *Am. J. Dis. Child., 131:* 1389, 1977.
68. Milkovich, L., and Van Den Berg, B. J. Effects of prenatal meprobamate and chlordiazepoxide hydrochloride on human embryonic and fetal development. *N. Engl. J. Med., 291:* 1268, 1974.
69. Barnes, A. C. The fetal environment: drugs and chemicals. In *Intrauterine Development,*

edited by A. C. Barnes, p. 362. Lea and Febiger, Philadelphia, 1968.
70. Corby, D. G. Aspirin in pregnancy: Maternal and fetal effects. *Pediatrics, 62:* 930, 1978.
71. Berlin, C. M., and Jacobsen, C. B. Congenital anomalies associated with parental LSD ingestion. *Soc. Ped. Res.*, Sec. Plenary Session, 1970.
72. Long, S. Y. Does LSD induce chromosomal damage and malformations? A review of literature. *Teratology, 6:* 75, 1972.
73. Dishotsky, N. T., Longhman, W. K., Mogar, R. E., and Lipscomb, W. R. LSD and genetic damage. Is LSD chromosome damaging, carcinogenic, mutagenic or teratogenic? *Science, 172:* 431, 1971.
74. Golden, N. L., Sokol, R. J., and Rubin, I. L. Angel dust: possible effects on the fetus. *Pediatrics, 65:* 18, 1980.
75. Jones, K. L., Smith, D. W., Ulleland, C. N. et al. Pattern of malformation in offspring of chronic alcoholic mothers. *Lancet, 1:* 1267, 1973.
76. Streissguth, A. P., Hanson, J. W., Streissguth, A. P. et al. The effects of moderate alcohol consumption during pregnancy on fetal growth and morphogenesis. *Fifth International Conference on Birth Defects,* p. 62. Excerpta Medica, Amsterdam, 1977.
77. Wilkins, L., Jones, H. W., Jr., Holman, G. H., and Stempfel, R. S., Jr. Masculinization of the female fetus association with administration of oral and intramuscular progestins during gestation; nonadrenal female pseudohermaphroditism. *J. Clin. Endocrinol. Metab., 18:* 559, 1958.
78. Revesz, C., Chappel, C. I., and Gandry, R. Masculinization of female fetuses in rat by progestational compounds. *Endocrinology, 66:* 140, 1960.
79. Greenwald, P., Barlow, J. J., Nasca, P. C., and Burnett, W. S. Vaginal cancer after maternal treatment with synthetic estrogens. *N. Engl. J. Med., 285:* 390, 1971.
80. Herbst, A. L., Ulfelder, H., and Poskanzer, D. C. Adenocarcinoma of the vagina. *N. Engl. J. Med., 284:* 878, 1971.
81. Herbst, A. L., Scully, R. E., Robboy, S. J., and Welch, W. R. Complications of prenatal therapy with diethylstilbestrol. *Pediatrics, 62:* 1151, 1978.
82. Rosenfeld, D. L., and Bronson, R. A. Reproductive problems in the DES-exposed female. *Obstet. Gynecol., 55:* 453, 1980.
83. Berger, M. J., and Goldstein, D. P. Impaired reproductive performance in DES-exposed women. *Obstet. Gynecol., 55:* 25, 1980.
84. Kaufman, R. H., Binder, G. L., Gray, P. M., and Adam, E. Upper genital tract changes associated with exposure in utero to diethylstilbestrol. *Am. J. Obstet. Gynecol., 128:* 51, 1977.
85. Bill, W. B., Schumacher, G. F. B., and Bibbo, M. Pathological semen and anatomical abnormalities of the genital tract in human male subjects exposed to diethylstilbestrol in utero. *J. Urol., 117:* 477, 1977.
86. Fraser, F. C., Kalter, H., Walker, B. E., and Fainstat, T. D. Experimental production of cleft palate with cortisone and other hormones. *J. Cell. Comp. Physiol. (Suppl. 1), 43:* 237, 1954.
87. Fainstat, T. D. Cortisone-induced congenital cleft palate in rabbits. *Endocrinology, 55:* 502, 1954.
88. White, P. *The Treatment of Diabetes Mellitus.* edited by E. P. Joslin, H. F. Root, P. White, and A. Marble. Lea and Febiger, Philadelphia, 1952.
89. Pedersen, R. M., Trygstrup, T., and Pedersen, J. Congenital malformations in infants of diabetic women. *Lancet, 1:* 1124, 1964.
90. Chung, C. S., and Myrianthopoulos, N. C. Factors affecting risks of congenital malformations. *Birth Defects, 11:* 23, 1975.
91. Soler, N. G., Walsh, C. H., and Malins, J. M. Congenital malformations in infants of diabetic mothers. *Q. J. Med., 45:* 303, 1976.
92. Blumel, J., Butler, M. C., Evans, E. B., and Eggers, G. W. N. Congenital anomaly of the sacrococcygeal spine. *Arch. Surg., 85:* 982, 1962.
93. Rusnak, S. L., and Driscoll, S. G. Congenital spinal anomalies in infants of diabetic mothers. *Pediatrics, 35:* 989, 1965.
94. Passarge, E., and Lenz, W. Syndrome of caudal regression in infants of diabetic mothers. *Pediatrics, 37:* 672, 1966.
95. Williamson, D. A. J. A syndrome of congenital malformations possibly due to maternal diabetes. *Dev. Med. Child. Neurol., 12:* 145, 1970.

96. Saxen, L., and Rapola, J. *Congenital Defects.* Holt, Rhinehart, and Winston, New York, 1969.
97. Landauer, W. Is insulin a teratogen? *Teratology, 5:* 129, 1972.
98. Sadler, T. W. Effects of maternal diabetes on early embryogenesis: II. Hyperglycemia-induced exencephaly. *Teratology, 21:* 349, 1980.
99. Cockroft, D. L., and Coppola, P. T. Teratogenic effects of excess glucose on headfold rat embryos in culture. *Teratology, 16:* 141, 1977.
100. Hurley, L. S. Nutritional Deficiencies and Excesses. In *Handbook of Teratology,* Vol. 1, edited by J. G. Wilson and F. C. Fraser, p. 261. Plenum Press, New York, 1977.
101. Ingalls, T. H., Curley, F. J., and Prindle, R. A. Experimental production of congenital abnormalities; timing and degree of anoxia as factors causing fetal deaths and congenital abnormalities in mouse. *N. Engl. J. Med., 247:* 758, 1952.
102. Lichty, J. A., Ting, R. Y., Bruns, P. D., and Dyar, E. Studies on babies born at high altitudes. I. Relation of altitude to birth and weight. *Am. J. Dis. Child., 93:* 666, 1957.
103. Harada, Y. Congenital Minamata Disease. In *Minamata Disease,* Kumanoto University, Study group of Minamata disease.
104. Matsumoto, H. G., Goyo, K., and Takevchi, T. Fetal Minamata disease. A neuropathological study of two cases of intrauterine intoxication by a methylmercury compound. *J. Neuropathol. Exp. Neurol., 24:* 563, 1965.
105. Snyder, R. D. Congenital mercury poisoning. *N. Engl. J. Med., 284:* 1014, 1971.
106. Bakir, F., Damluji, S. F., Amin-Zaki, L., Murtadha, M., Khalidi, A., Alrawi, N. Y., Tikkiti, S., Dhahir, H. I., Clarkson, T. W., Smith, J. C., and Doherty, R. A. Methylmercury poisoning in Iraq. *Science, 181:* 230, 1973.
107. Report of the 2,4,5-T Advisory Committee. Submitted to William D. Rickelshaus, Administrator, Environmental Protection Agency. Washington, D.C., 1971.
108. Hirschorn, K., and Cooper, M. L. Chromosomal aberrations in human disease. *Am. J. Med., 31:* 442, 1961.
109. Challacombe, D. N., and Taylor, A. Monosomy for a G autosome. *Arch. Dis. Child., 44:* 113, 1969.
110. Lejeune, J., Gautier, M., and Turpin, R. Les chromosomes humaines en culture de tissus. *C. R. Acad. Sci. (Paris), 248:* 602, 1959.
111. Jacobs, P. A., Baikie, A. G., Court Brown, W. M., and Strong, J. A. The somatic chromosomes in mongolism. *Lancet, 1:* 710, 1959.
112. Penrose, L. S. Relative aetiological importance of birth order and maternal age in mongolism. *Proc. R. Soc. (Biol.), 115:* 431, 1934.
113. Penrose, L. S. Mongolism. *Br. Med. Bull., 17:* 184, 1961.
114. Carter, C. O. The genetics of congenital malformations. In *Scientific Foundation of Obstetrics and Gynecology,* edited by E. E. Philipp, J. Barnes, and M. Newton. William Heinemann, London, 1970.
115. Carr, D. M. The chromosome abnormality in mongolism. *Can. Med. Assoc. J., 87:* 490, 1962.
116. Carter, C. O. et al. Chromosome translocation as a cause of familial mongolism. *Lancet, 2:* 678, 1960.
117. Fraccaro, M., Kaijser, K., and Lindsten, J. Chromosomal abnormalities in father and Mongol child. *Lancet, 1:* 724, 1960.
118. Sergovich, F. R., Soltan, H. C., and Carr, D. H. A 13-15/21 translocation chromosome in carrier father and mongol son. *Can. Med. Assoc. J., 87:* 852, 1962.
119. Koenig, E., Lubs, M., and Brandt, T. Congenital malformations and autosomal abnormalities. *Yale J. Biol. Med., 35:* 189, 1962.
120. Edwards, J. H., Harnden, D. G., Cameron, A. H., Crosse, J. M., and Wolff, O. W. A new trisomic syndrome. *Lancet, 1:* 787, 1960.
121. Patau, K., Smith, W. D., Therman, E., and Inhorn, S. L. Multiple congenital anomalies caused by an extra autosome. *Lancet, 1:* 790, 1960.
122. Carr, D. H. Chromosome studies in spontaneous abortions. *Obstet. Gynecol., 26:* 308, 1965.
123. Klinefelter, H. F., Reifenstein, F. C., and Albright, F. Syndrome characterized by gynecomastia, aspermatogenesis without a-leydigism and increased excretion of FSH. *J. Clin. Endocrinol. Metab., 2:* 615, 1942.
124. Jacobs, P. A., and Strong, J. A. A case of human intersexuality having a possible XXY sex

determining mechanism. *Nature (London), 183:* 302, 1959.
125. Moore, K. L. Sex reversal in newborn babies. *Lancet, 1:* 217, 1959.
126. Bergemann, E. Geschlechtschromatinbestimmungen am Neugeborenen. *Schweiz. Med. Wochenschr., 10:* 292, 1961.
127. Maclean, N., Harnden, D. G., Court Brown, W. M., Bond, J., and Mantle, D. J. Sex-chromosome abnormalities in newborn babies. *Lancet, 1:* 286, 1964.
128. Ferguson-Smith, M. A. Sex chromatin, Klinefelter's syndrome and mental deficiency. In *The Sex Chromatin,* edited by K. L. Moore, W. B. Saunders, Philadelphia, 1966.
129. Carr, D. H., Barr, M. L., Plunkett, E. R., Grumbach, M. M., Morishima, A., and Chu, E. H. Y. An XXXY sex chromosome complex in Klinefelter subjects with duplicated sex chromatin. *J. Clin. Endocrinol. Metab., 21:* 491, 1961.
130. Barr, M. L. et al. An XY/XXXY sex chromosome mosaicism in a mentally defective male patient. *J. Ment. Defic. Res., 6:* 65, 1962.
131. Moore, K. L. Sex chromatin and gonadal dysgenesis. In *The Sex Chromatin,* edited by K. L. Moore, W. B. Saunders, Philadelphia, 1966.
132. Ford, C. E., Jones, K. W., Polani, P. E., de Almeida, J. C., and Biggs, J. H. A sex chromosome anomaly in a case of gonodal dysgenesis (Turner's syndrome). *Lancet, 1:* 711, 1959.
133. Jacobs, P. A., Baikie, A. G., Court Brown, W. M., MacGregor, T. N., and MacLean, N. Evidence for the existence of the human "superfemale." *Lancet, 2:* 423, 1959.
134. Smith, D. W. *Recognizable Patterns of Human Malformation: Genetic, Embryologic and Clinical Aspects.* W. B. Saunders, Philadelphia, 1970.
135. Fraser, F. C. Genetics and congenital malformations. In *Progress in Medical Genetics,* edited by A. G. Steinberg, p. 38. Grune & Stratton, New York, 1961.
136. Stevenson, A. C. The load of hereditary defects in human populations. *Radiat. Res. (Suppl.), 1:* 306, 1959.
137. Francois, J. *L'Hérédité en Ophthalmologie,* Masson et Cie, Paris, 1958.
138. Hsia, D. Y. Phenylketonuria: a study of human biochemical genetics. *Pediatrics, 38:* 173, 1966.
139. Wilson, G. W. Experimental studies on congenital malformations. *J. Chron. Dis., 10:* 111, 1959.
140. Giroud, A., and Martinet, M. Action tératogène de l'hypervitaminose A chez la souris en fonction du stade embryonaire. *C. R. Soc. Biol. (Paris), 154:* 1353, 1960.
141. Hicks, S. P. The effects of ionizing radiation, certain hormones and radiometric drugs on the developing nervous system. *J. Cell. Physiol., 43:* 151, 1954.
142. Chang, M. C., and Hunt, D. M. Effects of *in vitro* radio-cobalt irradiation of rabbit ova on subsequent development *in vivo,* with special reference to the irradiation of maternal organism. *Anat. Rec., 137:* 511, 1960.
143. Eibs, H. G., Speilmann, H., and Hägele, M. Effects of sex steroid treatment during the preimplantation period on the development of mouse embryos in vivo and in vitro. In *Role of Pharmacokinetics in Prenatal and Perinatal Toxicology.* Edited by D. Neubert, H. J. Merker, H. Nau, and J. Langman. Georg Thieme Verlag, Stuttgart, 1978.
144. Nelson, M. M., Asling, C. W., and Evans, H. M. Production of multiple congenital malformations in young by maternal pteroyl-glutamic acid deficiency during gestation. *J. Nutr., 48:* 61, 1952.
145. Nelson, M. M., Wright, H. V., Asling, C. W., and Evans, H. M. Multiple congenital abnormalities resulting from transitory deficiency of pteroylglutamic acid during gestation in rats. *J. Nutr., 56:* 349, 1955.
146. Langman, J., and Welch, G. W. Excess vitamin A and the development of the cerebral cortex. *J. Comp. Neurol., 131:* 15, 1967.
147. Webster, W., Shimada, M., and Langman, J. Effect of fluorodeoxyuridine, colcemid and bromodeoxyuridine on developing neocortex of the mouse. *Am. J. Anat., 137:* 67, 1973.
148. Runner, M. N. Inheritance of susceptibility of congenital deformity. Metabolic clues provided by experiments with teratogenic agents. *Pediatrics, 23:* 245, 1959.
149. Fraser, F. C., and Fainstat, T. D. Production of congenital defects in offspring of pregnant mice treated with cortisone. *Pediatrics, 8:* 527, 1951.
150. Kalter, H. The inheritance of susceptibility to the teratogenic action of cortisone in mice (abstract). *Genetics, 39:* 185, 1954.

PART 2

Special Embryology

chapter 9

Skeletal System
(Skull; Limbs; Vertebral Column)

The skeletal system develops from the mesodermal germ layer, which appears during the third week of development. It forms a series of mesodermal tissue blocks, the **somites**, on each side of the neural tube (fig. 9-1*A*). Soon after its formation each somite becomes differentiated in a ventromedial part, the **sclerotome**, and a dorsolateral part, the **dermomyotome**. At the end of the fourth week the sclerotome cells become polymorphous and form a loosely woven tissue known as **mesenchyme** or embryonic connective tissue (fig. 9-1*B*). It is characteristic for the mesenchymal cells to migrate and to differentiate in many different ways. They may become fibroblasts, chondroblasts, or **osteoblasts**, the **bone forming cells**.

The bone forming capacity of mesenchyme is not restricted to the cells of the sclerotome, but occurs also in the somatic mesoderm layer of the body wall where the ribs are formed. More recently it has been shown that neural crest cells in the head region also differentiate into mesenchyme and participate in the formation of bones of the face.[1] In some bones, such as the flat

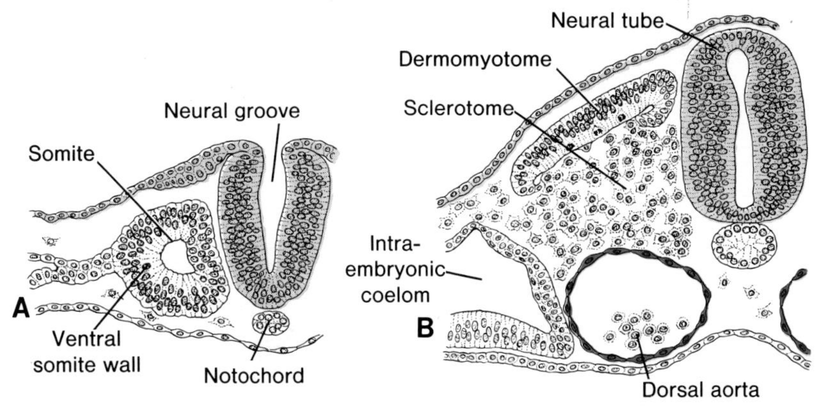

Figure 9-1. Development of the somite. *A*, The mesoderm has formed a somite and the cells are arranged around a small cavity. *B*, As a result of further differentiation the cells in the ventromedial wall lose their epithelial arrangement and become mesenchymal. They are collectively referred to as the scelerotome. The cells in the dorsolateral wall of the somite form the dermomyotome.

bones of the skull, the mesenchyme differentiates directly into bone, a process known as **membranous ossification** (fig. 9-2); in most bones, however, the mesenchymal cells first give rise to **hyaline cartilage models**, which in turn become ossified by **endochondral ossification** (fig. 9-7). (For histogenesis of bone, membranous and endochondral ossification, see regular histology textbooks.) In the following paragraphs the development of the most important bony structures and some of their abnormalities are discussed.

Skull

The skull can be divided into two parts: the **neurocranium** which forms a protective case around the brain, and the **viscerocranium**, which forms the skeleton of the face.

NEUROCRANIUM

The neurocranium is most conveniently divided into two portions: (1) the membranous part consisting of **flat bones**, which surround the brain as a vault: and (2) the **cartilaginous part** or **chondrocranium**, which forms the bones of the base of the skull.

Membranous Neurocranium. The sides and roof of the skull develop from mesenchyme investing the brain and undergo **membranous ossification**. As

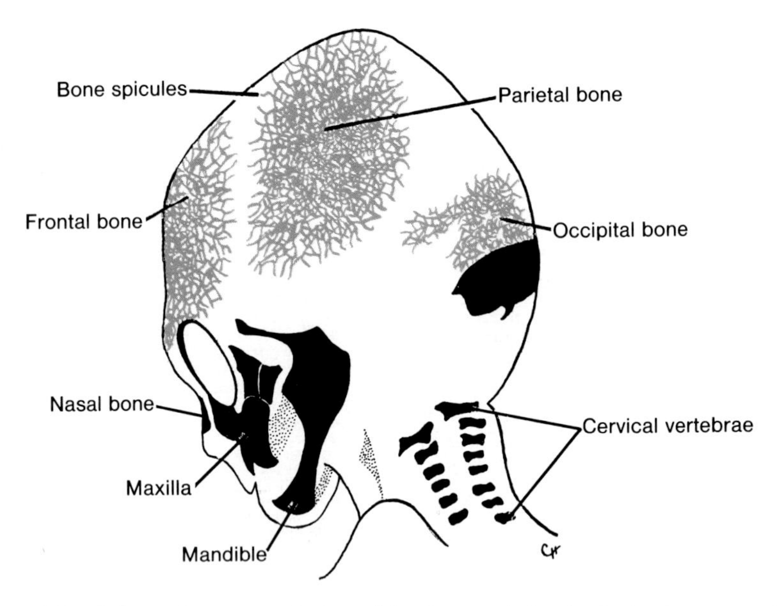

Figure 9-2. Schematic representation of the bones of the skull of a three-month-old fetus (drawn after a cleared preparation stained with alizarin). Note the spread of the bone spicules from the primary ossification centers in the flat bones of the skull.

a result a number of flat membranous bones are formed which are characterized by the presence of needle-like **bone spicules**. These spicules progressively radiate from the primary ossification centers toward the periphery (fig. 9-2). With further growth during fetal and postnatal life the membranous bones enlarge by apposition of new layers on the outer surface, and by simultaneous osteoclastic resorption from the inside.

NEWBORN SKULL

At birth the flat bones of the skull are separated from each other by narrow seams of connective tissue, the **sutures**. At points where more than two bones meet, the sutures are wide and known as the **fontanelles** (fig. 9-3). The most prominent of these is the **anterior fontanelle**, which is found where the two parietals and two frontals meet. The sutures and fontanelles allow the bones of the skull to overlap each other during the birth process. Soon after birth the membranous bones move back to their original position and give the skull a large round appearance. The size of the vault is strikingly large compared to the small facial region (fig. 9-3*B*).

Several of the sutures and the fontanelles remain membranous for a considerable time after birth. Growth of the bones of the vault is particularly rapid during growth and expansion of the flat bones which is caused mainly by the growth of the brain. Although a five- to seven-year-old child has nearly all of its cranial capacity, some of the sutures remain open until

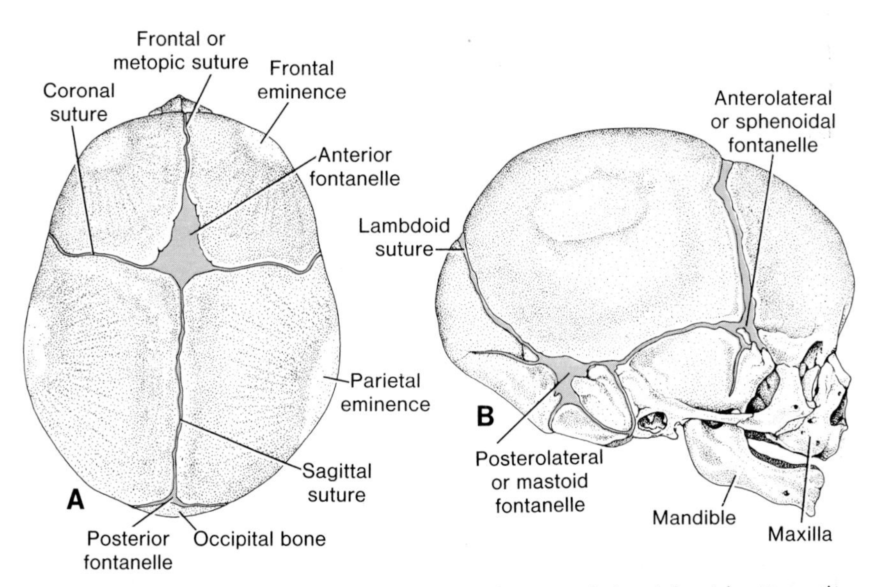

Figure 9-3. Skull of a newborn, seen from above and the right side. Note the anterior and posterior fontanelles and the sutures. The posterior fontanelle closes about three months after birth; the anterior fontanelle about the middle of the second year. Many of the sutures disappear during adult life.

adulthood. In the first few years after birth, palpation of the anterior fontanelle may give valuable information as to whether ossification of the skull is proceeding normally and whether intracranial pressure is normal.

Cartilaginous Neurocranium or Chondrocranium. This part of the skull consists initially of a number of separate cartilages. When these cartilages fuse and ossify by endochondral ossification the base of the skull is formed.

The base of the occipital bone is formed by the **parachordal cartilage** and the bodies of three **occipital sclerotomes**[2] (fig. 9-4). Rostral to the occipital base plate are found the **hypophyseal cartilages** and the **trabeculae cranii**. These cartilages soon fuse to form the body of the **sphenoid** and **ethmoid**, respectively. In this manner an elongated median plate of cartilage, extending from the nasal region to the anterior border of the **foramen magnum**, is formed.

A number of other mesenchymal condensations arise on either side of the median plate. The most rostral, the **ala orbitalis**, forms the lesser wing of the

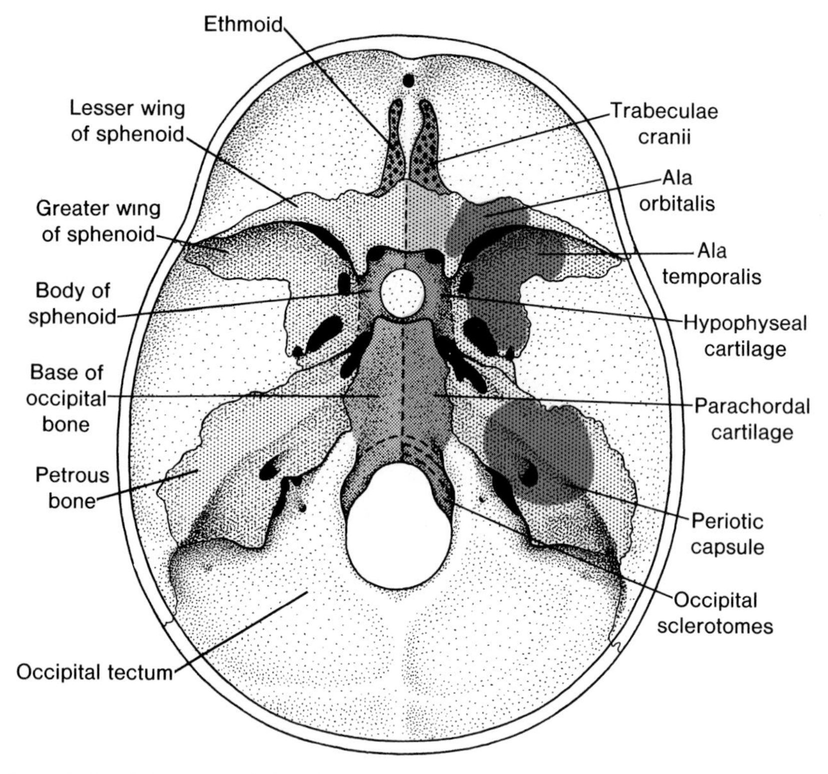

Figure 9-4. Schematized dorsal view of the chondrocranium or base of the skull in the adult. On the right are indicated in blue the various embryonic components participating in the formation of the median part of the chondrocranium; in red the components for the lateral part of the chondrocranium. On the left are indicated the names of the adult structures.

sphenoid bone. Caudally, it is followed by the **ala temporalis**, which gives rise to the greater wing of the sphenoid. A third component, the **periotic capsule**, gives rise to the petrous and mastoid parts of the temporal bone. These components later fuse with the median plate and with each other, except for the openings through which the cranial nerves leave the skull (fig. 9-4).

VISCEROCRANIUM

The viscerocranium consists of the bones of the face and is formed mainly by the cartilages of the first two pharyngeal arches (see figs 16-6 and 16-7). The first arch gives rise to a dorsal portion, the **maxillary process**, which extends forward beneath the region of the eye and gives rise to the **maxilla, the zygomatic bone** and **part of the temporal bone** (fig. 9-5). The ventral portion is known as Meckel's cartilage or the mandibular process. The mesenchyme around Meckel's cartilage condenses and ossifies by membranous ossification to give rise to the **mandible**. Meckel's cartilage disappears except in the **sphenomandibular** ligament. The dorsal tip of the mandibular process, along with that of the second pharyngeal arch, later give rise to the **incus**, the **malleus**, and the **stapes** (fig. 9-5). Ossification of the three ossicles begins in the fourth month, thus making these the first bones to become fully ossified (see also figs. 16-6 and 16-7).

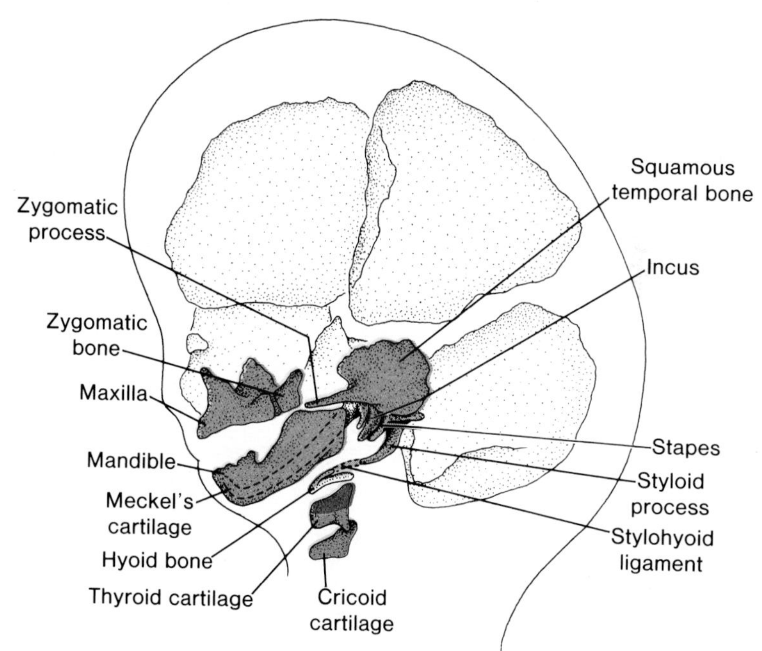

Figure 9-5. Lateral view of the head and neck region of an older fetus, showing the derivatives of the arch cartilages participating in the formation of the bones of the face.

At first the face is small in comparison with the neurocranium. This is caused by (1) the virtual absence of the paranasal air sinuses and (2) the small size of the bones, particularly of the jaws. With the appearance of teeth and the development of the air sinuses the face obtains its human characteristics.

Skull Abnormalities

CRANIOSCHISIS

Abnormalities of the skull vary from very large defects (**cranioschisis**) combined with gross brain abnormalities such as **anencephaly** (see fig 20-31), to small circumscribed defects detectable only by radiographic examination. Although children with severe skull and brain defects are not viable, those with relatively small defects in the skull through which brain tissue and/or meninges herniate (**encephalocele** or **cranial meningocele**) are frequently seen (fig. 9-6).

CRANIOSTENOSIS

Another important category of cranial abnormalities is caused by premature closure of one or more sutures. These abnormalities are collectively known as **craniosynostosis** or **craniostenosis**. The form of the skull depends

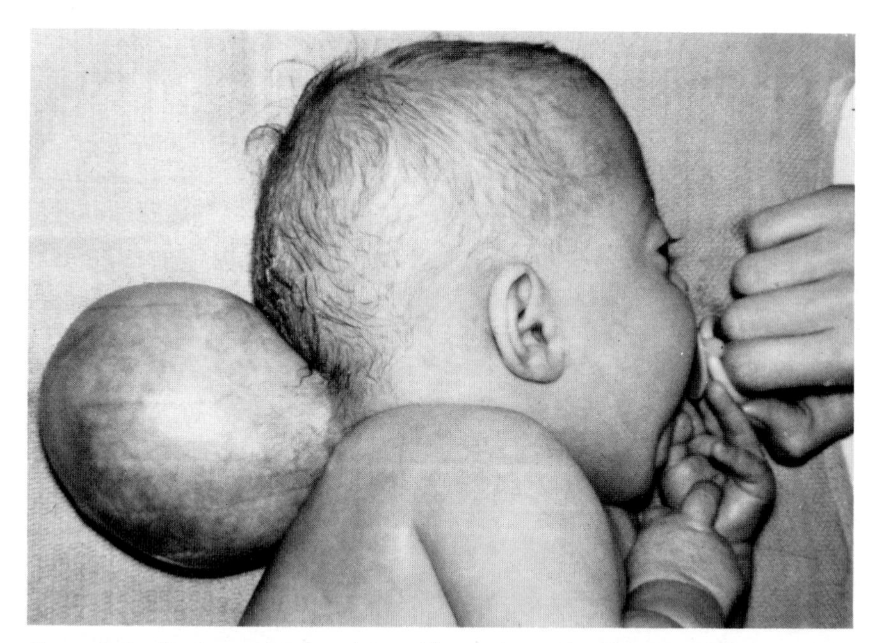

Figure 9-6. Photograph of patient with meningocele. This is a rather common abnormality, which frequently can be repaired successfully. (Courtesy Dr. J. Warkany. From Warkany, J. *Congenital Malformations*. Year Book Medical Publishers, Chicago, 1971. Used by permission.)

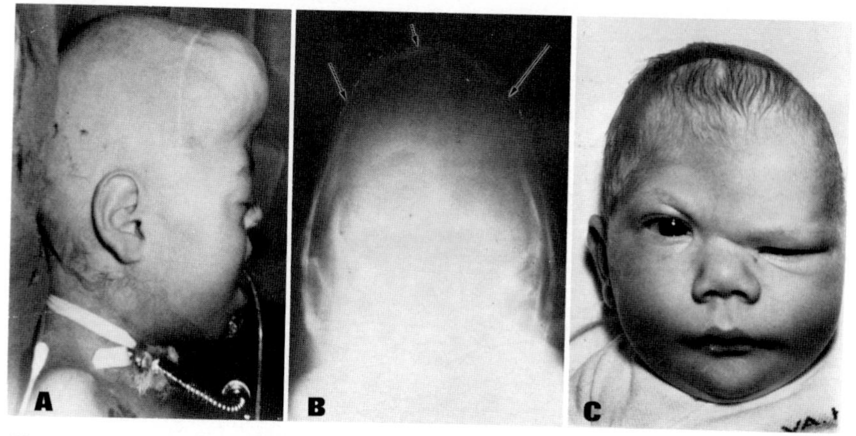

Figure 9-7. *A*, Photograph of a child with scaphocephaly caused by early closure of the sagittal suture. Note the frontal and occipital bossing. *B*, X-ray of child with acrocephaly caused by early closure of the coronal suture. *C*, Photograph of plagiocephaly resulting from early closure of coronal and lambdoid sutures on one side of the skull (see also fig. 9-3B). (Courtesy Dr. J. Jane, Department of Neurosurgery, University of Virginia.)

upon which of the sutures closed prematurely. Early closure of the sagittal suture results in frontal and occipital expansion and the skull becomes long and narrow (**scaphocephaly**) (fig. 9-7*A*). Premature closure of the coronal suture results in a short, high skull, known as **acrocephaly** or **tower skull** (fig. 9-7*B*). If the coronal and lambdoid sutures close prematurely on one side only, asymmetric craniostenosis, known as **plagiocephaly**, results (fig. 9-7*C*).

A great handicap in operating on patients with craniosynostosis is the tendency of the bones to reunite after the prematurely closed sutures have been opened.[3, 4]

Microcephaly is mainly an abnormality in which the brain fails to grow and in which consequently the skull fails to expand. Children with microcephaly are usually severely retarded.

Limbs

The limb buds become visible as paddle-shaped buds at the beginning of the fifth week (fig. 9-8*A*). Initially they consist of a core of mesenchyme and a covering layer of ectoderm.[5] At the apex of the buds the ectoderm is somewhat thickened and is known as the **apical ectodermal ridge**. This ridge exerts an inductive influence on the underlying mesenchyme which rapidly begins to grow and differentiate.[6, 7] In six-week-old embryos the terminal portion of the buds becomes flattened **(hand and foot plates)** and separated from the proximal segment by a circular constriction (fig. 9-8*B*). While the fingers and toes are formed by disappearance of the tissue in the radial grooves, a second constriction divides the proximal portion into two segments and the main parts of the extremities can be recognized (fig. 9-8*C*).[8]

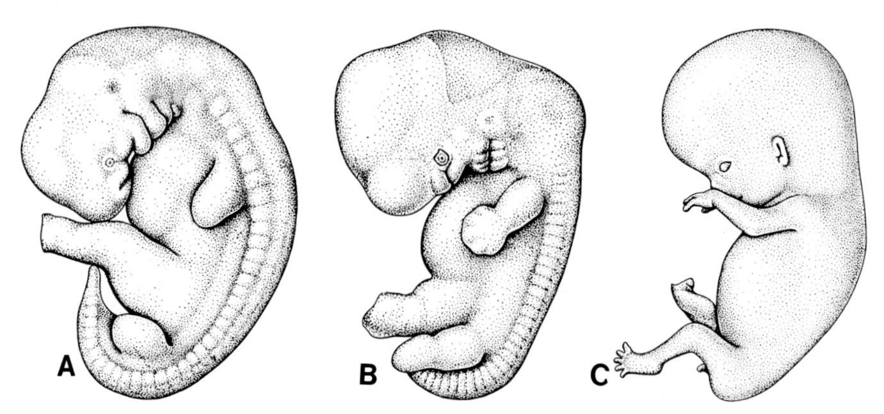

Figure 9-8. Schematic drawings of human embryos to demonstrate the development of the limb buds. *A*, At 5 weeks; *B*, at 6 weeks; *C*, at 8 weeks. Note that the hindlimb buds are somewhat behind in development in comparison with the forelimbs.

While the external shape is being established, the mesenchyme in the buds begins to condense and by the sixth week of development the first so-called **hyaline cartilage models**, foreshadowing the bones of the extremities, can be recognized (fig. 9-9). The ossification of the bones of the extremities, **endochondral ossification**, begins by the end of the embryonic period. Primary **ossification centers** are present in all long bones of the limbs by the twelfth week of development. From the primary center in the shaft or **diaphysis** of the bone, the endochondral ossification progresses gradually toward the ends of the cartilaginous "model."

At birth, the diaphysis of the bone is usually completely ossified but the two extremities, known as the **epiphyses**, are still cartilaginous. Shortly thereafter, however, ossification centers arise in the epiphyses.

A cartilage plate remains temporarily between the diaphyseal and epiphyseal ossification centers. This plate, known as the **epiphyseal plate**, plays an important role in the growth in length of the bone. On both sides of the plate endochondral ossification proceeds. When the bone has acquired its full length the epiphyseal plates disappear and the epiphyses unite with the shaft of the bone. (See your histology textbook for detailed information.)

In the long bones an epiphyseal plate is found on each extremity; in the smaller ones, such as the phalanges, only at one extremity; and in irregular bones, such as the vertebrae, one or more primary centers of ossification, and usually several secondary centers, are found.[9]

Knowledge about the appearance of various ossification centers is used by radiologists to determine whether a child has reached its proper maturation age.[10] Useful information of "bone age" is obtained from ossification studies in the hands and wrists of children.[11]

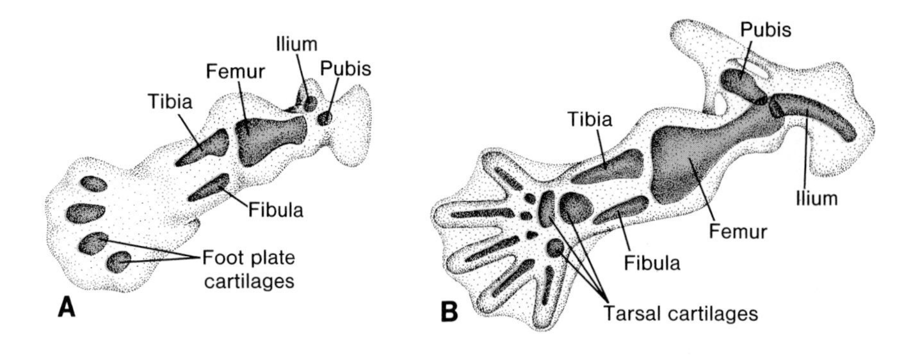

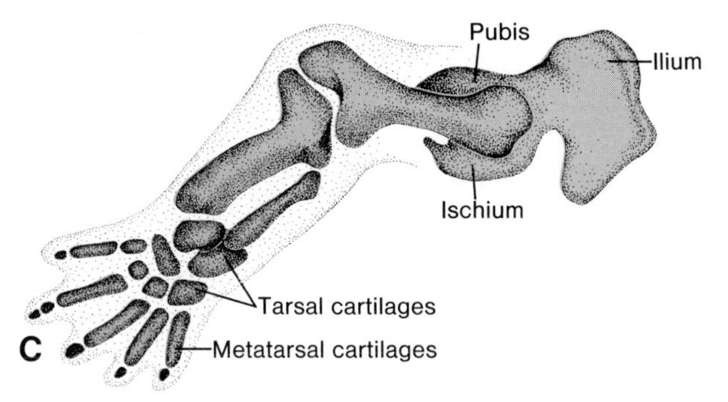

Figure 9-9. *A*, Schematic drawing of the lower extremity of an early six-week embryo illustrating the first hyaline cartilage models. *B* and *C*, Similar drawings showing the complete set of cartilage models at the end of the sixth and the beginning of the eighth week, respectively.

Limb Abnormalities

The abnormalities of the extremities vary greatly. In the most extreme form one or two extremities are absent (**amelia**) or represented only by hands and feet attached to the trunk by a small, irregularly shaped bone. Such a defect is known as **meromelia** (fig. 9-10*A*, *B*). Sometimes all segments of the extremities are present, but abnormally short (**micromelia**).

Although these abnormalities are rare and mainly of hereditary nature, the high incidence of children with these limb malformations born between 1957 and 1962 led to a review of the prenatal histories of the affected children.[12] It was thus noted that many of the mothers had taken **thalidomide**, a drug widely used as a sleeping pill and anti-nauseant. Presently it is well established that this drug may cause a characteristic syndrome of malformations, consisting of absence or gross deformities of the long bones, intestinal atresia, and cardiac anomalies (see Chapter 8).[13-16] After the drug

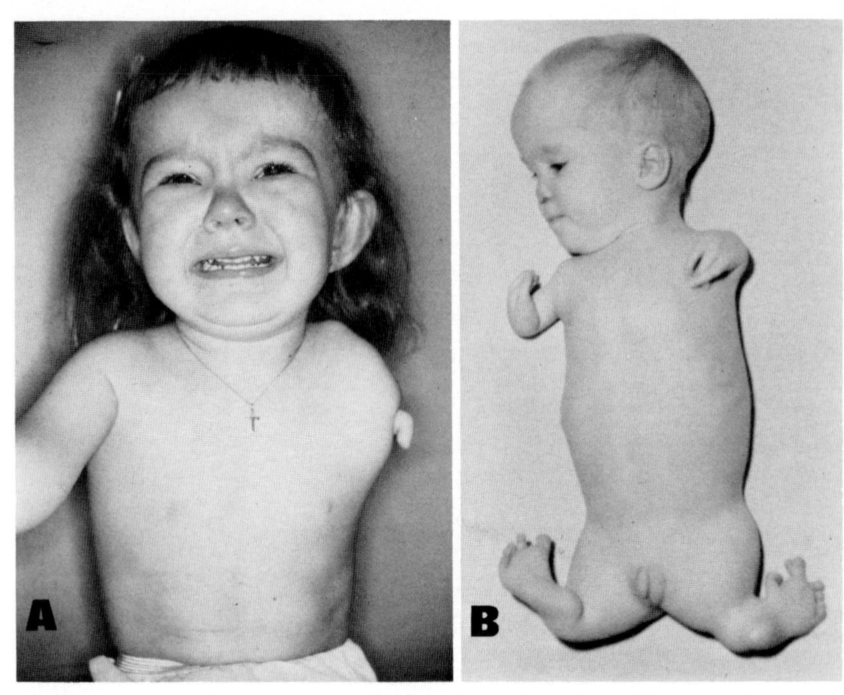

Figure 9-10. *A*, Photograph of child with unilateral amelia. *B*, Patient with meromelia. The hand is attached to the trunk by an irregularly shaped bone. (Courtesy Dr. M. Edgerton, Department of Plastic Surgery, University of Virginia.)

had been removed from the market, the abnormality was hardly ever observed.

POLYDACTYLY

A different category of limb abnormalities consists of the presence of extra fingers or toes (**polydactyly**). The extra digit frequently lacks proper muscular connections. Abnormalities with an excessive number of bones are mostly bilateral, while the absence of a digit such as a thumb is usually unilateral.

SYNDACTYLY

Abnormal fusion is usually restricted to the fingers or the toes (**syndactyly**). Normally the mesenchyme between the prospective digits in the hand and foot plate breaks down. Not infrequently this fails to occur and the result is fusion of one or more fingers and toes (fig. 9-11*A*). In some cases actual fusion of the bones occurs.

LOBSTER CLAW

This deformity consists of an abnormal cleft between the second and fourth metacarpal bones and soft tissues. The third metacarpal and phalangeal bones are almost always absent, and the thumb and index finger as well

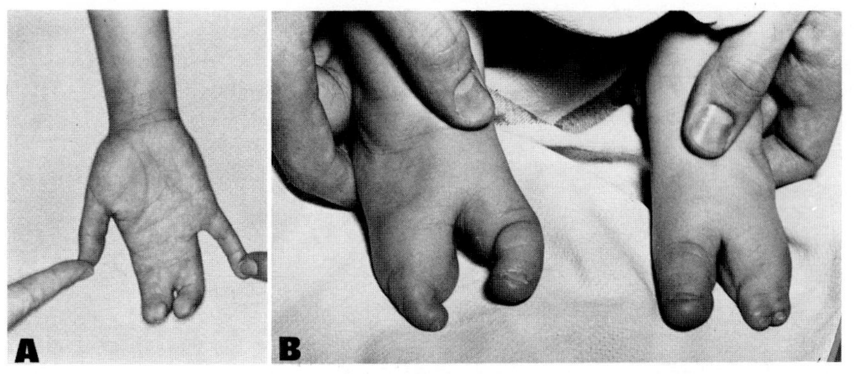

Figure 9-11. *A*, Photograph of syndactyly. *B*, Lobster claw hand. The third metacarpal and phalangeal bones are absent; the thumb and index finger as well as the fourth and fifth finger are fused. (Courtesy Dr. M. Edgerton, Department of Plastic Surgery, University of Virginia.)

as the fourth and fifth finger may be fused (fig. 9-11*B*). The two parts of the hand are somewhat opposed to each other and act like a lobster claw.

CLUB FOOT

This abnormality is usually seen in combination with syndactyly. The sole of the foot is turned inwards and the foot is adducted and plantar flexed. It is seen mainly in males and in some cases is hereditary. Whether abnormal positioning of the legs in utero may cause the abnormality is presently not known.[17]

CONGENITAL HIP DISLOCATION

This abnormality consists of an underdevelopment of the acetabulum and the head of the femur. The condition is rather common and occurs mostly in females. Although the dislocation usually occurs after birth, the abnormality of the bones develops prenatally. Since many babies with the abnormality are breech deliveries, it has been thought that breech posture may interfere with the development of the hip joint. The abnormality is frequently associated with laxity of the joint capsule.

Vertebral Column

During the fourth week of development the cells of the sclerotomes migrate medially to surround both the spinal cord and notochord (fig. 5-8*D*). The thus formed mesenchymal column retains traces of its segmental origin as the sclerotomic blocks are separated by less dense areas containing the **intersegmental arteries** (fig. 9-12*A*).

During further development the caudal portion of each sclerotome segment strongly proliferates and condenses (fig. 9-12*B*). This proliferation is so strong that it proceeds into the subjacent intersegmental tissue and in this

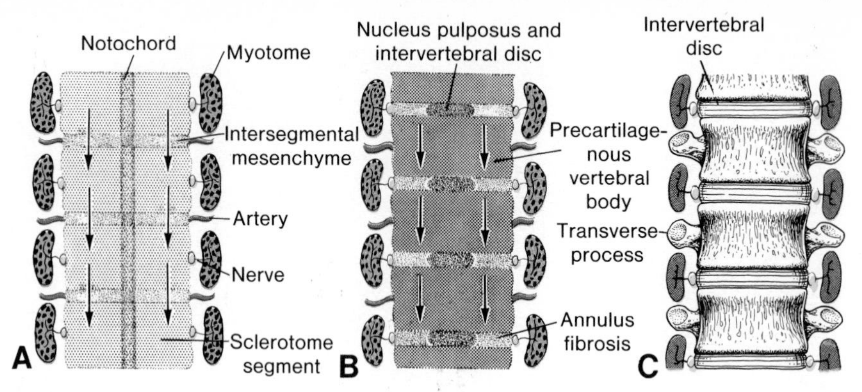

Figure 9-12. Scheme to show the formation of the vertebral column at various stages of development. *A*, At the fourth week of development the sclerotomic segments are separated by less dense intersegmental tissue. Note the position of the myotomes, intersegmental arteries, and segmental nerves. *B*, Condensation and proliferation of the caudal half of one sclerotome proceeds into the intersegmental mesenchyme and the cranial half of the subjacent sclerotome. Note the appearance of the intervertebral disc. Note the position of the arrows in A and *B*. *C*, The precartilaginous vertebral bodies are formed by the upper and lower halves of two successive sclerotomes and the intersegmental tissue. The myotomes overbridge the intervertebral discs.

manner binds the caudal half of one sclerotome to the cephalic half of the underlying sclerotome (see arrows in fig. 9-12*A* and *B*). Hence, by incorporation of the intersegmental tissue into the **precartilaginous vertebral body** (fig. 9-12*B*), the body of the vertebra becomes intersegmental in origin.[18]

The mesenchymal cells located between the cephalic and caudal parts of the original sclerotome segment do not proliferate and fill the space between two precartilaginous vertebral bodies. In this way they contribute to the formation of the **intervertebral disc**. (fig. 9-12*B*). While the notochord regresses entirely in the region of the vertebral bodies, it persists and enlarges in the region of the intervertebral disc. Here it undergoes mucoid degeneration and forms the **nucleus pulposus**, which is later surrounded by the circular fibers of the **annulus fibrosis**. Combined, these two structures form the **intervertebral disc** (fig. 9-12*C*).[19]

The rearrangement of the sclerotomes into the definitive vertebrae causes the myotomes to overbridge the intervertebral discs, and this alteration gives them the opportunity of moving the spine (fig. 9-12*C*). For the same reason, the intersegmental arteries, at first located between the sclerotomes, now pass midway over the vertebral bodies. The spinal nerves, however, come to lie near the intervertebral discs and leave the vertebral column through the intervertebral foramina.

Vertebral Column Abnormalities

The formation and subsequent rearrangement of the segmental sclerotomes into the definitive vertebrae is a complicated process, and it is not uncommon that two successive vertebrae fuse asymmetrically or that half a vertebra is missing. Likewise, it is not infrequently noted that the regular number of vertebrae is increased or decreased. A rather typical example of these abnormalities is formed in patients with the **Klippel-Feil syndrome**. These patients have a reduced number of cervical vertebrae, while the remaining vertebrae are fused or abnormal in shape. The condition is usually associated with other abnormalities.

One of the most serious vertebral defects, however, is the result of imperfect fusion or nonunion of the vertebral arches. Such an abnormality, known as **cleft vertebra**, is usually accompanied by abnormalities of the spinal cord, which herniates through the cleft and is thus exposed to the outside (For the various types of spina bifida see figs. 20-14 and 20-15).

General Skeletal Abnormalities

ACHONDROPLASIA

In addition to the abnormalities specifically affecting the skull, vertebral column, or limbs, in a number of diseases almost all the bones of the skeleton are affected.

One of the best known systemic abnormalities of the skeletal system is **achondroplasia**. This condition is caused by a disturbance of the endochondral ossification in the epiphysial plates of the long bones. The result is **dwarfism**. Both extremities are extremely short, the head has a normal size or is slightly enlarged and the center of the face is somewhat underdeveloped (fig. 9-13). Mental development is normal and occasionally high. The condition is inherited as a Mendelian dominant trait and occurs about once in 10,000 births.

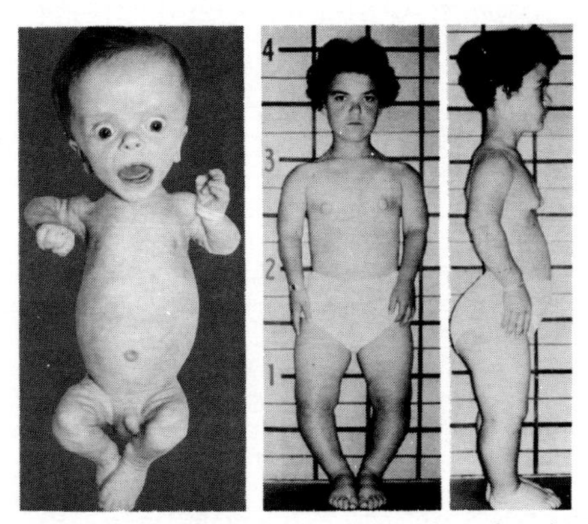

Figure 9-13. *A*, Three-month-old infant with achondroplasia. Note the large head, the short extremities, and the protruding abdomen. *B*, Achondroplasia in a 15-year-old girl. Note the dwarfism of the short limb type, the limbs being disproportionately shorter than the trunk. The limbs are bowed, there is an increase in the lumbar lordosis, and the face is small relative to the head.

ACROMEGALY

This condition is caused by congenital **hyperpituitarism** and is characterized by disproportional enlargement of the face, hands, and feet. Sometimes it causes an overall rapid growth resulting in **gigantism**.

In many patients skeletal abnormalities are accompanied by anomalies of other organ systems. These combinations of abnormalities are known as syndromes. Most syndromes are rare and the interested reader is referred to the books of Warkany,[17] Smith,[20] and Gorlin.[21]

SUMMARY

The skeletal system develops from mesenchyme. Some bones, such as the flat bones of the skull, undergo **membranous ossification**, that is, the mesenchyme cells are directly transferred into osteoblasts (fig. 9-2). In most bones, such as the long bones of the limbs, the mesenchyme condenses and forms hyaline cartilage models of bones (fig. 9-9). Ossification centers appear in these cartilage models and gradually the bone ossifies by **endochondral ossification**.

The **vertebral column** and **ribs** develop from the **sclerotome** compartments of the **somites**. A definitive vertebra is built up by condensation of the caudal half of one sclerotome and fusion with the cranial half of the subjacent sclerotome (fig. 9-12).

The **skull** has a complicated origin since it consists of the **neurocranium** and **viscerocranium**, each with its own membranous and cartilaginous components (figs. 9-2 to 9-4). Although many abnormalities occur in the skeletal system, limb abnormalities such as amelia, meromelia, and lobster claw hand (figs. 9-10 and 9-11) are the most common.

REFERENCES

1. Johnston, M. C., and Listgarten, M. A. Observations on the migration, interaction, and early differentiation of orofacial tissues. In *Developmental Aspects of Oral Biology*, edited by H. C. Slavkin and L. A. Baretta. Academic Press, New York, 1972.
2. Arey, H. B. The history of the first somite in human embryos. *Contrib. Embryol., 27:* 235, 1938.
3. Tessier, P. Ostéotomies totales de la face. Syndrome de Crouzon. Syndrome d'Apert. Oxycéphalies. Scaphocéphalies. Turricéphalies. *Ann Chir. Plast., 12:* 273, 1967.
4. Edgerton, M. T., Jane, J. A., and Berry, F. A. Craniofacial osteotomies and reconstruction in infants and children. *J. Plast. Reconstr. Surg., 54:* 13, 1974.
5. O'Rahilly, R. Normal development of the human embryo. In *Normal and Abnormal Embryological Development*, edited by C. H. Frantz. National Research Council, Washington, 1967.
6. Saunders, J. W. Control of growth patterns in limb development. In *Normal and Abnormal Embryological Development*, edited by C. H. Frantz. National Research Council, Washington, 1967.
7. Zwilling, E. Abnormal morphogenesis in limb development. In *Limb Development and*

Deformity: Problems of Evaluation and Rehabilitation, edited by C. A. Swinyard. Charles C. Thomas, Springfield, Ill., 1969.
8. Blechschmidt, E. The early stages of human limb development. In *Limb Development and Deformity: Problems of Evaluation and Rehabilitation*, edited by C. A. Swinyard. Charles C. Thomas, Springfield, Ill., 1969.
9. Gardner, E. Osteogenesis in the human embryo and fetus. In *The Biochemistry and Physiology of Bone*, edited by G. H. Bourne. Academic Press, New York, 1956.
10. Bayer, L. M., and Bayley, N. *Growth Diagnosis. Selected Methods for Interpreting and Predicting Physical Development from One Year to Maturity*. University of Chicago Press, Chicago, 1959.
11. Vaughan, V. C., and McKay, R. J. In *Nelson Textbook of Pediatrics*, ed. 10. W. B. Saunders, Philadelphia, 1975.
12. Lenz, W. Thalidomide and congenital abnormalities. *Lancet, 1:* 1219, 1962.
13. Weicker, H., and Hungerland, H. Thalidomid-embryopathie. I. Vorkommen inner and ausserhalb Deutschlands. *Dtsch. Med. Wochenschr., 87:* 922, 1962.
14. Vickers, T. H. Congenital abnormalities and thalidomide. *Med. J. Aust, 1:* 649, 1962.
15. Devitt, R. E. F., and Kenny, S. Thalidomide and congenital abnormalities. *Lancet, 1:* 430, 1962.
16. Knapp, K, Radiological aspects of thalidomide embryopathy. In *Limb Development and Deformity: Problems of Evaluation and Rehabilitation*, edited by C. A. Swinyard. Charles C. Thomas, Springfield, Ill., 1969.
17. Warkany, J. *Congenital Malformations: Notes and Comments*. Year Book Medical Publishers, Chicago, 1971.
18. Sensenig, E. C. The early development of the human vertebral column. *Contrib. Embryol., 33:* 21, 1949.
19. Peacock, A. Observations on the prenatal development of the intervertebral disc in man. *J. Anat., 85:* 260, 1951.
20. Smith, D. W. *Recognizable Patterns of Human Malformation: Genetic, Embryologic and Clinical Aspects*. W. B. Saunders, Philadelphia, 1976.
21. Gorlin, R. J. *Syndromes of the Head and Neck*, ed. 2. McGraw-Hill, New York, 1976.

chapter 10

Muscular System

Cross-Striated Musculature

The muscular system develops from the mesodermal germ layer and initially the somites play an important role. When the somite has differentiated into the sclerotome and dermomyotome, the cells of the **myotome** (fig. 10-1*A*) split off and become elongated and spindle-shaped. These cells, known as **myoblasts**, fuse together and form long multinucleated muscle fibers. Myofibrils soon appear in the cytoplasm and by the end of the third month **cross-striations** typical for **skeletal muscle** fibers appear. A similar process occurs in the ventrolateral body wall (fig. 10-1*B*), where the somatic mesoderm layer gives rise to cross-striated muscles for the body wall and limbs.[1] Since **smooth muscle** differentiates from the splanchnic mesoderm layer surrounding the gut and its derivatives (fig. 10-1*B*), and **cardiac muscle** is derived from splanchnic mesoderm surrounding the heart tube, all three types of muscle tissue are derived from mesoderm.

By the end of the fifth week the musculature in the body wall is divided into a small dorsal portion, the **epimere**, and a larger ventral part, the **hypomere** (fig. 10-2*A*). The nerve innervating the segmental muscles is also divided into a **dorsal primary ramus** for the epimere, and a **ventral primary ramus** for the hypomere (fig. 10-2*B*).

The muscles of the epimeres form the extensor muscles of the vertebral column, while those of the hypomeres give rise to the lateral and ventral flexor musculature (fig. 10-2*B*). The latter splits into three layers which, in the thorax, are represented by the **external intercostal**, the **internal intercostal**, and the **innermost intercostal** or **transverse thoracic muscle** (fig. 10-2*B*). In the abdominal wall these three muscle layers consist of the **external oblique**, the **internal oblique**, and the **transverse abdominis muscles**. The muscles in the wall of the thorax maintain their segmental character, owing to the ribs; in the abdominal wall, however, the muscles of the various segments fuse to form large sheets of muscle tissue.

In addition to the above three ventrolateral muscle layers, a ventral longitudinal column arises at the ventral tip of the hypomeres (fig. 10-2*B*). In the abdominal region this column is represented by the **rectus abdominis**

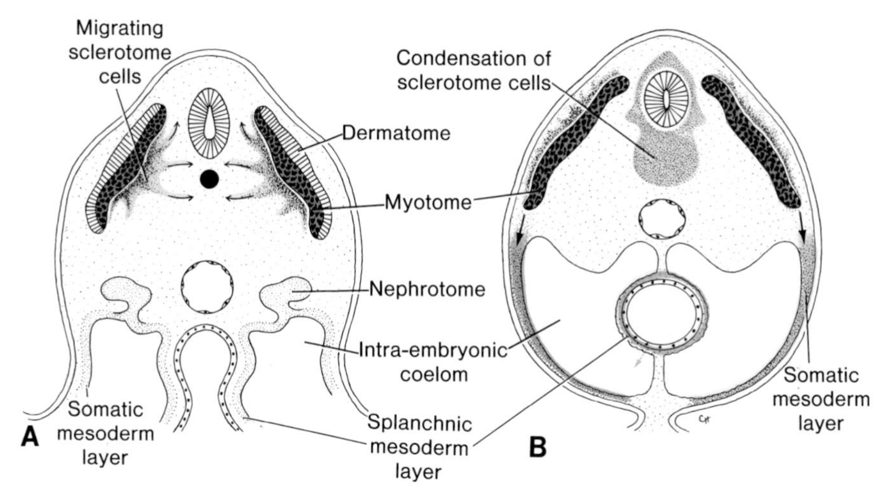

Figure 10-1. *A*, Diagrammatic transverse section through a four-week embryo, showing the cells of the myotome in close contact with the dermatome. *B*, Similar section as in *A*, showing the migration of the cells of the myotome in a ventral direction until they reach the intra-embryonic coelom.

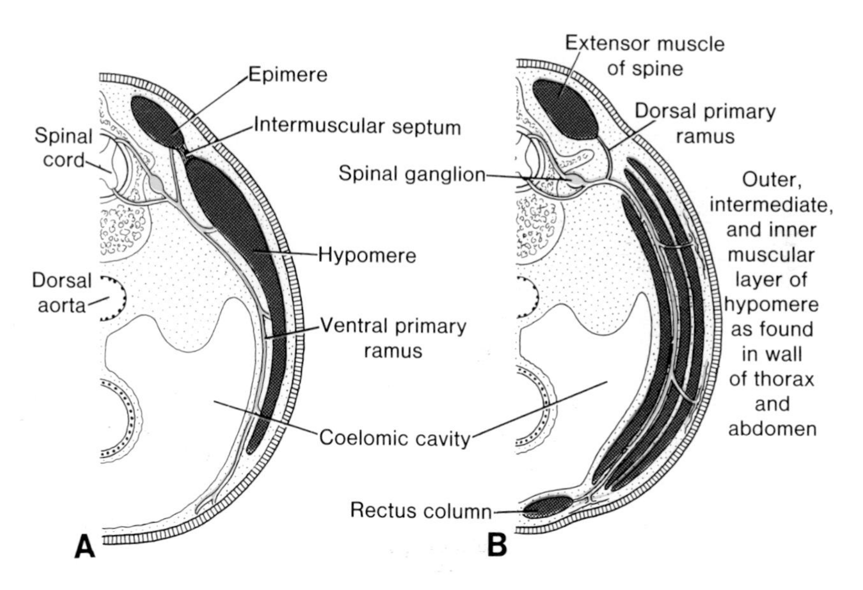

Figure 10-2. *A*, Transverse section through the thoracic region of a five-week embryo. The dorsal portion of the body wall musculature (epimere) and the ventral portion (hypomere) are innervated by a dorsal primary ramus and a ventral primary ramus, respectively. *B*, Similar sections as in *A*, at a later stage of development. The hypomere has formed three separate muscle layers and a ventral longitudinal muscle.

muscle, and in the cervical region by the **infrahyoid musculature**. In the thorax the longitudinal muscle normally disappears but is occasionally represented by the **sternalis muscle**.

TONGUE MUSCULATURE

In the region of the head the development of the myotomes is not so clear. Initially, four pairs of **occipital somites** can be distinguished, but the most cephalic of these disappears soon after its formation.[2] The myoblasts of the three remaining occipital myotomes are believed to migrate forward and to form the extrinsic and intrinsic musculature of the tongue (fig. 10-3*A*).[3] Indeed, their innervation by the hypoglossal nerve, which represents the occipital group of segmental nerves, gives considerable support to the theory that the tongue musculature is derived from the occipital somites.

EYE MUSCULATURE

Although the origin of the extrinsic muscles of the eye has not been traced in mammalian embryos, it has been suggested that these muscles originate from mesoderm surrounding the prochordal plate.[4] This mesoderm is thought to form three myotomes known as the **preotic myotomes** (fig. 10-3*A*).

PHARYNGEAL ARCH MUSCULATURE

Mesodermal cells located in the pharyngeal arches also differentiate into myoblasts and migrate subsequently in various directions (fig. 10-3*A*). Despite extensive migration, their origin can always be traced, as they remain innervated by the nerve of the arch of origin (see fig. 16-5)[5] In this manner

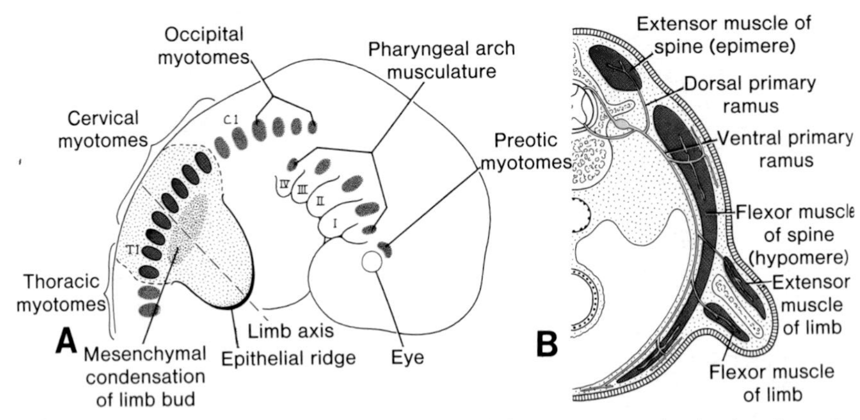

Figure 10-3. *A*, Schematic representation of the myotomes in the head, neck, and thorax region of a seven-week embryo. Note the localization of the preotic and occipital myotomes and the condensation of mesenchyme at the base of the limb bud. *B*, Transverse section through the region of attachment of the limb bud. Note the dorsal (extensor) and ventral (flexor) muscular component of the limb.

it has been possible to determine that the muscles of mastication, of the face, and of the pharynx and larynx originate from the different pharyngeal arches.

LIMB MUSCULATURE

The first indication of the limb musculature is found in the seventh week of development as a condensation of mesenchyme near the base of the buds (fig. 10-3*A*). In the human embryo this mesenchyme is derived from the somatic mesoderm. From here it migrates into the limb buds.[6]

With elongation of the limb buds the muscular tissue splits into flexor and extensor components (fig. 10-3*B*). Although initially the muscles of the limbs have a segmental character, with time they fuse and are then composed of muscle tissue derived from several segments.

The upper limb buds lie opposite the lower five cervical and upper two thoracic segments (fig. 10-4*A*, *B*) and the lower limb buds lie opposite the lower four lumbar and upper two sacral segments (fig. 10-4*C*). As soon as the buds are formed the various spinal nerves penetrate into the mesenchyme. At first they enter with isolated dorsal and ventral branches, but soon these branches unite to form large dorsal and ventral nerves. Thus the **radial nerve**, which supplies the extensor musculature, is formed by a combination of the dorsal segmental branches, whereas the **ulnar** and **median nerves**, which supply the flexor musculature, are formed by combination of the ventral branches. Immediately after the nerves have entered the limb buds, they establish an intimate contact with the differentiating mesodermal condensations, and the early contact between the nerve and the differentiating muscle cells is a prerequisite for their complete functional differentiation.

The spinal nerves not only play an important role in the differentiation and motor innervation of the limb musculature, but also provide the sensory innervation for the dermatomes. Although the original dermatomal pattern changes with growth of the extremities, an orderly sequence can still be recognized in the adult (fig. 10-4).

Muscle Abnormalities

Partial or complete absence of one or more muscles is a rather common occurrence. One of the best known examples is total or partial absence of the pectoralis major muscle. Similarly the palmaris longus, the serratus anterior and the quadratus femoris may be partially or entirely absent.

Excessive stretching of the sternocleidomastoid muscle during delivery may cause a hemorrhage in the muscle and a subsequent shortening. The condition is known as **congenital torticollis**.

Smooth Musculature

Smooth muscle tissue develops mainly from the splanchnic mesoderm layer surrounding the gastrointestinal tract and its derivatives. These mesodermal cells form the muscle coat of the gut, trachea, and bronchi, as well as

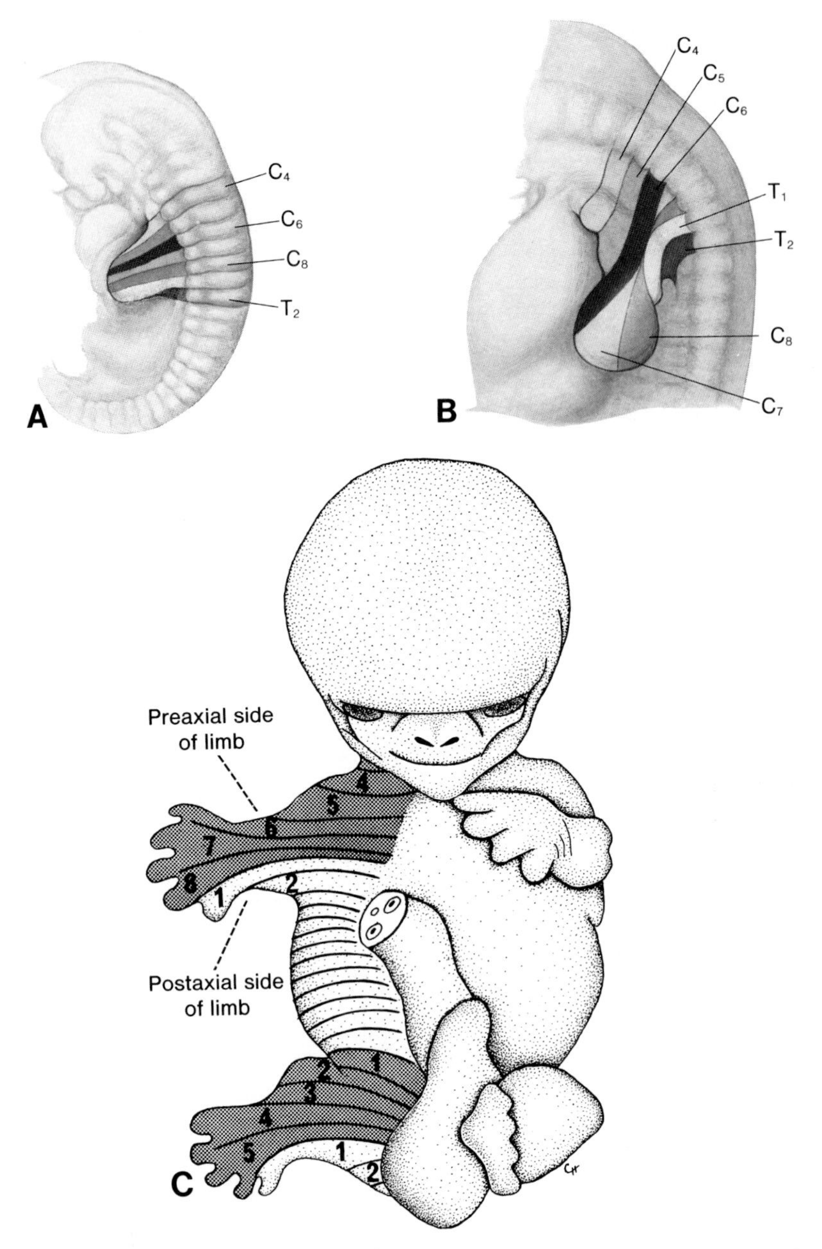

Figure 10-4. Schematic drawings of the limb buds indicating their segments of origin. With further development the segmental pattern disappears; an orderly sequence in the dermatome pattern, however, can still be recognized in the adult. *A*, Upper limb bud at five weeks; *B*, Upper limb bud at six weeks. *C*, Limb buds at seven weeks. (*A* and *B*, From Langman, J., and Woerdeman, M. W., *Atlas of Medical Anatomy*. W. B. Saunders Co., Philadelphia, 1978.)

that of the vessels found in the mesenteries. The vessels which develop in the limb buds, head, and body wall obtain their muscular coat from local mesenchyme. In fact, mesenchyme anywhere in the body is a potential source of smooth muscle tissue.

The smooth muscles of the iris form an exception. These muscles, the **sphincter** and **dilator pupillae**, are thought to differentiate from the ectoderm of the optic cup (see Chapter 18). Similarly the myoepithelial cells of the mammary and sweat glands are thought to originate from ectoderm.

Cardiac Muscle

Cardiac muscle in the embryo develops from the splanchnic mesoderm layer surrounding the endothelial heart tube. The myoblasts adhere to one another by special attachments that later develop into **inter-calated discs**. Myofibrils develop as in skeletal muscle, but the myoblasts do not fuse. During later development a few special bundles of muscle cells with few, irregularly distributed, myofibrils become visible. These bundles are **Purkinje fibers** and form the conducting system of the heart.

SUMMARY

All muscles are of **mesodermal origin**. Most of the skeletal muscles are derived from the **myotomes** which provide the myoblasts for the multinucleated muscle fibers. The myofibrils provide the cross-striations typical for skeletal muscle. Most of the head and neck muscles originate from the **branchial arch mesoderm**, and their source of origin can usually be traced by their nerve of origin. **All smooth muscles** as well as the **cardiac muscle fibers are derived from splanchnic mesoderm**.

REFERENCES

1. Strauss, W. L., and Rawles, M. E. An experimental study of the origin of the trunk musculature and ribs in the chick. *Am. J. Anat., 92:* 471, 1953.
2. Arey, H. B. The history of the first somite in human embryos. *Contrib. Embryol., 27:* 235, 1938.
3. Deuchar, E. M. Experimental demonstration of tongue muscle origin in chick embryos. *J. Embryol. Exp. Morph., 6:* 527, 1958.
4. Gilbert, P. W. Origin and development of the human extrinsic ocular muscles. *Contr. Embryol. Carneg. Inst., 36:* 59, 1957.
5. Gasser, R. F. The development of the facial muscles in man. *Am. J. Anat., 120:* 357, 1967.
6. O'Rahilly, R. Normal development of the human embryo. In *Normal and Abnormal Embryological Development*. National Research Council, Washington, D.C., 1967.

chapter 11

Body Cavities and Serous Membranes

Formation of Coelomic Cavity

At the end of the third week the intra-embryonic mesoderm on each side of the midline differentiates into a paraxial portion, an intermediate portion, and a lateral plate (fig. 11-1*A*). When intercellular clefts appear in the lateral mesoderm, the plates are divided into two layers: the **somatic mesoderm layer** and the **splanchnic mesoderm layer**, which is continuous with the mesoderm of the wall of the yolk sac (fig. 11-1*B*). The space bordered by these layers forms the **intra-embryonic coelom**.

At first the right and left intra-embryonic coelomic cavities are in wide open connection with the extra-embryonic coelom, but when the body of the embryo folds in cephalo-caudal and lateral directions, they lose this connection (fig 11-2*A*, *B*, and *C*). In this manner is formed a large intra-embryonic coelomic cavity extending from the thoracic to the pelvic region.

The cells of the somatic mesoderm lining the intra-embryonic coelomic cavity become mesothelial and form the **parietal layer of the serous membranes** lining the outside of the peritoneal, pleural, and pericardial cavities. In a similar manner the cells of the splanchnic mesoderm layer will form the **visceral layer of the serous membranes** covering the abdominal organs, lungs, and heart.

Diaphragm and Thoracic Cavity

The most important structure dividing the intra-embryonic coelomic cavity is formed by the **septum transversum**, a thick plate of mesodermal tissue occupying the space between the thoracic cavity and the stalk of the yolk sac (fig. 11-3*A*). This septum does not separate the thoracic and abdominal cavities entirely but leaves a large opening, the **pericardioperitoneal canal**, on each side of the foregut (fig. 11-3*A*).

When the lung buds begin to grow, they expand in caudo-lateral direction within the pericardioperitoneal canals (fig. 11-3*B*). As a result of the rapid growth of the lungs the pericardioperitoneal canals become too small, and they begin to expand into the mesenchyme of the body wall in dorsal, lateral, and ventral directions (fig. 11-3*B*) (see small arrows).

The expansion in ventral and lateral directions occurs in a plane lateral to

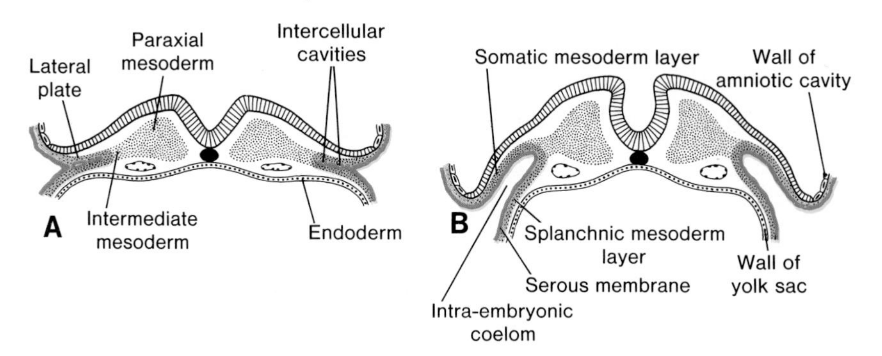

Figure 11-1. *A*, Transverse section through an embryo of approximately 19 days. Intercellular cavities are visible in the lateral plate. *B*, Section through an embryo of approximately 20 days. The lateral plate is divided into the somatic and splanchnic mesoderm layers which line the intra-embryonic coelomic cavities. The tissue bordering the intra-embryonic coelomic cavity differentiates into the serous membrane and is indicated by red lines.

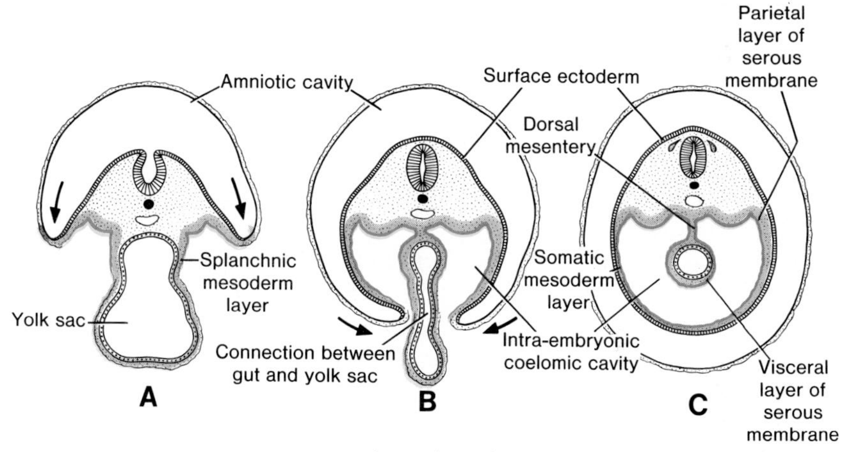

Figure 11-2. Transverse sections through embryos at various stages of development. *A*, The intra-embryonic coelom is in wide open communication with the extra-embryonic coelom. *B*, The intra-embryonic coelomic cavities are about to lose their contact with the extra-embryonic coelom. *C*, At the end of the fourth week. The splanchnic mesoderm layers are fused in the midline and form a double-layered membrane between the right and left intra-embryonic coelom. The ventral portion of this membrane, however, breaks down immediately after its formation.

the **pleuropericardial fold**. At first these folds appear as small ridges projecting into the primitive undivided thoracic cavity (fig. 11-3*B*). With the expansion of the lungs the mesoderm of the body wall is split into two components (fig. 11-4*A, B*): (1) the definitive wall of the thorax; and (2) a thin mesodermal membrane, the **pleuropericardial membrane**, which con-

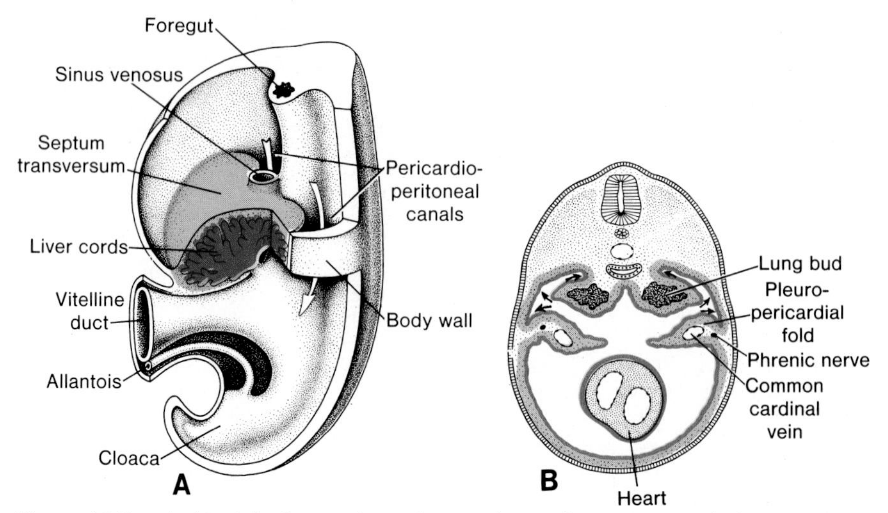

Figure 11-3. *A*, Model of a portion of an embryo of approximately five weeks. Parts of the body wall and the septum transversum have been removed to show the pericardioperitoneal canals. Note the size and thickness of the septum transversum and the liver cords penetrating in the septum. *B*, Drawing to show the growth of the lung buds into the pericardioperitoneal canals. Note the pleuropericardial folds. Arrows indicate the direction of expansion of the lung buds.

tains the common cardinal vein and phrenic nerve. When subsequently, as a result of the descent of the heart and the positional changes of the sinus venosus, the common cardinal veins shift toward the midline, the pleuropericardial membranes are drawn out in mesentery-like fashion (fig. 11-4*A*). They finally fuse with each other and with the root of the lungs, and the thoracic cavity is then divided into the definitive **pericardial cavity** and two **pleural cavities** (fig. 11-4*B*). In the adult the pleuropericardial membranes form the **fibrous pericardium.**

Although the pleural cavities are separated from the pericardial cavity, they remain temporarily in open communication with the abdominal cavity since the diaphragm is still incomplete. During further development the caudal border of the pleural cavities is delineated by a crescent-shaped fold, the **pleuroperitoneal fold** (fig. 11-5*A*). This fold projects into the caudal end of the pericardioperitoneal canal.[1] With further development the fold extends in medial and ventral directions and by the seventh week fuses with the mesentery of the esophagus and with the septum transversum (fig. 11-5*B*). Hence, **the connection between the thoracic and abdominal portions of the coelomic cavity is closed by the pleuroperitoneal membranes**. Further expansion of the pleural cavities into the mesenchyme of the body wall results in the addition of a peripheral rim to the pleuroperitoneal membranes (fig. 11-5*C*). Once this rim is established, myoblasts originating in the body wall penetrate the membranes to form the muscular part of the diaphragm.

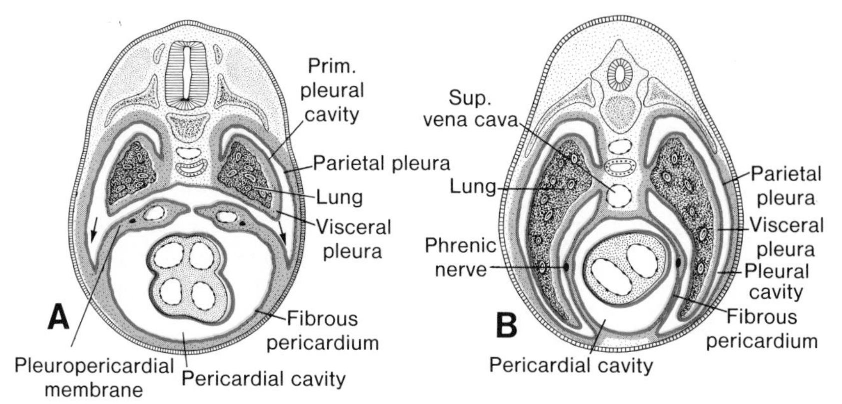

Figure 11-4. *A*, Schematic drawing showing the transformation of the pericardioperitoneal canals into the pleural cavities and the formation of the pleuropericardial membranes. Note the pleuropericardial folds with the common cardinal vein and phrenic nerve. The mesenchyme of the body wall is split into the pleuropericardial membranes and the definitive body wall. Arrows indicate direction of expansion of the primitive pleural cavity. *B*, Drawing through the thorax after fusion of the pleuropericardial folds with each other and with the root of the lungs. Note the position of the phrenic nerve which is now in the fibrous pericard. The right common cardinal vein has developed into the superior vena cava.

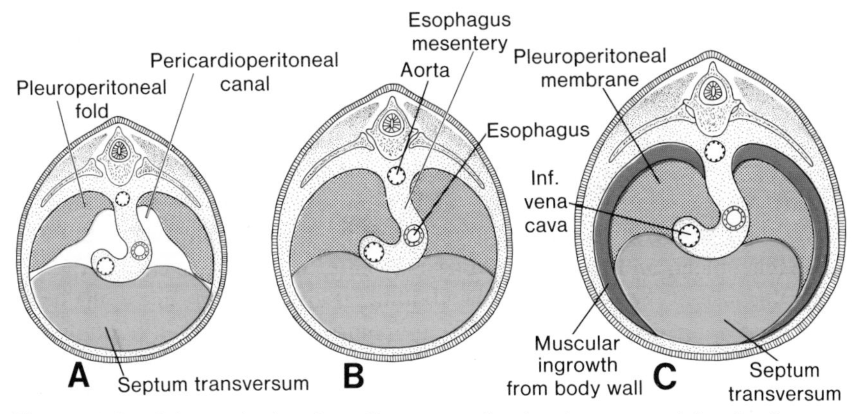

Figure 11-5. Schematic drawings illustrating the development of the diaphragm. *A*, The pleuroperitoneal folds appear at the beginning of the sixth week. *B*, The pleuroperitoneal folds have fused with the septum transversum and the mesentery of the esophagus in the seventh week, thus separating the thoracic cavity from the abdominal cavity. *C*, Transverse section at the fourth month of development. An additional rim derived from the body wall forms the most peripheral part of the diaphragm.

The diaphragm is thus derived from the following structures: (1) the septum transversum, which forms the tendinous part of the diaphragm; (2) the two pleuroperitoneal membranes; (3) muscular components from the lateral and dorsal body walls; and (4) the mesentery of the esophagus, in which the **crura of the diaphragm** develop (fig. 11-5*C*).

Position and Innervation of the Diaphragm. Initially the septum transversum lies opposite the cervical somites, and nerve components of the third, fourth, and fifth cervical segment[3] of the spinal cord grow into the septum. At first the nerves, known as the **phrenic nerves**, pass to the septum through the pleuropericardial folds (fig. 11-3*B*). This explains why, with the further expansion of the lungs and descent of the septum, they are located in the fibrous pericard (fig. 11-4*A, B*). Hence, in the adult the phrenic nerves reach the diaphragm via the fibrous pericard.

Although the septum transversum lies opposite the cervical segments during the fourth week, already by the sixth week the developing diaphragm is located at the level of the thoracic somites. This **descent of the diaphragm** is apparently caused by rapid growth of the dorsal part of the embryo (vertebral column) in comparison to the ventral part. By the beginning of the third month some of the dorsal bands of the diaphragm originate at the level of the first lumbar vertebra.

The phrenic nerves supply the diaphragm with its motor and sensory innervation. Since the most peripheral part of the diaphragm is derived from mesenchyme of the thoracic wall, it is generally accepted that some of the lower intercostal (thoracic) nerves contribute sensory fibers to the peripheral part of the diaphragm.

Diaphragmatic Hernia. A diaphragmatic hernia is one of the more common malformations in the newborn (1:2000) and is most frequently caused by failure of the pleuroperitoneal membrane(s) to close the pericardioperitoneal canal(s). The peritoneal and pleural cavities are then continuous with one another along the posterior body wall. Such a defect, known as the **congenital diaphragmatic hernia**, allows the abdominal viscera to enter the pleural cavity.[2, 3] Usually the hernia is on the left side, and intestinal loops and the stomach, spleen and part of the liver may enter the thoracic cavity (fig. 11-6*B*). Because of the presence of the abdominal viscera in the chest, the heart is pushed anteriorly, while the lungs are compressed and often hypoplastic.

Sometimes, the defective portion of the diaphragm is covered by a membrane composed of pleura and peritoneum without any tissue between the two. In such cases the intestinal contents penetrating the thoracic cavity are surrounded by the serous membranes.

Occasionally, a small part of the muscular fibers of the diaphragm fails to develop and a hernia may then remain undiscovered until the child is several years old. Such a defect is frequently seen in the anterior portion of the diaphragm and is then known as the **parasternal hernia**. A small peritoneal sac containing the intestinal loops then enters the chest between the sternal and costal portions of the diaphragm (fig. 11-6*A*).

Another type of diaphragmatic hernia, the **esophageal hernia**, is thought to be due to a congenital shortness of the esophagus. The cardia and upper

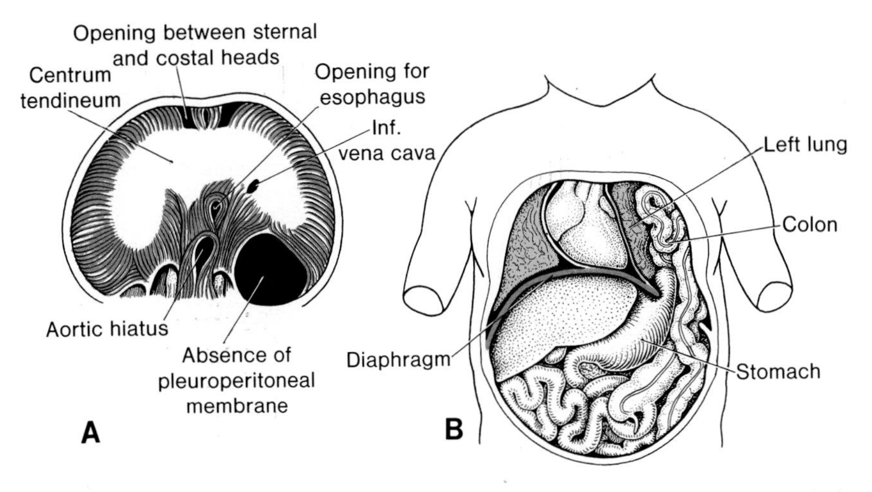

Figure 11-6. Congenital diaphragmatic hernia. *A*, Abdominal surface of the diaphragm, showing a large defect of the pleuroperitoneal membrane. *B*, Hernia of the intestinal loops and part of the stomach into the left pleural cavity. The heart and mediastinum are frequently pushed to the right, while the left lung is compressed.

part of the stomach are retained in the thorax and the stomach is then constricted at the level of the diaphragm.

Mesenteries and Abdominal Cavity

Initially the foregut, midgut, and hindgut are in broad contact with the mesenchyme of the posterior abdominal wall (fig. 11-2*A, B*). In the 8-mm embryo, however, the connecting tissue bridge has become narrow and the caudal part of the foregut, the midgut and the major part of the hindgut are suspended from the abdominal wall by the so-called **dorsal mesentery** (figs. 11-2*C* and 11-7). A ventral mesentery does not exist, except in the region of the terminal part of the esophagus, the stomach, and the upper part of the duodenum (fig. 11-7).

FORMATION OF VENTRAL MESENTERY

The stomach and upper part of the duodenum are initially in direct contact with the septum transversum, thus facilitating the growth of the liver cords into the mesenchyme of the septum (figs 11-3*A* and 14-7). As a result of the enormous growth of the liver, the septum transversum cannot contain the liver and the mesenchyme of the septum between the ventral abdominal wall and the liver becomes stretched and membranous, thus forming the **falciform ligament** (figs. 11-7 and 14-8*B*). In a similar manner the mesenchyme of the septum, extending between the liver and the ventral border of the stomach and duodenum, forms a membrane, known as the **lesser**

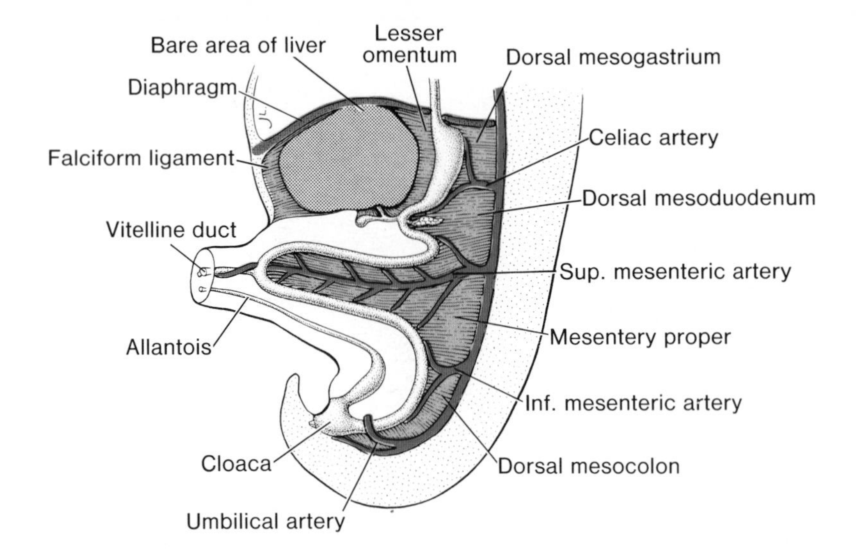

Figure 11-7. Schematic drawing showing the primitive dorsal and ventral mesenteries. Note that the liver is connected to the ventral abdominal wall and to the stomach by the falciform ligament and lesser omentum, respectively. The superior mesenteric artery runs through the mesentery proper and continues toward the yolk sac as the vitelline artery.

omentum. The free margin of the falciform ligament contains the umbilical vein (fig. 11-9*A*), which after birth is obliterated to form the **round ligament of the liver (ligamentum teres hepatis)**. The free margin of the lesser omentum contains the **bile duct**, **the portal vein** and the **hepatic artery**. In the adult the lower border of the lesser omentum forms the upper margin of the **epiploic foramen of Winslow**.

The liver is completely covered by peritoneum except for an area on its upper surface, where it remains in contact with the original portion of the septum transversum, the future **central tendon of the diaphragm** (fig. 11-7). This uncovered area is known as the **bare area of the liver**.

DORSAL MESENTERY

The dorsal mesentery extends from the lower end of the esophagus to the cloacal region of the hindgut. In the region of the stomach it is known as the **dorsal mesogastrium** or **greater omentum**; in the region of the duodenum, as the **dorsal mesoduodenum**; and in the region of the colon, as the **dorsal mesocolon**. The dorsal mesentery of the jejunal and ileal loops is known as the **mesentery proper**. Throughout its length the mesentery serves as a pathway for the blood vessels, nerves and lymphatics supplying the intestinal tract (fig. 11-7).

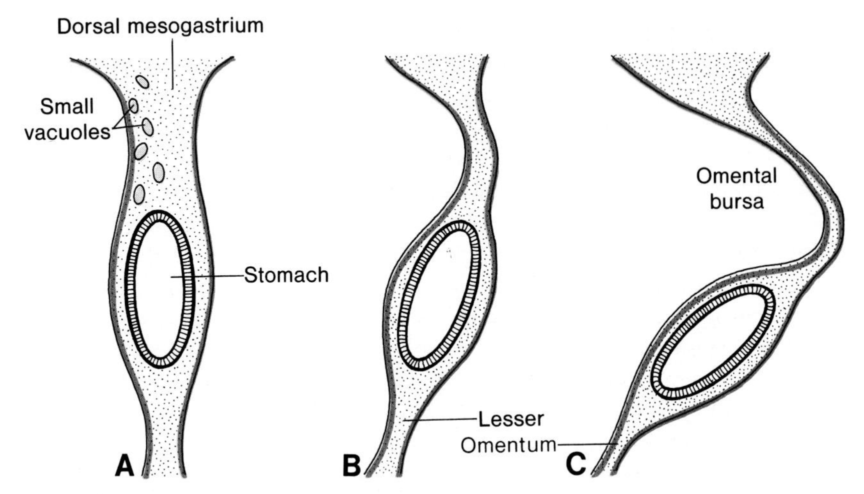

Figure 11-8. *A*, Schematic transverse section through a four-week embryo to show the intercellular clefts appearing in the dorsal mesogastrium. *B* and *C*, The clefts have fused and the omental bursa is formed as an extension of the right side of the coelomic cavity behind the stomach.

LESSER PERITONEAL SAC OR OMENTAL BURSA

In the fourth week of development small intercellular clefts appear in the mesenchyme dorsal to the stomach. These clefts fuse rapidly and a cavity, the future **omental bursa**, is formed behind the stomach (fig. 11-8).

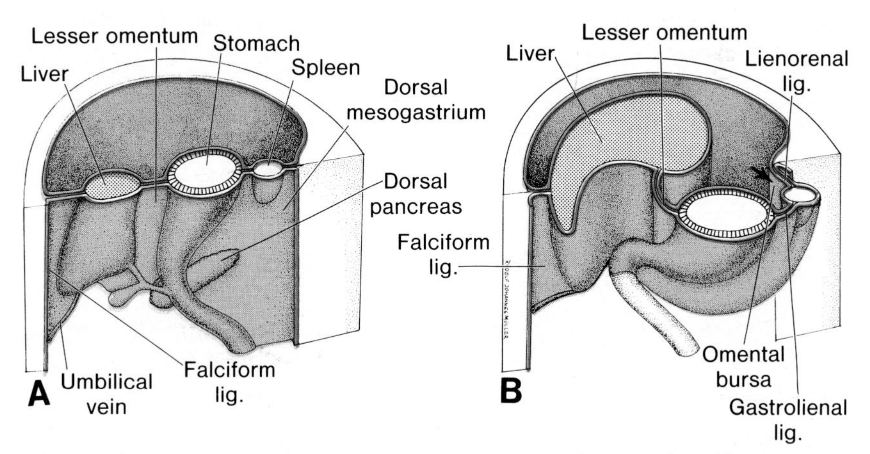

Figure 11-9. *A*, Drawing of the position of the spleen, stomach, and pancreas at the end of the 5th week. Note the position of the spleen and pancreas in the dorsal mesogastrium. *B*, Position of spleen and stomach at the 11th week. Note the formation of the omental bursa or lesser peritoneal sac.

With further extension in dorsal and lateral directions the dorsal mesogastrium begins to bulge to the left side (figs. 11-8*C*). This expansion has a profound effect on the stomach, the spleen, and the pancreas (figs. 11-9 and 11-10). The primordium of the spleen appears in the fifth week of development as a mesodermal proliferation between the two leaves of the dorsal mesogastrium (fig. 11-9*A*). With the formation of the omental bursa, a portion of the dorsal mesogastrium located between the spleen and dorsal midline swings to the left and fuses with the peritoneum of the posterior abdominal wall (figs. 11-9*B* and 11-10). The left leaf of this portion of the dorsal mesogastrium and the peritoneum subsequently disappears. The spleen, which always maintains an intraperitoneal position, is then connected to the dorsal body wall in the region of the left kidney by the **lienorenal ligament** and to the stomach by the **gastrolienal ligament**.

The formation of the omental bursa also influences the position of the pancreas. This organ initially grows into the dorsal mesoduodenum, but with time its tail expands into the dorsal mesogastrium (fig. 11-9*A*). Since the left leaf of this portion of the dorsal mesogastrium also fuses with the peritoneum of the dorsal body wall, the tail of the pancreas comes to lie in a retroperitoneal position (figs. 11-10*B* and 11-13).

As a result of the positional changes of the stomach, the dorsal mesogastrium, forming the left wall of the omental bursa, now bulges in a downward

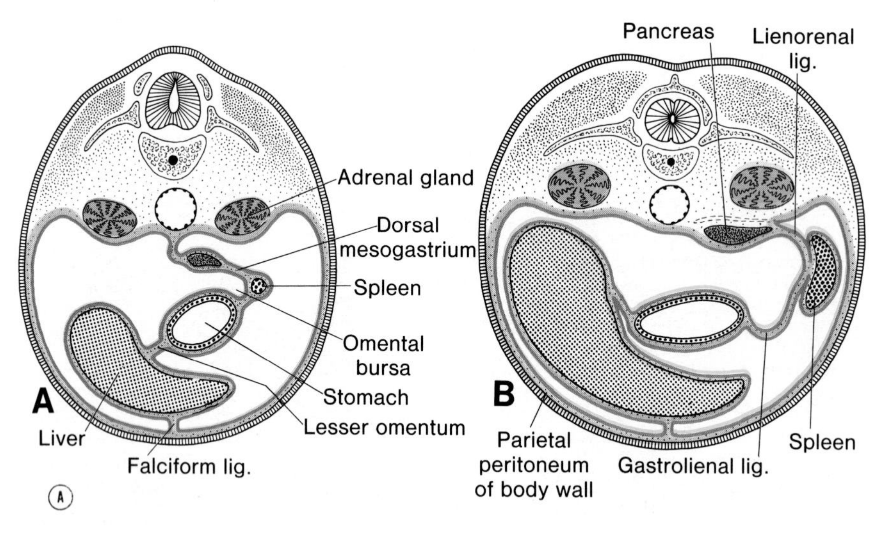

Figure 11-10. Diagrammatic transverse sections through the region of the stomach, liver, and spleen, showing the formation of the lesser peritoneal sac, the rotation of the stomach, and the position of the spleen and tail of the pancreas between the two leaves of the dorsal mesogastrium. With further development, the pancreas obtains a retroperitoneal position.

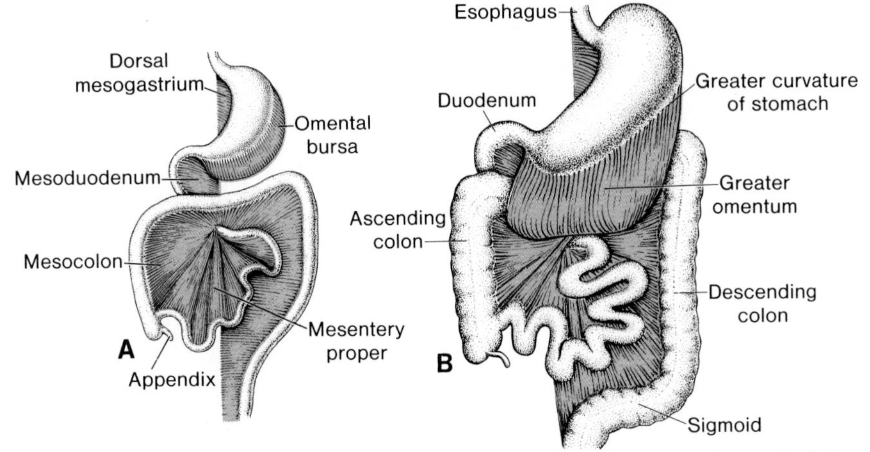

Figure 11-11. *A*, Schematic drawing of the dorsal mesentery at the end of the third month. The dorsal mesogastrium bulges out on the left side of the stomach, where it forms the omental bursa. *B*, The greater omentum hangs down from the greater curvature of the stomach in front of the transverse colon.

direction (fig. 11-11*A*). The dorsal mesogastrium continues to grow downward and forms a duplicature, extending over the transverse colon and small intestinal loops like an apron (figs. 11-11*B* and 11-12). This double-leaved apron is the **greater omentum**. Later its leaves fuse to form a single sheet

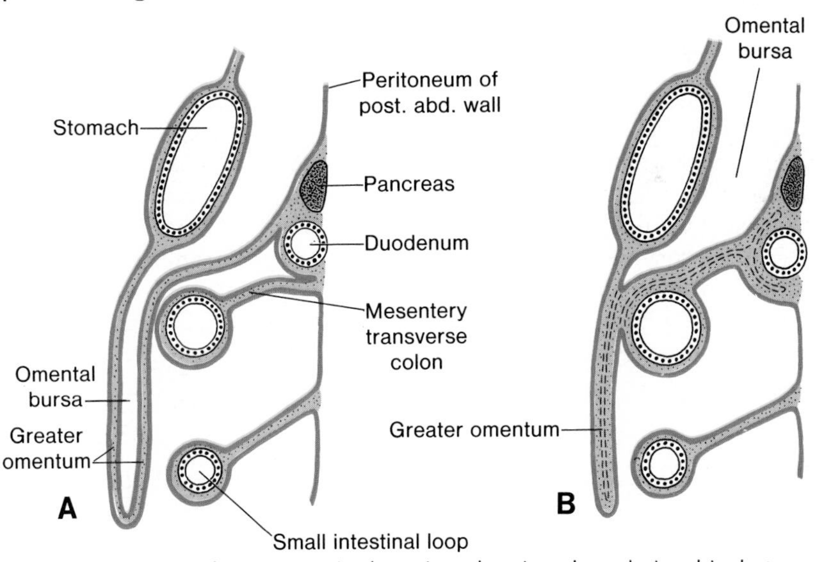

Figure 11-12. *A*, Schematic sagittal section showing the relationship between the greater omentum, stomach, transverse colon, and small intestinal loops at four months. The pancreas and duodenum have already acquired a retroperitoneal position. *B*, Similar section as in *A*, in the newborn. The leaves of the greater omentum have fused with each other and with the transverse mesocolon. The transverse mesocolon covers the already retroperitoneally located duodenum.

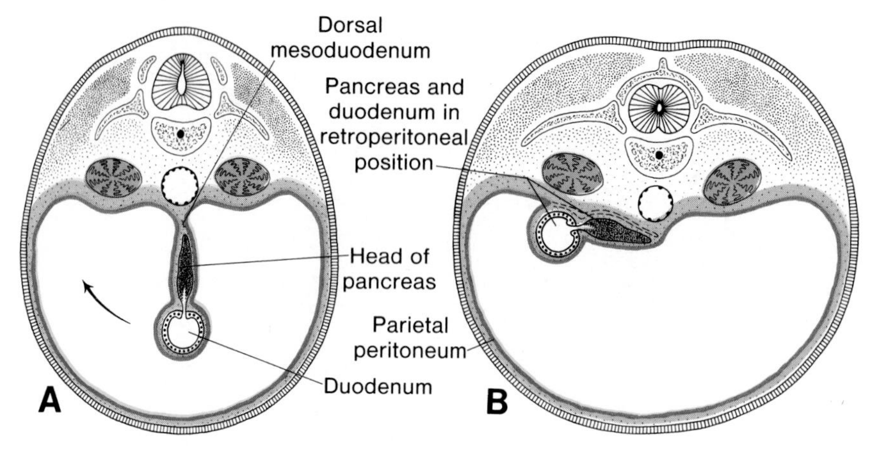

Figure 11-13. Transverse sections through the region of the duodenum at various stages of development. At first, the duodenum and the head of the pancreas are located in the median plane (*A*), but later they swing to the right and acquire a retroperitoneal position (*B*).

hanging from the greater curvature of the stomach. The posterior leaf of the greater omentum also fuses with the mesentery of the transverse colon (fig. 11-12*B*).

DORSAL MESODUODENUM

Rotation of the stomach and the duodenum, together with the rapid growth of the head of the pancreas, causes the duodenum to swing from its initial midline position to the right side of the peritoneal cavity (fig 11-13). The duodenum and the head of the pancreas are pressed against the dorsal body wall, and the right surface of the dorsal mesoduodenum fuses with the adjacent peritoneum. Both layers subsequently disappear, and the duodenum and head of the pancreas become fixed in a **retroperitoneal position**. The entire pancreas thus obtains a retroperitoneal position. The dorsal mesoduodenum disappears entirely except in the region of the pylorus of the stomach, where a small portion of the duodenum remains intraperitoneal.

MESENTERY PROPER

The mesentery of the primitive intestinal loop, **the mesentery proper**, undergoes profound changes with rotation and coiling of the loops (see figs. 14-13 to 14-15). When the caudal limb moves to the right side of the abdominal cavity, the dorsal mesentery twists around the origin of the superior mesenteric artery (figs. 11-7 and 11-11*A*). Later, when the ascending and descending portions of the colon obtain their definitive positions, their mesenteries are pressed against the peritoneum of the posterior abdominal wall (fig. 11-14). After fusion of these layers, the ascending and descending

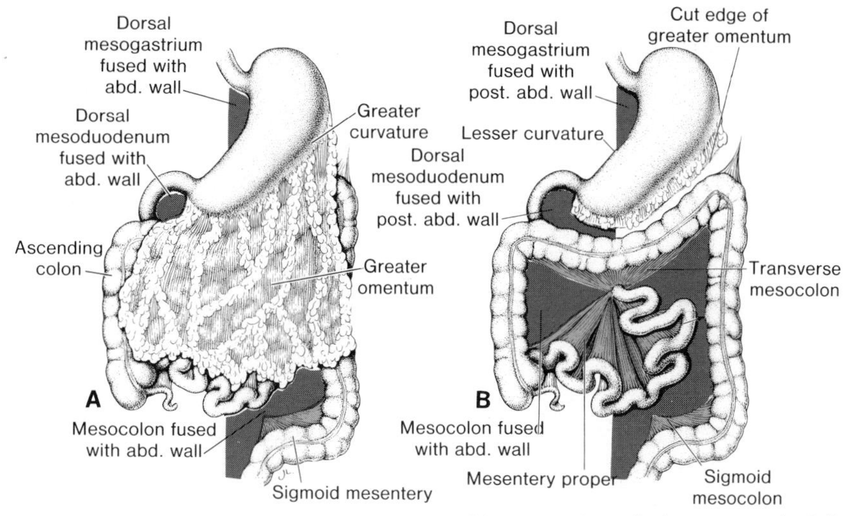

Figure 11-14. Frontal view of the intestinal loops with and after removal of the greater omentum. The cross-hatched areas indicate the parts of the dorsal mesentery which fuse with the posterior abdominal wall. Note the line of attachment of the mesentery proper.

colons are permanently anchored in a retroperitoneal position. The appendix and lower end of the cecum, however, retain their free mesentery.

The fate of the transverse mesocolon is different. It fuses with the posterior wall of the omental bursa (fig. 11-12), but maintains its mobility. Its line of attachment finally extends from the hepatic flexure of the ascending colon to the splenic flexure of the descending colon (fig. 11-14*B*).

The mesentery of the jejuno-ileal loops is at first continuous with that of the ascending colon (fig. 11-11*A*). When the mesentery of the ascending mesocolon fuses with the posterior abdominal wall, the mesentery of the jejuno-ileal loops obtains a new line of attachment which extends from the area where the duodenum becomes intraperitoneal to the ileocecal junction (fig. 11-14*B*).

Mobile Cecum and Colon. Normally the ascending colon, except for its most caudal part (approximately one inch), is fused to the posterior abdominal wall and covered by peritoneum on its anterior surface and sides. Persistence of a portion of the mesocolon gives rise to what is usually termed a **mobile cecum**. In its most extreme form, the mesentery of the ascending colon has failed to fuse with the posterior body wall, so that the root of the common mesentery is limited to a small area around the origin of the superior mesenteric artery. Such an unusually long mesentery allows for abnormal movements of the gut or even volvulus of the cecum and colon. Similarly, retrocolic pockets may occur behind the ascending mesocolon, and a **retrocolic hernia** represents the entrapment of portions of the small intestinal loops behind the mesocolon.

SUMMARY

At the end of the third week intercellular clefts appear in the mesoderm on each side of the midline. When these spaces fuse the **intra-embryonic coelom** bordered by a **somatic mesoderm** and **a splanchnic mesoderm** layer is formed (figs. 11-1 and 11-2). With the cephalo-caudal and transverse folding of the embryo the intra-embryonic coelomic cavity extends from the thoracic to the pelvic region. The somatic mesoderm will form the **parietal layer** of the **serous membranes** lining the outside **of the peritoneal, pleural and pericardial cavities**. The **splanchnic layer** will form the **visceral layer of the serous membranes** covering the lungs, heart and abdominal organs.

The **diaphragm** divides the coelomic cavity into the **thoracic cavity** and the **peritoneal cavity**. It develops from four components: (1) the septum transversum, (2) the pleuroperitoneal membranes, (3) dorsal mesentery of esophagus; and (4) muscular components of the body wall (fig. 11-5). Congenital diaphragmatic hernias occur frequently as a result of a defect of the pleuroperitoneal membrane on the left side.

The **thoracic cavity** is divided into the **pericardial cavity** and two **pleural cavities** for the lungs by the **pleuropericardial membranes** (see fig. 11-4).

With the transverse folding of the embryonic disc the two lateral parts of the intra-embryonic coelom come together (fig. 11-2) and fuse in the midline. The splanchnic mesoderm then encloses the primitive gut and suspends it from the dorsal body wall as a double layered membrane, the **dorsal mesentery** (fig. 11-7). The ventral mesentery is short and connects the stomach and duodenum with the liver (**lesser omentum**) and the liver with the anterior abdominal wall (**falciform ligament**). As a result of further rotation of the stomach and other gastrointestinal loops the greater omentum, lienorenal, and gastrolienal ligaments are formed (figs. 11-9 to 11-11).

REFERENCES

1. Wells, L. J. Development of the human diaphragm and pleural sacs. *Contrib. Embryol., 35:* 107, 1954.
2. Tarnay, T. J. Diaphragmatic hernia. *Ann. Thorac. Surg., 5:* 66, 1968.
3. Gray, S. W., and Skandalakis, J. E. The diaphragm. In *The Embryological Basis for the Treatment of Congenital Defects.* W. B. Saunders, Philadelphia, 1972.

chapter 12

Cardiovascular System

The vascular system of the human embryo appears in the middle of the third week, when the embryo is no longer able to satisfy its nutritional requirements by diffusion alone. At this stage mesenchymal cells in the splanchnic mesoderm layer of the late presomite embryo proliferate and form isolated cell clusters known as the **angiogenetic clusters** (fig. 12-1*B*). (For blood and blood vessel formation see Chapter 5.)

At first the clusters are located on the lateral sides of the embryo, but they rapidly spread in cephalic direction (fig. 12-1*A*). With time they acquire a

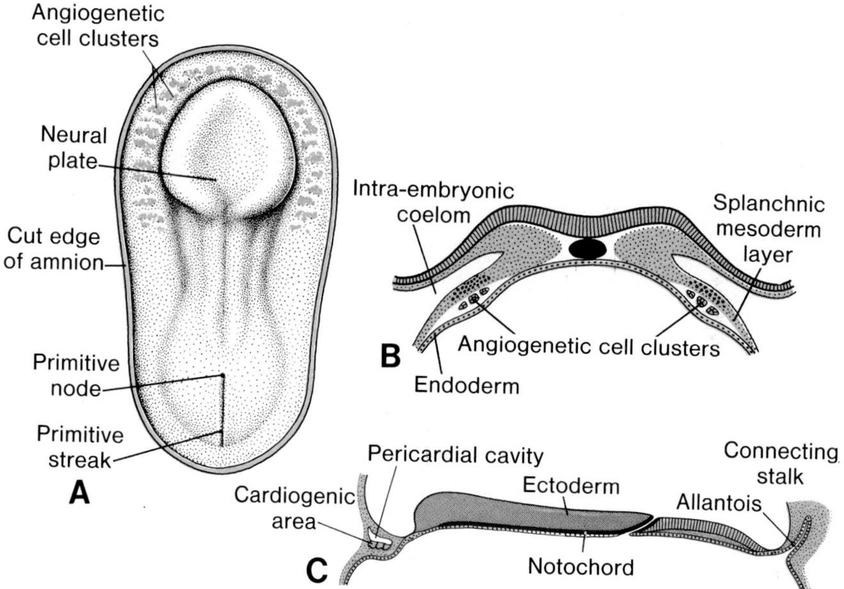

Figure 12-1. *A*, Dorsal view of a late presomite embryo (about 18 days) after removal of the amnion. The angiogenetic cell clusters formed in the splanchnic mesoderm in front of the neural plate and on each side of the embryo are visible through the overlying ectoderm and somatic mesoderm layer. *B*, Transverse section through similar embryo to show the position of the angiogenetic cell clusters in the splanchnic mesoderm layer. *C*, Cephalo-caudal section through similar embryo showing the position of the pericardial cavity and cardiogenic area.

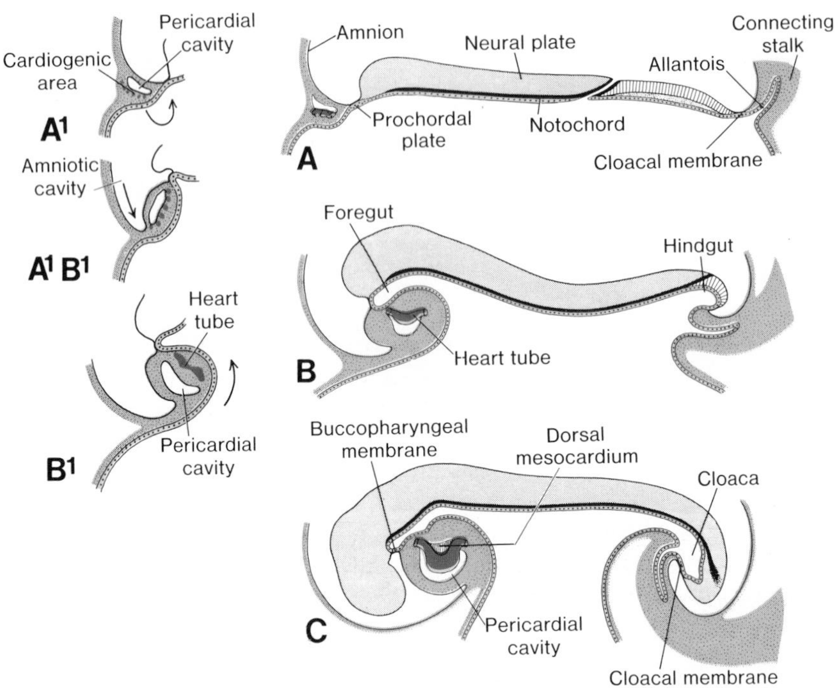

Figure 12-2. Drawings to show the result of the rapid growth of the brain vesicles on the position of the pericardial cavity and the developing heart tube. Initially the cardiogenic area and the pericardial cavity are located in the front of the prochordal plate. As a result of the rotation along a transverse axis through the prochordal plate, the cardiogenic area (heart tube) finally comes to lie dorsally to the pericardial cavity. *A*, 18 days; *B*, 21 days; *C*, 22 days.

lumen, unite, and form a **horseshoe-shaped** plexus of small blood vessels. The anterior central portion of this plexus is known as the **cardiogenic area** and the intra-embryonic coelomic cavity located over this region later develops into the **pericardial cavity** (fig. 12-2*A*).

In addition to the horseshoe-shaped plexus, other clusters of angiogenetic cells appear bilaterally, parallel and close to the midline of the embryonic shield. These clusters also acquire a lumen and form a pair of longitudinal vessels, the **dorsal aortae**. At a later stage these vessels gain connections with the horseshoe-shaped plexus which will form the heart tube.

Formation and Position of Heart Tube

Initially the central portion of the cardiogenic area is located anterior to the prochordal plate and the neural plate (figs. 12-2*A*). With the closure of the neural plate and the subsequent formation of the brain vesicles, however, the central nervous system grows so rapidly in cephalic direction that it

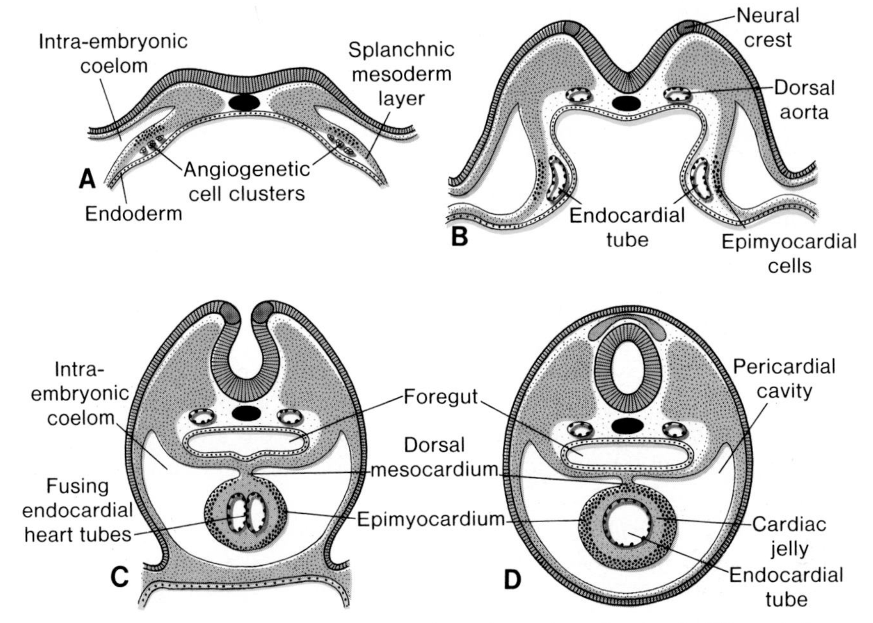

Figure 12-3. Schematic transverse sections through embryos at different stages of development, showing the formation of a single heart tube from paired primordia. *A*, Early presomite embryo (approximately 17 days); *B*, late presomite embryo (approximately 18 days); *C*, at four somites (approximately 21 days); *D*, at eight somites (approximately 22 days) (adapted from several sources).

extends over the central cardiogenic area and the future pericardial cavity (fig. 12-2). As a result the prochordal plate (future buccopharyngeal membrane) and the central portion of the cardiogenic plate are pulled forward, while, in addition, they are rotated about 180° along a transverse axis. Hence, the central portions of the cardiogenic plate and pericardial cavity become located ventrally and caudally to the buccopharyngeal membrane (fig. 12-2*C*).

Simultaneous with the cephalo-caudal flexion the originally flat embryonic shield folds in transverse direction. As a result the two lateral endothelial heart tubes come closer to each other and fuse (fig. 12-3). This fusion begins at the cephalic end of the tubes and extends in caudal direction. Hence, a single endocardial tube is formed (fig. 12-4).[1]

The developing primitive heart tube bulges gradually more and more into the pericardial cavity. Initially, however, the tube remains attached to the dorsal side of the pericardial cavity by a fold of mesodermal tissue, the **dorsal mesocardium** (figs. 12-3 and 12-5). A ventral mesocardium is never formed. With further development the dorsal mesocardium disappears also.

While these events occur the mesoderm adjacent to the endocardial tubes gradually thickens, and forms the **epimyocardial mantle** (figs. 12-3 and 12-5). This layer is at first separated from the endothelial tube by the gelatinous substance, the **cardiac jelly**.[2] Later the jelly is invaded by mesenchymal cells. Finally the wall of the heart tube consists of three layers: (1) the **endocardium**, forming the internal endothelial lining of the heart; (2) the **myocardium**, forming the muscular wall; and (3) the **epicardium** or **visceral pericardium**, covering the outside of the tube.

FORMATION OF HEART LOOP

At first the heart forms a straight tube (fig. 12-4*B*) inside the pericardial cavity (fig. 12-5). The intrapericardial part consists of the future **bulboventricular portion**. The atrial portion and the sinus venosus are still paired and lie outside the pericardium in the mesenchyme of the septum transversum (fig. 12-6*A*).

Since the bulboventricular portion of the heart tube grows much more rapidly than the pericardial cavity, elongation of the tube cannot be accomplished in a longitudinal direction and the tube is forced to bend. The cephalic portion of the loop bends in ventral and caudal direction and slightly to the right (fig. 12-6*B*, *C*), while the caudal atrial portion shifts in dorsocranial direction and slightly to the left (figs. 12-6*C* and 12-7*A*).

While the cardiac loop is being formed, local expansions become visible throughout the length of the tube. The **atrial portion**, initially a paired

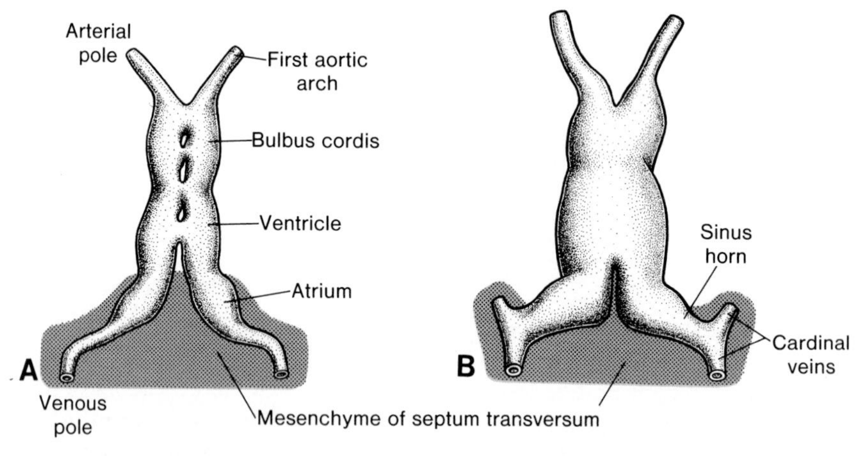

Figure 12-4. Fusion of the two heart tubes. *A*, Ventral view at approximately the 21st day; *B*, at the 22nd day. Note that the atrial region is the last to fuse and that the sinus horns are embedded in the tissue of the septum transversum.

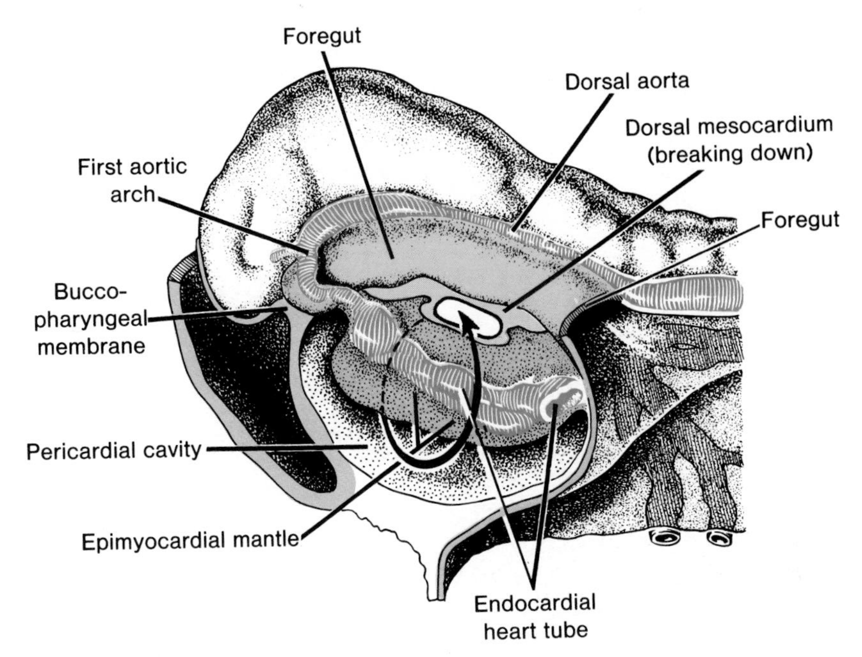

Figure 12-5. Drawing of the cephalic end of an early somite embryo. The developing endocardial heart tube and its investing layer bulge into the pericardial cavity. Note that the dorsal mesocardium is breaking down.

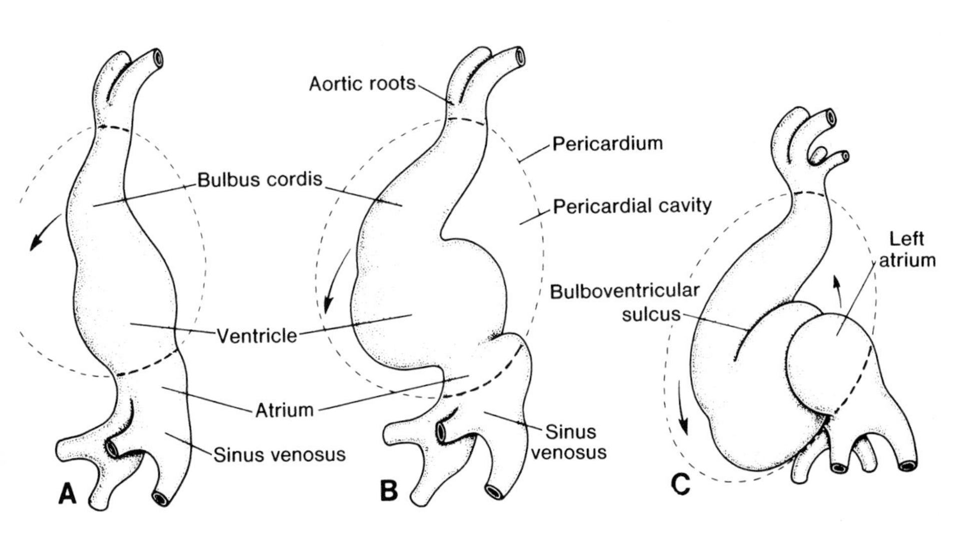

Figure 12-6. Formation of the cardiac loop. *A*, At 8 somites; *B*, 11 somites; *C*, 16 somites. Broken line indicates pericardium. Note how the atrium gradually assumes an intra-pericardial position (modified from Kramer).

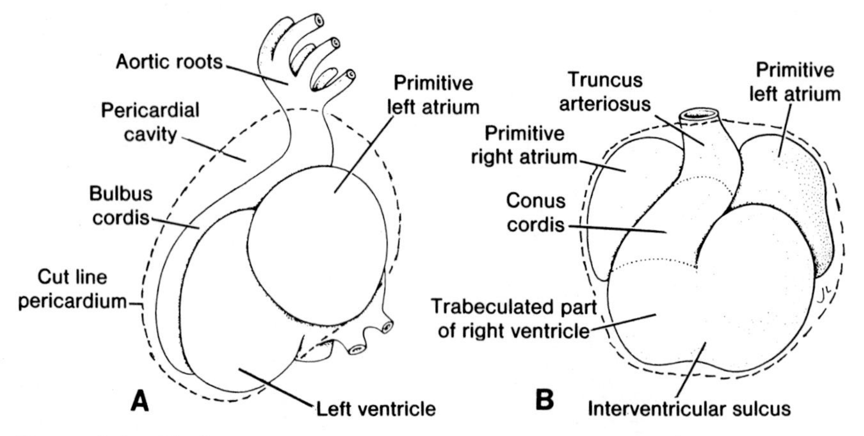

Figure 12-7. The heart of a 5-mm embryo (approximately 28 days). *A*, Seen from the left; *B*, in frontal view (modified after Kramer[2]). Note that the bulbus cordis is divided into: (1) the truncus arteriosus; (2) the conus cordis; and (3) the trabeculated part of the right ventricle.

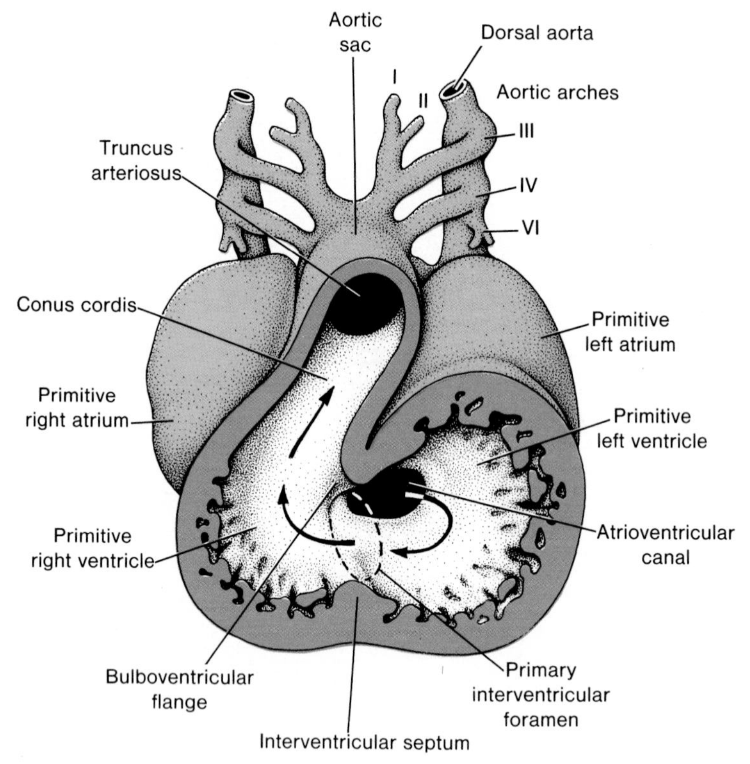

Figure 12-8. Frontal section through the heart of a 6-mm embryo showing the primary interventricular foramen and the entrance of the atrium into the primitive left ventricle. Note the bulboventricular flange. Arrow indicates direction of blood flow.

structure located outside the pericardial cavity, forms a common atrium and becomes incorporated into the pericardial cavity (fig. 12-7*A*).[3] The **atrioventricular junction** remains narrow and forms the **atrioventricular canal**, which connects the common atrium and the early embryonic ventricle (fig. 12-8). The **bulbus cordis** is narrow except for its proximal third. This portion will form the **trabeculated part of the right ventricle** (figs. 12-7*B* and 12-8). The midportion, known as the **conus cordis**, will form the outflow tracts of both ventricles. The distal part of the bulbus, the **truncus arteriosus**, will form the roots and proximal portion of the aorta and pulmonary artery (fig. 12-8). The junction between the ventricle and the bulbus cordis, externally indicated by the **bulboventricular sulcus** (fig. 12-6*C*), likewise remains narrow and is called the **primary interventricular foramen** (fig. 12-8).

At the end of the loop formation, the smooth-walled heart tube begins to form primitive trabeculae in two sharply defined areas just proximal and distal to the **primary interventricular foramen** (fig. 12-8). The atrial portion and the other portions of the bulbus remain temporarily smooth-walled. The primitive ventricle, which is now trabeculated, is called the **primitive left ventricle**, since it will form the major portion of the definitive left ventricle. Likewise, the trabeculated proximal one-third of the bulbus cordis may be referred to as the **primitive right ventricle** (fig. 12-8).

The truncoconal portion of the heart tube, initially located on the right side of the pericardial cavity, shifts gradually to a more medial position. This change in position is the result of the formation of two transverse dilatations of the atrium, bulging on each side of the bulbus cordis (figs. 12-7*B* and 12-8).

DEVELOPMENT OF SINUS VENOSUS

Since the sinus venosus greatly contributes to the definitive form of the atrium, it will be necessary to describe its development briefly.

The **sinus venosus** maintains its paired condition longer than any other portion of the heart tube. In the middle of the fourth week it consists of a small **transverse portion** and the **right** and **left sinus horns** (fig. 12-9*A*). Each horn receives blood from three important veins: (1) the **vitelline** or **omphalomesenteric vein**; (2) the **umbilical vein**; and (3) the **common cardinal vein** (see also fig. 12-40). At first the communication between the sinus and the atrium is wide. Soon, however, the entrance of the sinus shifts to the right (fig. 12-9*B*). This is primarily caused by left to right shunts of blood, which occur in the venous system during the fourth and fifth week of development.

With the obliteration of the left umbilical vein at the 5-mm stage and the left vitelline vein at the 7-mm stage, the left sinus horn rapidly loses its importance (fig. 12-9*B*). When finally the left common cardinal vein is obliterated at the 60-mm stage (10 weeks), all that remains of the left sinus horn is the **oblique vein of the left atrium** and the **coronary sinus** (fig. 12-10).

As a result of the left-to-right shunts, the right sinus horn and veins enlarge greatly. The right horn, which now forms the only communication

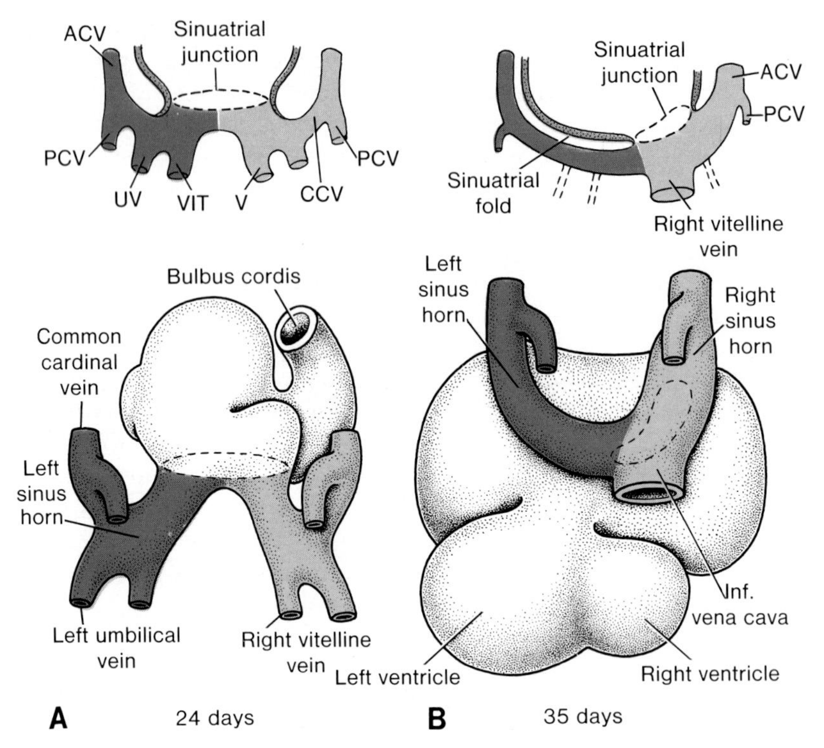

Figure 12-9. Two stages in the development of the sinus venosus, seen from dorsal. Broken line in *A* and *B* indicates the entrance of the sinus venosus into the atrial cavity. *A*, At 18 somites (approximately 24 days); *B*, approximately 35 days. Each drawing is accompanied by a scheme to show in transverse section the great veins and their relation to the atrial cavity. ACV, anterior cardinal vein; PCV, posterior cardinal vein; CCV, common cardinal vein; UV, umbilical vein; VIT V, vitelline vein (see also figs. 12-10 and 12-40).

between the original sinus venosus and the atrium, is then gradually incorporated into the right atrium to form the smooth-walled part of the right atrium (figs. 12-11 and 12-14). Its entrance, the **sinu-atrial orifice**, is flanked on each side by a valvular fold, the **right** and **left venous valves** (fig. 12-11*A*). Dorsocranially, the valves fuse, thereby forming a ridge, known as the **septum spurium** (fig. 12-11*A*). Initially the valves are large, but when the right sinus horn is incorporated into the wall of the atrium, the left venous valve and the septum spurium fuse with the developing atrial septum (fig. 12-11*B*). The superior portion of the right venous valve disappears entirely; the inferior portion develops into two parts: (1) the **valve of the inferior vena cava**, and (2) the **valve of the coronary sinus** (fig. 12-11*B*). The crista terminalis forms the dividing line between the original trabeculated part of the right atrium and the smooth walled part (sinus venarum) which originates from the right sinus horn (fig. 12-11*B*).

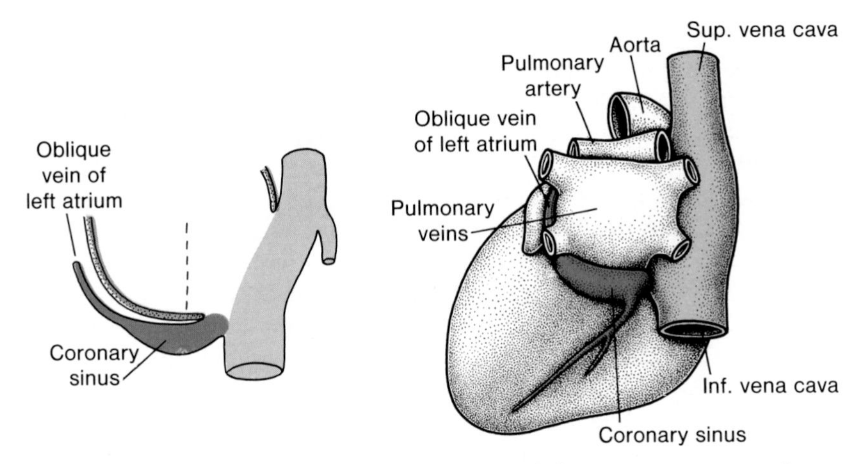

Figure 12-10. Final stage in the development of the sinus venosus and great veins.

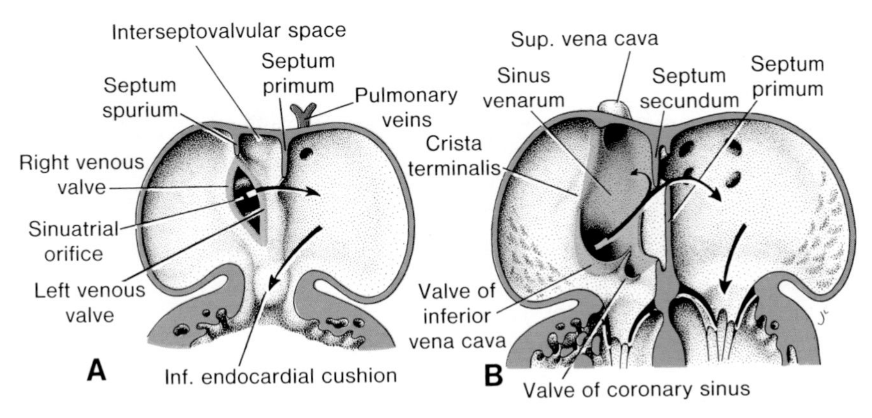

Figure 12-11. Coronal sections through the heart at the level of the atrioventricular canal seen from ventral to show the development of the venous valves. *A*, At 7- to 8-mm stage (5 weeks); *B*, newborn (modified after His). The sinus venarum, indicated in blue, is smooth walled and derived from the right sinus horn. The arrows indicate the blood flow.

Formation of Cardiac Septa

The major septa of the heart are formed between the 27th and 37th days of development, when the embryo grows in length from 5 mm to approximately 16 to 17 mm.

One method by which a septum may be formed involves two actively growing masses of tissue which approach each other until they fuse, thereby dividing the lumen into two separate canals (fig. 12-12*A*, *B*). Such a septum may also be formed by active growth of a single cell mass which continues expanding until it reaches the opposite side of the lumen (fig. 12-12*C*). **Cell proliferation is the basic prerequisite for such a septum formation.**

The other manner in which a septum may be formed does not involve cell proliferation. When, for example, a narrow strip of tissue in the wall of the atrium or ventricle would fail to grow, while the areas on each side of it would expand rapidly, a narrow ridge would be formed between the two expanding portions (fig. 12-12*D*, *E*). When growth of the expanding portions continues on either side of the narrow portion, the two walls approach each other and eventually fuse, thus forming a septum (fig. 12-12*F*). Such a septum will never completely divide the original lumen, but will leave a narrow communicating canal between the two expanded sections. It is usually closed secondarily by tissue contributed by neighboring proliferating tissues.

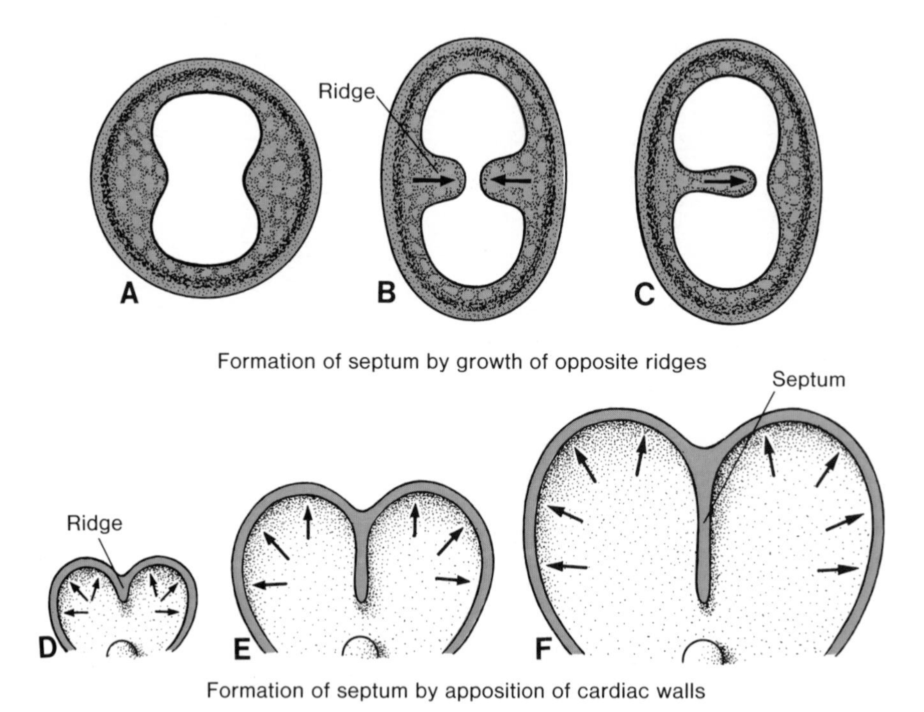

Figure 12-12. Septum formation by two actively growing ridges which approach each other until they fuse (*A* and *B*). Sometimes the septum is formed by an actively growing single cell mass (*C*). *D–F*, Septum formation by fusion of two expanding portions of the wall of the heart. Such a septum will never completely separate two cavities.

Septum Formation in Common Atrium

At the end of the fourth week a sickle-shaped crest grows from the roof of the common atrium into the lumen. This crest is considered to represent the first portion of the **septum primum** (figs. 12-11*A* and 12-13*A*, *B*). The two limbs of this septum extend in the direction of the endocardial cushions in the atrioventricular canal. The opening between the lower rim of the septum primum and the endocardial cushions is the **ostium primum** (fig. 12-13*A*, *B*). With further development extensions of the superior and inferior endocardial cushions grow along the edge of the septum primum, thereby gradually

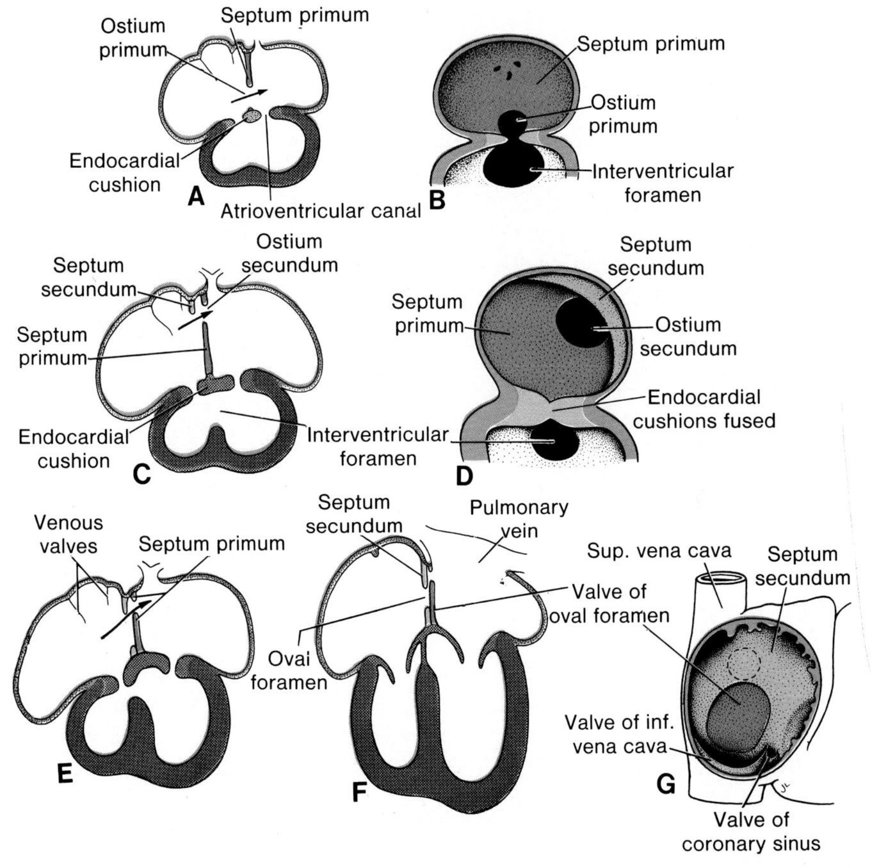

Figure 12-13. Schematic representation of the atrial septa at various stages of development. *A*, At 6 mm (approximately 30 days); *B*, same stage as in *A*, but seen from the right; *C*, 9 mm (approximately 33 days); *D*, same stage as in *C*, but seen from the right; *E*, 14 mm (approximately 37 days); *F*, newborn; *G*, view of the atrial septum seen from the right, same stage as in *F*.

closing the ostium primum (fig. 12-13*C, D*). Before closure is completed, however, perforations appear in the septum primum. When these perforations coalesce, the **ostium secundum** is formed, thus ensuring a free blood flow from the right to the left primitive atrium (fig. 12-13*B, D*).

When the lumen of the right atrium expands as a result of the incorporation of the sinus horn, a new crescent-shaped fold appears. This new fold, the **septum secundum** (fig. 12-13*C, D*), never forms a complete partition in the atrial cavity (fig. 12-13*G*). Its anterior limb extends downward to the septum in the atrioventricular canal.[4] When the left venous valve and the septum spurium fuse with the right side of the septum secundum, the free concave edge of the septum secundum begins to overlap the ostium secundum (fig. 12-11*A, B*). The opening left by the septum secundum is called the **oval foramen.** When the upper part of the septum primum gradually disappears, the remaining part of the septum primum becomes the **valve of the oval foramen**. The passage between the two atrial cavities consists of an obliquely elongated cleft (fig. 12-13*E–G*), and blood from the right atrium flows to the left side through this cleft (see arrows in figs. 12-11*B* and 12-13*E*).

After birth, when the lung circulation begins and the pressure in the left atrium increases, the valve of the oval foramen is pressed against the septum secundum, thus obliterating the oval foramen and separating the right and left atria. In about 20 per cent of the cases fusion of the septum primum and septum secundum is incomplete, and a narrow oblique cleft remains between the two atria.
This condition is known as **probe patency** of the oval foramen; it does not allow infracardiac shunting of blood.[5]

FURTHER DIFFERENTIATION OF ATRIA

While the primitive right atrium enlarges by incorporation of the right sinus horn, the primitive left atrium likewise expands greatly. Initially a single embryonic pulmonary vein develops as an outgrowth of the posterior left atrial wall, just to the left of the septum primum (fig. 12-14*A*).[3] This vein gains connection with the veins of the developing lung buds. During further development, the primitive pulmonary vein and progressively more and more of its branches become incorporated into the left atrium, thus forming the large **smooth-walled** part of the adult atrium. While initially one vein enters the left atrium, ultimately four pulmonary veins enter (fig. 12-14*B*).

In the fully developed heart, the original embryonic left atrium is represented by little more than the **trabeculated atrial appendage** while the smooth-walled part originates from the pulmonary veins (fig. 12-14). On the right side, the original embryonic right atrium becomes the trabeculated right atrial appendage containing the pectinate muscles, while the smooth-walled part (**sinus venarum**) originates from the right sinus horn.

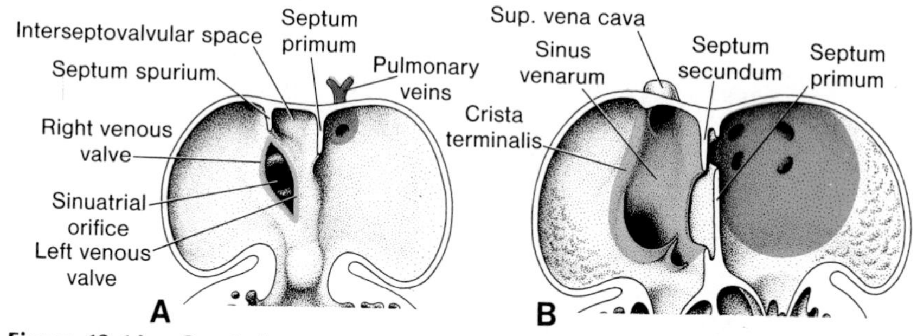

Figure 12-14. Coronal sections through the heart to show the development of the smooth-walled portions of the right and left atrium. Both the wall of the right sinus horn (blue) and the pulmonary veins (red) are incorporated into the heart to form the smooth-walled parts of the atria.

Septum Formation in Atrioventricular Canal

At the end of the fourth week two mesenchymal cushions, the **atrioventricular endocardial cushions**, appear at the superior and inferior borders of the atrioventricular canal (figs. 12-15 and 12-16). Initially the atrioventricular canal gives access only to the primitive left ventricle and is separated from the bulbus cordis by the **bulbo-(cono-) ventricular flange** (fig. 12-8). By the 9-mm stage (33rd day), however, the posterior extremity of the flange terminates almost midway along the base of the superior endocardial cushion and is much less prominent than before (fig. 12-16). Since the atrioventricular canal simultaneously enlarges to the right, blood passing through the atrioventricular orifice now has direct access to the primitive left as well as the primitive right ventricle.

In addition to the inferior and superior endocardial cushions, two other cushions, the **lateral atrioventricular cushions**, appear on the right and left borders of the canal (figs. 12-15 and 12-16). The main cushions, in the meantime, project further into the lumen and by the 10-mm stage they fuse

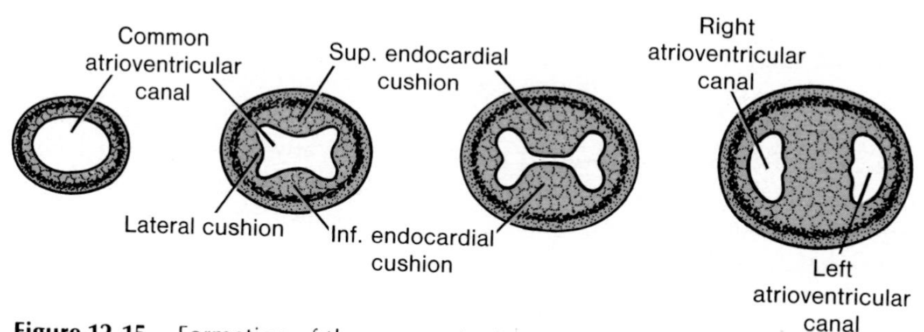

Figure 12-15. Formation of the septum in the atrioventricular canal. From left to right, 4-, 6-, 9-, and 12-mm stages, respectively. The initial circular opening becomes gradually widened in transverse direction.

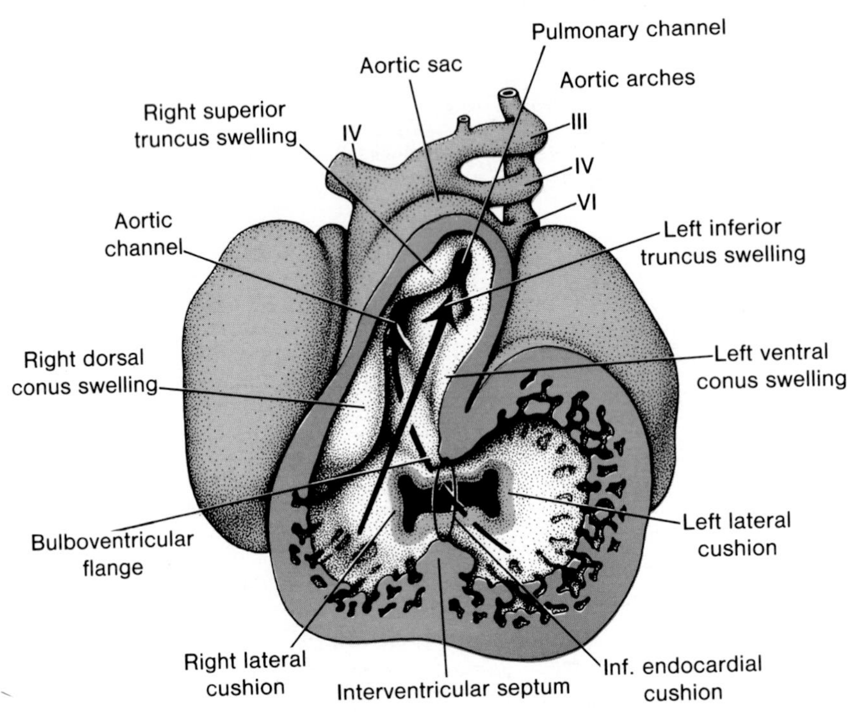

Figure 12-16. Frontal section through the heart of a 9-mm embryo. At this stage of development blood from the atrial cavity enters the left primitive ventricle as well as the right primitive ventricle. Note the development of the cushions in the A-V canal. The swellings in the truncus and conus are clearly visible. The ring indicates the primitive interventricular foramen. Arrows indicate blood flow.

with each other, resulting in a complete division of the canal into right and left atrioventricular orifices (fig. 12-15).

ATRIOVENTRICULAR VALVES

After the endocardial cushions have fused, each atrioventricular orifice is surrounded by localized proliferations of mesenchymal tissue (fig. 12-17*A*). When the tissue located on the ventricular surface of these proliferations becomes hollowed out by the bloodstream, the newly formed valves remain attached to the ventricular wall by muscular cords only (fig. 12-17*B*). Finally the muscular tissue in the cords degenerates and is replaced by dense connective tissue. The valves then consist of connective tissue covered by endocardium and are connected to thickened trabeculae in the wall of the ventricle, the **papillary muscles**, by means of the **chordae tendineae** (fig. 12-17*C*). In this manner two valve leaflets are formed in the left atrioventricular canal: the **bicuspid** or **mitral valve;** and three on the right side, the **tricuspid valve.**

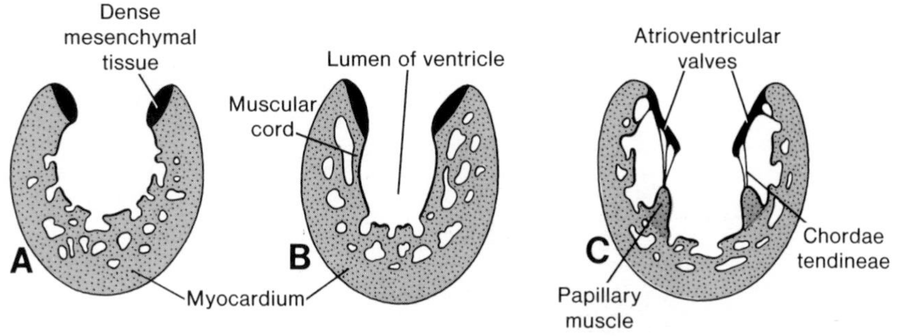

Figure 12-17. Formation of the atrioventricular valves and chordae tendineae. Note how the valves are hollowed out from the ventricular side, but remain attached to the ventricular wall by the chordae tendineae.

ABNORMALITIES OF ATRIAL SEPTUM

Atrial septal defects belong to the most common congenital heart abnormalities. One of the most significant defects is the **ostium secundum** defect. This anomaly is characterized by a large opening between the left and right

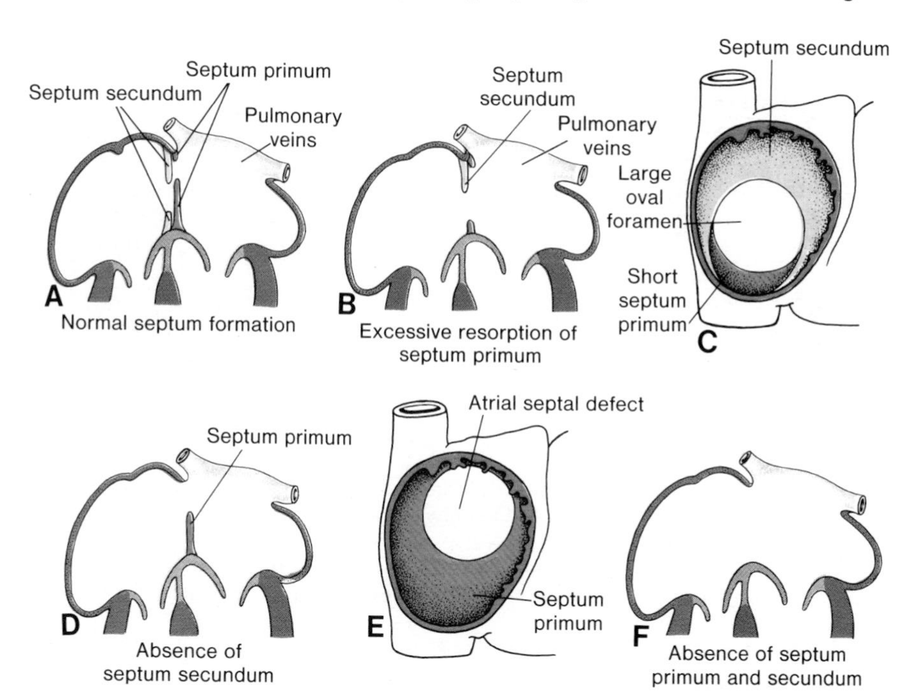

Figure 12-18. *A*, Normal atrial septum formation, *B*, *C*, Ostium secundum defect caused by excessive resorption of the septum primum. *D*, *E*, Similar defect caused by failure of development of the septum secundum. *F*, Common atrium or cor trilocular biventriculare—complete failure of the septum primum and septum secundum to form.

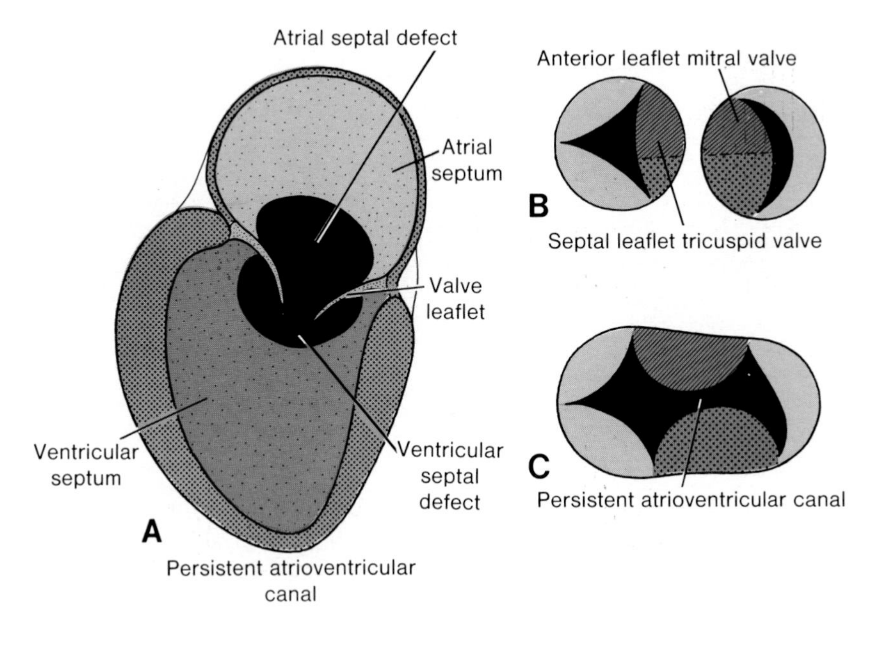

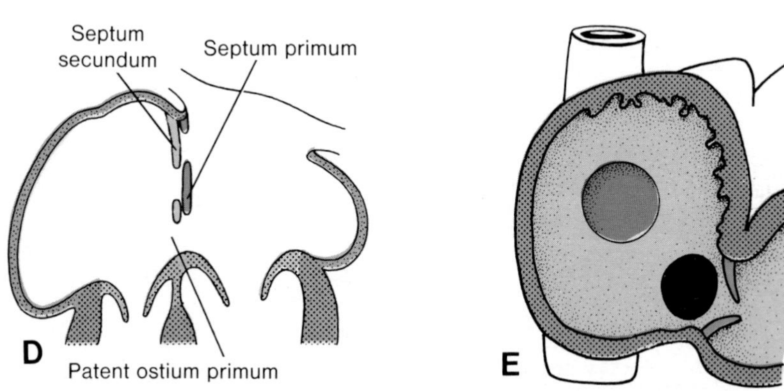

Figure 12-19. *A*, Persistent common atrioventricular canal. This abnormality is always accompanied by a septum defect in the atrial as well as in the ventricular portion of the cardiac partition. *B*, Valves in the atrioventricular orifices under normal conditions. *C*, Split valves in case of a persistent atrioventricular canal. *D*, and *E*, Ostium primum defect caused by incomplete fusion of the atrioventricular endocardial cushions.

atria and is caused either by excessive resorption of the septum primum (fig. 12-18*B*, *C*) or by inadequate development of the septum secundum (fig. 12-18*D*, *E*). Depending on the size of the opening, considerable intracardiac shunting may occur from left to right.[6]

The most serious abnormality in this group is the complete absence of the atrial septum (fig. 12-18*F*). This condition, known as **common atrium** or **cor triloculare biventriculare**, is always associated with serious defects elsewhere in the heart.[7]

Occasionally the oval foramen closes during prenatal life. This abnormality, known as **premature closure of the oval foramen**, leads to massive hypertrophy of the right atrium and ventricle, and underdevelopment of the left side of the heart.[8] Death usually occurs shortly after birth.

ABNORMALITIES OF ATRIOVENTRICULAR CANAL

The endocardial cushions of the atrioventricular canal not only divide this canal into a right and left orifice, but also participate in the formation of the membranous portion of the interventricular septum and in the closure of the ostium primum. Hence, whenever the cushions fail to fuse, the result is a **persistent atrioventricular canal**, combined with a defect in the cardiac septum (fig. 12-19*A*). This septal defect has an atrial and ventricular component, separated by abnormal valve leaflets in the single atrioventricular orifice (fig. 12-19*C*).

Occasionally the endocardial cushions in the atrioventricular canal fuse only partially. The defect in the atrial septum is then similar to that in the above-described abnormality, but the interventricular septum is closed (fig. 12-19*D*, *E*). This defect, known as the **ostium primum defect**, is usually combined with a cleft in the anterior leaflet of the tricuspid valve (fig. 12-19*C*).[10]

Another important abnormality is the obliteration of the right atrioventricular orifice. This abnormality, known as **tricuspid atresia** (fig. 12-20*B*), is characterized by the absence or fusion of the tricuspid valves. The defect

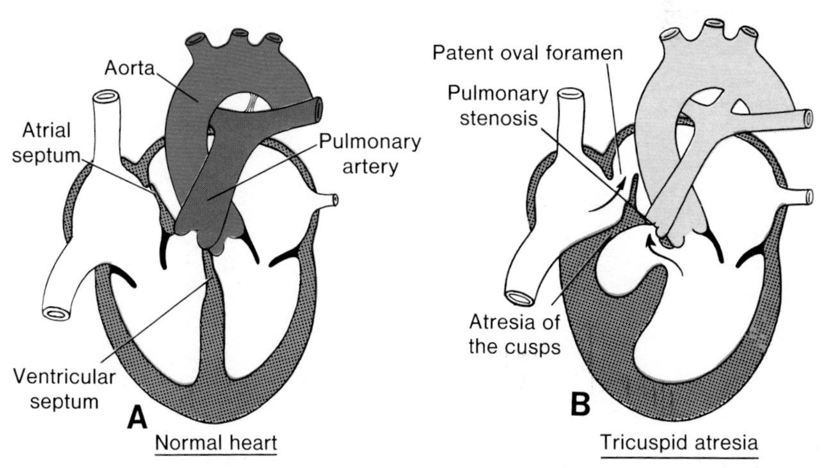

Figure 12-20. *A*, Drawing of normal heart. *B*, Tricuspid atresia. Note the small right ventricle and the large left ventricle.

is always associated with: (1) patency of the oval foramen; (2) ventricular septal defect; (3) underdevelopment of the right ventricle; and (4) hypertrophy of the left ventricle.[11]

Septum Formation in Ventricles

By the end of the fourth week, the two primitive ventricles begin to dilate. This is accomplished by continuous growth of the myocardium on the outside and continuous diverticulation and trabecula formation on the inside (figs. 12-8, 12-16, and 12-24).

The medial walls of the expanding ventricles become apposed and gradually fuse together, thus forming the **muscular interventricular septum** (fig. 12-24). Sometimes the fusion between the two walls is not complete, which manifests itself as a more or less deep apical cleft between the two ventricles. The space between the free rim of the muscular ventricular septum and the fused endocardial cushions permits communication between the two ventricles. This opening is later closed by the **membranous interventricular septum.**

Septum Formation in Truncus Arteriosus and Conus Cordis

During the fifth week a pair of opposing ridges appear in the cephalic part of the truncus. These ridges, the **truncus swellings**, are located on the right superior wall (**right superior truncus swelling**) and on the left inferior wall (**left inferior truncus swelling**) (fig. 12-16). The right superior truncus swelling grows distally and to the left, while the left inferior swelling grows distally and to the right. Hence, while growing in the direction of the aortic sac, the swellings twist around each other thus foreshadowing the spiral course of the future septum (fig. 12-21). After complete fusion the ridges form a septum known as the **aorticopulmonary septum**, dividing the truncus into an **aortic** and **pulmonary channel.**

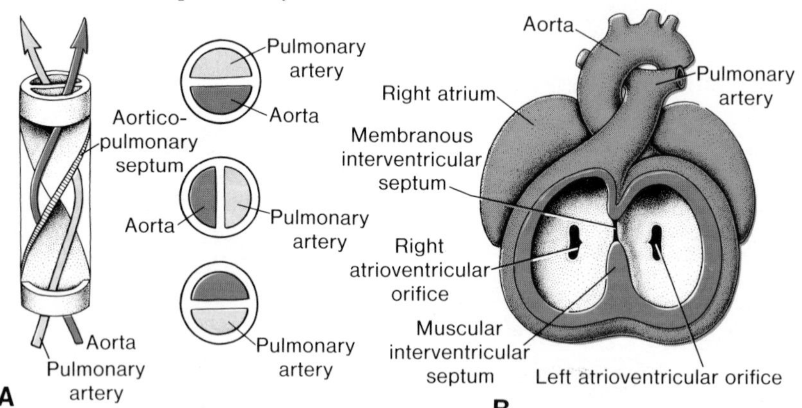

Figure 12-21. *A*, Diagram to show the spiral shape of the aorticopulmonary septum. *B*, Position of aorta and pulmonary artery at 25-mm stage (eighth week). Note how the aorta and pulmonary artery twist around each other.

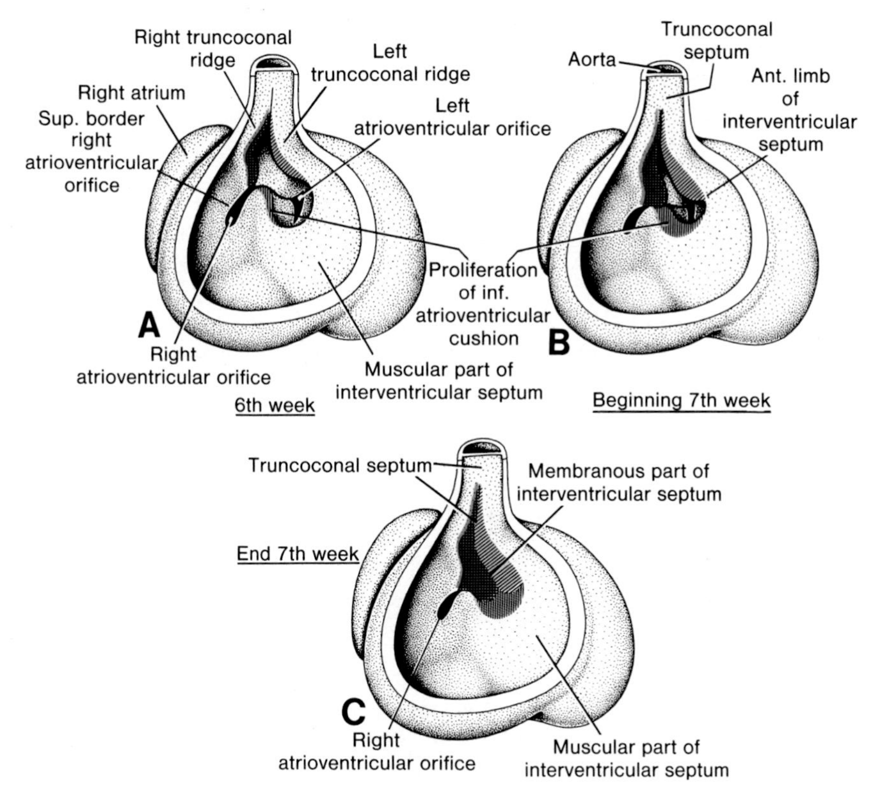

Figure 12-22. Schematic drawings showing the development of the truncoconal ridges and the closure of the interventricular foramen. The proliferations of the right and left conus swellings, combined with the proliferation of the inferior atrioventricular cushion, eventually close the interventricular foramen and form the membranous portion of the interventricular septum. *A*, At six weeks (12 mm); *B*, beginning of seventh week (14.5 mm); *C*, end of seventh week (20 mm) (modified after Hamilton, Boyd, and Mossman).

At about the time that the truncus swellings appear, similar swellings develop along the right dorsal and left ventral walls of the **conus cordis** (fig. 12-16). After the truncus septum has been completed, the conus swellings grow toward each other and in a distal direction to unite with the truncus septum.

The proximal extremity of the right conus swelling terminates at the superior border of the right atrioventricular orifice (figs. 12-22 and 12-23). The left conus swelling extends proximally along the right side of the anterior limb of the muscular interventricular septum (fig. 12-22).

When the two conus swellings have fused, the septum divides the conus into an anterolateral portion the outflow tract of the right ventricle (fig. 12-23), and a posteromedial portion, which forms the outflow tract of the left ventricle (fig. 12-24).

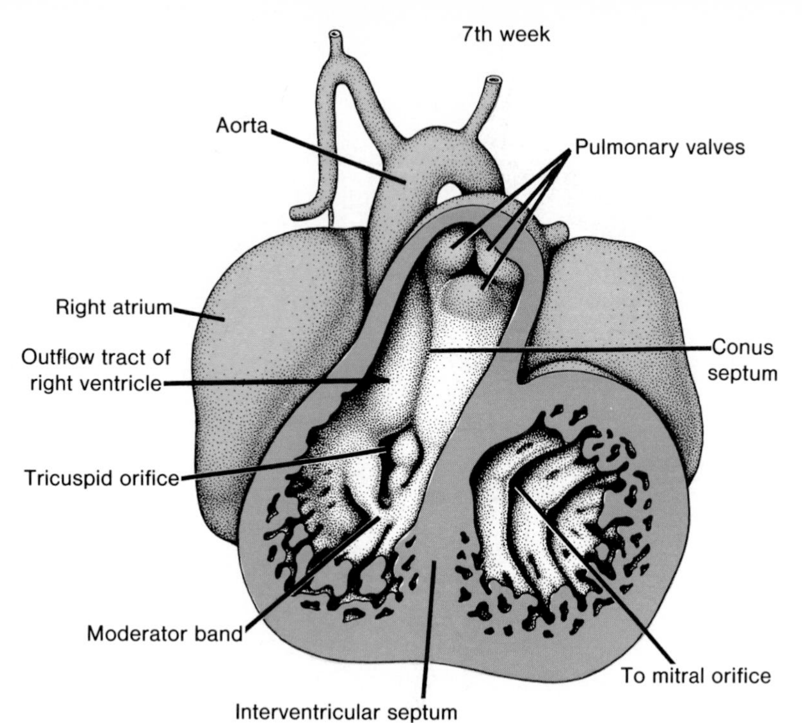

Figure 12-23. Frontal section through the heart of a 16-mm embryo. Note the conus septum and the position of the pulmonary valves.

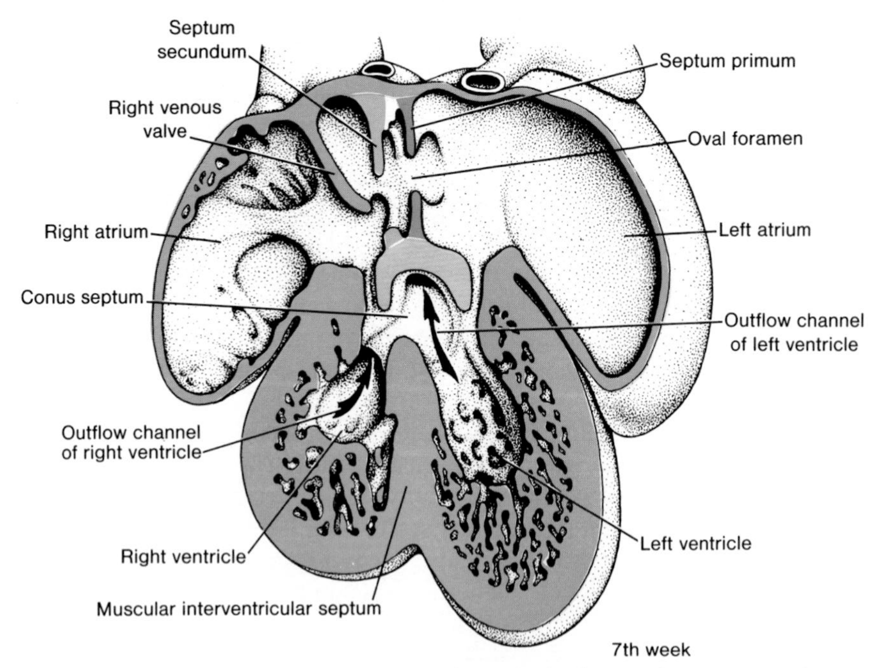

Figure 12-24. View of a frontal section through the heart of a 15-mm embryo. The conus septum is completed and blood from the left ventricle enters the aorta. Note the septa in the atrial region.

The interventricular foramen, found above the muscular interventricular septum, becomes reduced in size with the completion of the conus septum (fig. 12-22). During further development, closure of the foramen is accomplished by the outgrowth of tissue from the inferior endocardial cushion along the top of the muscular interventricular septum (fig. 12-22).[12] This tissue fuses with the abutting parts of the conus septum. After complete closure the interventricular foramen becomes the **membranous part of the interventricular septum.**

Semilunar Valves. When the partitioning of the truncus has almost been completed, the primordia of the semilunar valves become visible as small tubercles. These tubercles are found on the main truncus swellings and one of each pair is assigned to the pulmonary and aortic channels, respectively (fig. 12-25). Opposite the fused truncus swellings, a third tubercle appears in both channels. Gradually the tubercles are hollowed out at their upper surface, thus forming the **semilunar valves** (fig. 12-26). This process is already well advanced at the 16-mm stage and virtually completed in the 40-mm embryo.

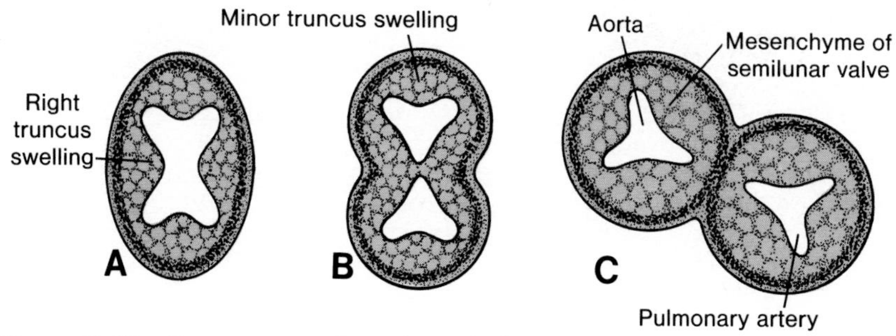

Figure 12-25. Transverse sections through the truncus arteriosus at the level of the semilunar valves at five, six, and seven weeks of development (*A*, *B*, and *C*, respectively) (after Kramer[2]).

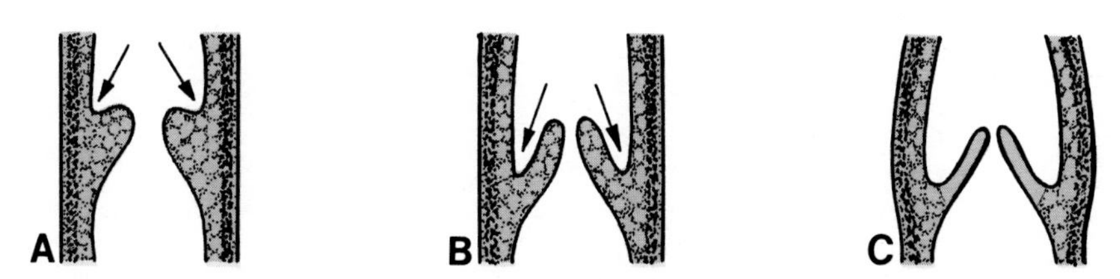

Figure 12-26. Longitudinal sections through the semilunar valves. From left to right, sixth, seventh, and ninth weeks of development, respectively.

ABNORMALITIES OF INTERVENTRICULAR SEPTUM

Considering the complicated formation of the membranous part of the interventricular septum, it is not surprising that defects easily arise. Indeed, a **defect of the membranous septum** is frequently seen (fig. 12-27). Although it may be found as an isolated lesion, the defect is also often associated with abnormalities in the partition of the truncoconal region. Depending on the size of the opening, the blood carried by the pulmonary artery may be 1.2 to 1.7 times more abundant than that carried by the aorta.[9] Occasionally the

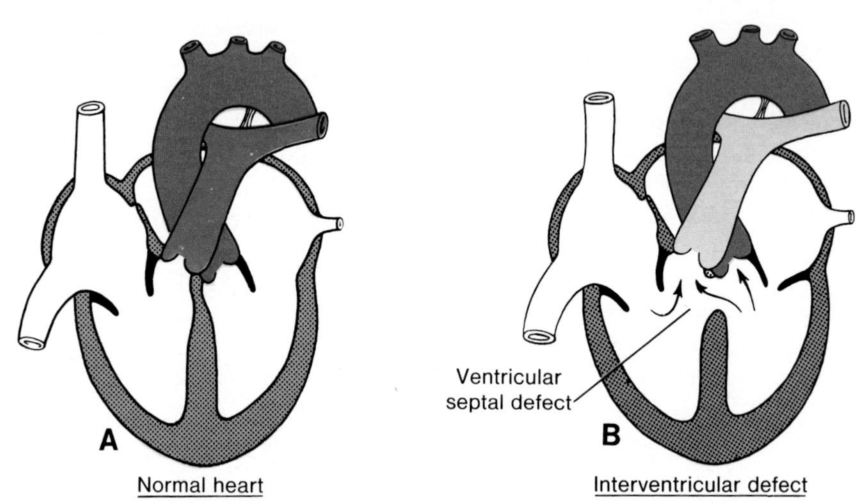

Figure 12-27, *A*, Drawing of normal heart. *B*, Isolated defect in the membranous portion of the interventricular septum. Note the blood of the left ventricle flows to the right through the interventricular foramen.

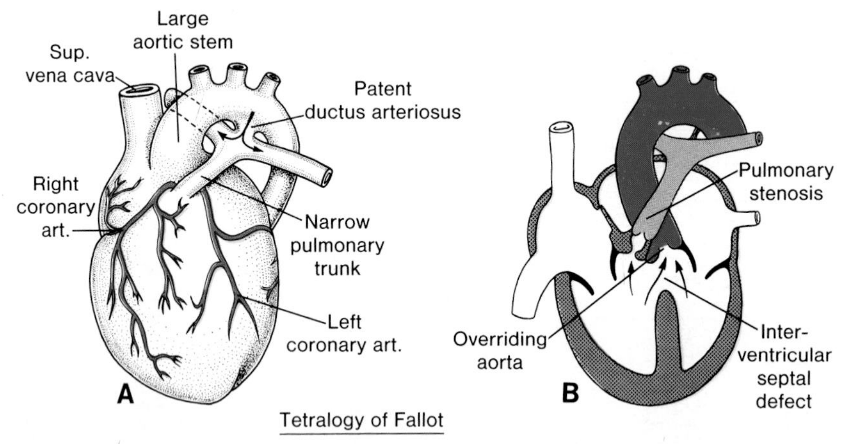

Figure 12-28. Tetralogy of Fallot. *A*, Surface view. *B*, Schematic drawing to show: (1) stenosis of the pulmonary tract; (2) overriding aorta; (3) interventricular septal defect; and (4) hypertrophy of the right ventricle.

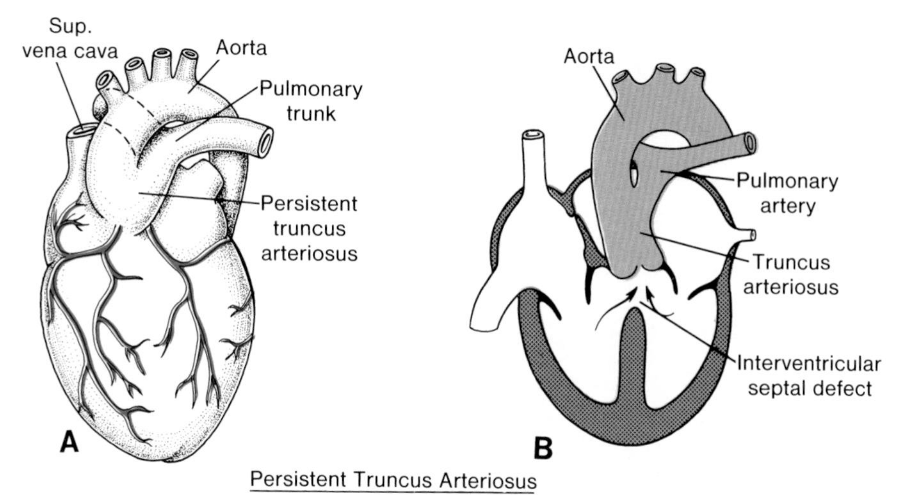

Figure 12-29. Persistent truncus arteriosus. The pulmonary artery originates from the common truncus. The septum in truncus and conus has failed to form. This abnormality is always combined with an interventricular septal defect.

defect is not restricted to the membranous part, but involves the muscular part of the septum as well.[13]

ABNORMALITIES OF TRUNCUS AND CONUS

The most frequently seen abnormality of this region is an unequal division of the conus, due to an anterior displacement of the truncoconal septum. This results in a narrow right ventricular outflow region, that is, a **pulmonary infundibular stenosis,** and a large defect of the interventricular septum (fig. 12-28). The aorta arises directly above the septal defect from both ventricular cavities, and the resulting higher pressure on the right side causes hypertrophy of the right ventricular wall. This abnormality known as the **tetralogy of Fallot**, is the most important type of malformations causing cyanosis. The defect is compatible with life.[14, 15]

When the truncoconal ridges fail to fuse and to descend toward the ventricles, a **persistent truncus arteriosus** results (fig. 12-29). In such a case the pulmonary artery arises some distance above the origin of the undivided truncus. Since the ridges also participate in the formation of the interventricular septum, the persistent truncus is always accompanied by a defective interventricular septum. The undivided truncus thus overrides both ventricles and receives blood from both sides.

Sometimes the truncoconal septum fails to follow its normal spiral course, and descends straight downward (fig. 12-30*A*). As a consequence the aorta originates from the right ventricle, and the pulmonary artery from the left. This condition, known as **transposition of the great vessels**, is one of the most common cardiac abnormalities.[16] Sometimes it is associated with a

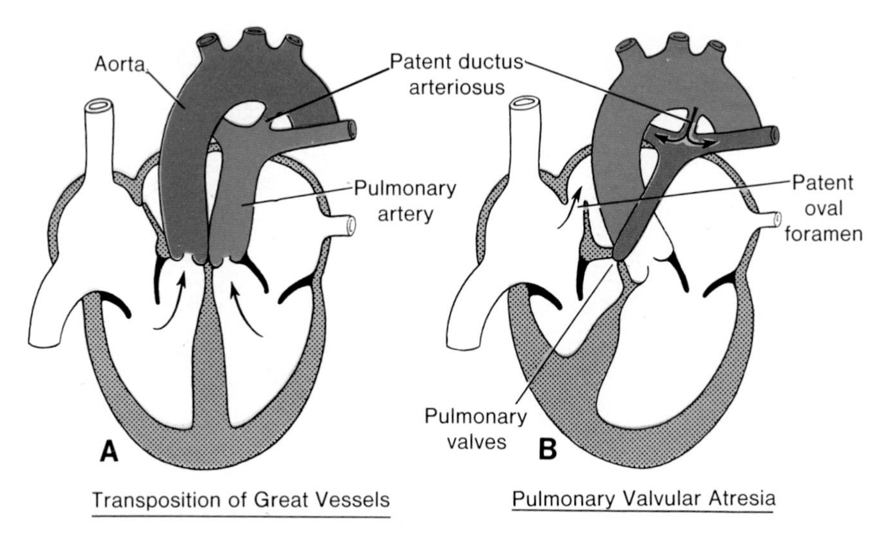

Figure 12-30. *A*, Transposition of the great vessels. *B*, Pulmonary atresia with normal aortic root. The only access route to the lungs is by way of the patent ductus arteriosus.

defect in the membranous part of the interventricular septum. It is usually combined with an open ductus arteriosus.

ABNORMALITIES OF SEMILUNAR VALVES

In this important group of abnormalities the semilunar valves of the pulmonary artery or aorta are fused for a variable distance and may even form an imperforated diaphragm. In case of a **valvular stenosis of the pulmonary artery**, the trunk of the pulmonary artery is narrow or even atretic (fig. 12-30*B*)[17] The patent oval foramen then forms the only outlet for blood from the right side of the heart. The ductus arteriosus is always patent and represents the only access route to the pulmonary circulation.

In case of **aortic valvular stenosis** (fig. 12-31*A*), the fusion of the thickened valves may be so complete that only a pinhole opening remains. The size of the aorta itself, however, is usually normal.[18]

When fusion of the semilunar aortic valves is complete—a condition known as **aortic valvular atresia** (fig. 12-31*B*)—the aorta, left ventricle, and left atrium are markedly underdeveloped. The abnormality is usually accompanied by a wide opening ductus arteriosus, which delivers blood into the aorta.[19]

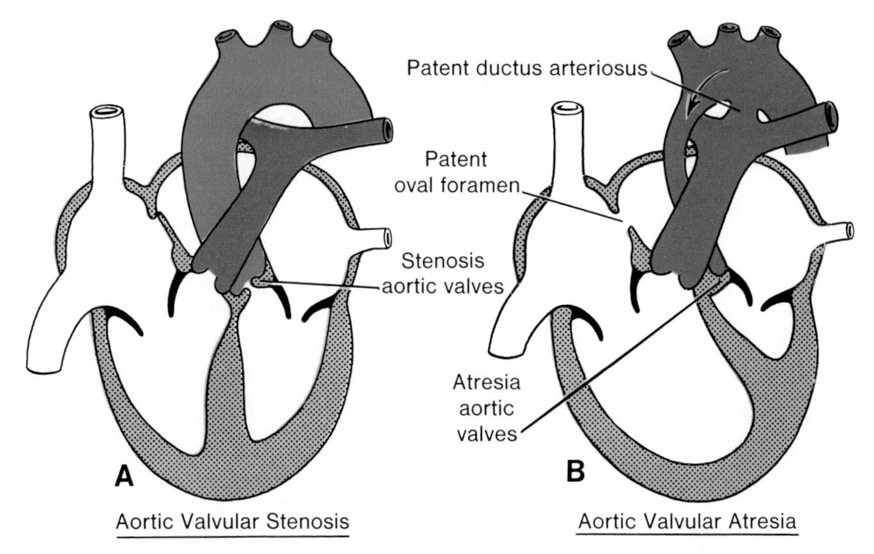

Figure 12-31. *A*, Aortic valvular stenosis. *B*, Aortic valvular atresia. Arrow in arch of the aorta indicates direction of blood flow. The coronary arteries are supplied by this retroflux. Note the small left ventricle and the large right ventricle.

ABNORMALITIES IN HEART POSITION

In addition to the malformations described above, the position of the heart itself may be abnormal. Most frequently seen is **dextrocardia.** In this condition the heart is located in the right side of the thorax and the abnormality is usually associated with a total or partial **situs inversus.**

A rare anomaly is **ectopia cordis.** The heart is then located on the surface of the chest. Basically this malformation is caused by failure of the embryo to close in the midline (sternal cleft).

CAUSES OF CARDIOVASCULAR ABNORMALITIES

Fetal Environment. German measles affecting pregnant women in the first 10 weeks of gestation is frequently followed by cataract, deafness, and congenital heart malformations in the newborn. The risk of giving birth to a defective infant following infection has been estimated at approximately 10 per cent.[20] In order of frequency the following heart and vessel abnormalities are seen: **persistent ductus arteriosus**, ventricular septal defect, Fallot's tetralogy, atrial septal defect, and pulmonary valvular stenosis.[21–22]

Genetic Factors. There are many reports of family pedigrees showing more than one member affected with congenital heart malformations; this suggests that these malformations may be transmitted genetically. In contrast to these reports, however, recent authors are less inclined to incriminate genetic factors. Even though the familial incidence of cardiac defects has been noted in some cases, no definite mode of inheritance has so far been determined.[23–26]

SUMMARY

The entire cardiovascular system, the heart, blood vessels, and blood cells originate from the mesodermal germ layer. Although initially paired by the 22nd day of development the two tubes (figs. 12-4 and 12-5) form a single, slightly bent heart tube (fig. 12-6) consisting of an inner endocardial tube and a surrounding epimyocardial mantle. During the fourth to the seventh week the heart becomes divided in a typical four-chambered structure.

Septum Formation in the Atrium. The **septum primum**, a sickle-shaped crest descending from the roof of the atrium, never divides the atrium in two, but leaves a lumen "**ostium primum**" for communication between the two (fig. 12-13). Later a **septum secundum** is formed, but it also fails to split the common atrium into two parts. Only **at birth**, when pressure in the left atrium increases, are the two septa pressed against each other and is any communication between the two closed. Abnormalities in the atrial septum may consist of total absence (fig. 12-18) to a small opening known as the oval foramen.

Septum Formation in the Atrioventricular (A-V) Canal. Two large endocardial cushions divide the A-V canal into the right tricuspid and the left bicuspid or mitral A-V canal (fig. 12-16). Persistence of the common atrioventricular canal (fig. 12–19) or abnormal division such as atresia of the tricuspid canal (fig. 12-20B) are well-known defects.

Septum Formation in the Ventricles. The interventricular septum consists of a thick **muscular** part and a thin **membranous** portion (fig. 12-21) formed by (1) an inferior endocardial A-V cushion; (2) the right conus swelling; and (3) the left conus swelling (fig. 12-22). In many cases these three components fail to fuse, resulting in an open interventricular foramen. Although this abnormality may be isolated, it is frequently combined with other compensatory defects (fig. 12-27 and 12-28).

Septum Formation in the Bulbus. The bulbus is divided into (1) the truncus (aorta and pulmonary trunk); (2) the conus (outflow tract of the aorta and pulmonary trunk); and (3) the trabeculated portion of the right ventricle. The truncus region is divided by the spiral-shaped **aorticopulmonary septum** into the two main arteries (fig. 12-21). The conus swellings divide the outflow tracts of the aortic and pulmonary channels and close the interventricular foramen (fig. 12-23). Many vascular abnormalities such as **transposition of the great vessels** and **pulmonary valvular atresia** result from abnormal division of the truncoconal region.

REFERENCES

1. Goss, C. M. The development of the median coordinated ventricle from the lateral hearts in rat embryos with three to six somites. *Anat. Rec., 112:* 761, 1952.
2. Kramer, T. C. The partitioning of the truncus and conus and the formation of the membranous portion of the interventricular septum in the human heart. *Am. J. Anat., 71:* 343, 1942.
3. Los, J. A. The development of the pulmonary veins and the coronary sinus in the human embryo. Doctoral thesis, Unviersity of Leyden, The Netherlands, 1958.
4. Van Mierop, L. H. S., Alley, R. D., Kausel, M. W., and Stranahan, A. The anatomy and embryology of endocardial cushion defects. *J. Thorac. Cardiovasc. Surg., 43:* 71, 1962.
5. Wright, R. R., Anson, B. J., and Cleveland, H. C. The vestigial valves and the interatrial foramen of the adult human heart. *Anat. Rec., 100:* 331, 1948.
6. Dexter, L. Atrial septal defects. *Br. Heart J., 18:* 209, 1956.
7. Rogers, H. M., and Edwards, J. E. Cor triloculare biventriculare. *Am. Heart J., 45:* 623, 1953.
8. Campbell, M. Natural history of atrial septal defect. *Br. Heart J., 32:* 820, 1970.
9. Edwards, J. E. Congenital malformations of the heart and great vessels. In *Pathology of the Heart,* edited by S. E. Gould, p. 266. Charles C Thomas, Springfield, Ill., 1953.
10. Rogers, H. M., and Edwards, J. E. Incomplete division of the atrioventricular canal with patent interatrial foramen primum. *Am. Heart J., 36:* 28, 1948.
11. Edwards, J. E., and Burchell, H. B. Congenital tricuspid atresia; a classification. *Med. Clin. North Am., 33:* 1177, 1949.
12. Odgers, P. N. B. The development of the parts membranacea septi in the human heart. *J. Anat., 72:* 247, 1939.
13. Mason, D. G., and Hunter, W. C. Localized congenital defects of the cardiac interventricular septum. *Am. J. Pathol., 13:* 835, 1937.
14. Brinton, W. D., and Campbell, M. Necropsies in some congenital disease of the heart, mainly Fallot's tetralogy. *Br. Heart J., 15:* 335, 1953.
15. Baffes, T. G., Johnson, F. R., Pott, W. J., and Gibson, S. Anatomic variations in tetralogy of Fallot. *Am. Heart J., 46:* 657, 1953.
16. Van Mierop, L. H. S. Transposition of the great arteries. I. Clarification of further confusion. *Am. J. Cardiol., 28:* 735, 1971.
17. Greenwold, P. Congenital pulmonary atresia with intact ventricular septum. In *Proceedings of the 29th Scientific Session of the American Heart Association,* p. 51, 1956.
18. Campbell, M., and Kauntze, R. Congenital valvular stenosis. *Br. Heart J., 15:* 179, 1953.
19. Monie, J. W., and De Pape, A. D. J. Congenital aortic atresia. *Am. Heart J., 40:* 595, 1950.
20. Warkany, J. Etiologic factors of congenital heart disease. In *Congenital Heart Disease,* edited by H. W. Kaplan and S. J. Robinson, p. 83. American Association for the Advancement of Science, Washington, D.C., 1960.
21. Campbell, M. Place of maternal rubella in the aetiology of congenital heart diseae. *Br. Med. J., 1:* 5227, 1961.
22. Bell, J. On rubella in pregnancy. *Br. Med. J., 1:* 1302, 1959.
23. McKeown, T., MacMahon, B., and Parsons, C. G. The familial incidence of congenital malformations of the heart. *Br. Heart J., 15:* 273, 1953.
24. Polani, P. E., and Campbell, M. An etiological study of congenital heart disease. *Ann. Hum. Genet., 19:* 209, 1955.
25. Uchida, J. A., and Rowe, R. D. Discordant heart anomalies in twins. *Am. J. Hum. Genet., 9:* 133, 1957.
26. Book, J. A. Heredity and heart disease. *Am. J. Public Health, 50:* 1, 1960.

Arterial System

AORTIC ARCHES

When the branchial arches are formed during the fourth and fifth week of development (see fig. 16-2), each arch receives its own cranial nerve (see fig. 16-5) and its own artery (see fig. 16-2). These arteries are known as **aortic arches** and arise from the **aortic sac**, the most distal part of the truncus arteriosus (figs. 12-8 and 12-32). They are embedded in the mesenchyme of the branchial arches and terminate in the dorsal aortae. With the formation of the successive branchial arches the aortic sac contributes a branch to each new arch, thus giving rise to a total of six pairs of arteries (fig. 12-32). During further development this arterial pattern becomes greatly modified and some vessels retrogress altogether. The fifth arch is often underdeveloped or not formed at all.

In a 4-mm embryo the **first aortic arch** has largely disappeared (fig. 12-33). A small portion, however, persists to form the **maxillary artery**. The **second aortic arch** similarly will disappear soon. The remaining portions of this arch are the **hyoid** and **stapedial arteries**.[1] The third arch is large; the fourth and sixth arches are in the process of formation. Even though the sixth arch is not completed, the **primitive pulmonary artery** is already present as a major branch (fig. 12-33*A*).

In a 10-mm embryo the first two aortic arches have disappeared (fig. 12-33*B*). The third, fourth, and sixth arches are large. The trunco-aortic sac has

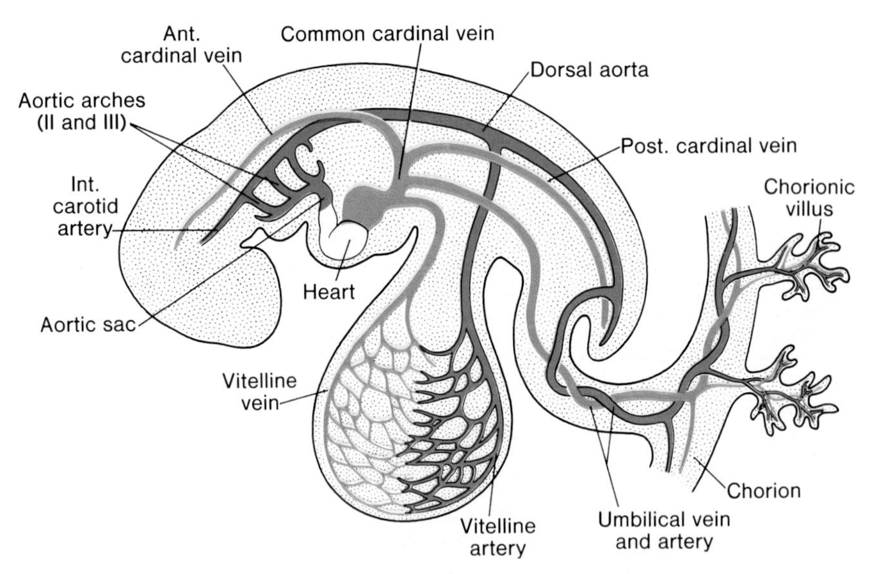

Figure 12-32. Schematic drawing of the main intra- and extra-embryonic arteries (red) and veins in a 4-mm embryo (end of the fourth week). Only the vessels on the left side of the embryo are represented.

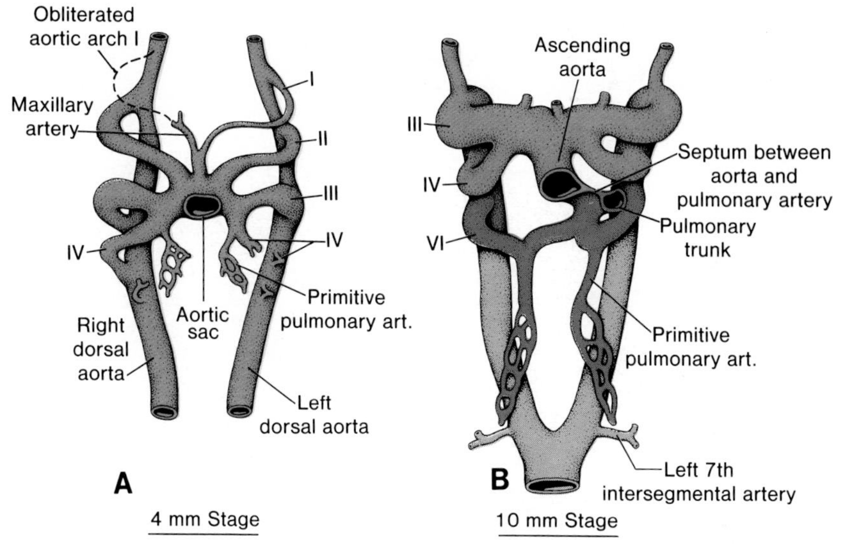

Figure 12-33. *A,* Drawing of the aortic arches at the 4-mm stage. The first aortic arch is obliterated before the sixth arch has fully developed. *B,* Aortic arch system at the 10-mm stage. Note the aorticopulmonary septum and the large primitive pulmonary arteries.

been divided so that the sixth arches are now continuous with the pulmonary trunk.

Since during further development the aortic arch system gradually loses its original symmetrical form, the basic vascular pattern is schematically shown in figure 12-34*A*, while the definitive pattern is given in figure 12-34*B* and *C*. This representation may be helpful in understanding the transformation from the original into the adult arterial system.

The following changes occur:

The **third aortic arch** forms the **common carotid artery** and the first part of the **internal carotid artery**. The remainder of the internal carotid is formed by the cranial portion of the dorsal aorta. The **external carotid artery** is a sprout of the third aortic arch.

The **fourth aortic arch** persists on both sides, but its ultimate fate is different on the right and left sides. On the left it forms part of the arch of the aorta, between the left common carotid and the left subclavian arteries. On the right it forms the most proximal segment of the right subclavian artery, the distal part of which is formed by a portion of the right dorsal aorta and the seventh intersegmental artery (fig. 12-34*B*). The **fifth aortic arch** is transient and is never well developed.

The **sixth aortic arch,** also known as the **pulmonary arch**, gives off an important branch which grows toward the developing lung bud (fig. 12-33*B*). On the right side the proximal part becomes the proximal segment of the right pulmonary artery. The distal portion of this arch loses its connection

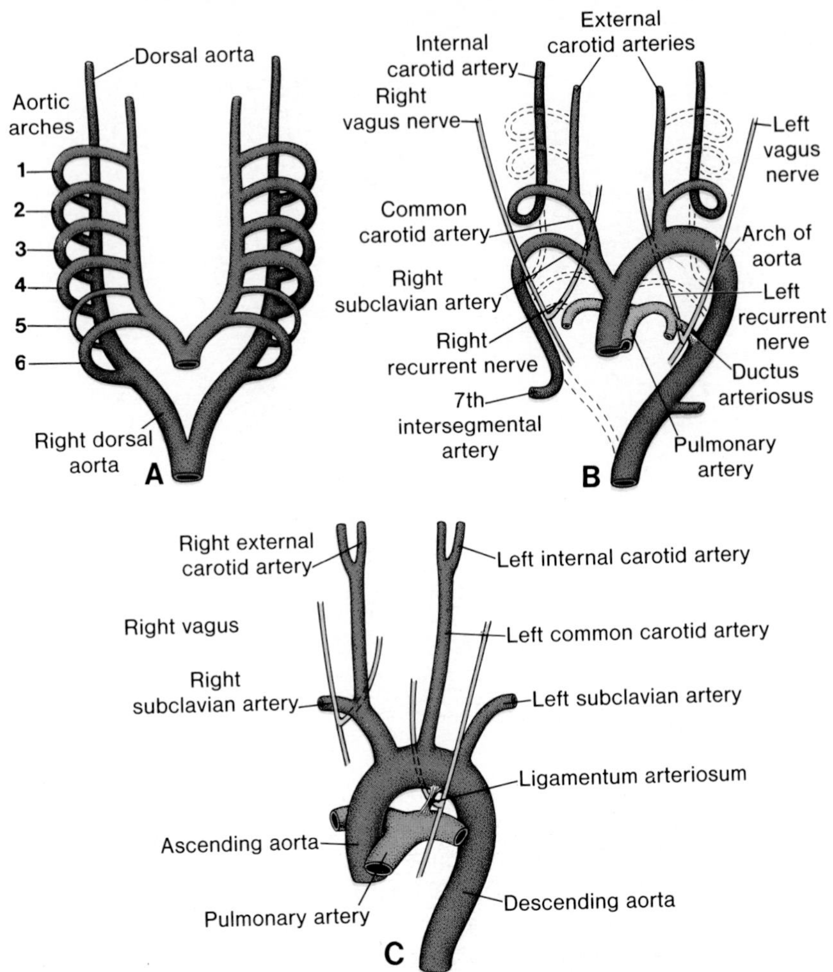

Figure 12-34. *A*, Diagram of the aortic arches and dorsal aortas before transformation into the definitive vascular pattern. *B*, Diagram of the aortic arches and dorsal aortas after the transformation. The obliterated components are indicated by broken lines. Note the patent ductus arteriosus and the position of the seventh intersegmental artery on the left. *C*, The great arteries in the adult. Compare the distance between the place of origin of the left common carotid artery and the left subclavian in *B* and *C*. After disappearance of the distal part of the sixth aortic arch and the fifth arch on the right, the right recurrent laryngeal nerve hooks around the right subclavian artery. On the left the nerve remains in place and hooks around the ligamentum arteriosum.

with the dorsal aorta and disappears. On the left, the distal part persists during intra-uterine life as the **ductus arteriosus**.

OTHER CHANGES IN ARCH SYSTEM

Simultaneously with the alterations in the aortic arch system, a number of other changes occur. (1) The dorsal aorta located between the entrance of

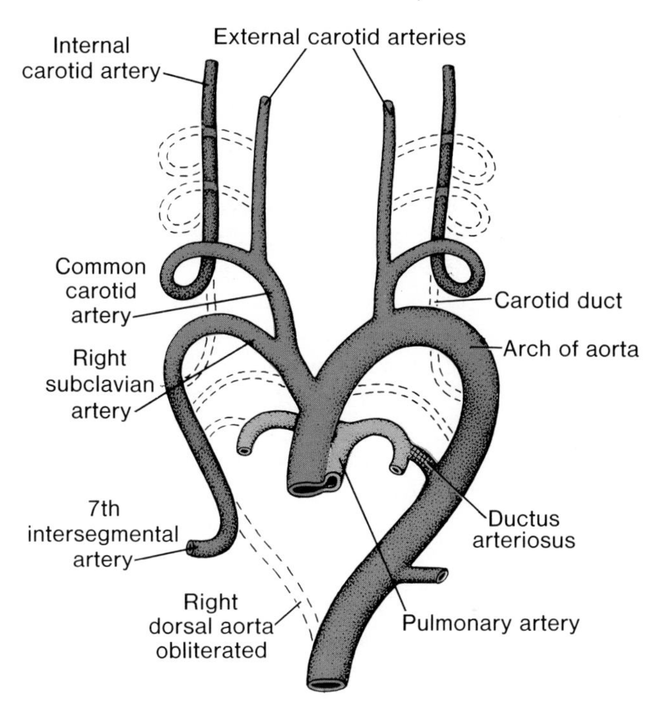

Figure 12-35. Drawing demonstrating the changes in the original aortic arch system.

the third and fourth arches, known as the **carotid duct**, is obliterated (fig. 12-35). (2) The right dorsal aorta disappears between the origin of the seventh intersegmental artery and the junction with the left dorsal aorta (fig. 12-35). The formation of the neck causes the heart to descend from its initial cervical position into the thoracic cavity. Hence, the carotid and innominate arteries elongate considerably (fig. 12-34*C*). As a further result of this caudal shift, the left subclavian artery, distally fixed in the arm bud, shifts its point of origin from the aorta at the level of the seventh intersegmental artery (fig. 12-34*B*) to an increasingly higher point, until it comes close to the origin of the left common carotid artery (fig. 12-34*C*).[2] (4) As a result of the caudal shift of the heart and the disappearance of various portions of the aortic arches, the course of the **recurrent laryngeal nerves** becomes different on the right and left sides. Initially these nerves, branches of the vagus, supply the sixth branchial arches. When the heart descends, they hook around the sixth aortic arches and then ascend again to the larynx. Hence, their recurrent course. When on the right the distal part of the sixth aortic arch and the fifth aortic arch disappear, the recurrent laryngeal nerve moves up and hooks around the right subclavian artery. On the left the nerve does not move up, since the distal part of the sixth aortic arch persists as the ligamentum arteriosum.

VITELLINE AND UMBILICAL ARTERIES

The vitelline arteries, initially a number of paired vessels supplying the yolk sac (fig. 12-32), gradually fuse and form the arteries located in the dorsal mesentery of the gut. In the adult they are represented by the **coeliac, superior mesenteric**, and **inferior mesenteric arteries**. These vessels supply the derivatives of the foregut, midgut, and hindgut, respectively (fig. 11-7).

The umbilical arteries, initially paired ventral branches of the dorsal aortas, course to the placenta in close association with the allantois (fig. 12-32). During the fourth week, however, each artery acquires a secondary connection with the dorsal branch of the aorta, the **common iliac artery**, and loses its original origin. After birth the proximal portions of the umbilical arteries persist as the **internal iliac** and the **superior vesical arteries**, while the distal parts are obliterated to form the **medial umbilical ligaments**.

Abnormalities of Great Arteries

PATENT DUCTUS ARTERIOSUS

Under normal conditions the ductus arteriosus is functionally closed through contraction of its muscular wall shortly after birth.[3] Anatomical

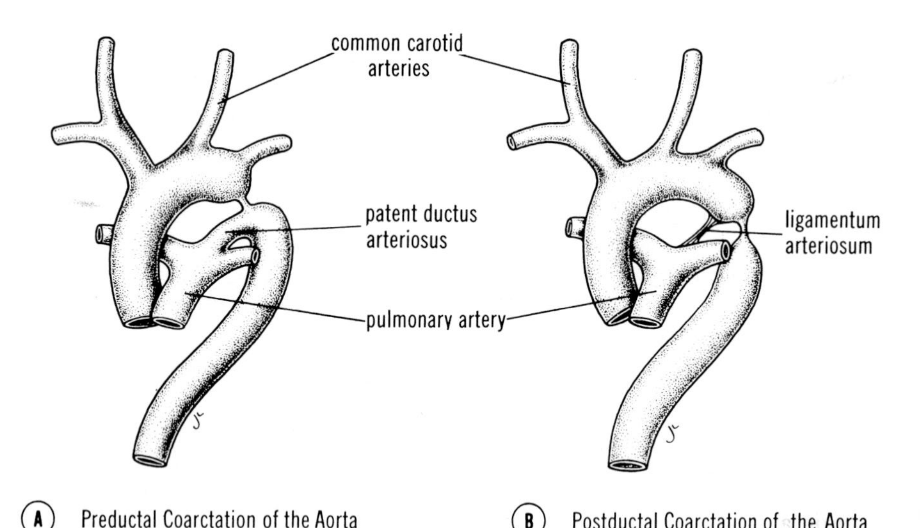

Figure 12-36. Coarctation of the aorta. *A*, Preductal type; *B*, postductal type. The caudal part of the body is supplied by large, hypertrophied intercostal and internal thoracic arteries.

closure, however, by means of intima proliferation takes from one to three months.[4] A patent ductus arteriosus, more common in females than in males, is one of the most frequently seen abnormalities of the great vessels and may occur either as an isolated abnormality or in combination with other heart defects (figs. 12-28*A* and 12-30). Particularly defects which cause large differences between aortic and pulmonary pressures may cause heavy blood flow through the ductus, thus preventing its normal closure.

COARCTATION OF AORTA

Coarctation of the aorta (fig. 12-36*A*, *B*) is a condition in which the aortic lumen below the origin of the left subclavian artery is significantly narrowed. Since the constriction may be located above or below the entrance of the ductus arteriosus, two types of coarctation may be distinguished: the **preductal** and **postductal** types. The cause of the aortic narrowing is primarily an abnormality in the media of the aorta followed by intima proliferations.[5] In the preductal type the ductus arteriosus persists, whereas in the postductal type this channel usually is obliterated. In the latter case, a collateral circulation between the proximal and distal parts of the aorta is established by way of large intercostal and internal thoracic arteries.[6] In this manner the lower part of the body is supplied with blood.

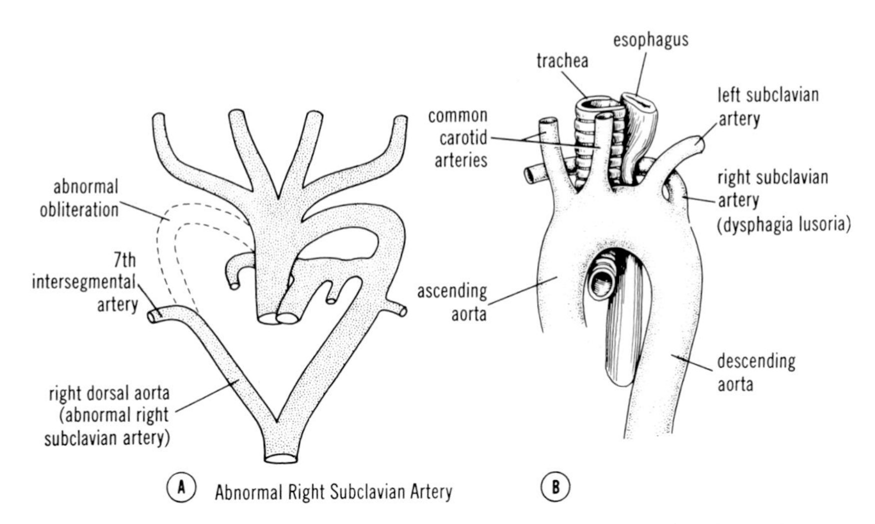

Figure 12-37. Abnormal origin of the right subclavian artery (dysphagia lusoria). *A,* Scheme to show the obliteration of the right fourth aortic arch and the proximal portion of the right dorsal aorta, and the persistence of the distal portion of the right dorsal aorta. *B,* The abnormal right subclavian artery crosses the midline behind the esophagus and may compress the latter.

ABNORMAL ORIGIN OF RIGHT SUBCLAVIAN ARTERY

In this abnormality (fig. 12-37*A*, *B*) the right subclavian artery is formed by the distal portion of the right dorsal aorta and the seventh intersegmental artery. The right fourth aortic arch and the proximal part of the right dorsal aorta have been obliterated. With shortening of the aorta between the left common carotid and the left subclavian arteries, the origin of the abnormal right subclavian artery is finally found just below that of the left subclavian artery[2] Since its stem is derived from the right dorsal aorta, it must cross the midline behind the esophagus to reach the right arm. This abnormality, which is frequently seen, may occasionally cause difficulties in swallowing and respiration. In such instances, the recurrent laryngeal nerve does not hook around the right subclavian artery, but passes directly from the vagus to the larynx musculature.

DOUBLE AORTIC ARCH

In this abnormality the right dorsal aorta persists between the origin of the seventh intersegmental artery and its junction with the left dorsal aorta (fig. 12-38). A **vascular ring** is thus formed which surrounds the trachea and esophagus and frequently compresses these structures, causing difficulties in breathing and swallowing.[7, 8]

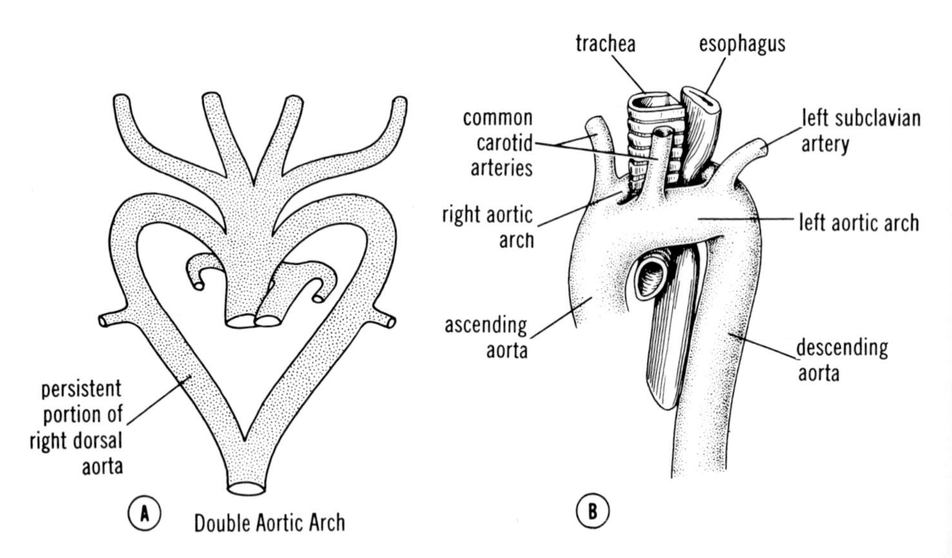

Figure 12-38. Double aortic arch. *A,* Scheme showing the persistence of the distal portion of the right dorsal aorta. *B,* The double aortic arch forms a vascular ring around the trachea and the esophagus.

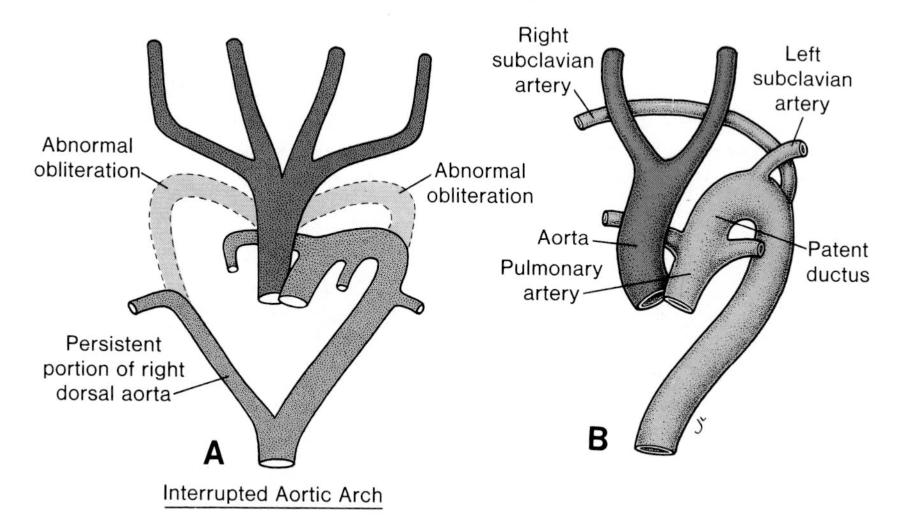

Figure 12-39. *A,* Scheme showing the obliteration of the fourth aortic arch on the right as well as on the left and persistence of the distal portion of the right dorsal aorta. *B,* Case of interrupted aortic arch. The aorta supplies the head; the pulmonary artery by way of the ductus arteriosus supplies the remaining parts of the body.

RIGHT AORTIC ARCH

In such a case the left fourth aortic arch and left dorsal aorta have been completely obliterated and are replaced by the corresponding vessels on the right side. Occasionally, when the ligamentum arteriosum is situated on the left side and passes behind the esophagus, it may cause complaints with swallowing.

INTERRUPTED AORTIC ARCH

This interesting anomaly is caused by obliteration of the fourth aortic arch on the left side (fig. 12-39*A*, *B*). It is frequently combined with an abnormal origin of the right subclavian artery. The ductus arteriosus remains wide open and the descending aorta and subclavian arteries are supplied with blood of low oxygen content. The aortic trunk supplies the two common carotid arteries.[12]

Venous System

In the fifth week three pairs of major veins can be distinguished: (1) the **vitelline** or **omphalomesenteric veins**, carrying blood from the yolk sac to the sinus venosus; (2) the **umbilical veins**, originating in the chorionic villi and carrying oxygenated blood to the embryo; and (3) the **cardinal veins**, draining the body of the embryo proper (fig. 12-40).

VITELLINE VEINS

Before entering the sinus venosus, the vitelline veins form a plexus around the duodenum and pass through the septum transversum. The liver cords growing into the septum interrupt the course of the veins and an extensive vascular network is formed, known as the **hepatic sinusoids** (fig. 12-41*B*).

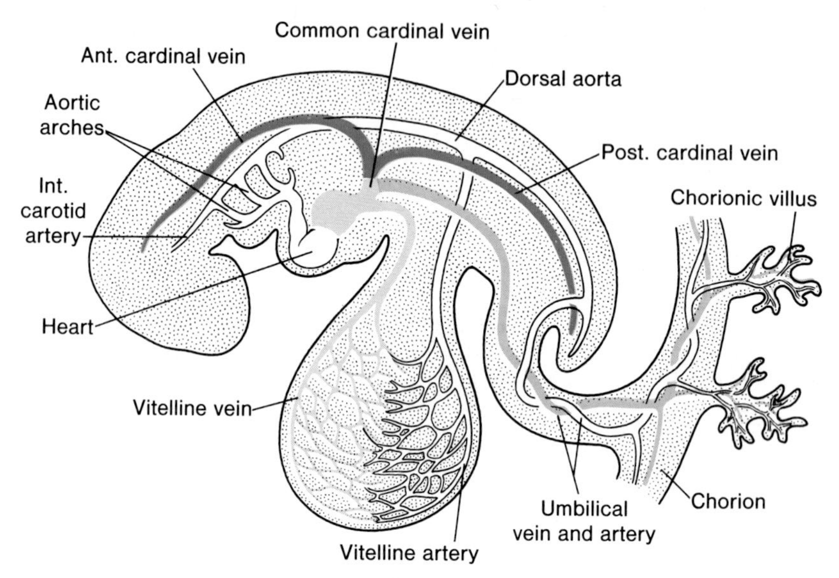

Figure 12-40. Schematic drawing of the main components of the venous system in a 4-mm embryo (end of the fourth week).

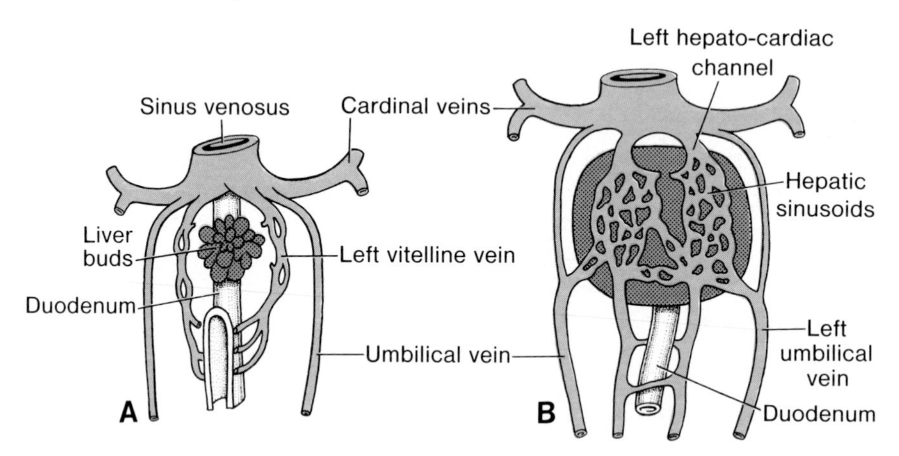

Figure 12-41. Schematic drawings to show the development of the vitelline and umbilical veins during the fourth (*A*) and fifth weeks (*B*). Note the plexus around the duodenum, the formation of the hepatic sinusoids and the beginning of the left-to-right shunt between the vitelline veins.

With the reduction of the left sinus horn, blood from the left side of the liver is rechanneled toward the right, resulting in an enlargement of the right vitelline vein (right hepato-cardiac channel). Ultimately the right hepato-cardiac channel forms the **posthepatic portion of the inferior vena cava**. The proximal part of the left vitelline vein disappears (fig. 12-42*A*, *B*). The anastomotic network around the duodenum develops into a single vessel, the **portal vein** (fig. 12-42*B*). The **superior mesenteric vein**, which drains the primitive intestinal loop, is considered as the successor of the right vitelline vein. The distal portion of the left vitelline vein also disappears (fig. 12-42*A*, *B*).

UMBILICAL VEINS

The umbilical veins pass initally on each side of the liver, but soon become connected to the hepatic sinusoids (fig. 12-41*A*, *B*). The proximal part of both umbilical veins, as well as the remainder of the right umbilical vein, then disappear, so that the left vein is the only one to carry blood from the placenta to the liver (fig. 12-42). With the increase of the placental circulation, a direct communication is formed between the left umbilical vein and the right hepato-cardiac channel, the **ductus venosus** (fig. 12-42*A*, *B*). This vessel bypasses the sinusoidal plexus of the liver. After birth the left umbilical vein and ductus venosus are obliterated and form the **ligamentum teres hepatis** and **ligamentum venosum,** respectively (see "Circulatory Changes at Birth").

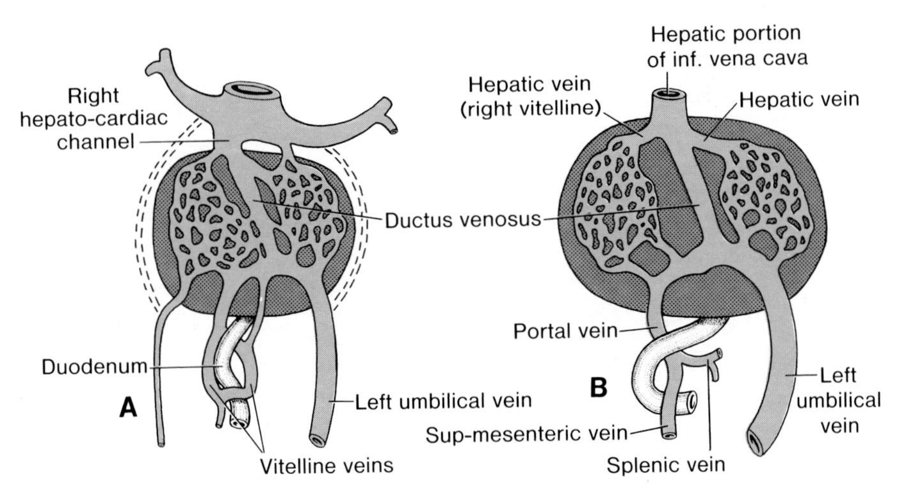

Figure 12-42. Development of vitelline and umbilical veins in the second (*A*) and third months (*B*). Note the formation of the ductus venosus, the portal vein, and the hepatic portion of the inferior vena cava. The splenic and superior mesenteric veins enter into the portal vein.

CARDINAL VEINS

Initially the cardinal veins form the main venous drainage system of the embryo. It consists of the **anterior cardinal veins**, which drain the cephalic part of the embryo, and the **posterior cardinal veins**, which drain the remaining part of the body of the embryo. The anterior and posterior veins join before entering the sinus horn and form the short **common cardinal veins**. During the fourth week the cardinal veins form a symmetrical system (fig. 12-43*A*).

During the fifth to the seventh week a number of additional veins are formed: (1) the **subcardinal veins**, which mainly drain the kidneys; (2) the **sacrocardinal veins**, which drain the lower extremities; and (3) the **supracardinal veins**, which drain the body wall by way of the intercostal veins, thereby taking over the function from the posterior cardinal veins (fig. 12-43*A*).

Characteristic for the formation of the vena cava system is the appearance of anastomoses between left and right in such a manner that the blood from the left is channeled to the right side.

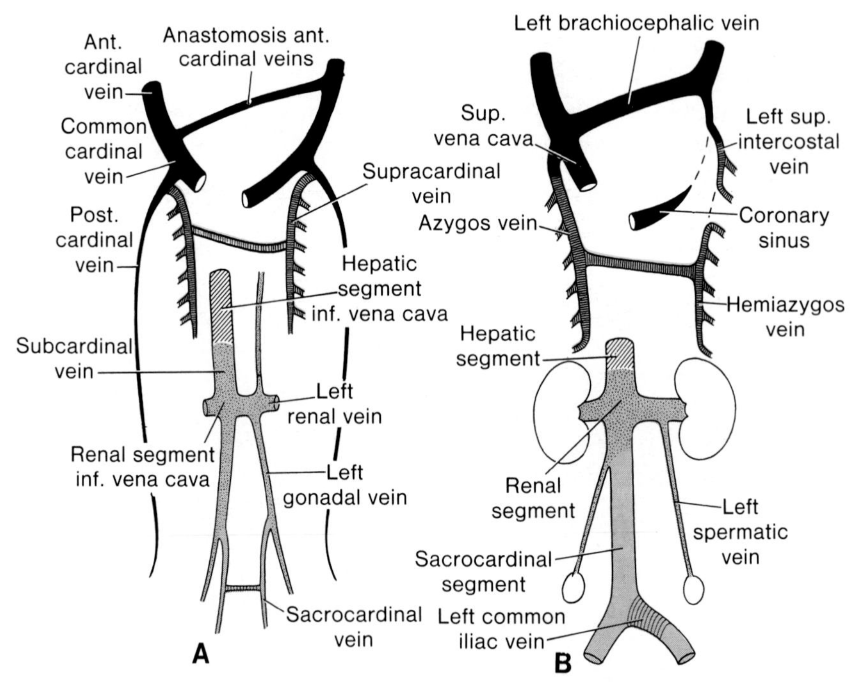

Figure 12-43. Diagrams showing the development of the inferior vena cava, the azygos vein, and the superior vena cava. *A*, In the seventh week. Note the anastomoses which are formed between the subcardinals, the supracardinals, the sacrocardinals, and the anterior cardinals. *B*, The venous system at birth. Note the three components of the inferior vena cava.

The **anastomosis between the anterior cardinal veins** develops into the **left brachiocephalic vein** (fig. 12-43*A*, *B*). Most of the blood from the left side of the head and the left upper extremity is then channeled to the right. The terminal portion of the left posterior cardinal vein entering into the left brachiocephalic vein is retained as a small vessel, the **left superior intercostal vein** (fig. 12-43*B*). This vessel receives blood from the second and third intercostal spaces. The **superior vena cava** is formed by the right common cardinal vein and the proximal portion of the right anterior cardinal vein.[9, 10]

The **anastomosis between the subcardinal veins** is formed by the **left renal vein**. When this communication has been established, the left subcardinal vein disappears, and only its distal portion remains as the **left gonadal vein**. Hence, the right subcardinal vein becomes the main drainage channel and develops into the **renal segment of the inferior vena cava** (fig. 12-43*B*).

The **anastomosis between the sacrocardinal veins** is formed by the **left common iliac vein** (fig. 12-43*B*). The right sacrocardinal vein finally becomes the sacrocardinal segment of the inferior vena cava. When the renal segment of the inferior vena cava connects with the hepatic segment, which is derived from the right vitelline vein, the inferior vena cava is complete. It consists then of a hepatic segment, a renal segment, and a sacrocardinal segment.

With obliteration of the major portion of the posterior cardinal veins, the supracardinal veins gain in importance. The fourth to eleventh right intercostal veins empty into the right supracardinal vein, which, together with a portion of the posterior cardinal vein, forms the **azygos vein** (fig. 12-43). On the left, the fourth to seventh intercostal veins enter into the left supracardinal vein. After development of a communicating vessel between the two supracardinals, the left supracardinal empties into the azygos vein and is then known as the **hemiazygos vein** (fig. 12-43*B*).

Abnormal Venous Drainage

The complicated development of the venae cavae accounts for the fact that deviations from the normal pattern are frequently seen.

DOUBLE INFERIOR VENA CAVA AT LUMBAR LEVEL

In this abnormality (fig. 12-44*A*), the left sacrocardinal vein has failed to lose its connection with the left subcardinal, and the left common iliac vein may or may not be present. The left gonadal vein, however, is present as in normal conditions.

ABSENCE OF INFERIOR VENA CAVA

In this case (fig. 12-44*B*) the right subcardinal vein has failed to make its connection with the liver and shunts its blood directly into the right supracardinal vein (figs. 12-43 and 12-44*B*). Hence, the blood stream from the caudal part of the body reaches the heart by way of the azygos and superior vena cava. The hepatic vein enters into the right atrium at the site of the

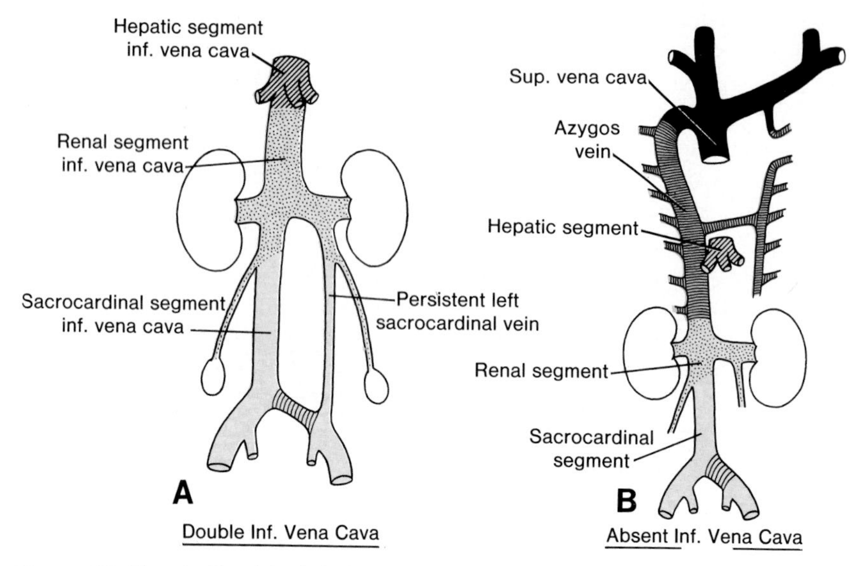

Figure 12-44. *A*, Double inferior vena cava at the lumbar level due to the persistence of the left sacrocardinal vein. *B*, Absent inferior vena cava. The lower half of the body is drained by the azygos vein, which enters the superior vena cava. The hepatic vein enters the heart at the site of the inferior vena cava.

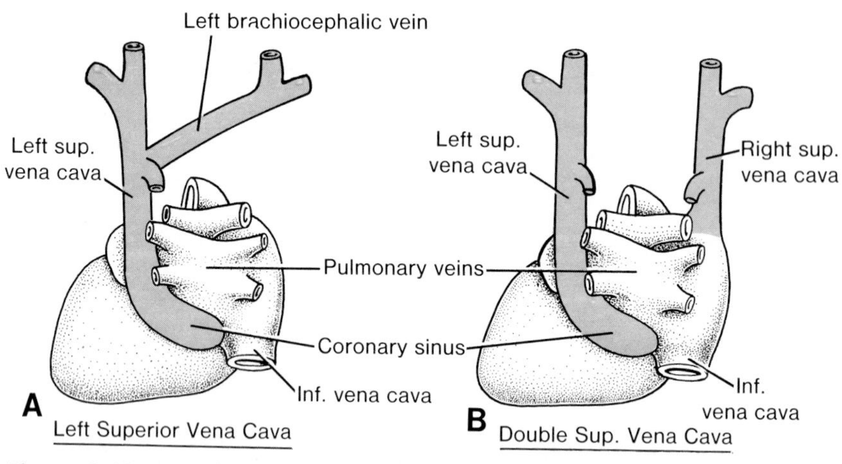

Figure 12-45. *A*, Left superior vena cava draining into the right atrium by way of the coronary sinus (dorsal view). *B*, Double superior vena cava. The communicating (brachiocephalic) vein between the two anterior cardinals has failed to develop (dorsal view).

inferior vena cava.[11] Usually this abnormality is associated with other heart malformations.

LEFT SUPERIOR VENA CAVA

This abnormality (fig. 12-45*A*) is caused by the persistence of the left anterior cardinal vein and the obliteration of the common cardinal and proximal part of the anterior cardinal vein on the right. In such a case, the blood from the right is channeled toward the left by way of the brachiocephalic vein.[12] The left superior vena cava drains into the right atrium by way of the left sinus horn, that is, the coronary sinus.

DOUBLE SUPERIOR VENA CAVA

This condition (fig. 12-45*B*) is characterized by the persistence of the left anterior cardinal vein and the failure of the left brachiocephalic vein to form. The persistent left anterior cardinal vein, which is called **left superior vena cava**, drains into the right atrium by way of the coronary sinus.

Circulatory Changes at Birth

FETAL CIRCULATION

Our present knowledge of the fetal circulation is based largely on angiocardiographic investigations in the fetal lamb and in the human fetus.[13–15]

Before birth, blood from the placenta—about 80% saturated with oxygen—returns to the fetus by way of the umbilical vein. On approaching the liver, the main portion of this blood flows through the ductus venosus directly into the inferior vena cava, thereby short-circuiting the liver. A smaller portion enters the liver sinusoids and mixes here with blood from the portal circulation (fig. 12-46). A sphincter mechanism in the ductus venosus, close to the entrance of the umbilical vein, regulates the flow of umbilical blood through the liver sinusoids. It is thought that this sphincter closes when, because of a uterine contraction, the venous return is too high, thus preventing a sudden overloading of the heart.[16]

After a short course in the inferior vena cava, where the placental blood mixes with deoxygenated blood returning from the lower limbs, it enters the right atrium. Here it is guided toward the oval foramen by the valve of the inferior vena cava, and the major portion of the blood stream passes directly into the left atrium. A small portion, however, is prevented from so doing by the lower edge of the septum secundum, the **crista dividens**, and remains in the right atrium. Here it mixes with the desaturated blood returning from the head and arms by way of the superior vena cava.

From the left atrium, where it mixes with a small amount of blood returning from the lungs, the blood stream enters the left ventricle and ascending aorta. Since the coronary and carotid arteries are the first branches of the ascending aorta, the heart musculature and the brain are supplied

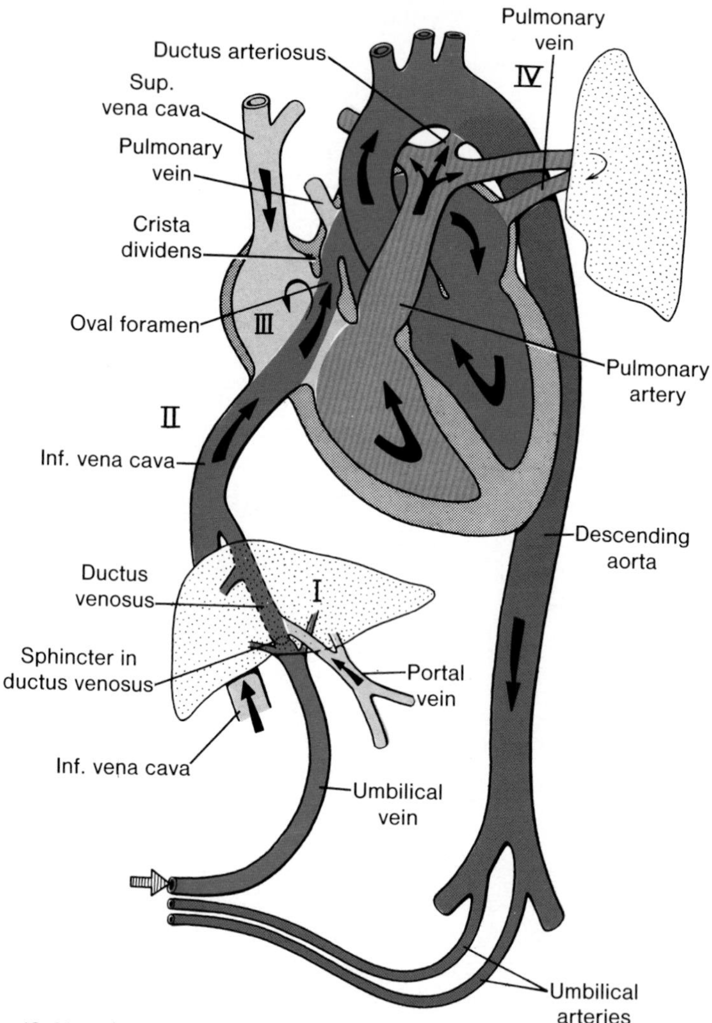

Figure 12-46. Plan of the human circulation before birth. Arrows indicate the direction of the blood flow. Note where the oxygenated blood mixes with deoxygenated blood: (1) in the liver; (2) in the inferior vena cava; (3) in the right atrium; and (4) at the entrance of the ductus arteriosus into the descending aorta.

with well-oxygenated blood. The desaturated blood from the superior vena cava flows by way of the right ventricle into the pulmonary trunk. Since the resistance in the pulmonary vessels during fetal life is high, the main portion of this blood passes directly through the **ductus arteriosus** into the descending aorta, where it mixes with blood from the proximal aorta. From here on the blood stream flows toward the placenta by way of the two umbilical arteries. The oxygen saturation in the umbilical arteries is about 58 per cent.

During its course from the placenta to the organs of the fetus, the high oxygen content of the blood in the umbilical vein gradually decreases by mixing with desaturated blood. Theoretically this may occur in the following places: (1) in the liver by mixture with a small amount of blood returning from the portal system; (2) in the inferior vena cava, which carries deoxygenated blood returning from the lower extremities, pelvis, and kidneys; (3) in the right atrium by mixture with blood returning from the head and limbs; and (4) at the entrance of the ductus arteriosus into the descending aorta.

CHANGES AT BIRTH

The sudden changes occurring in the vascular system at birth are caused by the cessation of the placental blood flow and the beginning of the lung respiration. Since, at the same time, the ductus arteriosus closes by muscular contraction of its wall, the amount of blood flowing through the lung vessels increases rapidly. This in turn results in a rise in pressure in the left atrium. Simultaneously with these changes on the left, the pressure in the right atrium decreases as a result of interruption of the placental blood flow. The septum primum is then apposed to the septum secundum, and the oval foramen closes functionally.

Summarizing, the following changes occur in the vascular system after birth (fig. 12-47):

① **Closure of the umbilical arteries** is accomplished by contraction of the smooth musculature in their walls and is probably caused by thermal and mechanical stimuli and a change in oxygen tension.[17] Functionally, the arteries are closed a few minutes after birth. The actual obliteration of the lumen by fibrous proliferation, however, may take from two to three months.[18, 19] The distal parts of the umbilical arteries then form the **medial umbilical ligaments**, while the proximal portions remain open as the **superior vesical arteries** (fig. 12-47).

② **Closure of the umbilical vein and ductus venosus** occurs shortly after that of the umbilical arteries. Hence, blood from the placenta may enter the newborn for some time after birth. After obliteration, the umbilical vein forms the **ligamentum teres hepatis** in the lower margin of the falciform ligament. The ductus venosus, which courses from the ligamentum teres to the inferior vena cava, is also obliterated and forms the **ligamentum venosum**.

③ **Closure of the ductus arteriosus** by contraction of its muscular wall occurs almost immediately after birth and is probably mediated by bradykinin, a substance released from the lungs during initial inflation. Angiocardiography and cardiac catheterization, however, have revealed that during the first days after birth a left-to-right shunt is not unusual.[17, 20] Complete anatomical obliteration by proliferation of the intima is thought to take from one to three months.[18] In the adult, the obliterated ductus arteriosus forms the **ligamentum arteriosum**.

④ **Closure of the oval foramen** is caused by an increased pressure in the

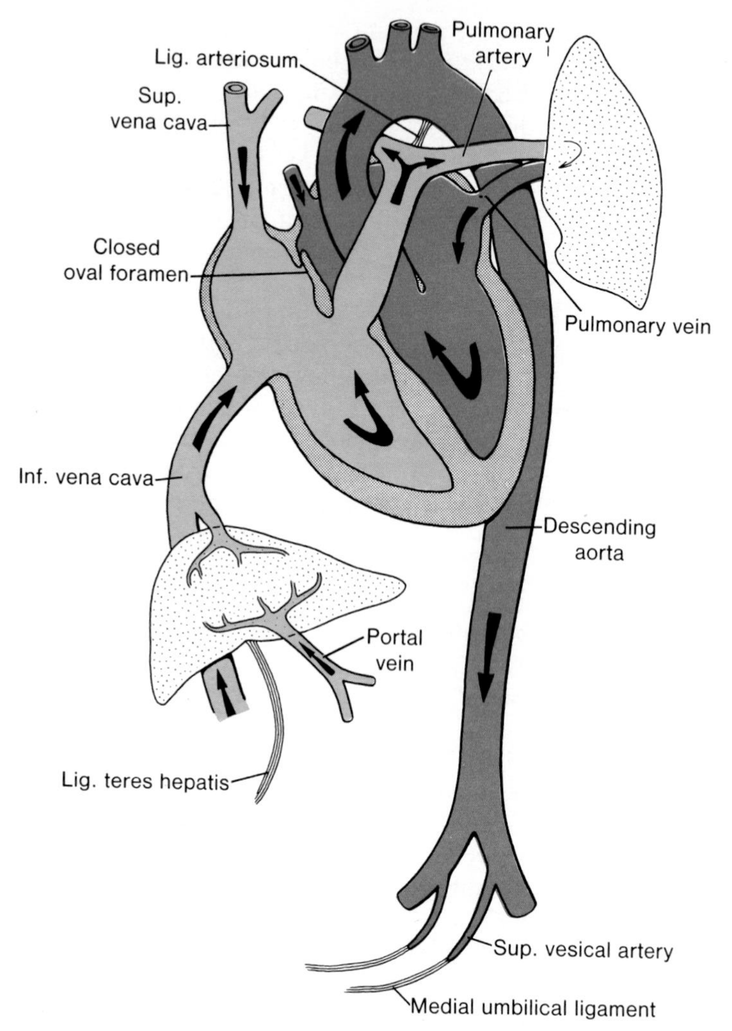

Figure 12-47. Plan of the human circulation after birth. Note the changes occurring as a result of the beginning of respiration and the interruption of the placental blood flow.

left atrium combined with a decrease in pressure on the right side. With the first good breath the septum primum is pressed against the septum secundum. During the first days of life, however, this closure is reversible. Crying of the baby creates a shunt from right to left, thus accounting for the cyanotic periods in the newborn. Constant apposition gradually leads to fusion of the two septa in about one year. In 20 to 25 per cent of all individuals, however, perfect anatomical closure may never be obtained.[19]

SUMMARY

Arterial System. Although each of the six branchial arches has its own aortic artery (figs. 16-2 and 12-34), most obliterate entirely or partially (figs. 12-34*C*, and 12-35*A*). Two important components are (1) the arch of the aorta (right fourth aortic arch); (2) pulmonary artery (sixth aortic arch) which during fetal life is connected to the aorta through the ductus arteriosus; (3) right subclavian artery formed by the left fourth aortic arch, distal portion of the right dorsal aorta, and the seventh intersegmental artery (fig. 12-34*B*). The most common vascular aortic arch abnormalities include: (1) open ductus arteriosus and coarctation of the aorta (fig. 12-36); (2) persistent right aortic arch and abnormal right subclavian artery (figs. 12-37, and 12-38), both causing respiratory and swallowing complaints.

Venous System. Three systems can be recognized: (1) the vitelline system which develops into the **portal system**; (2) the cardinal system, which forms the **caval system**; (3) the umbilical venous system, which disappears after birth. The complicated caval system, in particular, is characterized by many abnormalities such as double inferior and superior vena cava and left superior vena cava (fig. 12-45).

Changes at Birth. During prenatal life the placental circulation provides the fetus with its oxygen, but after birth the lungs take over the gas exchange. In the circulatory system the following changes take place at birth and in the first postnatal months: (1) the ductus arteriosus closes; (2) the oval foramen closes; (3) the umbilical vein and ductus venosus close and remain as the ligamentum teres hepatis and ligamentum venosum; (4) the umbilical arteries close and form the medial umbilical ligaments.

REFERENCES

1. Padget, D. H. The development of the cranial arteries in the human embryo. *Contrib. Embryol., 32:* 205, 1948.
2. Barry, A. The aortic arch derivatives in the human adult. *Anat. Rec., 111:* 221, 1951.
3. Adams, F.·H., and Lind, J. Physiologic studies on the cardiovascular status of the normal newborn infant (with special reference to the ductus arteriosus). *Am. J. Dis. Child., 93:* 13, 1957.
4. Jager, V. V., and Wollenman, O. J. An anatomical study of the closure of the ductus arteriosus. *Am. J. Pathol., 18:* 595, 1942.
5. Wielenga, G. The relationship between coarctation of the aorta and the ligamentum arteriosum. Doctoral thesis, University of Leyden, The Netherlands, 1959.
6. Edwards, J. E., Christensen, N. A., Cladgett, O. T., and McDonald, J. R. Pathologic considerations in coarctation of the aorta. *Proc. Mayo Clin., 23:* 324, 1948.
7. Griswold, H. E., and Young, M. D. Double aortic arch. *Pediatrics, 4:* 751, 1949.
8. Ekstrom, G., and Sandblom, P. Double aortic arch; embryonic development. *Acta Chir. Scand., 102:* 183, 1951.
9. Grünwald, P. Die Entwicklung der Vena Cava Caudalis. *Z. Mikrosk. Anat. Forsch., 43:* 275, 1938.

10. McClure, F. W., and Butler, E. G. The development of the vena cava inferior in man. *Am. J. Anat., 35:* 331, 1925.
11. Anderson, R. C., Heilig, W., Novick, R., and Jarvis, C. Anomalous inferior vena cava with azygos drainage. *Am. Heart J., 49:* 318, 1955.
12. Winter, F. S. Persistent left superior vena cava. *Angiology, 5:* 90, 1954.
13. Lind, J., and Wegelius, C. Human fetal circulation in the mammalian fetus; physiological aspects of development. *Symp. Quant. Biol., 19:* 109, 1954.
14. Barcroft, J. Foetal and neonatal physiology. *Br. Med. Bull., 17:* 247, 1961.
15. Dawes, J. S. Changes in the circulation at birth and the effects of asphyxia. In *Recent Advances in Pediatrics*, edited by D. Gairdner. Little, Brown, & Co., Boston, 1958.
16. Gribbe, G., Hirvonen, L., Lind, J., and Wegelius, C. Cineangiocardiographic recordings of the cyclic changes in volume of the left ventricle. *Cardiology, 34:* 348, 1959.
17. Adams, F. H., and Lind, J. Physiologic studies on the cardiovascular status of normal newborn infants with special reference to the ductus arteriosus. *Pediatrics, 19:* 431, 1957.
18. Ode, E. De ductus arteriosus. Doctoral thesis, University of Leyden, The Netherlands, 1951.
19. Patten, B. M. The development of the heart. In *Pathology of the Heart*, edited by S. E. Gould, p. 20. Charles C Thomas, Springfield, Ill., 1953.
20. Lind, J., Boesen, T., and Wegelius, C. Selective angiocardiography in congenital heart disease. *Prog. Cardiovasc. Dis., 2:* 293, 1959–1960.

chapter 13

Respiratory System

When the embryo is approximately four weeks old the primordium of the respiratory system appears as an **outgrowth** from the ventral wall **of the foregut** (fig. 13-1*A*). Hence, the **epithelium** of the internal lining of the larynx, the trachea, and bronchi, as well as that of the lungs, is entirely of endodermal origin.[1] The **cartilaginous** and **muscular** components of the trachea and lungs, however, are derived from the splanchnic mesoderm surrounding the foregut.

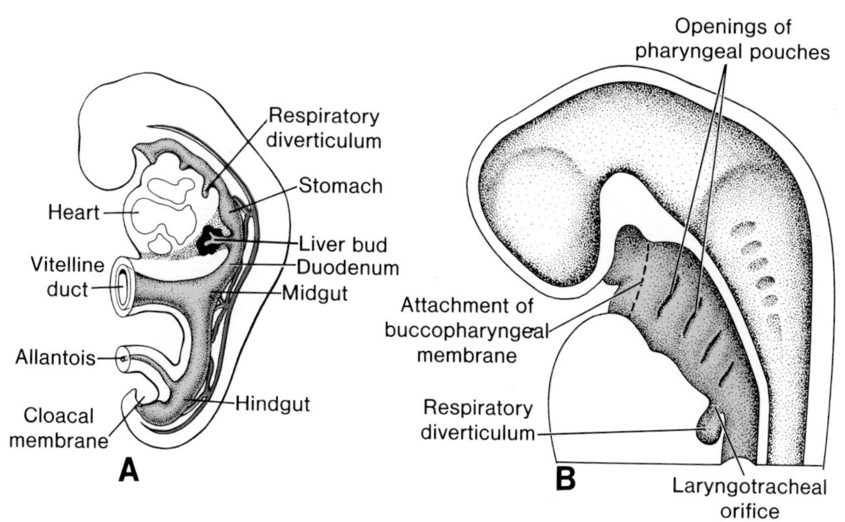

Figure 13-1. *A*, Drawing of an embryo of approximately 25 days to show the relation of the respiratory diverticulum to the heart, the stomach, and the liver. *B*, Sagittal section through the cephalic end of a five-week embryo showing the openings of the pharyngeal pouches and the laryngotracheal orifice.

Initially the respiratory diverticulum is in wide open communication with the foregut (fig. 13-1*B*), but when the diverticulum expands in caudal direction it becomes separated from the foregut by the development of two longitudinal ridges, the esophagotracheal ridges (fig. 13-2*A*). When these ridges subsequently fuse to form a septum, known as the **esophagotracheal septum**, the foregut is divided in a dorsal portion, the **esophagus**, and a

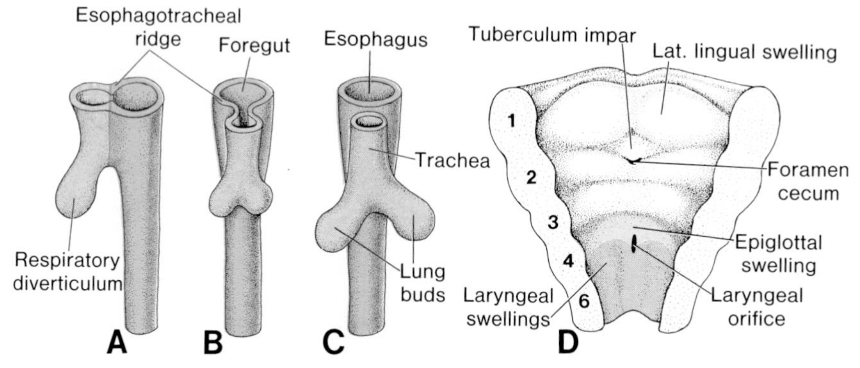

Figure 13-2. *A–C*, Successive stages in the development of the respiratory diverticulum. Note the esophagotracheal ridges and the formation of the septum, splitting the foregut in the esophagus and trachea with lung buds. *D*, The ventral portion of the pharynx seen from above. Note the laryngeal orifice and the surrounding swellings.

ventral portion, the **trachea** and **lung buds** (fig. 13-2*B*,*C*).[2-4] The respiratory primordium, however, maintains its open communication with the pharynx through the **laryngeal orifice** (fig. 13-2*D*).

Larynx

The internal lining of the larynx is of endodermal origin but the cartilages and muscles originate from mesenchyme of the branchial arches (see fig. 16-7). As a result of the rapid proliferation of this mesenchyme the laryngeal orifice changes in appearance from a sagittal slit to a T-shape (fig. 13-3*A*). When subsequently the mesenchyme of the two arches is transformed into the **thyroid**, **cricoid**, and **arytenoid cartilages**, the characteristic adult shape of the laryngeal orifice can be recognized (fig. 13-3*B*).

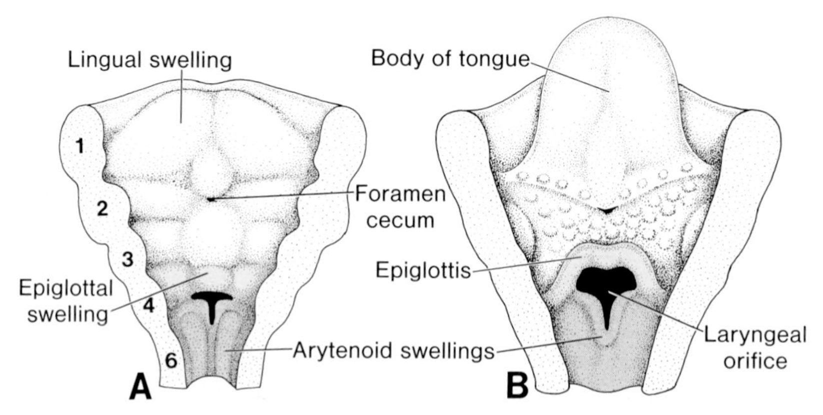

Figure 13-3. Drawings showing the laryngeal orifice and surrounding swellings at successive stages of development. *A*, At 6 weeks; *B*, at 12 weeks.

At about the time that the cartilages are formed the laryngeal epithelium also proliferates rapidly, resulting in a temporary occlusion of the lumen. When, subsequently, vacuolization and recanalization occur, a pair of lateral recesses, the **laryngeal ventricles**, are formed. These recesses are bounded by folds of tissue which do not disappear, but differentiate into the **false** and **true vocal cords**.

Since the musculature of the larynx is derived from mesenchyme of the fourth and sixth branchial arches, all laryngeal muscles are innervated by branches of the tenth cranial nerve, the **vagus nerve**. The **superior laryngeal nerve represents the fourth branchial arch**, and **the recurrent laryngeal nerve the sixth branchial arch**. (For further detail on the laryngeal cartilages see "Head and Neck," Chapter 16.)

Trachea, Bronchi, and Lungs

During its separation from the foregut, the respiratory primordium forms a midline structure, the trachea, and two lateral outpocketings, the **lung buds** (fig. 13-2*B*,*C*). The right lung bud subsequently divides into three branches, the **main bronchi**, and the left into two main bronchi (fig. 13-4*A*), thus foreshadowing the presence of three lobes on the right side and two on the left (fig. 13-4*B*,*C*).

With subsequent growth in caudal and lateral directions, the lung buds penetrate into the coelomic cavity (fig. 13-5*A*,*B*). This space is rather narrow and is known as the **pericardioperitoneal canal**. It is found on either side of the foregut (figs. 13-5*A* and 11-3*B*), and is gradually filled up by the expanding lung buds. When subsequently the pericardioperitoneal canals are separated from the peritoneal and pericardial cavities by the pleuropericardial folds, the remaining spaces are the **primitive pleural cavities** (see Fig. 11-4). The mesoderm, which covers the outside of the lung, develops into

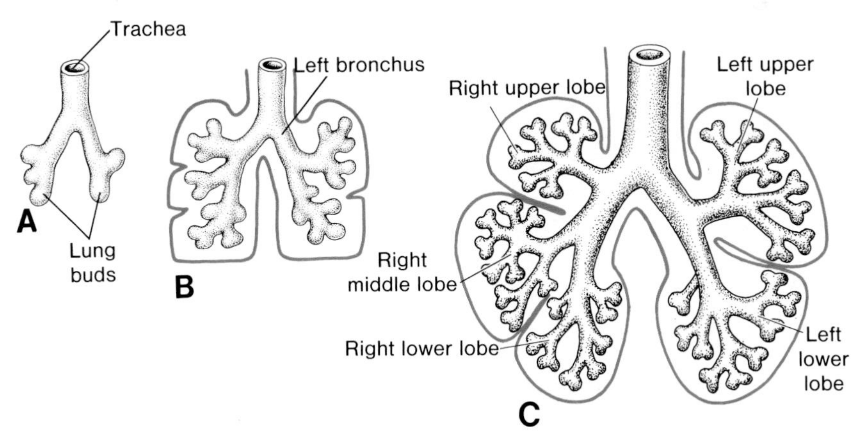

Figure 13-4. Successive stages in the development of the trachea and lungs. *A*, At five weeks; *B*, at six weeks; *C*, at eight weeks (adapted from several sources).

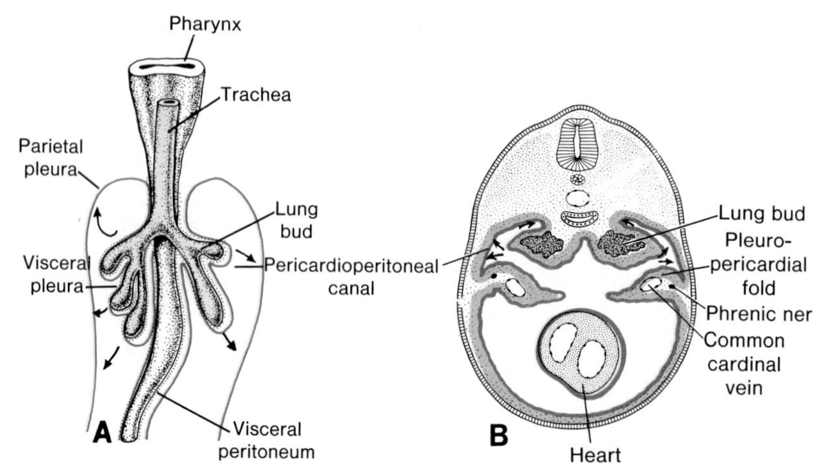

Figure 13-5. Drawings to show the expansion of the lung buds in the pericardioperitoneal canals. At this stage of development the canals are in wide communication with the peritoneal and pericardial cavities (see also fig. 11-3). *A*, Lung buds seen from ventral. *B*, Transverse section through the lung buds. Note the pleuropericardial folds which will divide the thoracic coelomic cavity into the pleural and pericardial cavities (see also fig. 11-3*B*).

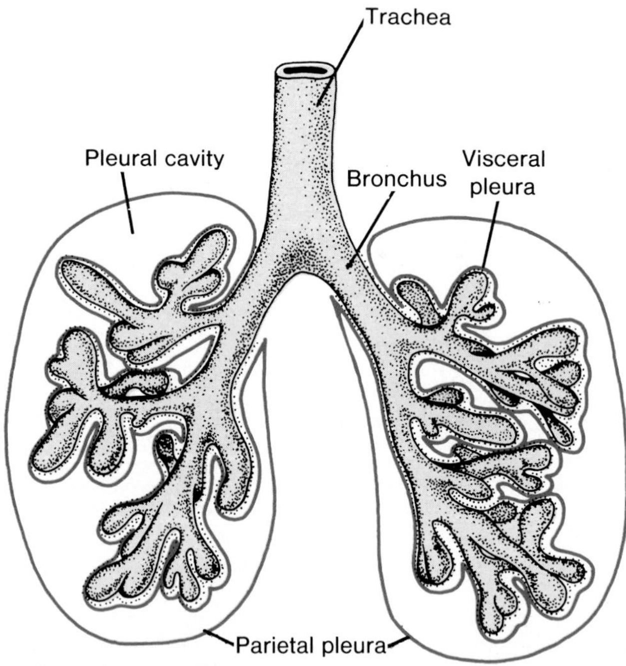

Figure 13-6. Once the pericardioperitoneal canals are closed off from the pericardial and peritoneal cavity, respectively, the lungs expand in the pleural cavities. Note the visceral and parietal pleura and the definitive pleural cavity. The visceral pleura extends between the lobes of the lungs.

the **visceral pleura**. The somatic mesoderm layer, covering the body wall from the inside, becomes the **parietal pleura** (Fig. 13-5 *A*). The space between the parietal and visceral pleura is the **pleural cavity** (fig. 13-6).

During further development the main bronchi divide repeatedly in a dichotomous fashion, and by the end of the sixth month approximately 17 generations of subdivisions have been formed. Before the bronchial tree has reached its final shape, however, **an additional six divisions** will be formed. These **appear during postnatal life**. While all these new subdivisions are formed and the bronchial tree develops, the lungs migrate caudally and by the time of birth the bifurcation of the trachea is located opposite the fourth thoracic vertebra.

MATURATION OF LUNGS

Up to the seventh prenatal month the bronchioli continuously divide into more and smaller canals (canalicular phase) (fig. 13-7*A*) and the vascular supply increases steadily. Respiration becomes possible when some of the cells of the cuboidal respiratory bronchioli change into thin flat cells (fig. 13-7*B*). These cells are intimately associated with numerous blood and lymph capillaries and the surrounded spaces are now known as **terminal sacs** or **primitive alveoli**. During the seventh month sufficient capillaries are present to guarantee adequate gas exchange and the premature infant is able to survive.

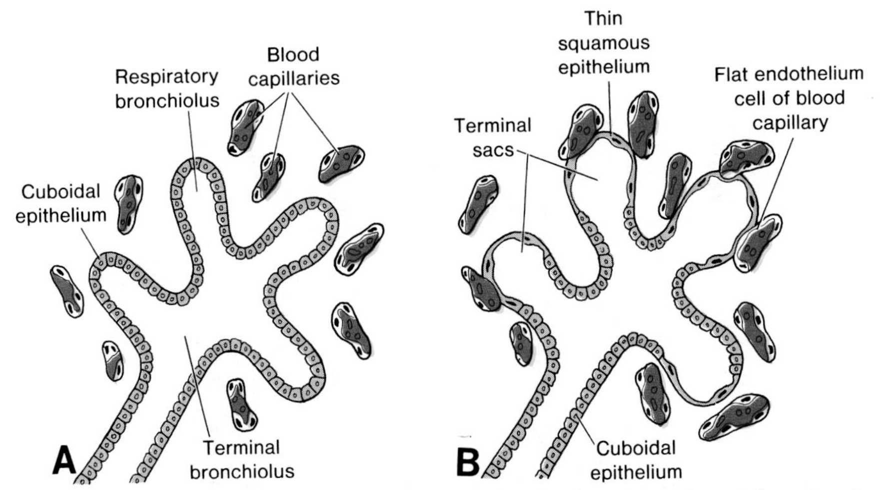

Figure 13-7. Schematic drawings representing the histological and functional development of the lung. *A*, The canalicular phase which lasts from 15 to 25 weeks. Note the cuboidal cells lining the respiratory bronchioli. *B*, The terminal sac period begins at the end of the sixth and beginning of the seventh premature month. Several cuboidal cells become very thin and intimately associated with the endothelium of the blood and lymph capillaries. The canaliculi are now known as *terminal sacs* or *primitive alveoli*.

During the last two months of prenatal life and for several years of postnatal life the number of terminal sacs increases steadily. In addition, the cells lining the sacs, known as **alveolar epithelial cells** (type I), become gradually thinner so that surrounding capillaries protrude into the alveolar sacs (fig. 13-8). The thus formed intimate contact makes up the **blood-air barrier**. Characteristic **mature alveoli** are not present before birth.[5–7] In addition to the endothelial cells and the flat alveolar epithelial cells, another cell type develops at the end of the sixth month. These cells, the **alveolar epithelial cells** (type II), are the prospective **surfactant** producers, a substance capable of lowering the surface tension at the air-alveolar (blood) interface.[8,9]

Before birth the lungs are filled with fluid which contains a high Cl- concentration, little protein, some mucus from the bronchial glands, and surfactant from the alveolar epithelial cells (type II).[10] The amount of surfactant in the fluid increases, particularly during the last two weeks before birth.

When respiration begins at birth most of the lung fluid is rapidly resorbed by the blood and lymph capillaries, while a small amount is probably expelled via the trachea and bronchi during the delivery process.[11] When the fluid is resorbed from the alveolar sacs, the surfactant, however, remains deposited as a thin, phospholipid coat on the alveolar cell membranes. With air entering the alveoli during the first breath, the surfactant coat prevents

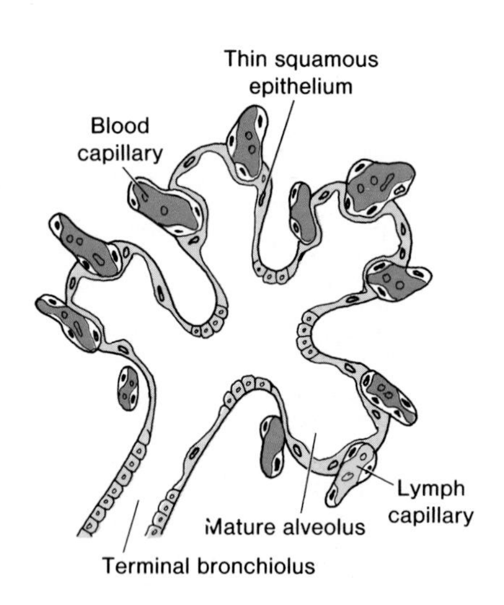

Figure 13-8. Lung tissue in infant. Note the thin squamous epithelial cells (also known as alveolar epithelial cells-type I) and the surrounding capillaries protruding into the alveolar sacs.

the development of an air-water (blood) interphase with high surface tension. Without the fatty surfactant layer the alveoli would collapse during expiration (atelectasis).

LUNGS IN PREMATURE, STILLBORN, AND NORMAL POSTNATAL SITUATIONS

Surfactant appears to be particularly important for survival of the **premature infant**. When insufficient amounts of surfactant are present, the air-water (blood) surface membrane tension becomes high and the risk is great that part of the alveoli will collapse during expiration. As a result **respiratory distress** may develop, and it is believed that this is a common cause of death in the premature infant. In these cases the partially collapsed alveoli contain a fluid with high protein content and many hyaline membranes, and lamellar bodies, probably derived from the surfactant layer. The disease is therefore also known as **hyaline membrane disease**.

In a **stillborn** infant the lung fluid is not resorbed and consequently when the lungs are placed in water at autopsy, they will sink. Once the newborn has inhaled, the lungs float on water, since the fluid has been replaced by air.

Respiratory movements after birth cause the air to enter the lungs, which subsequently expand and fill the pleural cavity. Although the alveoli increase somewhat in size, growth of the lungs after birth is mainly due to an increase in the number of respiratory bronchioli and alveoli. It is estimated that only one-sixth of the adult number of alveoli are present at birth. The remaining alveoli are formed during the first 10 years of postnatal life through the continous formation of new primitive alveoli.[12]

LUNG ABNORMALITIES

Although many abnormalities of the lung and bronchial tree have been described, e.g., blind-ending trachea with absence of lungs and agenesis of one lung, most of these gross abnormalities are rare. More frequently seen are abnormal divisions of the bronchial tree, sometimes resulting in the presence of supernumerary lobules. These variations of the bronchial tree are of little functional significance, but may cause unexpected difficulties in bronchoscopy.

More interesting are **ectopic lung lobes** which arise from the trachea or esophagus. It is believed that these lobes are formed from additional respiratory buds of the foregut which develop independently of the main respiratory system.

Most important clinically are the **congenital cysts of the lung**.[13] Such cysts, formed by dilation of the terminal or larger bronchi, may be multiple, giving the lung a honeycomb appearance on x-ray, or they may be restricted to one or more larger ones. Since cystic structures of the lung usually drain poorly, they frequently cause chronic infections.

SUMMARY

The **respiratory system** is an outgrowth of the ventral wall of the foregut and the epithelium of the larynx, trachea, bronchi, and alveoli is of endodermal origin. The cartilaginous and muscular components are of mesodermal origin. In the fourth week of development the trachea and lung buds are split off from the foregut by the **esophagotracheal septum**, thus dividing the foregut in the respiratory diverticulum anteriorly and the esophagus posteriorly. Contact between the two is maintained through the larynx, which is formed by tissue of the lower branchial arches (see Fig. 16-7). Distally the left lung bud develops into two main bronchi and two lobes; on the right in three main bronchi and three lobes. Faulty partitioning of the foregut by the esophagotracheal septum causes fistulas between the esophagus and trachea (see fig. 14-4).

After a pseudoglandular (5 to 17 weeks) and a canalicular (15 to 25 weeks) phase (fig. 13-7), cells of the cuboidal lined bronchioli change into thin, flat cells, intimately associated with blood and lymph capillaries. In the seventh month gas exchange between the blood and air in the **primitive alveolus** is possible. Before birth the lungs are filled with fluid with little protein, some mucus, and **surfactant**. This substance is produced by **alveolar epithelial cells** (type II) and forms a phospholipid coat on the alveolar membranes. At the beginning of respiration the lung fluid is resorbed, with exception of the surfactant coat which prevents the collapse of the alveoli during expiration by reducing the surface tension at the air-blood capillary interphase. Absence or insufficient amount of the surfactant in the premature baby causes respiratory distress due to collapse of the primitive alveoli (**hyaline membrane disease**).

Growth of the lungs after birth is mainly due to an increase in the **number** of respiratory bronchioli and alveoli and not to an increase in **size** of the alveoli. New alveoli are formed at least during the first 10 years of postnatal life.

REFERENCES

1. Low, F. N., and Sampaio, M. M. The pulmonary alveolar epithelium as an entodermal derivative. *Anat. Rec.*, *127:* 51, 1957.
2. Waterson, D. J., Carter, R. E., and Aberdeen, E. Oesophageal atresia: tracheo-oesophageal fistula, a study of survival in 218 infants. *Lancet*, *1:* 819, 1962.
3. Langman, J. Esophagus atresia accompanied by vessel anomalies. *Arch. Chir. Neerl.*, *4:* 39, 1952.
4. Avery, M. E. *The Lung and its Disorders in the Newborn Infant*. W. B. Saunders, Philadelphia, 1968.
5. Boyden, E. A. The pattern of terminal airspaces in a premature infant of 30–32 weeks that lived nineteen and a quarter hours. *Am. J. Anat.*, *126:* 31, 1969.
6. Emery, J. *The Anatomy of the Developing Lung*. William Heinemann, London, 1969.

7. Boyden, E. A. Development of the human lung. In *Practice of Pediatrics*, Vol. 4, edited by J. Brenneman. Harper and Row, Hagerstown, Md., 1972.
8. Williams, M. C., and Mason, R. J. Development of the type II cell in the fetal rat lung. *Am. Rev. Resp. Dis., 115:* 37–47, 1977.
9. Rosenbaum, R. M., Picciano, P., Kress, Y., and Wittner, M. Ultrastructure of in vitro type II epithelial cell cysts derived from adult rabbit lung cells. *Anat. Rec., 188:* 241, 1977.
10. Oliver, R. E. Fetal lung liquids. *Fed. Proc., 36:* 2669, 1977.
11. Davis, J. A. The first breath and development of lung tissue. In *Scientific Foundations of Obstetrics and Gynecology*, edited by E. E. Philipp, J. Barnes and M. Newton. William Heinemann, London, 1970.
12. Hislop, A. and Reid, L. Growth and development of the respiratory system. In *Scientific Foundations of Paediatrics*, edited by J. A. Davis and J. Dobbing. W. B. Saunders, Philadelphia, 1974.
13. Salzberg, A. M. Congenital malformations of the lower respiratory tract. In *Disorders of the Respiratory Tract in Children. Vol. I. Pulmonary Disorders*, ed. 2. W. B. Saunders, Philadelphia, 1972.

chapter 14

Digestive System

As a result of the cephalo-caudal and lateral folding of the embryo, the endoderm lined cavity is partially incorporated into the embryo to form the **primitive gut**. Two other portions of the endoderm lined cavity, the **yolk sac** and the **allantois** remain temporarily outside the embryo (fig. 14-1*A*-*D*).

In the cephalic as well as the caudal part of the embryo the primitive gut forms a blind ending tube, the **foregut** and **hindgut**, respectively. The middle part, the **midgut**, remains temporarily connected to the yolk sac by means of the **vitelline duct** or **yolk stalk** (fig. 14-1*D*) (see Chapter 5).

The development of the primitive gut and its derivatives is usually discussed in four sections: (1) the **pharyngeal gut** or **pharynx**, which extends from the buccopharyngeal membrane to the tracheobronchial diverticulum (fig. 14-1*D*). Since this section is of particular importance for the development of the head and neck, it will be discussed in Chapter 16. (2) The **foregut**, lying caudal to the tracheobronchial diverticulum and extending as far caudally as the liver outgrowth; (3) the **midgut**, beginning caudal to the liver bud (the anterior intestinal portal) and extending to a point where, in the adult, the junction of the right two-thirds and left one-third of the transverse colon is found (known, in the embryo, as the posterior intestinal portal); (4) the **hindgut**, extending from the posterior intestinal portal to the cloacal membrane (fig. 14-1).

The endoderm forms the epithelial lining of the digestive tract and gives rise to the glands such as the liver and the pancreas. The muscular and peritoneal components of the wall of the gut are derived from the splanchnic mesoderm.

Figure 14-1. Schematic drawings of sagittal sections through embryos at various stages of development to demonstrate the effect of the cephalo-caudal and lateral flexion on the position of the endoderm lined cavity. Note the formation of the foregut, midgut, and hindgut. *A*, Presomite embryo; *B*, 7-somite embryo; *C*, 14-somite embryo; *D*, at the end of the first month.

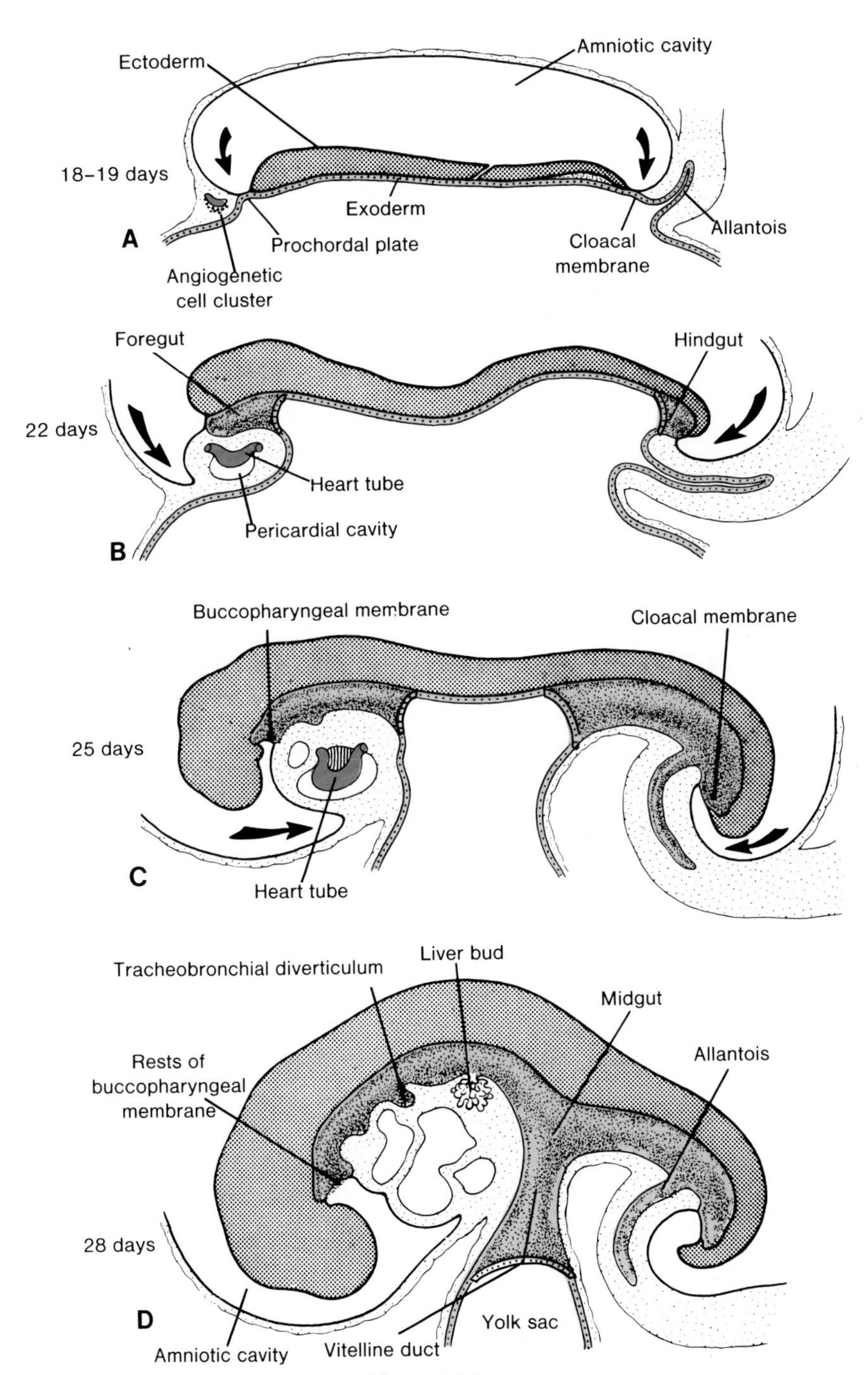
Ectoderm
Amniotic cavity
18–19 days
A
Exoderm
Prochordal plate
Angiogenetic
cell cluster
Cloacal
membrane
Allantois
Foregut
Hindgut
22 days
Heart tube
Pericardial cavity
B
Buccopharyngeal membrane
Cloacal membrane
25 days
C
Heart tube
Tracheobronchial diverticulum
Liver bud
Midgut
Rests of
buccopharyngeal
membrane
Allantois
28 days
D
Amniotic cavity
Vitelline duct
Yolk sac

Figure 14-1.

Foregut

ESOPHAGUS

When the embryo is approximately four weeks old, a small diverticulum appears at the ventral wall of the foregut at the border with the pharyngeal gut (fig. 14-2*A*). This **respiratory** or **tracheobronchial diverticulum** is gradually separated from the dorsal part of the foregut through a partition, which is known as the **esophagotracheal septum** (fig. 14-2*B*, *C*). In this manner the foregut is divided into a ventral portion, the **respiratory primordium**, and a dorsal portion, the **esophagus.**

Initially the esophagus is very short (fig. 14-3*A*) but with the descent of the heart and lungs, it lengthens rapidly (fig. 14-3*B*). The muscular coat, which is formed by the surrounding mesenchyme, is striated in its upper two-thirds and innervated by the vagus; in the lower one-third the muscle coat is smooth and innervated by the splanchnic plexus.

Atresia of Esophagus and Esophagotracheal Fistula. This abnormality is thought to result either from a spontaneous deviation of the esophagotracheal septum in a posterior direction, or from some mechanical factor pushing the dorsal wall of the foregut anteriorly. In its most common form the proximal part of the esophagus ends as a blind sac, whereas the distal part is connected to the trachea by a narrow canal at a point just above the bifurcation (fig. 14-4*A*). Occasionally, the fistulous canal between the trachea and the distal portion of the esophagus is replaced by a ligamentous cord (fig. 14-4*B*). Rarely do both the proximal and distal portions of the esophagus open into the trachea (fig. 14-4*C*).

Atresia of the esophagus prevents the normal passage of amniotic fluid into the intestinal tract; this results in the accumulation of excess fluid in the

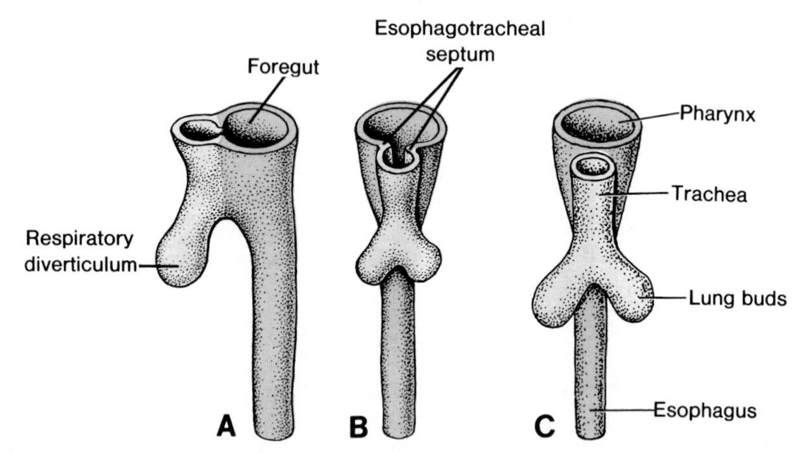

Figure 14-2. Successive stages in the development of the respiratory diverticulum and the esophagus through the partitioning of the foregut. *A*, At the end of the third week (lateral view). *B* and *C*, During the fourth week (ventral view).

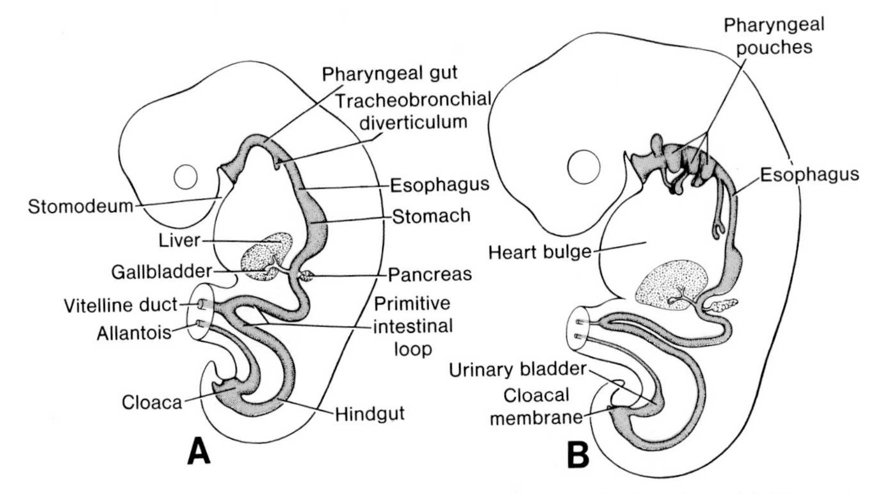

Figure 14-3. Schematic drawings of embryos during the fourth and fifth weeks of development to show the formation of the gastrointestinal tract and the various derivatives originating from the endodermal germ layer.

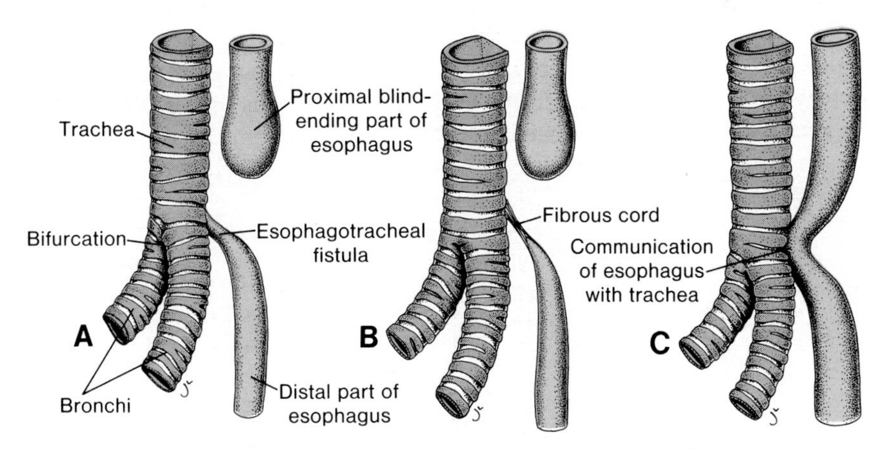

Figure 14-4. *A*, Atresia of the esophagus and esophagotracheal fistula. *B*, Atresia of esophagus. The connection between the distal part of the esophagus and trachea is formed by a fibrous cord. *C*, The proximal and distal parts of the esophagus are both connected to the trachea by a narrow canal.

amniotic sac (**polyhydramnios**) and consequently an enlarged uterus. Although a newborn child with atresia of the esophagus may appear normal, at its first attempt to drink the proximal portion of the esophagus will fill rapidly and milk will flow over into the trachea and lungs.

STOMACH

The stomach appears as a fusiform dilation of the foregut in the fourth week of development (fig. 14-3). During the following weeks its appearance and position change greatly as a result of the different rate of growth in various regions of its wall, and the changes in position of the surrounding organs.[1] The positional changes of the stomach are most easily explained by assuming that it rotates around (1) a longitudinal and (2) an anteroposterior axis (figs. 14-5 and 14-6).

Around its longitudinal axis, the stomach rotates 90° clockwise, causing its left side to face anteriorly and its right side, posteriorly (fig. 14-5*A–C*). Hence, the left vagus nerve, initially innervating the left side of the stomach, now innervates the anterior wall; similarly the right vagus innervates the posterior wall. During this rotation the original posterior wall of the stomach grows faster than the anterior portion and this results in the formation of the **greater** and **lesser curvatures** (fig. 14-5*C*). Since at this stage of development the stomach is attached to the posterior and anterior body wall by the dorsal and ventral mesogastrium, respectively (see Chapter 11, p. 150), rotation around the longitudinal axis is thought to pull the dorsal mesogastrium to the left, thus helping in the formation of the **omental bursa**, a pouch of the peritoneal cavity located behind the stomach (fig. 14-5*D–F*).

The cephalic and caudal ends of the stomach are originally located in the midline, but during further growth the caudal or **pyloric part** moves to the right and upward, and the cephalic or **cardiac portion** to the left and slightly downward (fig. 14-6*A*, *B*). The stomach thus assumes its final position, and its length axis runs from above left to below right.

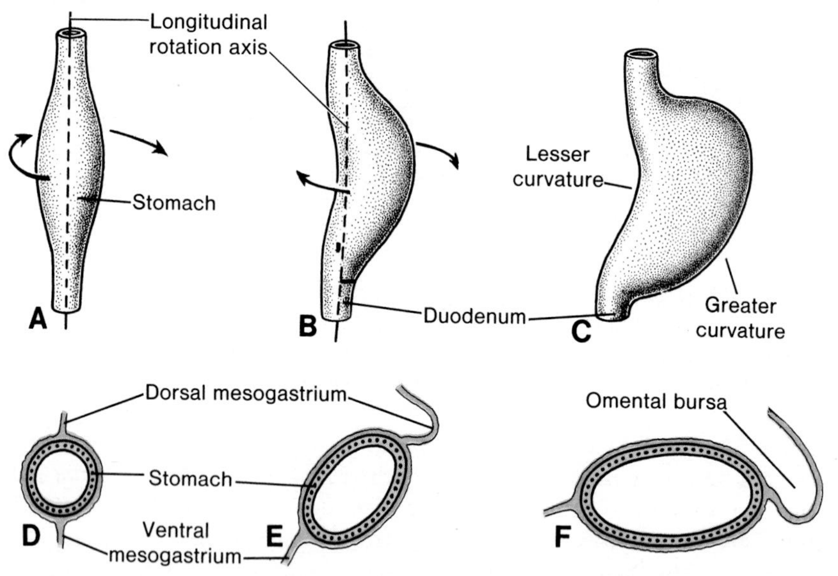

Figure 14-5. Schematic representation of the positional changes of the stomach. *A*, *B*, and *C* show the rotation of the stomach along its longitudinal axis as seen from anterior; *D*, *E*, and *F* show, in transverse section, the effect of rotation on the peritoneal attachments.

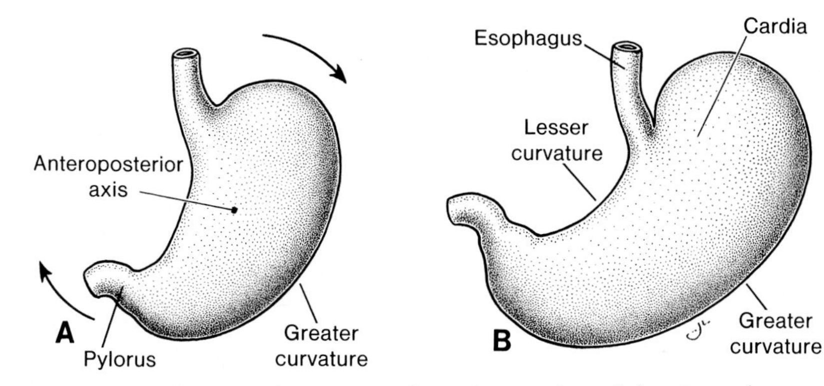

Figure 14-6. Schematic drawings to show the rotation of the stomach around the anteroposterior axis. Note the change in position of the pylorus and cardia.

Pyloric Stenosis. Sometimes the circular and, to a lesser degree, the longitudinal musculature of the stomach in the region of the pylorus is hypertrophied. This is one of the most common abnormalities of the stomach in infants and is believed to develop during fetal life. There is an extreme narrowing of the pyloric lumen, and the passage of food is obstructed, resulting in severe progressive vomiting. A few cases have been described in which the pylorus was atretic.[1]

Other malformations of the stomach, such as duplications and the presence of a prepyloric septum, are rare.[2, 3]

DUODENUM

This portion of the intestinal tract is formed by the terminal part of the foregut and the cephalic part of the midgut. The junction of the two parts is located directly distal to the origin of the liver bud (fig. 14-7). As the stomach rotates, the duodenum takes on the form of a C-shaped loop, rotates to the right, and finally comes to lie retroperitoneally (fig. 14-7*B*) (see also figs 11-12 and 11-13). During the second month the lumen of the duodenum may temporarily be obliterated. Under normal conditions, however, the lumen is reestablished shortly afterward. Since the foregut is supplied by the celiac artery and the midgut by the superior mesenteric artery, the duodenum is supplied by branches of both arteries (fig. 11-7).

LIVER AND GALLBLADDER

The liver primordium appears in the middle of the third week as an outgrowth of the endodermal epithelium at the distal end of the foregut (fig. 14-7*A*, *B*).[4] This outgrowth, known as the **hepatic diverticulum** or **liver bud**, consists of rapidly proliferating cell strands which penetrate the **septum transversum**, that is, the mesodermal plate between the pericardial cavity and the stalk of the yolk sac (fig. 14-7*A*, *B*). While the hepatic cell strands continue to penetrate in the septum, the connection between the hepatic

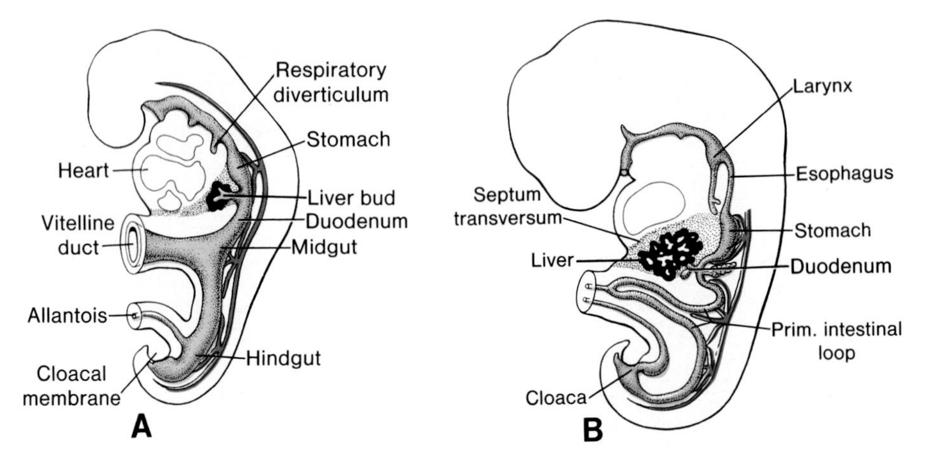

Figure 14-7. *A*, Drawing of a 3-mm embryo (approximately 25 days) to show the primitive gastrointestinal tract and the formation of the hepatic diverticulum. The liver bud is formed by the endodermal epithelial lining of the foregut. *B*, Drawing of a 5-mm embryo (approximately 32 days). The epithelial liver cords penetrate the mesenchyme of the septum transversum.

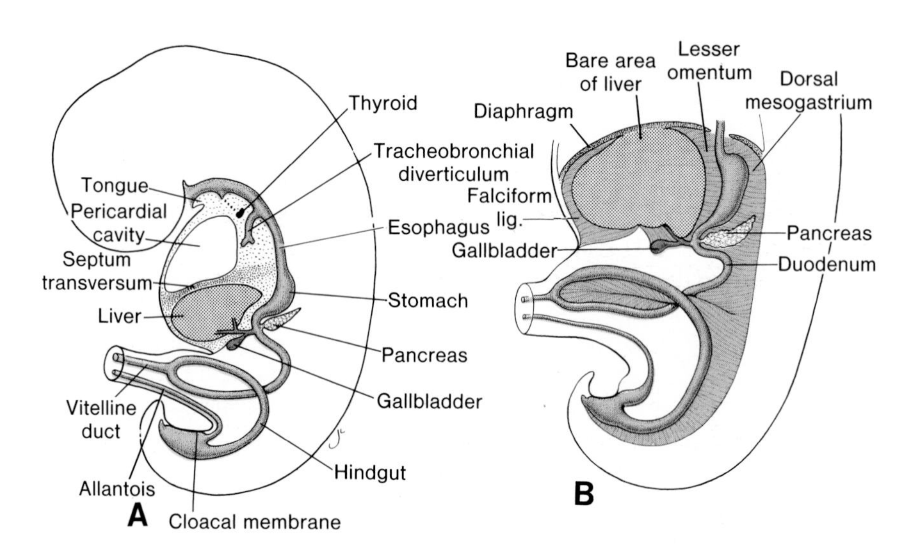

Figure 14-8. *A*, Drawing of a 9-mm embryo (approximately 36 days). The liver expands caudally into the abdominal cavity. Note the condensation of mesenchyme in the area between the liver and the pericardial cavity, foreshadowing the formation of the diaphragm. *B*, Drawing of a slightly older embryo. Note the falciform ligament extending between the liver and the anterior abdominal wall, and the lesser omentum between the liver and the foregut (stomach and duodenum). The liver is entirely surrounded by peritoneum, except in its contact area with the diaphragm. This area is known as the bare area of the liver.

diverticulum and the foregut (duodenum) narrows, thus forming the **bile duct**. A small ventral outgrowth is formed by the bile duct and this outgrowth gives rise to the **gallbladder** and the **cystic duct** (figs. 14-8 and 14-10). During further development the epithelial liver cords intermingle with the vitelline and umbilical veins forming the hepatic sinusoids. The liver cords differentiate into the parenchyma and form the lining of the biliary ducts. The **hemopoietic cells**, **Kupffer cells**, and connective tissue cells are derived from the mesoderm of the septum transversum.

Liver and its Relation to Peritoneum. As a result of its continuous rapid growth, the liver becomes too large for the confines of the septum transversum and begins gradually to protrude into the abdominal cavity. The mesoderm of the septum between the ventral abdominal wall and the liver becomes stretched and very thin, thus forming a thin membrane, known as the **falciform ligament** (fig. 14-8*B*). The umbilical vein, originally found in the mesoderm of the septum, now occupies a position in the free, caudal margin of the falciform ligament. Similarly the mesoderm of the septum between the liver and the foregut (stomach and duodenum) becomes stretched and membranous, thereby forming the **lesser omentum (gastrohepatic and duodenohepatic ligaments)** (fig. 14-8*B*). In the free margin of the lesser omentum are found the **bile duct**, the **portal vein**, and the **hepatic artery**. Hence, when the liver bulges caudally into the abdominal cavity, the mesoderm of the septum transversum located between the liver and the foregut, and the liver and the ventral abdominal wall, becomes membranous, thus forming the **lesser omentum** and **falciform ligament**, respectively. Together they form the peritoneal connection between the foregut and the ventral abdominal wall and are known as the **ventral mesogastrium**.

The mesoderm on the surface of the liver differentiates into visceral peritoneum except on its cranial surface (fig. 14-8*B*). In this region the liver remains in contact with the rest of the original septum transversum. This portion of the septum consists of densely packed mesoderm and will form the tendinous portion of the diaphragm. The surface of the liver, which is in contact with the future diaphragm, is never covered by peritoneum and is known as the **bare area of the liver** (fig. 14-8*B*).

Function of Liver in Fetus. In the 10th week of development the weight of the liver is approximately 10 per cent of the total body weight. Though this may be attributed partly to the presence of a large number of sinusoids another important factor is its **hemopoietic function**. Large nests of proliferating cells, which produce red and white blood cells, are found between the hepatic cells and the walls of the vessels. This activity subsides gradually during the last two months of intra-uterine life and only small hemopoietic islands remain at birth. The weight of the liver is then only 5 per cent of the total body weight.

Another important function of the liver begins approximately at the 12th week. At this time bile is formed by the hepatic cells. Since, meanwhile, the

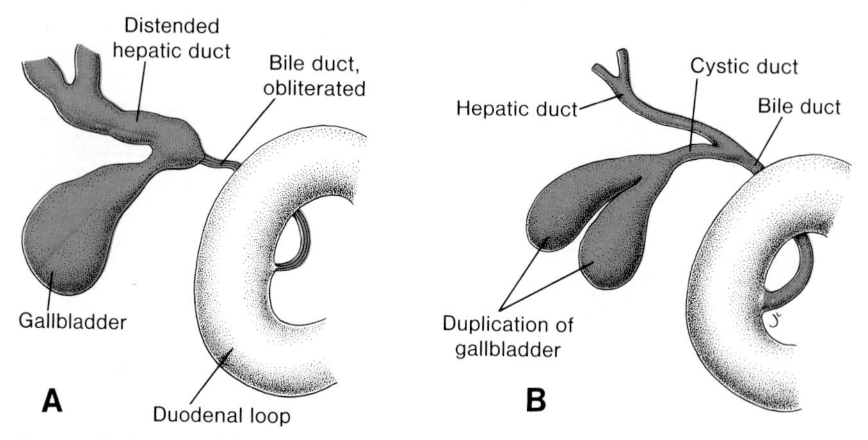

Figure 14-9. *A*, Obliteration of the bile duct, resulting in a distention of the gallbladder and hepatic ducts distal to the obliteration. *B*, Bifid gallbladder.

gallbladder and **cystic duct** have developed and the cystic duct has joined the hepatic duct to form the **bile duct** (fig. 14-8*B*), the bile can enter the gastrointestinal tract. As a result its contents obtain a dark green color. As a result of the positional changes of the duodenum, the entrance of the bile duct gradually shifts from its initial anterior position to a posterior one, and consequently the bile duct is found passing behind the duodenum (figs. 14-10 and 14-11).

Atresia of Gallbladder and Bile Ducts. Initially the gallbladder is a hollow organ, but as a result of proliferation of its epithelial lining it becomes temporarily solid. The definitive lumen develops by vacuolization of the epithelium. When this fails to occur the gallbladder remains atretic and does not develop.[5]

The intra- and extrahepatic ducts also go through a solid stage in their development. If the lumen fails to reopen, the ducts will appear as narrow, fibrous cords. Occasionally, such an atresia is limited to a small portion of the bile duct only (fig. 14-9*A*). The gallbladder and the hepatic duct proximal to the atresia are then considerably distended, and severe, steadily increasing jaundice will become obvious after birth.

In addition to atresia of the gallbladder, duplication, partial subdivision, and diverticula of the gallbladder have frequently been described (fig. 14-9*B*).

PANCREAS

The pancreas is formed by two buds originating from the endodermal lining of the duodenum (fig. 14-10). While the **dorsal pancreatic bud** is located in the dorsal mesentery, the **ventral pancreatic bud** is closely related to the bile duct.

When the duodenum rotates to the right and becomes C-shaped, the ventral pancreatic bud migrates dorsally in a manner similar to the shifting of the entrance of the bile duct (fig. 14-10*B*). Finally the ventral pancreas comes to lie immediately below and behind the dorsal pancreas (fig. 14-11).

Later the parenchyma as well as the duct systems of the dorsal and ventral pancreatic buds fuse (fig. 14-11*B*). The ventral bud forms the **uncinate process** and the inferior part of the head of the pancreas. The remaining part of the gland is derived from the dorsal bud. The **main pancreatic duct** is formed by the distal part of the dorsal pancreatic duct and the entire ventral pancreatic duct (fig. 14-11*B*). The proximal part of the dorsal pancreatic duct either is obliterated or persists as a small channel, the **accessory pancreatic duct**. The main pancreatic duct, together with the bile duct, enters the duodenum at the site of the major papilla; the entrance of the accessory duct is at the site of the minor papilla. In about 10 per cent of all cases the duct system fails to fuse and the original double system persists.[6]

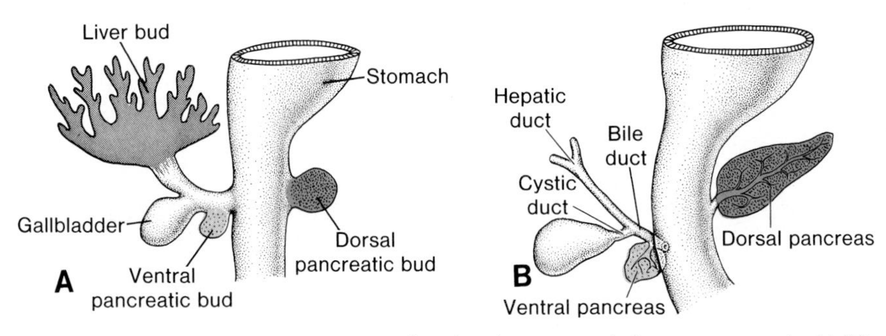

Figure 14-10. Successive stages in the development of the pancreas. *A*, At 30 days (approximately 5 mm); *B*, 35 days (approximately 7 mm). The ventral pancreatic bud is initially located close to the hepatic diverticulum, but later migrates posteriorly around the duodenum in the direction of the dorsal pancreatic bud.

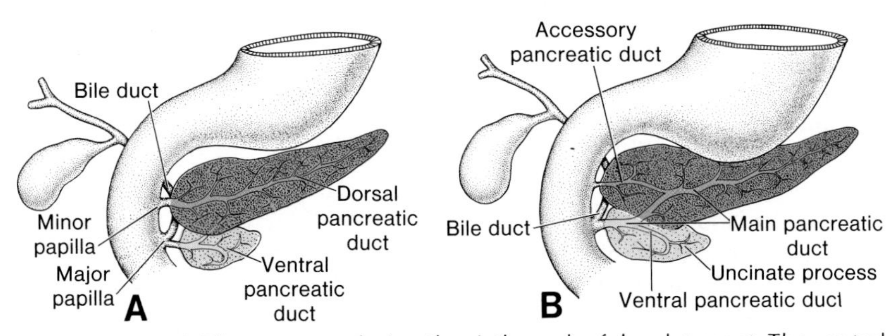

Figure 14-11. *A*, The pancreas during the sixth week of development. The ventral pancreatic bud is in close contact with the dorsal pancreatic bud. *B*, Drawing showing the fusion of the pancreatic ducts. The main pancreatic duct enters the duodenum in combination with the bile duct at the major papilla. The accessory pancreatic duct enters the duodenum at the minor papilla.

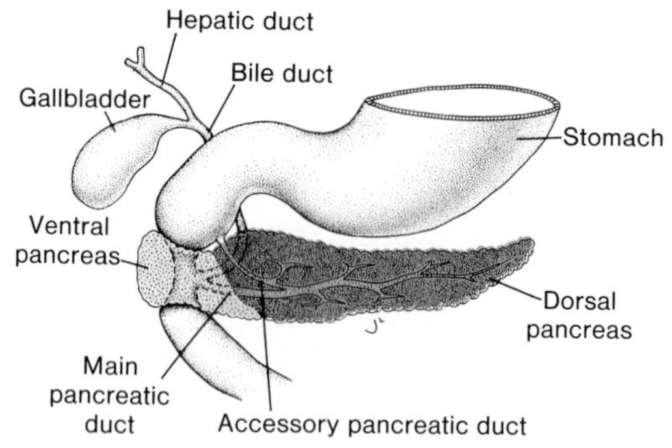

Figure 14-12. Annular pancreas. The ventral pancreas forms a ring around the duodenum, occasionally resulting in duodenal stenosis.

The **pancreatic islets** or **islets of Langerhans** develop from the parenchymatous pancreatic tissue in the third month of fetal life and are scattered throughout the gland. **Insulin secretion** begins at approximately the fifth month.[7] Since the fetal insulin level is independent from the maternal insulin level, it is unlikely that much insulin crosses the placenta.[8]

Annular Pancreas. The ventral pancreatic bud probably consists of two components which, under normal conditions, fuse and rotate around the duodenum in such a manner that they come to lie below the dorsal pancreatic bud. Occasionally, however, the right portion of the ventral bud migrates along its normal route, but the left part migrates in an opposite direction. In this manner the duodenum is surrounded by pancreatic tissue, and an **annular pancreas** is formed (fig. 14-12).[9] The malformation sometimes constricts the duodenum and causes complete obstruction.

Heterotopic Pancreatic Tissue. Heterotopic pancreatic tissue may be found anywhere from the distal end of the esophagus to the tip of the primary intestinal loop. Most frequently it is found in the mucosa of the stomach and in Meckel's diverticulum.[10] Here it may show all the pathological changes characteristic of the pancreas itself.

Midgut

In the 5-mm embryo the midgut is suspended from the dorsal abdominal wall by a short mesentery, and communicates with the yolk sac by way of the **vitelline duct** or **yolk stalk** (figs. 14-1 and 14-7). In the adult, the midgut begins immediately distal to the entrance of the bile duct into the duodenum (fig. 14-12) and terminates at the junction of the proximal two-thirds of the transverse colon with the distal one-third. Over its entire length the midgut is supplied by the **superior mesenteric artery** (fig. 14-13*A*).

The development of the midgut is characterized by a rapid elongation of the gut and its mesentery, resulting in the formation of the **primary intestinal loop** (figs. 14-7*B*, 14-8, and 14-13). At its apex, the loop remains in open

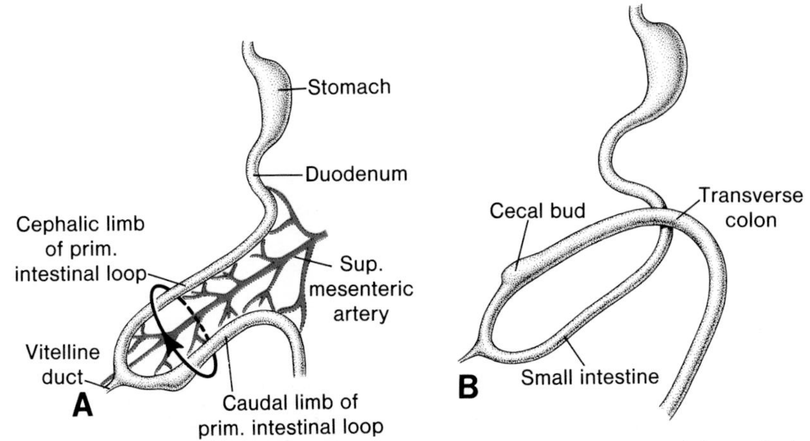

Figure 14-13. *A*, Schematic drawing of the primitive intestinal loop before rotation (lateral view). The superior mesenteric artery forms the axis of the loop. Arrow indicates the direction of the counterclockwise rotation. *B*, Similar view as in *A*, showing the primitive intestinal loop after 180° counterclockwise rotation. Note that the transverse colon passes in front of the duodenum.

connection with the yolk sac by way of the narrow **vitelline duct** (fig. 14-13). The cephalic limb of the loop develops into the distal part of the duodenum, the jejunum, and part of the ileum. The caudal limb becomes the lower portion of the ileum, the cecum, and the appendix, the ascending colon and the proximal two-thirds of the transverse colon. The junction of the cranial and caudal limbs can, in the adult, only be recognized if a portion of the vitelline duct persists as **Meckel's** or **ileal diverticulum** (fig. 14-18).

PHYSIOLOGICAL HERNIATION

Development of the primary intestinal loop is characterized by rapid elongation, particularly of the cephalic limb. As a result of this rapid growth and the simultaneous expansion of the liver, the abdominal cavity temporarily becomes too small to contain all the intestinal loops and they enter the extra-embryonic coelom in the umbilical cord during the sixth week of development (**physiological umbilical herniation**) (figs. 14-14 and 6-1).

ROTATION OF MIDGUT

Coincident with the growth in length, the primitive intestinal loop rotates around an axis formed by the **superior mesenteric artery** (fig. 14-13). When viewed from the front this rotation occurs in a counterclockwise direction and amounts to approximately 270° when it is completed (figs. 14-13 and 14-15). Even during the rotation movement, the elongation of the small intestinal loop continues and the jejunum and ileum form a number of coiled loops (fig. 14-14). The large intestine likewise grows considerably in length, but fails to participate in the coiling phenomenon. The rotation

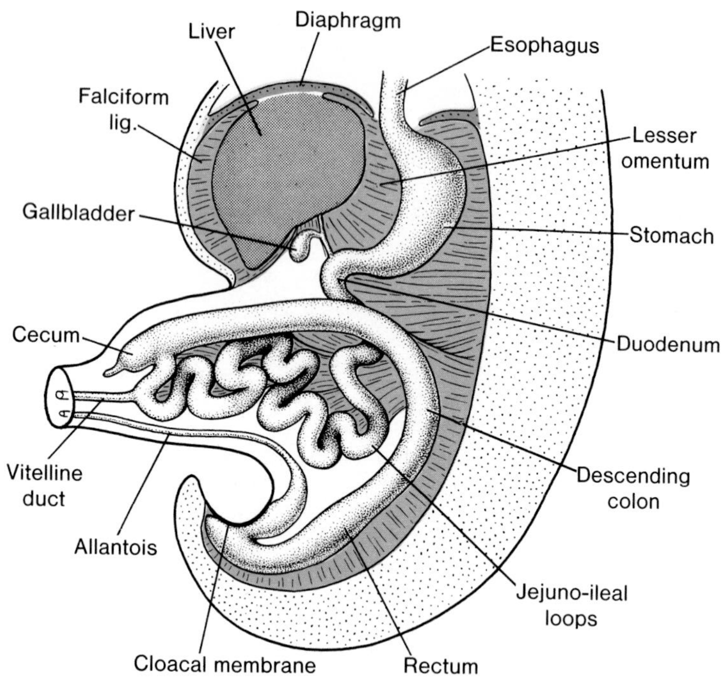

Figure 14-14. Umbilical herniation of the intestinal loops in an embryo of approximately eight weeks (crown-rump length, 35 mm). Coiling of the small intestinal loops and formation of the cecum occur during the herniation.

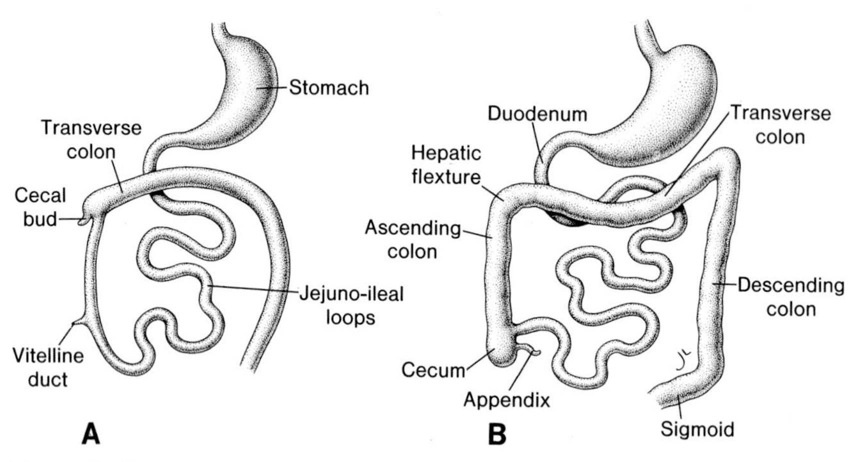

Figure 14-15. *A*, Anterior view of the intestinal loops after 270° counterclockwise rotation. Note the coiling of the small intestinal loops and the position of the cecal bud in the right upper quadrant of the abdomen. *B*, Similar view as in *A*, with the intestinal loops in the final position. Cecum and appendix are located in the right lower quadrant of the abdomen.

occurs during the herniation (about 90°) as well as during the return of the intestinal loops into the abdominal cavity (remaining 180°).

RETRACTION OF HERNIATED LOOPS

At about the end of the third month the herniated intestinal loops begin to return to the abdominal cavity. Though the factors responsible for this return are not precisely known, it is thought that (1) regression of the mesonephroi, (2) reduced growth of the liver, and (3) actual expansion of the abdominal cavity play important roles.

The proximal part of the jejunum is the first part to reenter the abdominal cavity and it comes to lie on the left side (fig. 14-15*B*). The later returning loops gradually settle more and more to the right. The **cecal swelling**, which appears at about the 12-mm stage as a small conical dilation of the caudal limb of the primitive intestinal loop, is the last part of the gut to reenter the abdominal cavity. It is temporarily located in the right upper quadrant directly below the right lobe of the liver (fig. 14-15*A*). From here it descends into the right iliac fossa, thereby forming the **ascending colon** and the **hepatic flexure** (fig. 14-15*B*). During this process the distal end of the cecal swelling forms a narrow diverticulum, the **primitive appendix** (fig. 14-16).

Since the appendix develops during the descent of the colon, it is understandable that its final position frequently is posterior to the cecum or colon. This position of the appendix is referred to as **retrocecal** or **retrocolic**, respectively (fig. 14-17).

FIXATION OF INTESTINAL LOOPS

As the intestines return to the abdominal cavity their mesenteries are pressed against the posterior abdominal wall and in several areas they fuse with the parietal peritoneum, thus fixing certain loops against the posterior abdominal wall (see Chapter 11, p. 155).

Remnants of Vitelline Duct. In 2 to 4 per cent of people, a small portion of the vitelline duct persists, forming an outpocketing of the ileum, known as **Meckel's** or **ileal diverticulum** (fig. 14-18*A*). In the adult, this diverticulum

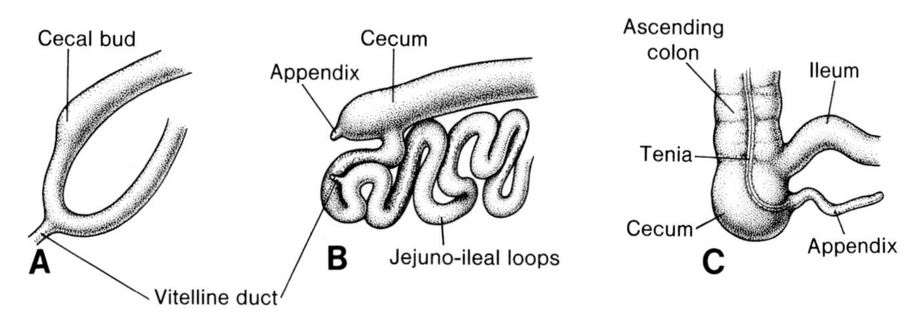

Figure 14-16. Successive stages in the development of the cecum and appendix. *A*, At seven weeks; *B*, Eight weeks; *C*, Newborn.

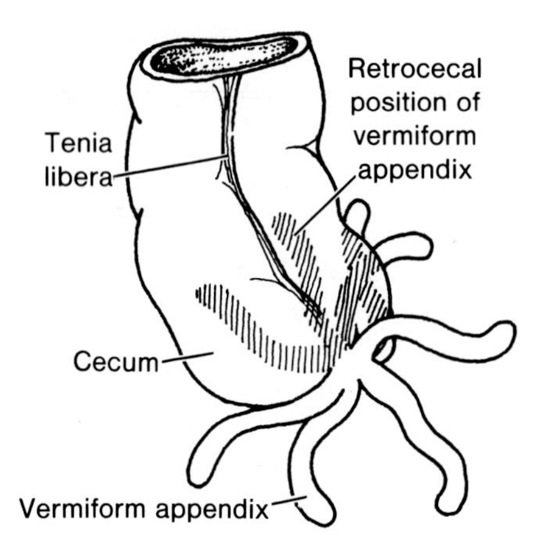

Figure 14-17. Various positions of appendix. In about 50 per cent of cases the appendix is found in a retrocecal or retrocolic position. (From Anderson, J. *Grant's Atlas of Anatomy*, 7th ed. Williams & Wilkins, Baltimore, 1978.)

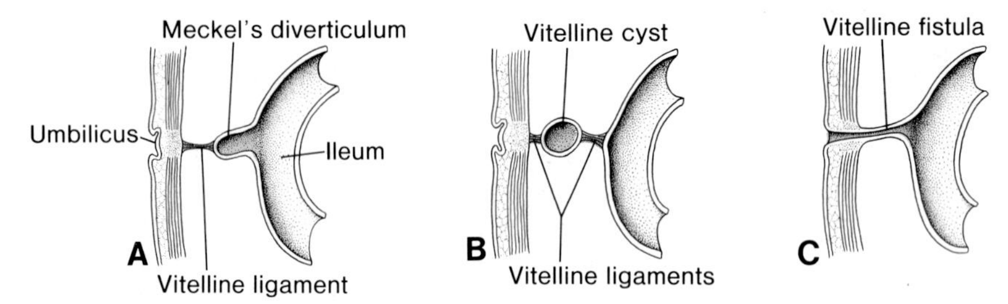

Figure 14-18. Remnants of the vitelline duct. *A*, Meckel's or ileal diverticulum combined with fibrous cord (vitelline ligament). *B*, Vitelline cyst attached to the umbilicus and the wall of the ileum by vitelline ligaments. *C*, Vitelline fistula connecting the lumen of the ileum with the umbilicus.

is located about 40 to 60 cm from the ileocecal valve on the antimesenteric border of the ileum and does not usually cause any complaints.[11] When it contains heterotopic pancreatic tissue or gastric mucosa, however, ulceration, bleeding, or even perforation may occur.

Sometimes the vitelline duct remains patent over its entire length, thus forming a direct communication between the umbilicus and the intestinal tract. This abnormality is known as **umbilical** or **vitelline fistula** (fig. 14-18*C*). A fecal discharge may then be found at the umbilicus. In another variation, both ends of the vitelline duct are transformed into fibrous cords, while the middle portion forms a large cyst, the **enterocystoma** or **vitelline cyst** (fig. 14-18*B*). Since the fibrous cords traverse the peritoneal cavity, they may easily cause intestinal strangulation or **volvulus**. Intestinal loops twist easily around such fibrous strands and become obstructed.

Omphalocele. Sometimes the intestinal loops fail to return from the umbilical cord into the abdominal cavity. When this occurs the loops remain in the extra-embryonic coelom of the umbilical cord. At birth the herniated

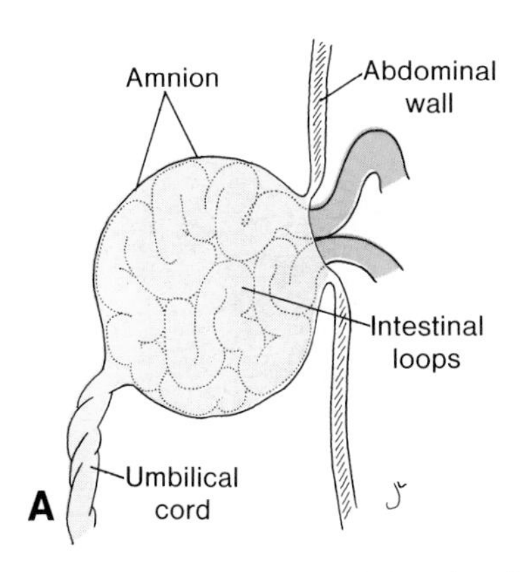

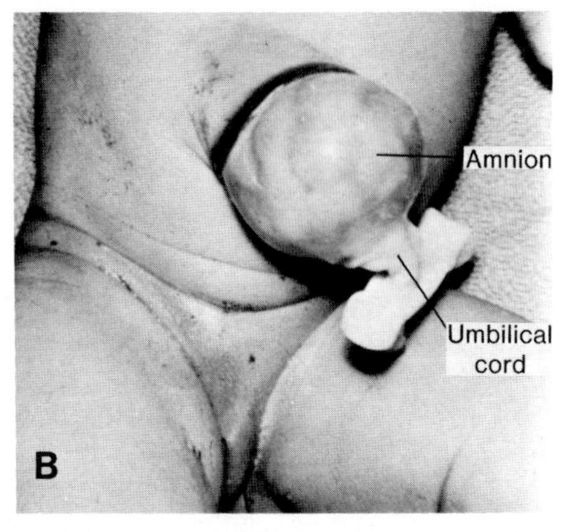

Figure 14-19. *A*, Omphalocele—failure of the intestinal loops to return to the abdominal cavity. The herniated loops are surrounded by a membranous sac formed by the amnion. *B*, Photograph of omphalocele in newborn child. (Courtesy Dr. S. Shaw, Department of Surgery, University of Virginia.)

loops cause a large swelling in the umbilical cord and are covered only by the amnion. This defect is known as an **omphalocele** (fig. 14-19).[13]

Congenital Umbilical Hernia. In this abnormality the muscular layers and skin around the umbilicus are absent and the surface layer is formed by the amnion. The viscera return to the abdomen but herniate again during the fetal period. They protrude then outside the abdominal cavity and are covered by **peritoneum and amnion**, but not by skin. The sac is extremely thin and often ruptures during birth.

In most severe cases, all the viscera, including the liver, may be found outside the abdominal cavity. This abnormality is known as **eventration of the abdominal viscera**. The defect is usually associated with exstrophy of the urinary bladder (see Chapter 15, p. 257).

Abnormal Rotation of Intestinal Loop. The primitive intestinal loop normally rotates 270° counterclockwise. Occasionally, however, rotation amounts to 90° only. When this occurs, the colon and cecum are the first portions of the gut to return from the umbilical cord and they settle on the left side of the abdominal cavity (fig. 14-20*A*). The later returning loops then become located more and more to the right.[14] The abnormality is often called **left-sided colon**.

In some cases, there is **reversed rotation of the intestinal loop**. The primitive loop then rotates 90° in a clockwise direction. In such an abnormality the transverse colon passes behind the duodenum (fig. 14-20*B*) and lies behind the superior mesenteric artery. The main danger in all cases of abnormal rotation is that the twisting of the intestinal loops causes a kink in the arteries and thus a vascular obstruction of a loop. The other possibility is that abnormal peritoneal bands cause a direct obstruction of a loop.

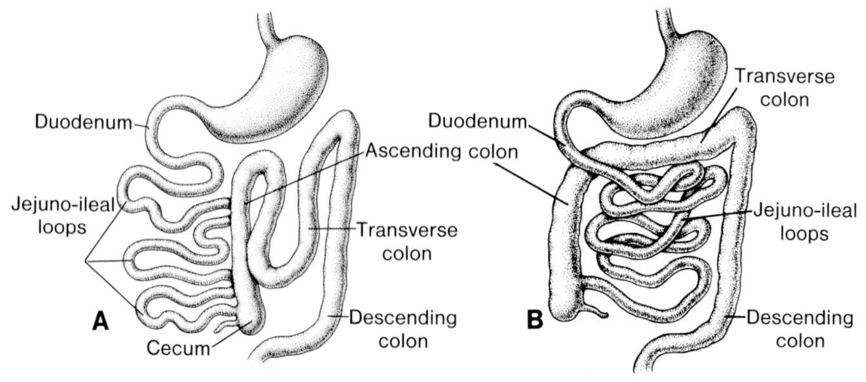

Figure 14-20. *A*, Abnormal rotation of the primitive intestinal loop. The colon is located on the left side of the abdomen and the small intestinal loops on the right. Note that the ileum enters the cecum from the right. *B*, The primitive intestinal loop is rotated 90° in a clockwise direction (reversed rotation). The transverse colon passes behind the duodenum.

Duplications of Gastrointestinal Tract. Duplications of intestinal loops may occur anywhere along the length of the alimentary canal.[15] They are most frequently located in the region of the ileum, where they may vary from a small diverticulum to a large cyst (fig. 14-21*C*, *D*). Although duplications of the gut always remain attached to the segment of origin, their mucosa may be greatly different. A duplication of the rectum has been found to be lined with gastric mucosa.[16]

During development the alimentary canal goes through a transient solid

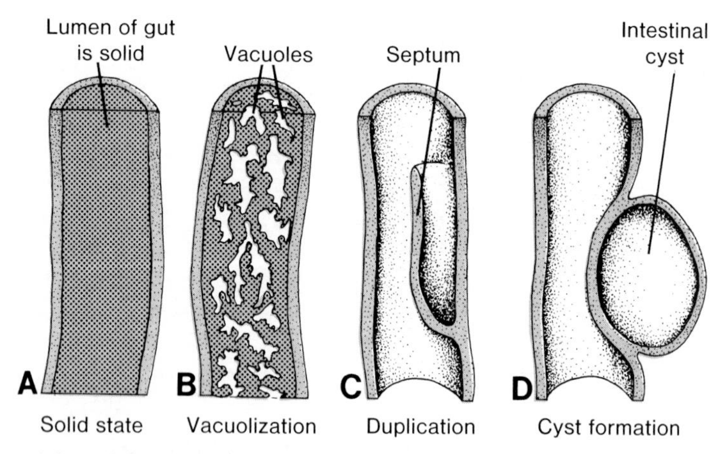

Figure 14-21. Schematic drawings of a portion of the ileum showing the solid state (*A*), vacuole formation (*B*), persistence of a septum in the vacuolated lumen (*C*). This abnormality leads frequently to diverticulum formation, particularly in the loops of the small intestine, *D*, Cyst formation. Frequently communication with the lumen of the gut is absent.

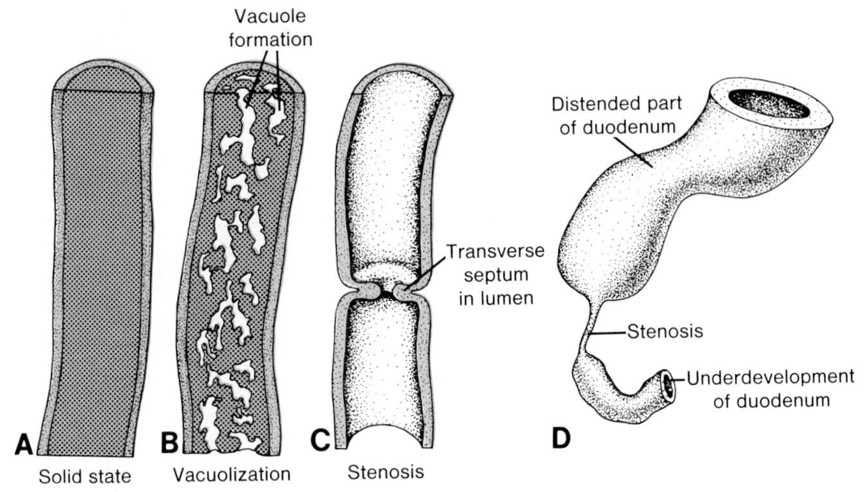

Figure 14-22. Schematic drawings of a portion of the duodenum showing the solid state (*A*), vacuole formation (*B*), and stenosis caused by insufficient vacuolization (*C*). *D*, Stenosis of the duodenum. Note the distention of the proximal part of the duodenum and the narrowing of the distal part.

state, which is followed by reestablishment of the lumen by means of vacuolization (fig. 14-21*A*, *B*). If, however, some isolated vacuoles do not properly fuse with those forming the main lumen, a duplication may arise (fig. 14-21*C*). Similarly when no communication with the main lumen is established, a cyst will be formed. They are usually located on the mesenteric side of the intestinal loop (fig. 14-21*D*).

Atresia and Stenosis of Gut. Atresia and stenosis may likewise occur anywhere along the length of the primitive intestinal loop. In case of atresia (complete absence of the lumen), a thin diaphragm is usually found across the lumen of the gut (fig. 14-22 *A*, *B*, *C*). This diaphragm is thought to result from an incomplete vacuolization of the lumen. The duodenum is particularly known for the presence of a **stenosis** or narrowing of the lumen. Such a stenosis frequently results in a distention of the proximal part and a marked narrowing of the intestinal loops below the level of the stenosis (fig. 14-22*D*). The proximal part of the duodenum may exceed the stomach in size. Duodenal stenosis in the newborn is usually accompanied by severe vomiting of bile colored fluid.

Hindgut

The hindgut gives rise to the distal third of the transverse colon, the descending colon, the sigmoid, the rectum, and the upper part of the anal canal. The endoderm of the hindgut also forms the internal lining of the bladder and urethra (see Chapter 15).

The terminal portion of the hindgut enters into the cloaca, an endoderm lined cavity which is in direct contact with the surface ectoderm. In the contact area between the endoderm and ectoderm, the **cloacal membrane** is formed (fig. 14-23*A*, *B*).

During further development a transverse ridge, the **urorectal septum**, arises in the angle between the allantois and the hindgut (fig. 14-23). This septum gradually grows caudad, thereby dividing the cloaca into an anterior portion, the **primitive urogenital sinus**, and a posterior part, the **anorectal canal** (fig. 14-23*C*). When the embryo is seven weeks old the urorectal septum reaches the cloacal membrane, at which point the **perineum** is formed. The cloacal membrane is then divided into the posterior **anal membrane**, and the anterior **urogenital membrane** (for discussion of further development of the urogenital sinus, see Chapter 15, p. 242).

In the meantime, the anal membrane is surrounded by mesenchymal swellings, and in the eighth week it is found at the bottom of an ectodermal depression, known as **anal pit** or **proctodeum** (fig. 14-24). In the ninth week

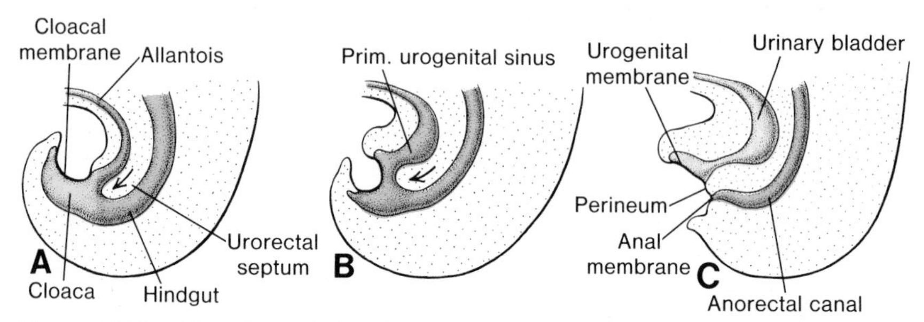

Figure 14-23. Drawings of the cloacal region in embryos at successive stages of development. Arrow indicates the route of descent followed by the urorectal septum. Note the anorectal canal and the perineum.

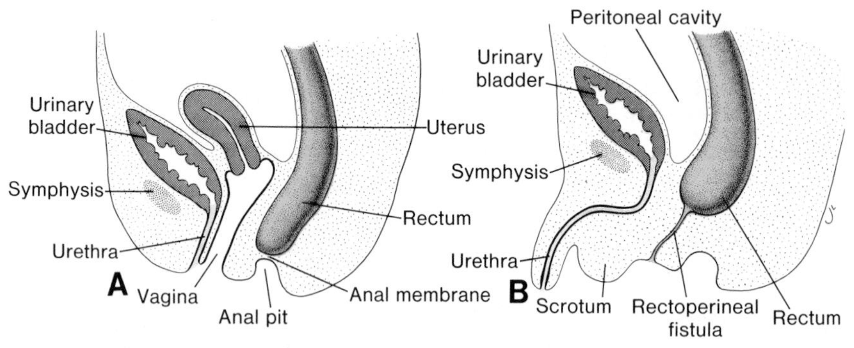

Figure 14-24. *A*, Imperforate anus. The anal membrane persists as a diaphragm between the upper and lower portions of the anal canal. *B*, Rectal atresia. The ampulla of the rectum has failed to develop. Usually a rectoperineal fistula is present.

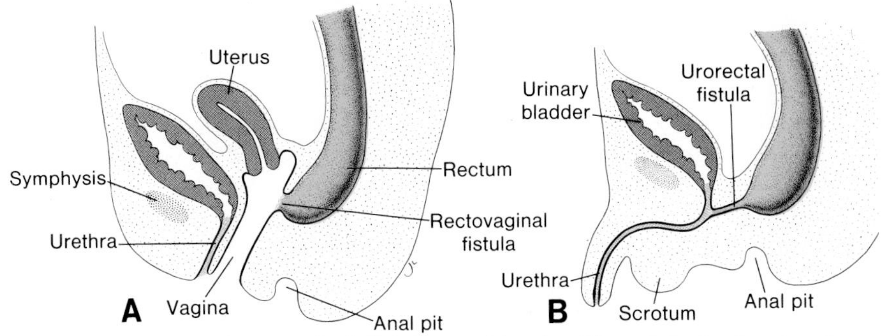

Figure 14-25. *A*, Rectovaginal fistula combined with rectal atresia, due to a defect in the formation of the urorectal septum. *B*, Urorectal fistula combined with rectal atresia.

the anal membrane ruptures and an open pathway is formed between the rectum and the outside. The upper part of the anal canal is thus endodermal in origin and is vascularized by the artery of the hindgut, the **inferior mesenteric artery**. The lower third of the anal canal, however, is of ectodermal origin and is supplied by the systemic rectal arteries, branches of the **internal pudendal artery**. The junction between the endodermal and ectodermal parts is formed by the **pectinate line**, which is found just below the anal columns. Approximately at this line the epithelium changes from columnar into stratified squamous epithelium.

Imperforate Anus, Rectal Atresia. In simple cases, the anal canal ends blind at the anal membrane, which then forms a diaphragm between the endodermal and ectodermal portions of the anal canal (fig. 14-24*A*).[7] In more serious cases, a thick layer of connective tissue may be found between the terminal end of the rectum and the surface, either due to a failure of the anal pit to develop, or due to an atresia of the ampullar part of the rectum, **rectal atresia** (fig. 14-24*B*). **Deviation of the urorectal septum** in dorsal direction probably **causes** many **rectal and anal abnormalities**.[18]

Rectal Fistulas. Rectal fistulas are frequently observed in association with an imperforate anus, and may be found between the rectum and the vagina, the urinary bladder, or the urethra (fig. 14-25).[19] Frequently, such a fistula opens to the surface in the perineal region (fig. 14-24*B*).

SUMMARY

The epithelium of the digestive system and its derivatives are of endodermal origin; the muscular and peritoneal components are of mesodermal origin. The system extends from the buccopharyngeal membrane to the cloacal membrane (fig. 14-1) and is divided in the (1) pharyngeal gut, (2) foregut, (3) midgut, and (4) hindgut. Since the pharyngeal gut gives rise mainly to the pharynx and related glands it is discussed under Chapter 16—"Head and Neck."

The **foregut** gives rise to the esophagus; trachea and lung buds; the stomach; and the duodenum proximal to the entrance of the bile duct. In addition, the liver, pancreas, and biliary apparatus develop as outgrowths of the endodermal epithelium of the upper part of the duodenum (fig. 14-7). Since the upper part of the foregut is divided by a septum (the esophagotracheal septum) into the esophagus posteriorly and the trachea and lung buds anteriorly, deviation of the septum may result in abnormal openings between the trachea and esophagus (fig. 14-4). The epithelial liver cords and biliary system growing out into the septum transversum (fig. 14-7) differentiate into the parenchyma and form the lining of the biliary ducts. The hemopoietic cells, present in great numbers before birth, the Kupffer, and connective tissue cells are of mesodermal origin. The pancreas develops from a ventral bud and a dorsal bud which later fuse to form the definitive pancreas (figs. 14-10 and 14-11). Sometimes the two parts surround the duodenum—annular pancreas—causing constriction of the gut (fig. 14-12).

The **midgut** forms the primary intestinal loop (fig. 14-13) and gives rise to the duodenum distal to the entrance of the bile duct till the junction of the proximal two-thirds of the transverse colon with the distal one-third (fig. 14-13). At its apex the primary loop remains temporarily in open connection with the yolk sac through the vitelline duct. During the sixth week the loop grows so rapidly that it protrudes into the umbilical cord—physiological herniation (fig. 14-14). During the 10th week it returns into the abdominal cavity. While these processes occur the midgut loop rotates 270° counterclockwise (fig. 14-15). Remnants of the vitelline duct (fig. 14-18); failure of the midgut to return to the abdominal cavity (fig. 14-19); malrotation; stenosis; and duplications of parts of the gut (figs. 14-20 to 14-22) are common abnormalities.

The **hindgut** gives rise to the distal one-third of the transverse colon till the upper part of the anal canal (the distal part of the anal canal originates from the ectodermal anal pit (fig. 14-24). The caudal part of the hindgut is divided by the urorectal septum into the rectum and anal canal posteriorly and urinary bladder and urethra anteriorly (fig. 14-23). Deviation of the urorectal septum may result in rectal atresia and abnormal openings between the rectum and the urethra, bladder, or vagina (figs. 14-24 and 14-25).

REFERENCES

1. Salebury, A. M., and Collins, R. E. Congenital pyloric atresia. *Arch. Surg., 80:* 501, 1960.
2. Muller, G. S. Intrathoracic duplications of the foregut. II. *S. Afr. Med. J., 34:* 259, 1960.

3. Rawling, J. T. A prepyloric septum. *Br. J. Surg.*, *47:* 162, 1960.
4. Severn, C. B. A morphological study of the development of the human liver. I. Development of the hepatic diverticulum. *Am. J. Anat.*, *131:* 133, 1971.
5. Houle, M. P., and Hill, P. S. Congenital absence of the gallbladder. *J. Maine Med. Assoc.*, *51:* 108, 1960.
6. Dawson, W., and Langman, J. An anatomical-radiological study on the pancreatic duct pattern in man. *Anat. Rec.*, *139:* 59, 1961.
7. Falin, L. T. The development and cytodifferentiation of the islets of Langerhans in human embryos and foetuses. *Acta Anat.*, *68:* 147, 1967.
8. Coltart, T. M., Beard, R. W., Turner, R. C., and Oakley, N. W. Blood glucose and insulin relationships in the human mother and fetus before onset of labour. *Br. Med. J.*, *4:* 17, 1969.
9. Weatherill, D., Forgrave, E. G., and Carpenter, W. S. Annular pancreas producing duodenal obstruction in the newborn. *Am. J. Dis. Child.*, *95:* 202, 1958.
10. Martinez, N. S., Morlach, C. G., Dockerty, B., Waugh, J. M., and Weber, H. Heterotopic pancreatic tissue involving the stomach. *Ann. Surg.*, *147:* 1, 1958.
11. Brookes, V. B. Meckel's diverticulum in children. *Br. J. Surg.*, *42:* 57, 1954.
12. Smith, J. R. Accessory enteric formations; a classification and nomenclature. *Arch. Dis. Child.*, *35:* 87, 1960.
13. McKeown, T., MacMahon, B., and Record, R. G. An investigation of 69 cases of exomphalos. *Am. J. Hum. Genet.*, *5:* 168, 1953.
14. Estrada, R. L. *Anomalies of Intestinal Rotation and Fixation.* Charles C Thomas, Springfield, Ill., 1968.
15. Christensen, C. R. Duplications in the gastro-intestinal tract in children. *Dan. Med. Bull.*, *5–6:* 281, 1959.
16. Clift, M. M. Duplication of the small intestine. *J. Am. Med. Wom. Assoc.*, *9:* 396, 1954.
17. Kiesemetter, W. B., and Nixon, H. H. Imperforate anus. I. Its surgical anatomy. *J. Pediatr. Surg.*, *2:* 60, 1967.
18. Gourgh, M. H. Congenital abnormalities of the anus and rectum. *Arch. Dis. Child.*, *36:* 146, 1961.
19. Stephens, F. D. *Congenital Malformations of the Rectum, Anus and Genito-Urinary Tracts.* E. and S. Livingstone, Ltd. Edinburgh, 1963.

chapter 15

Urogenital System

Functionally, the urogenital system can be divided into two entirely different components: (1) the **urinary system** and (2) the **genital system.** Embryologically and anatomically, however, they are intimately interwoven. Both develop from a common mesodermal ridge along the posterior wall of the abdominal cavity, and the excretory ducts of both systems initially enter a common cavity, the cloaca.

With further development, the overlapping of the two systems is particularly evident in the male. The primitive excretory duct first functions as a urinary duct, but later is transformed into the main genital duct. Moreover, in the adult, the urinary as well as the genital organs discharge urine and semen through a common duct, the penile urethra.

Urinary System

FORMATION OF EXCRETORY UNIT

In the beginning of the fourth week the intermediate mesoderm in the cervical region loses its contact with the somite and forms segmentally arranged cell clusters, known as the **nephrotomes** (fig. 15-1*A*). The nephrotomes grow in a lateral direction and obtain a lumen (fig. 15-1*B*). The newly established tubules, the **nephric tubules**, open medially into the intra-embryonic coelom, while at their lateral ends they grow in a caudal direction. During the caudal growth the tubules of succeeding segments unite and form a longitudinal duct on each side of the embryo (fig. 15-2). While this occurs small branches of the dorsal aorta cause invaginations in the wall of the nephric tubule as well as in that of the coelomic cavity, thus forming the **internal** and **external glomeruli**, respectively (fig. 15-1*B*). Together the glomeruli and the nephric tubule form the **excretory unit.**

In the thoracic, lumbar, and sacral regions the intermediate mesoderm (1) loses its contact with the coelomic cavity; (2) loses its segmentation; and (3) forms two, three, or even more excretory tubules per original segment (fig. 15-2*B*). As a result the external glomeruli fail to develop and the unsegmented mesoderm forms the **nephrogenic tissue cords.** They give rise to the excretory (renal) tubules of all kidney systems and produce bilateral longitudinal ridges, the **urogenital ridges**, on the dorsal wall of the coelomic cavity (fig. 15-3).

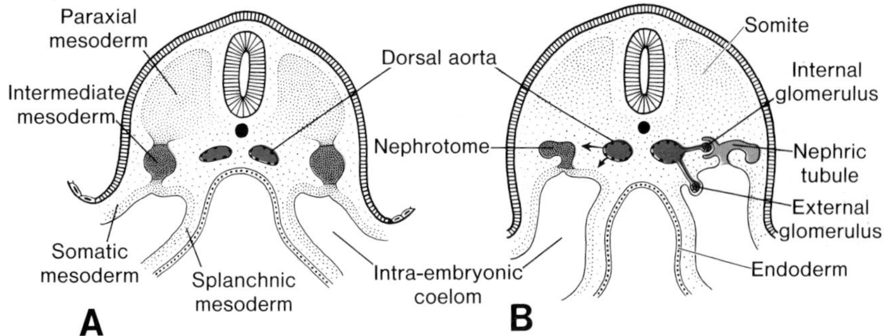

Figure 15-1. Schematic transverse sections through embryos at various stages of development to show the formation of the nephric tubule. *A*, At 21 days; *B*, at 25 days. Note the formation of the external and internal glomeruli, and the open connection between the coelomic cavity and the nephric tubule (modified after Heuser).

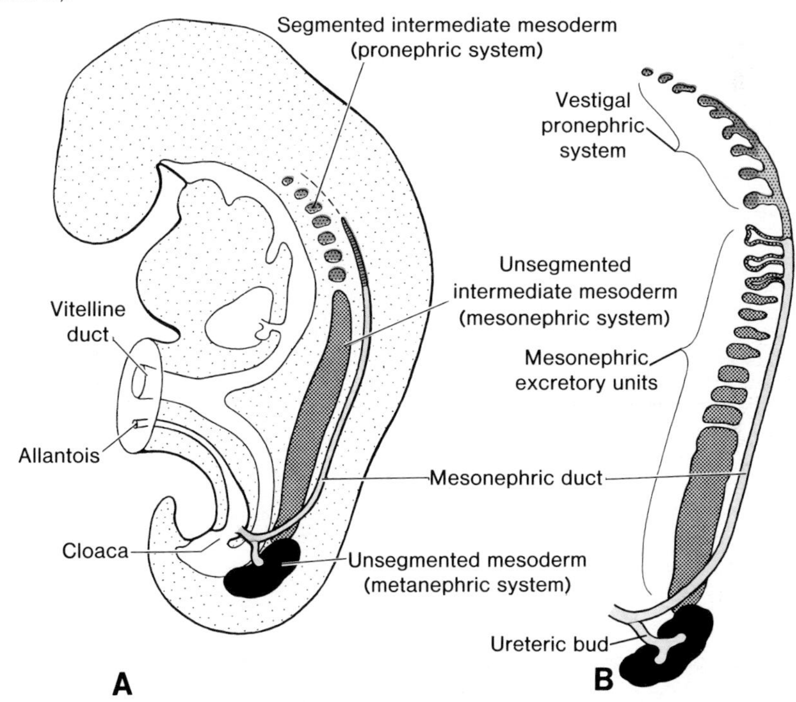

Figure 15-2. *A*, Schematic diagram showing the relation of the intermediate mesoderm of the pronephric, mesonephric, and metanephric systems. In the cervical and upper thoracic regions the intermediate mesoderm is segmented; in the lower thoracic, lumbar, and sacral regions it forms a solid, unsegmented mass of tissue, the nephrogenic cord. Note the longitudinal collecting duct, initially formed by the pronephros but later taken over by the mesonephros. *B*, Schematic representation of the excretory tubules of the pronephric and mesonephric systems in a five-week-old embryo. Note the remnant of the pronephric excretory tubules and longitudinal collecting duct.

Kidney Systems

Three different, slightly overlapping kidney systems are formed during intra-uterine life in man: the **pronephros**, the **mesonephros**, and **metanephros** or **permanent kidney.**

PRONEPHROS

In the human embryo the pronephros is represented by 7 to 10 solid cell groups in the cervical region (fig. 15-2*B*). The first formed vestigial nephrotomes regress before the last ones are formed, and at the end of the fourth week all indications of the pronephric system have disappeared.

MESONEPHROS

During regression of the pronephric system, the first excretory tubules of the mesonephros appear. They lengthen rapidly, form an "S"-shaped loop, and acquire a glomerulus at their medial extremity (fig. 15-3*A*). Here the tubule forms **Bowman's capsule.** The capsule and glomerulus form together a **mesonephric (renal) corpuscle.** At the opposite end the tubule enters the longitudinal collecting duct, known as the **mesonephric** or **Wolffian duct** (figs. 15-2 and 15-3).

In the middle of the second month, the mesonephros forms a large ovoid organ on each side of the midline (fig. 15-3*B*). Since the developing gonad

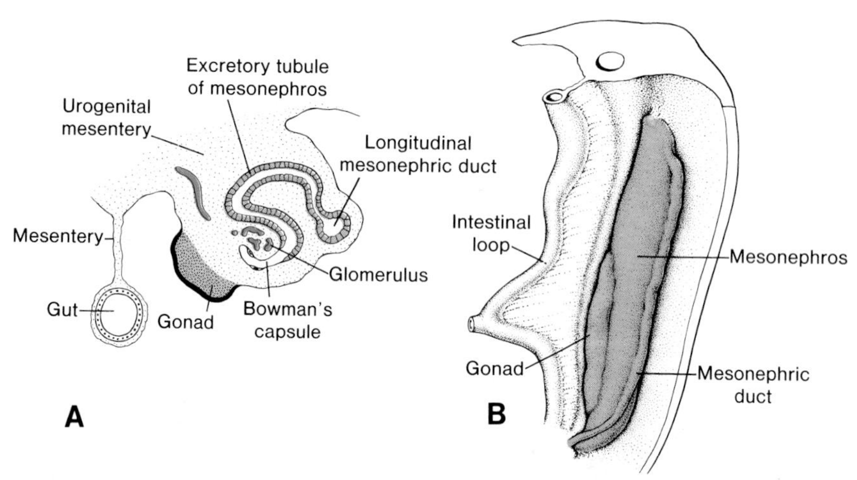

Figure 15-3. *A*, Transverse section through the urogenital ridge in the lower thoracic region of a five-week embryo, showing the formation of an excretory tubule of the mesonephric system. Note the appearance of the capsule of Bowman and the gonadal ridge. The mesonephros and gonad are attached to the posterior abdominal wall by a broad urogenital mesentery. *B*, Drawing to show the relationship of the gonad and the mesonephros. Note the size of the mesonephros. The mesonephric duct (Wolffian duct) runs along the lateral side of the mesonephros.

is located on its medial side, the ridge formed by both organs is known as the **urogenital ridge** (fig. 15-3*A*). While the caudal tubules are still differentiating, the cranial tubules and glomeruli show degenerative changes and by the end of the second month the majority have disappeared. A few of the caudal tubules and the mesonephric duct, however, persist in the male but disappear in the female (see "Genital System").

Although functional activity of the mesonephric system has been shown to exist in embryos of the cat, rabbit, and pig, and great resemblances in ultrastructure exist between the mesonephros and the metanephros, functional activity of the mesonephros has not been demonstrated in the human embryo.[1]

METANEPHROS OR PERMANENT KIDNEY

The third urinary organ, the **metanephros** or **permanent kidney**, appears in the fifth week. Its excretory units develop from the **metanephric mesoderm** (fig. 15-4) in the same manner as in the mesonephric system. The development of the duct system, however, differs from the other kidney systems.

Collecting System. The collecting ducts of the permanent kidney develop from the **ureteric bud**, an outgrowth of the mesonephric duct close to its entrance into the cloaca (figs. 15-2 and 15-4). The bud penetrates the metanephric tissue, which, as a cap, is molded over its distal end (fig. 15-4). Subsequently the bud dilates, forming the **primitive renal pelvis**; simultaneously it splits into a cranial and caudal portion, the future **major calyces** (fig. 15-5*A*, *B*).

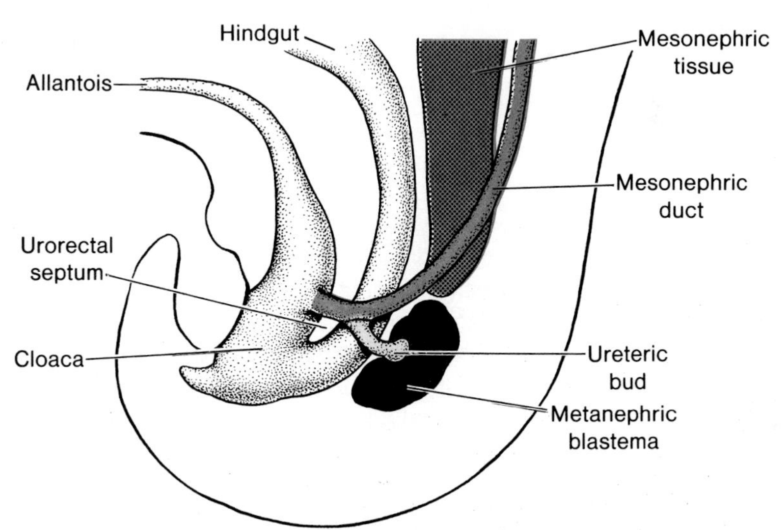

Figure 15-4. Schematic drawing to show the relationship of the hindgut and cloaca at the end of the fifth week. The ureteric bud begins to penetrate the metanephric mesoderm or blastema.

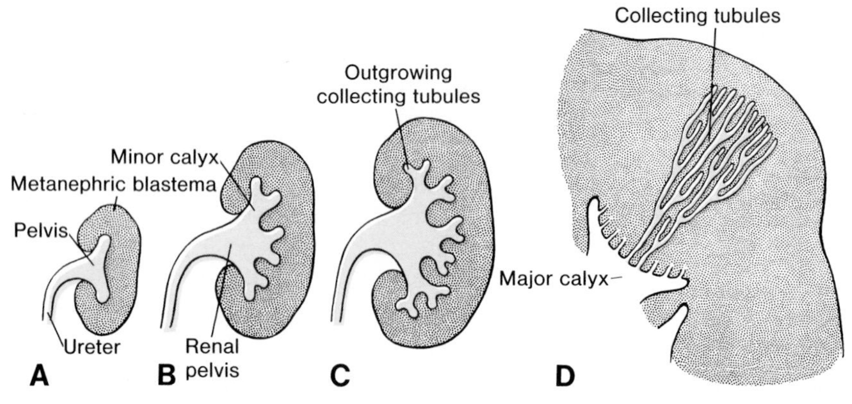

Figure 15-5. Schematic drawings showing the development of the renal pelvis, calyces, and collecting tubules of the metanephros. *A*, At six weeks; *B*, end of sixth week; *C*, seven weeks *D*, newborn. Note the pyramid form of the collecting tubules entering the minor calyx.

Each calyx, while penetrating into the metanephric tissue, forms two new buds. These buds continue to subdivide until 12 or more generations of tubules have been formed (fig. 15-5). While at the periphery more tubules are formed until the end of the fifth month, the tubules of the second order enlarge and absorb those of the third and fourth generations, thus forming the **minor calyces** of the renal pelvis. During further development the collecting tubules of the fifth and successive generations elongate considerably and converge on the minor calyx, thereby forming the **renal pyramid** (fig. 15-5*D*). Hence, the **ureteric bud gives rise to the ureter, the renal pelvis, the major and minor calyces, and approximately one to three million collecting tubules.**

Excretory System. Each newly formed collecting tubule is covered at its distal end by a so-called **metanephric tissue cap** (fig. 15-6*A*). Under the inductive influence of the tubule, cells of the tissue cap form small vesicles, the **renal vesicles**, which in turn give rise to small tubules (fig. 15-6*B*,*C*).[2, 3] These tubules form the **nephrons or excretory units**. The proximal end of the nephron forms the **Bowman's capsule** of the renal glomerulus (fig. 15-6*C*, *D*). The distal end forms an open connection with one of the collecting tubules, thus establishing a passageway from the glomerulus to the collecting unit. Continuous lengthening of the **excretory tubule** results in the formation of the **proximal convoluted tubule**, the **loop of Henle**, and the **distal convoluted tubule** (fig. 15-6*E*, *F*). Hence, the kidney develops from two different sources: (1) **the metanephric mesoderm which provides the excretory units**, and (2) **the ureteric bud which gives rise to the collecting system**.

At birth the kidneys have a lobulated appearance. During infancy the lobulation disappears as a result of further growth of the nephrons. Their numbers, however, do not increase.

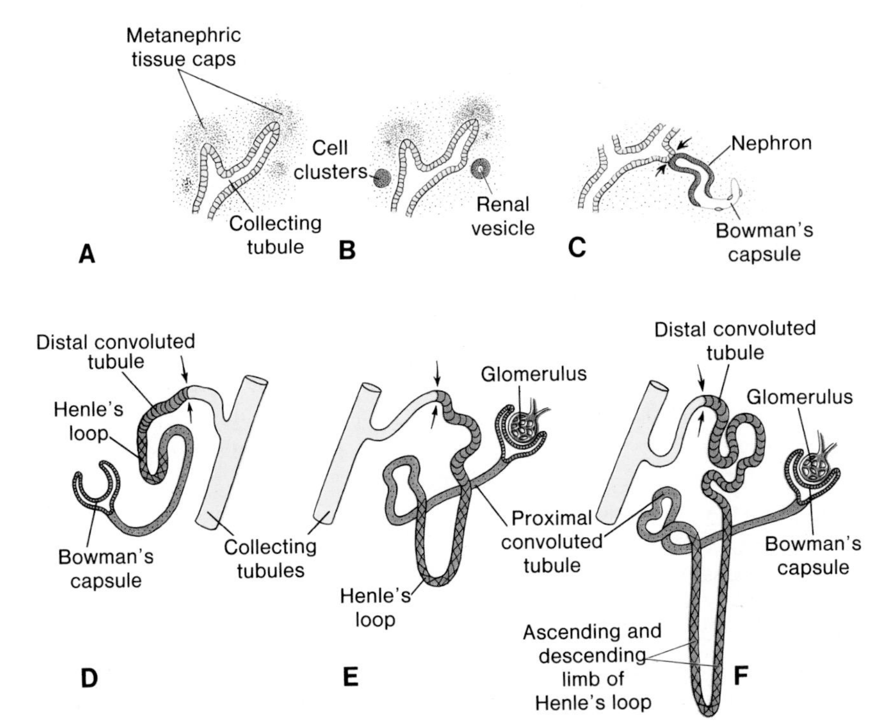

Figure 15-6. Schematic representation of the development of a metanephric excretory unit. Arrows indicate the place where the excretory unit (red) establishes an open communication with the collecting system (blue), thus allowing for the flow of urine from the glomerulus into the collecting ducts.

Congenital Cystic Kidney. According to the "nonunion" theory of the formation of renal cysts, the collecting and excretory tubules occasionally fail to join. The excretory units develop then in a normal manner and may even form functional glomeruli. Accumulation of urine in the convoluted tubules, however, causes them to dilate and gradually to form cysts. These cysts may be so numerous that insufficient active renal tissue remains.

Sometimes one or more cysts are found close to the pelvis of the kidney. These cysts are thought to be remnants of the nephrons of the second, third, or fourth order. When their corresponding collecting tubules are absorbed into the minor calyces, the nephrons usually disappear. It is presently thought that the main cause for cyst formation in the kidney is an abnormal development of the collecting system.[4, 5] In some cases cyst formation was found to result from hyperplasia of the wall of the collecting tubules, while in others abnormal differentiation of the ureteric bud, resulting in dilated, constricted, or sometimes atretic tubules, was held responsible.

Renal Agenesis. Bilateral or unilateral renal agenesis is presumably caused by an early degeneration of the ureteric bud. When the ureteric bud does not reach the metanephric tissue cap, the latter fails to proliferate. Unilateral renal agenesis is seen in about one of 1500 individuals; bilateral agenesis is rare.

In case of renal agenesis the amount of amniotic fluid may be small (**oligohydramnos**). The fetus will survive since the kidneys are not necessary for exchange of waste products. Hence, the fetus will develop, but after birth will die within a few days.[6]

Double and Ectopic Ureter. Early splitting of the ureteric bud may result in partial or complete duplication of the ureter (fig. 15-7*A*, *B*). The metanephric tissue may then be divided into two parts, each with its own renal pelvis and ureter. More frequently, however, the two parts have a number of lobes in common, as a result of the intermingling of the collecting tubules. In rare cases one ureter opens into the bladder, while the other enters the vagina, urethra, or vestibule (fig. 15-7*C*). This abnormality results from the development of two ureteric buds. One of the buds usually has a normal position, while the abnormal bud moves downwards together with the mesonephric duct and hence its low, abnormal entrance in the bladder, urethra, vagina, or epididymal region (fig. 15-7*C*).[7]

POSITION OF KIDNEY

The kidney, initially located in the pelvic region, later shifts to a more cranial position in the abdomen. This so-called **ascent of the kidney** is caused by a diminution of the body curvature as well as by the growth of the body in the lumbar and sacral regions (fig. 15-8). In the pelvis the metanephros receives its arterial supply from a pelvic branch of the aorta. During its

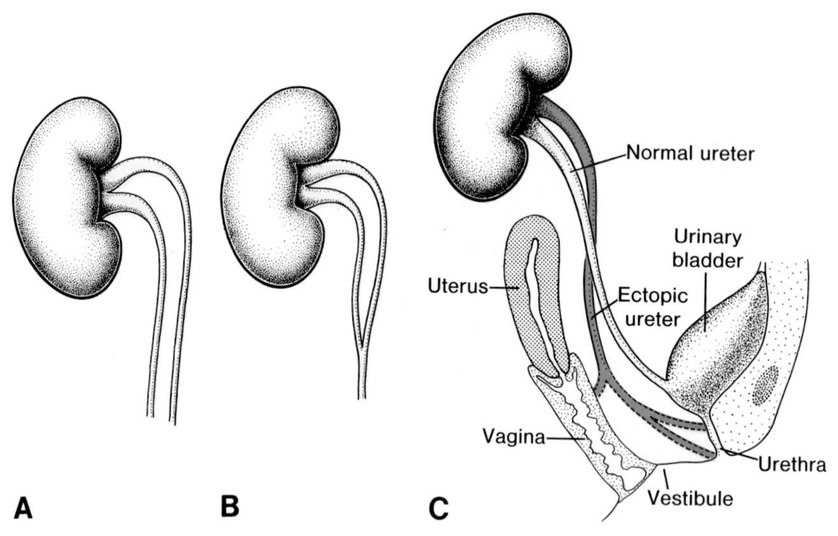

Figure 15-7. *A*, *B*, Complete and partial double ureter. *C*, Possible sites of ectopic ureteral openings in the vagina, urethra, and vestibule.

"ascent" to the abdominal level it is vascularized by arteries which originate from the aorta at continuously higher levels. The lower vessels usually degenerate but vascular variations, such as two or three **supernumerary renal arteries**, may result from the persistence of the embryonic vessels.

Pelvic and Horseshoe Kidney. During their ascent the kidneys pass through the arterial fork formed by the umbilical arteries, but occasionally one of them fails to do so. It then remains in the pelvis close to the common iliac artery and is known as a **pelvic kidney** (fig 15-9*A*). Sometimes both kidneys are pushed so close together during their passage through the arterial fork that the lower poles fuse. This results in the formation of a **horseshoe kidney** (fig. 15-9*B*). The horseshoe kidney is usually located at the level of the lower lumbar vertebrae, since its ascent is prevented by the root of the

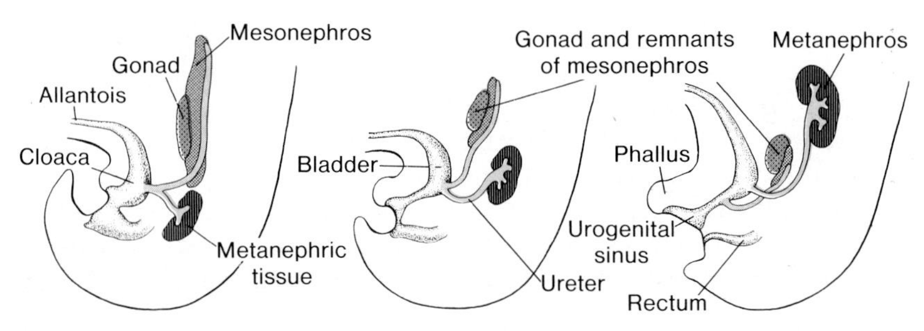

Figure 15-8. Ascent of the kidney. Note the change in position between the metanephros and mesonephric system. The mesonephric system degenerates almost entirely and only a few remnants persist in close contact with the gonad. In both the male and female embryo the gonad descends from its original level to a much lower position.

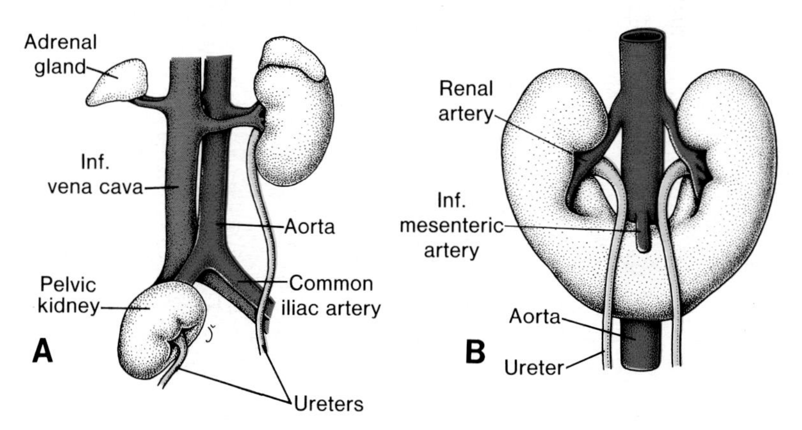

Figure 15-9. *A*, Unilateral pelvic kidney. Note the position of the adrenal gland on the affected side. *B*, Horseshoe kidney, ventral view. Note the position of the inferior mesenteric artery.

inferior mesenteric artery (fig. 15-9*B*). The ureters arise from the anterior surface of the kidney and pass ventral to the isthmus in caudal direction. Horseshoe kidney is a rather common abnormality and is found in 1 of about 600 people.

FUNCTION OF KIDNEY

The metanephros or definitive kidney becomes functional during the second half of pregnancy. The urine is passed into the amniotic cavity and mixes with the amniotic fluid. This fluid is swallowed by the fetus and enters the intestinal tract, where it is absorbed into the bloodstream. It thus enters the placenta which during fetal life functions as a kidney system to transfer metabolic waste products to the mother.

BLADDER AND URETHRA

During the fourth to seventh weeks of development, the **urorectal septum** divides the cloaca into the **anorectal canal**, and the **primitive urogenital sinus** (fig. 15-10). The cloacal membrane itself is then divided into the **urogenital membrane**, anteriorly, and the **anal membrane**, posteriorly (fig. 15-10*C*). (For a more detailed description see Chapter 14 and fig. 14-23.)

Three portions of the primitive urogenital sinus can be distinguished: (1) the upper and largest part is the **urinary bladder** (fig. 15-11*A*). Initially the bladder is continuous with the allantois, but when the lumen of the allantois is obliterated, a thick fibrous cord, the **urachus**, remains connecting the apex of the bladder with the umbilicus (fig. 15-11*B*).[13] In the adult the ligament is known as the **median umbilical ligament.** (2) A rather narrow canal, the **pelvic part of the urogenital sinus**, which in the male gives rise to the prostatic and membranous parts of the urethra. (3) The **definitive urogenital sinus**, also known as the **phallic part of the urogenital sinus.** It is considerably flattened from side to side and is separated from the outside by the urogenital membrane (fig. 15-11*A*).

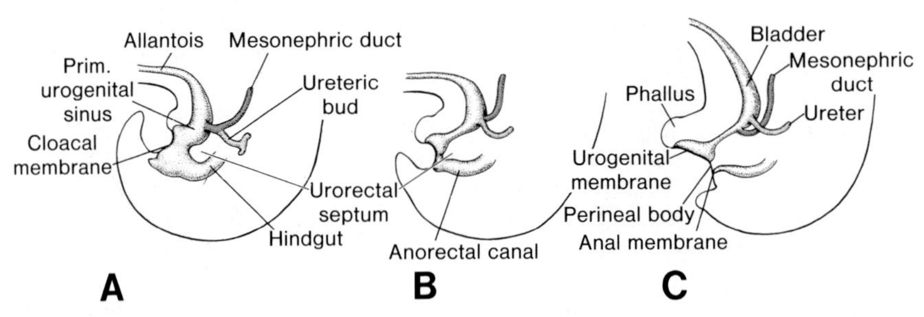

Figure 15-10. Diagrams showing the division of the cloaca into the urogenital sinus and anorectal canal. Note that the mesonephric duct is gradually absorbed into the wall of the urogenital sinus and that the ureters enter separately. *A*, End of the fifth week; *B*, seven weeks; *C*, eight weeks.

The development of the definitive urogenital sinus differs greatly between the two sexes (see "Genital System," p. 255).

During division of the cloaca, the caudal portions of the mesonephric ducts are gradually absorbed into the wall of the urinary bladder (fig. 15-12). Consequently the ureters, initially outbuddings of the mesonephric ducts, enter the bladder separately (fig. 15-12*B*). As a result of the ascent of the kidneys the orifices of the ureters move further cranially; those of the mesonephric ducts move close together to enter the prostatic urethra and in the male become the **ejaculatory ducts** (fig. 15-12*C*, *D*).[8] Since both the mesonephric ducts and the ureters are of mesodermal origin, the mucosa of the bladder formed by incorporation of the ducts, the **trigone of the bladder**, is mesodermal in origin. The remaining part of the bladder is derived from the urogenital sinus and is endodermal in origin. With time, the mesodermal

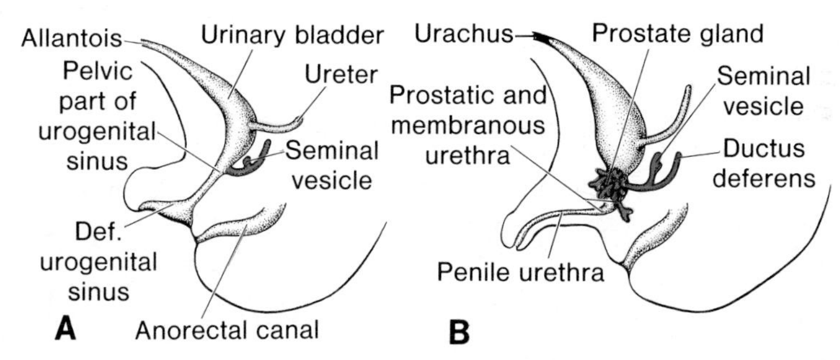

Figure 15-11. *A*, Development of the urogenital sinus into the urinary bladder, the pelvic part of the urogenital sinus, and the definitive urogenital sinus. *B*, In the male the definitive urogenital sinus develops into the penile urethra. The prostate gland is formed by outbuddings of the urethra, while the seminal vesicles are formed by an outbudding of the ductus deferens.

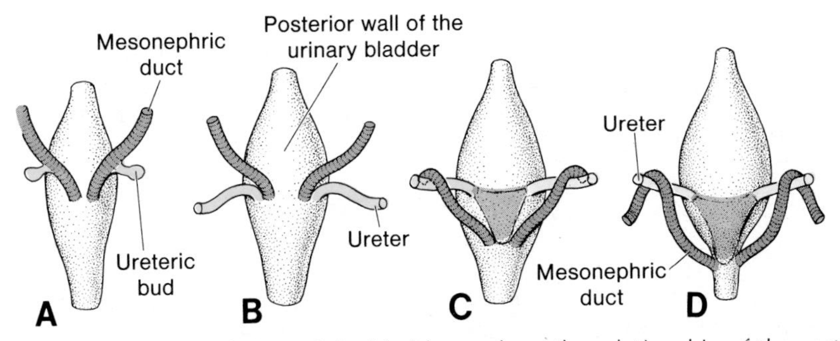

Figure 15-12. Dorsal view of the bladder to show the relationship of the ureters and mesonephric ducts during development. Initially the ureter is formed by an outgrowth of the mesonephric duct, but with time it obtains a separate entrance into the urinary bladder. Note the trigone of the bladder, formed by incorporation of the mesonephric ducts.

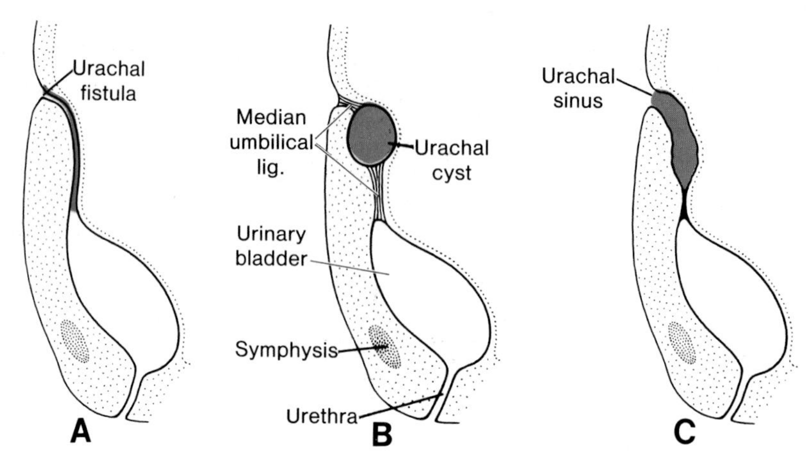

Figure 15-13. *A*, Urachal fistula. *B*, Urachal cyst. *C*, Urachal sinus. The sinus may or may not be in open communication with the urinary bladder.

lining of the trigone is replaced by endodermal epithelium, so that finally the inside of the bladder is completely lined with epithelium of endodermal origin.[8]

Urethra. The epithelium of the male and female urethra is of endodermal origin, while the surrounding connective and smooth muscle tissue are derived from splanchnic mesoderm. At the end of the third month, the epithelium of the prostatic urethra begins to proliferate and forms a number of outbuddings which penetrate the surrounding mesenchyme. In the male these buds form the **prostate gland** (fig. 15-11*B*). In the female the cranial part of the urethra gives rise to the **urethral** and **paraurethral glands.**

Urachal Fistula, Cyst, and Sinus. When the lumen of the intra-embryonic portion of the allantois persists, urine may drain from the umbilicus (fig. 15-13*A*). This abnormality is known as a **urachal fistula.**[9] If only a localized area of the allantois persists, secretory activity of its lining results in a cystic dilation, the **urachal cyst** (fig. 15-13*B*). When the lumen in the upper part persists, it forms a **urachal sinus.** This sinus is usually continuous with the urinary bladder (fig 15-13*C*).[9]

Genital System

GONADS

Although the sex of the embryo genetically is determined at the time of fertilization, the gonads do not acquire male or female morphological characteristics until the seventh week of development.

The gonads appear initially as a pair of longitudinal ridges, the **genital** or **gonadal ridges** (fig. 15-14) and are formed by proliferation of the coelomic epithelium and a condensation of the underlying mesenchyme (fig. 15-14*B*).

Germ cells do not appear in the genital ridges until the sixth week of development.[10]

In human embryos the primordial germ cells appear at an early stage of development among the endoderm cells in the wall of the yolk sac close to the allantois (fig. 15-15*A*).[11, 12] They migrate by ameboid movement along the dorsal mesentery of the hindgut (fig. 15-15*B*) and in the sixth week of development invade the genital ridges. If they fail to reach the ridges, the

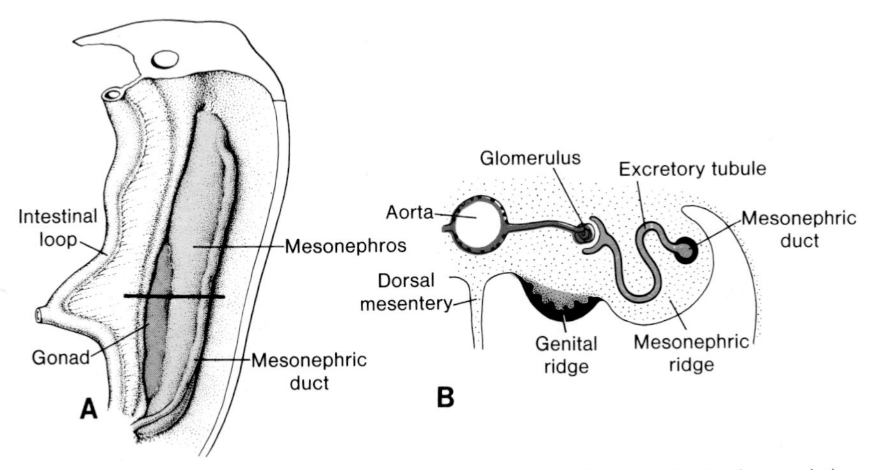

Figure 15-14. *A*, Drawing to show the relationship of the genital ridge and the mesonephros. Note the location of the mesonephric duct. *B*, Transverse section through the mesonephros and genital ridge at a level indicated in *A*.

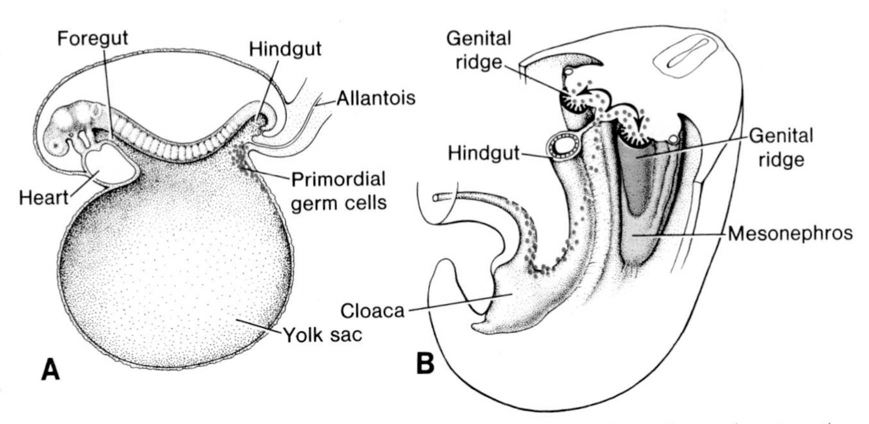

Figure 15-15. *A*, Schematic drawing of a three-week-old embryo showing the primordial germ cells in the wall of the yolk sac, close to the attachment of the allantois (after Witchi). *B*, Drawing to show the migration path of the primordial germ cells along the wall of the hindgut and the dorsal mesentery into the genital ridge.

gonads do not develop. Hence, the primordial germ cells have an inductive influence on the development of the gonad into ovary or testis.

Indifferent Gonad. Shortly before and during the arrival of the primordial germ cells the coelomic epithelium of the genital ridge proliferates and epithelial cells penetrate the underlying mesenchyme. Here they form a number of irregularly shaped cords, the **primitive sex cords** (fig. 15-16). In both male and female embryos these cords are connected to the surface epithelium, and it is impossible to differentiate between the male and female gonad. Hence, the gonad is known as the **indifferent gonad.**

Testis. If the embryo is genetically male the primordial germ cells carry an XY sex chromosome complex. Under influence of the Y-chromosome, which has a testis-determining effect, the primitive sex cords continue to proliferate and penetrate deep into the medulla to form the **testis** or **medullary cords** (table 15-1) (fig. 15-17*A*). Toward the hilus of the gland the cords break up into a network of tiny cell strands which later give rise to the tubules of the **rete testis** (fig. 15-17*A*, *B*).

During further development, the testis cords lose contact with the surface epithelium, and they become separated from it by a dense layer of fibrous connective tissue, the **tunica albuginea**, a typical characteristic of the testis (fig. 15-17).

In the fourth month the testis cords become horseshoe-shaped, and their extremities are continuous with those of the rete testis (fig. 15-17*B*).[13] The testis cords are now composed of primitive germ cells and **sustentacular cells of Sertoli**[14] derived from the surface of the gland (fig. 1-11*A*).

The **interstitial cells of Leydig** develop from the mesenchyme located between the testis cords and are particularly abundant in the fourth to sixth

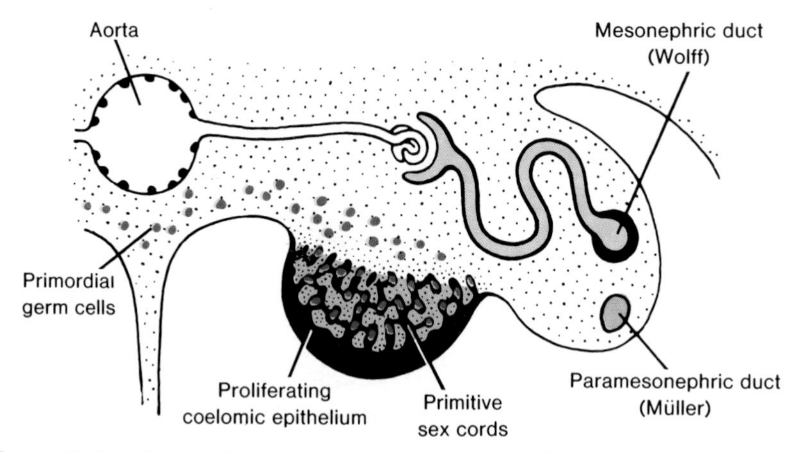

Figure 15-16. Schematic transverse section through the lumbar region of a six-week embryo, showing the indifferent gonad with the primitive sex cords. Some of the primordial germ cells are surrounded by cells of the primitive sex cords.

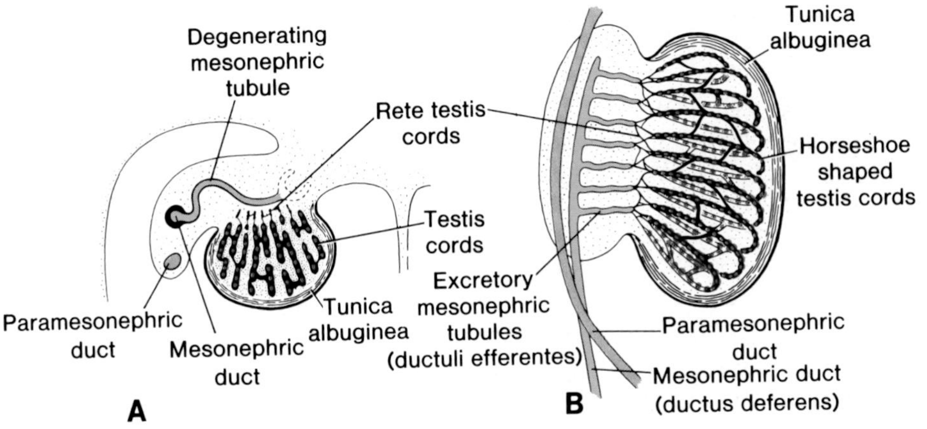

Figure 15-17. *A*, Transverse section through the testis in the eighth week of development. Note the tunica albuginea, the testis cords, the rete testis, and the primordial germ cells. The glomerulus and Bowman's capsule of the mesonephric excretory tubule are in regression. *B*, Schematic representation of the testis and the genital ducts in the fourth month of development. The horseshoe-shaped testis cords are continuous with the rete testis cords. Note the ductuli efferentes (excretory mesonephric tubules) which enter the mesonephric duct.

Table 15-1. **Influence of Primordial Germ Cells on Indifferent Gonad**

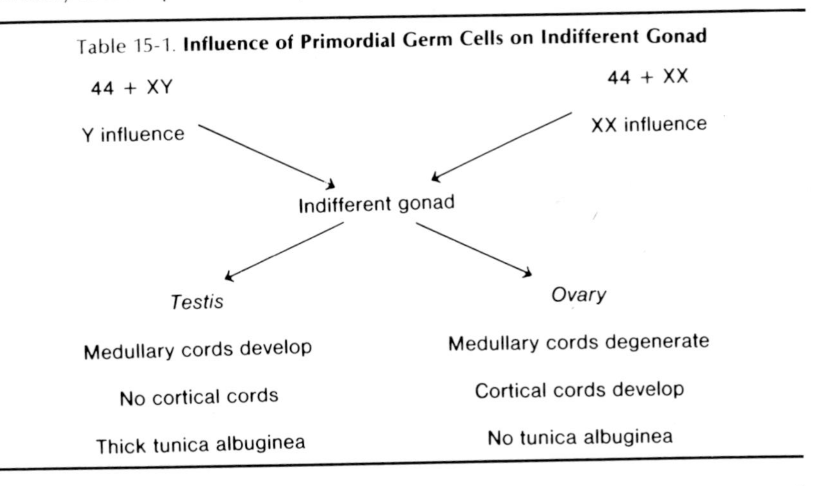

months of development. The testis is now able to influence the sexual differentiation of the genital ducts and external genitalia.

The cords remain solid until puberty, when they acquire a lumen, thus forming the **seminiferous tubules.** Once the seminiferous tubules are canalized, they join the rete testis tubules, which in turn enter the **ductuli efferentes.** These efferent ductules are the remaining parts of the excretory tubules of the mesonephric system. They function as the link between the rete testis and the **mesonephric** or **Wolffian duct**, which is known as the **ductus deferens** (fig. 15-17*B*).

Ovary. In female embryos with an XX sex chromosome complement, the primitive sex cords are broken up into irregular cell clusters (fig. 15-18*A*). These clusters, containing groups of primitive germ cells, are mainly located in the medullary part of the ovary. Later they disappear and are replaced by a vascular stroma which forms the **ovarian medulla** (table 15-1).

The surface epithelium of the female gonad, unlike that of the male, continues to proliferate. In the seventh week it gives rise to a second generation of cords, the **cortical cords**, which penetrate the underlying mesenchyme but remain close to the surface (fig. 15-18*A*). In the fourth month these cords are also split into isolated cell clusters, each surrounding one or more primitive germ cells (fig. 15-18*B*). The germ cells subsequently develop into the oogonia, while the surrounding epithelial cells, descendants of the surface epithelium, form the **follicular cells** (see Chapter 1).

It may thus be stated that the sex of an embryo is determined at the time of fertilization and depends on whether the spermatocyte carries an X- or a Y-chromosome. In embryos with an XX sex chromosome configuration, the medullary cords of the gonad regress and a secondary generation of cortical cords develops (fig. 15-18). In embryos with an XY sex chromosome complex, the medullary cords develop into testis cords and the secondary cortical cords fail to develop (fig. 15-17).

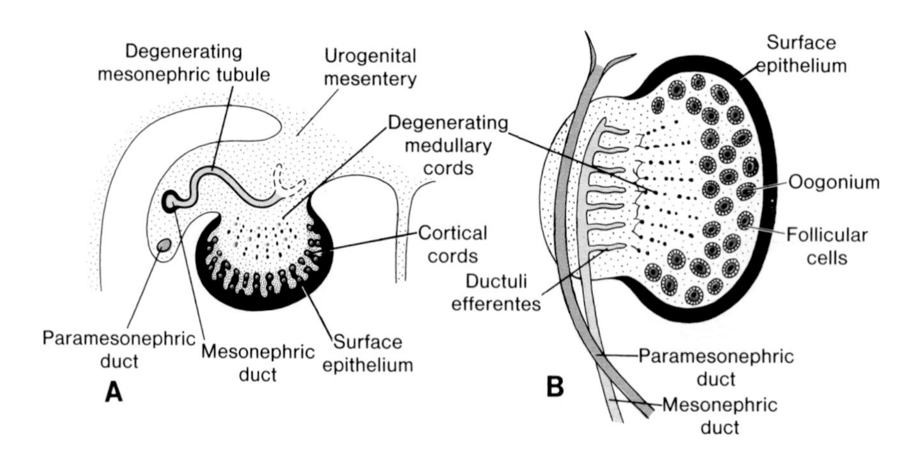

Figure 15-18. *A*, Transverse section through the ovary at the seventh week of development, showing the degeneration of the primitive (medullary) sex cords and the formation of the cortical cords. *B*, Schematic drawing of the ovary and genital ducts in the fifth month of development. Note the degeneration of the medullary cords. The excretory mesonephric tubules (ductuli efferentes) do not communicate with the rete. The cortical zone of the ovary contains groups of oogonia surrounded by follicular cells.

GENITAL DUCTS

Indifferent Stage. Both male and female embryos have initially two pair of genital ducts: (1) the **mesonephric ducts**, and (2) the **paramesonephric ducts.** The paramesonephric duct arises as a longitudinal invagination of the coelomic epithelium on the anterolateral surface of the urogenital ridge (fig. 15-19). Cranially the duct opens in the coelomic cavity with a funnel-like structure; caudally, it runs first lateral to the mesonephric duct, but then crosses it ventrally to grow in caudomedial direction (fig. 15-19). In the midline it comes in close contact with the paramesonephric duct from the opposite side. The two ducts are initially separated by a septum, but later fuse to form the **uterine canal** (fig. 15-21*A*). The caudal tip of the combined ducts projects into the posterior wall of the urogenital sinus, where it causes a small swelling, the **paramesonephric or Müllerian tubercle** (fig. 15-21*A*). The mesonephric ducts open into the urogenital sinus on either side of the Müllerian tubercle.

DIFFERENTIATION OF DUCT SYSTEM

Development of the genital duct system and the external genitalia occurs under influence of hormones circulating in the fetus during intra-uterine life. The fetal testes produce a nonsteroidal substance, known as the **inducer** substance, which induces differentiation and growth of the mesonephric duct system and inhibits the development of the paramesonephric ducts. Because

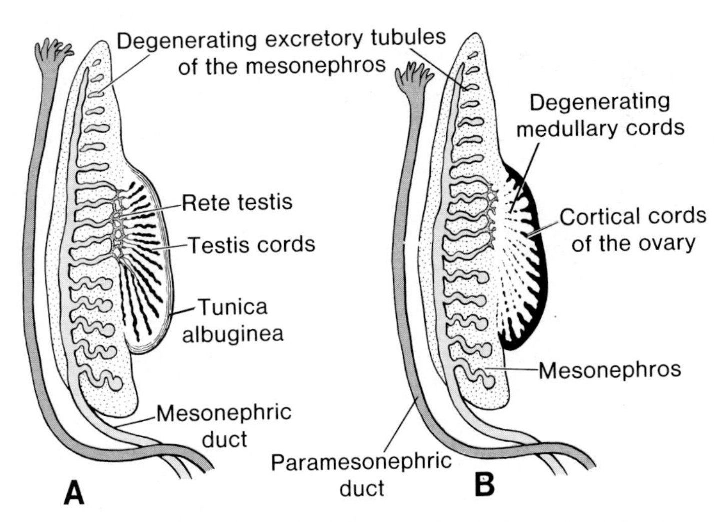

Figure 15-19. Diagram of the genital ducts in the sixth week of development in the male (*A*) and in the female (*B*). The mesonephric and paramesonephric ducts are present in both the male and female. Note the excretory tubules of the mesonephros and their relationship to the developing gonad in both sexes.

of its inhibiting influence on the paramesonephric system, the substance has also been called the **suppressor** substance. In addition to the inducer substance, the testes produce **androgens** which stimulate the growth of the penis, formation of the penile urethra, fusion of the scrotal swellings, and development of the prostate and seminal vesicles[15] (table 15-2).

In the female the paramesonephric duct system develops into the uterine tubes and uterus, probably in response to **maternal and placental estrogens** circulating in the fetus. Since the male inducer substance is absent the mesonephric duct system regresses. In the absence of androgens, the indifferent external genitalia are stimulated by maternal and placental estrogens and differentiate into labia majora, labia minora, clitoris, and part of the vagina (table 15-2).

Genital Ducts in Male. As the mesonephros regresses, a few excretory tubules, the **epigenital tubules**, establish contact with the cords of the rete testis and finally form the **ductuli efferentes** of the testis (fig. 15-20). The excretory tubules along the caudal pole of the testis, the **paragenital tubules**, do not join the cords of the rete testis (fig. 15-20*B*). Their vestiges are collectively known as the **paradidymis.**

The mesonephric duct persists (except for its most cranial portion, the **appendix epididymis**) and forms the main genital duct (fig. 15-20). Immediately below the entrance of the ductuli efferentes it elongates greatly and

Table 15-2. **Influence of Sex Gland on Further Sex Differentiation**

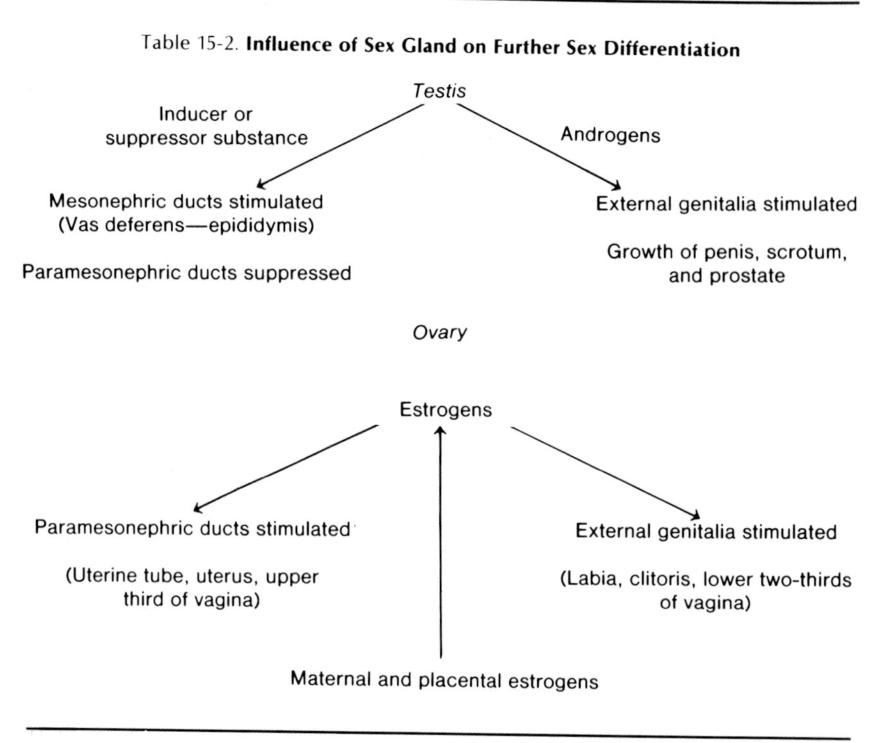

becomes highly convoluted, thus forming the (**ductus**) **epididymis.** From the tail of the epididymis to the outbudding of the **seminal vesicle**, the mesonephric duct obtains a thick muscular coat and is known as the **ductus deferens.** Beyond the seminal vesicle the duct is known as **ejaculatory duct.**

The paramesonephric duct in the male degenerates entirely, except for a small portion at its cranial end, the **appendix testis.** The fate of its caudal part is not precisely known. According to some authors it develops into the **utriculus prostaticus**, a small diverticulum in the wall of the prostatic urethra. Others believe that it is formed by an outpocketing of the urogenital sinus (fig. 15-20*B*).

Genital Ducts in Female. The paramesonephric duct develops into the main genital duct of the female. Initially three parts can be recognized: (1) a cranial vertical portion which opens into the coelomic cavity; (2) a horizontal part which crosses the mesonephric duct; and (3) a caudal vertical part which fuses with its partner from the opposite side (fig. 15-21*A*).

With the descent of the ovary, the first two parts develop into the **uterine tube** (fig. 15-21*B*) and the caudal fused parts into the **uterine canal.** When the second part of the paramesonephric duct courses in mediocaudal direction, the urogenital ridges gradually come to lie in a transverse plane (fig. 15-22*A*, *B*). After the ducts have fused in the midline, a broad transverse

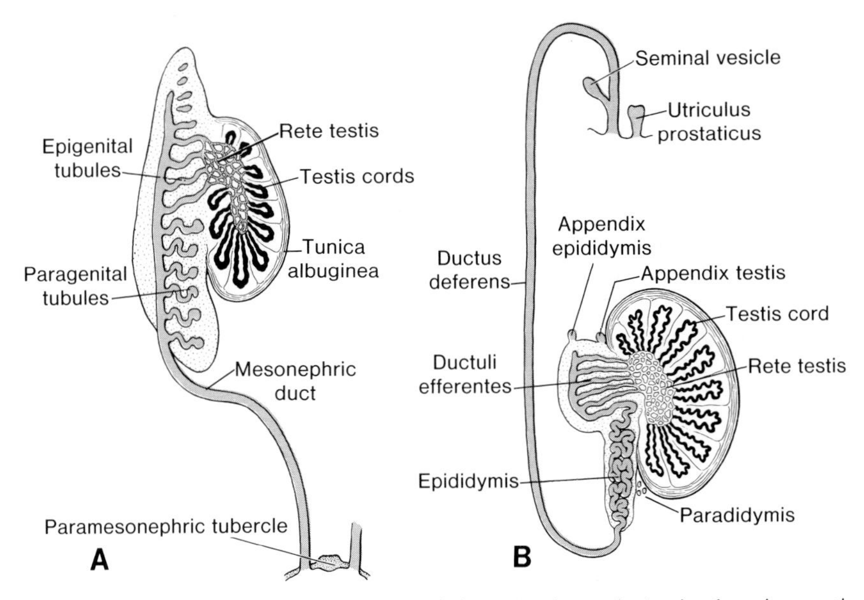

Figure 15-20. *A*, Diagram of the genital ducts in the male in the fourth month of development. The paramesonephric duct has degenerated except for the appendix testis and the utriculus prostaticus. *B*, The genital duct after descent of the testis. Note the horseshoe-shaped testis cords, the rete testis, and the ductuli efferentes entering the ductus deferens. The paradidymis is formed by the remnants of the paragenital mesonephric tubules.

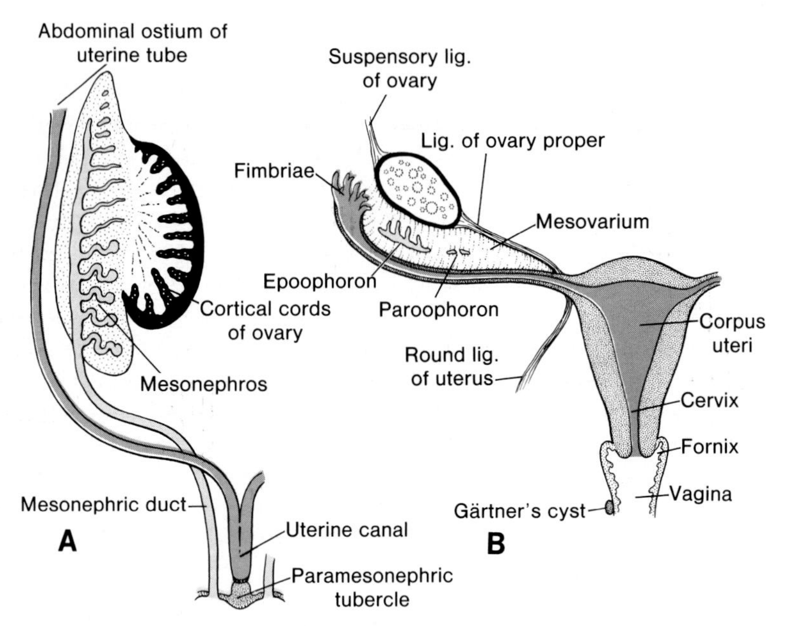

Figure 15-21. *A*, Schematic drawing of the genital ducts in the female at the end of the second month of development. Note the paramesonephric or Müllerian tubercle and the formation of the uterine canal. *B*, The genital ducts after descent of the ovary. The only parts remaining of the mesonephric system are the epoophoron, paroophoron, and Gärtner's cyst. Note the suspensory ligament of the ovary, the ligament of the ovary proper, and the round ligament of the uterus.

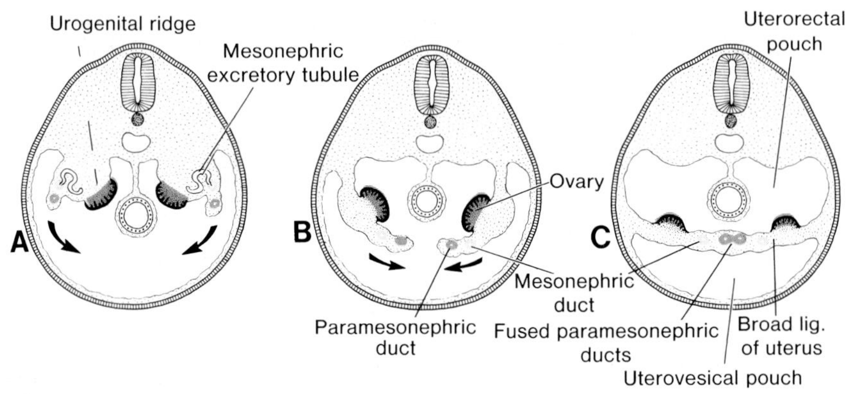

Figure 15-22. Transverse sections through the urogenital ridge at progressively lower levels. Note that the paramesonephric ducts approach each other in the midline to fuse. As a result of the fusion a transverse fold, the broad ligament of the uterus, is formed in the pelvis. The gonads come to lie at the posterior aspect of the transverse fold.

pelvic fold is established (fig. 15-22*C*). This fold, which extends from the lateral sides of the fused paramesonephric ducts towards the wall of the pelvis, is known as the **broad ligament of the uterus.** In its upper border is found the uterine tube and on its posterior surface the ovary (fig. 15-22*C*). The uterus and broad ligaments divide the pelvic cavity in the **uterorectal pouch** and the **uterovesical pouch.** The fused paramesonephric ducts give rise to the **corpus** and the **cervix of the uterus.** They are surrounded by a layer of mesenchyme. With time this mesenchyme forms the muscular coat of the uterus, the **myometrium** and its peritoneal covering, the **perimetrium.**

Vagina. Initially it was believed that the paramesonephric duct gave rise to both the uterus and the entire vagina. This view was abandoned since it was observed that solid evaginations of the posterior wall of the urogenital sinus became canalized and participated in the formation of the vagina. According to the most recent theory the upper one-third of the vagina is derived from the uterine canal, while the lower two-thirds is of urogenital sinus origin. Hence, the vagina is of double origin.[16, 17]

Shortly after the solid tip of the paramesonephric ducts has reached the urogenital sinus (figs. 15-23*A* and 15-24*A*), two solid evaginations grow out from the pelvic part of the sinus (figs. 15-23*B* and 15-24*B*). These evaginations, the **sinovaginal bulbs**, proliferate strongly and form a solid plate. The

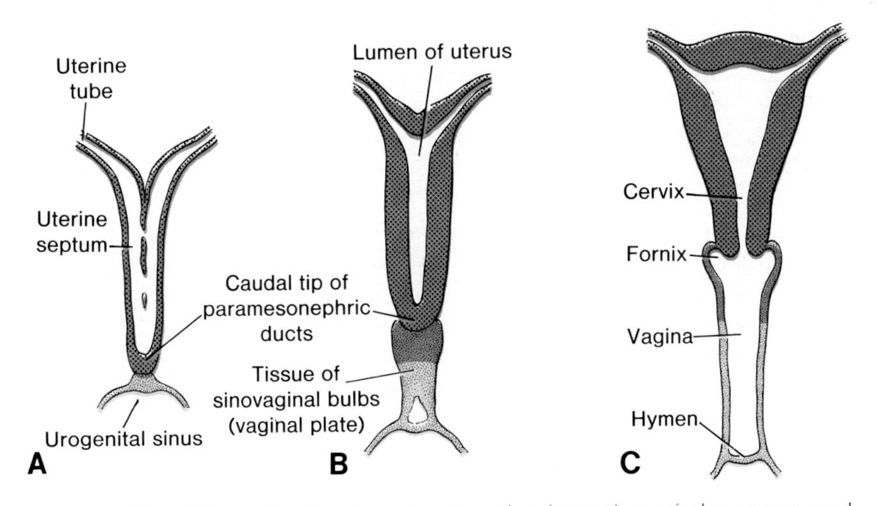

Figure 15-23. Schematic drawing showing the formation of the uterus and vagina. *A*, At nine weeks. Note the disappearance of the uterine septum. *B*, At the end of the third month. Note the tissue of the sinovaginal bulbs. *C*, Newborn. The upper third of the vagina and the fornices are formed by vacuolization of the paramesonephric tissue and the lower two-thirds by vacuolization of the sinovaginal bulbs.

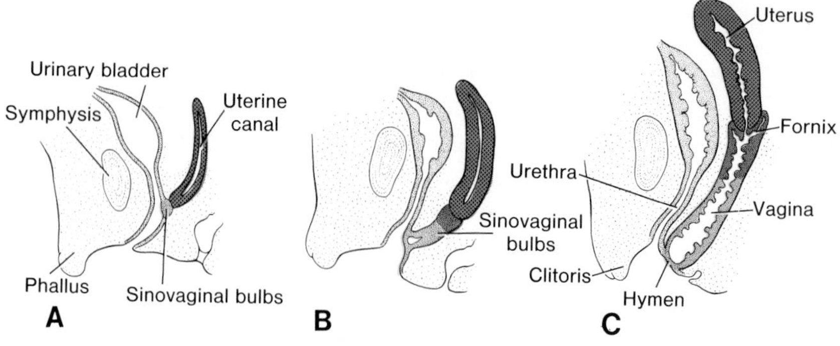

Figure 15-24. Schematic sagittal sections showing the formation of the uterus and vagina at various stages of development.

proliferation continues at the cranial end of the plate, thus increasing the distance between the uterus and the urogenital sinus. By the fifth month the vaginal outgrowth is entirely canalized. The wing-like expansions of the vagina around the end of the uterus, the **vaginal fornices** are of paramesonephric origin (fig. 15-24*C*).

The lumen of the vagina remains separated from that of the urogenital sinus by a thin tissue plate, known as the **hymen** (figs. 15-23*C* and 15-24*C*). It consists of the epithelial lining of the sinus and a thin layer of vaginal cells and usually ruptures during perinatal life.

Some remnants of the cranial and caudal excretory tubules may be found in the female. They are located in the mesovarium, where they form the epoophoron and paroophoron, respectively (fig. 15-21*B*). The mesonephric duct disappears entirely except for a small cranial portion found in the epoophoron and occasionally a small caudal portion, which may be found in the wall of the uterus or vagina. Later in life it may form a cyst, known as Gärtner's cyst (fig. 15-21*B*).

Duplication and Atresia of Uterine Canal. Lack of fusion of the paramesonephric ducts in a localized area or throughout the length of the ducts may explain all different types of duplication of the uterus. In its extreme form the uterus is entirely double (**uterus didelphys**) (fig. 15-25*A*), in the least severe form only slightly indented in the middle (**uterus arcuatus**) (fig. 15-25*B*). One of the more common anomalies is the **uterus bicornis**, in which the uterus has two horns entering a common vagina (fig. 15-25*C*). This condition is normal in many of the mammals below the primates.

In patients with complete or partial atresia of one of the paramesonephric ducts, the rudimentary part lies as an appendage to the well-developed side, but since its lumen usually does not communicate with the vagina, complications frequently ensue (**uterus bicornis unicollis with one rudimentary horn**) (fig. 15-25*D*). If the atresia involves both sides partially, an **atresia of the cervix** may result (fig. 15-25*E*). If the sinovaginal bulbs fail to fuse or do not

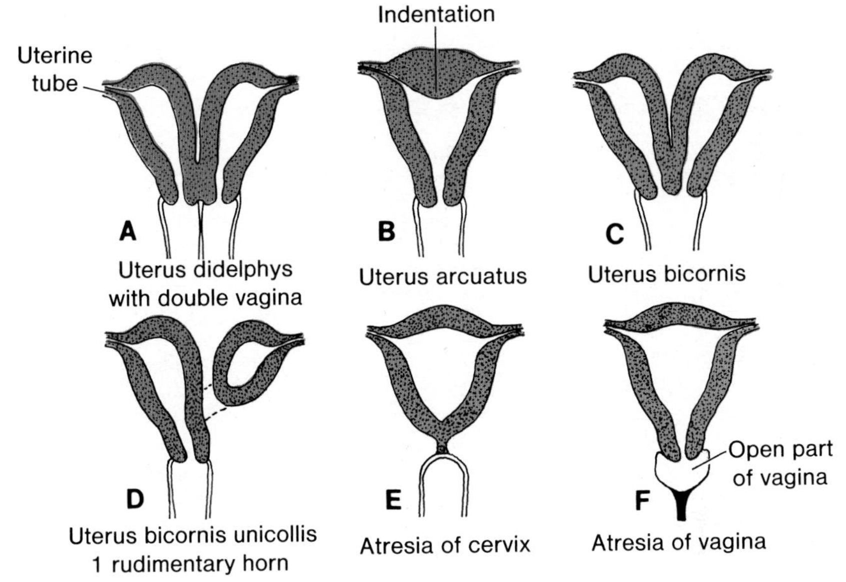

Figure 15-25. Schematic representation of the main abnormalities of the uterus and vagina, caused by persistence of the uterine septum or obliteration of the lumen of the uterine canal.

develop at all, a **double vagina** or **atresia of the vagina**, respectively, results (fig. 15-25*A*, *F*). In the latter case a small vaginal pouch, originating from the paramesonephric ducts, usually surrounds the opening of the cervix.

EXTERNAL GENITALIA

Indifferent Stage. In the third week of development mesenchyme cells, originating in the region of the primitive streak, migrate around the cloacal membrane to form a pair of slightly elevated folds, the **cloacal folds** (fig. 15-26*A*). Directly cranial to the cloacal membrane the folds unite to form the **genital tubercle.** When, in the sixth week, the cloacal membrane is subdivided into the urogenital and anal membranes, the cloacal folds are likewise

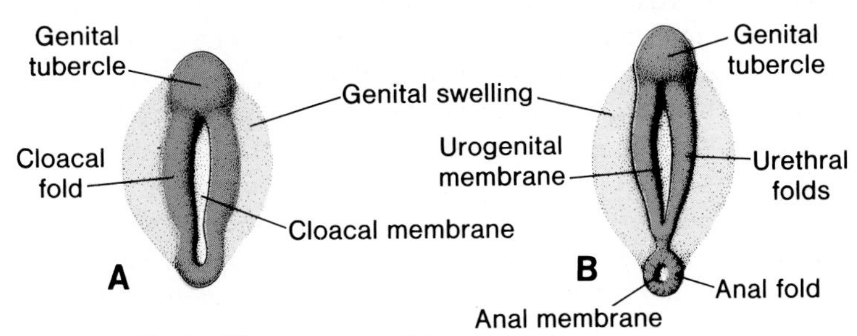

Figure 15-26. The indifferent stage of the external genitalia. *A*, At approximately four weeks; *B*, at approximately six weeks.

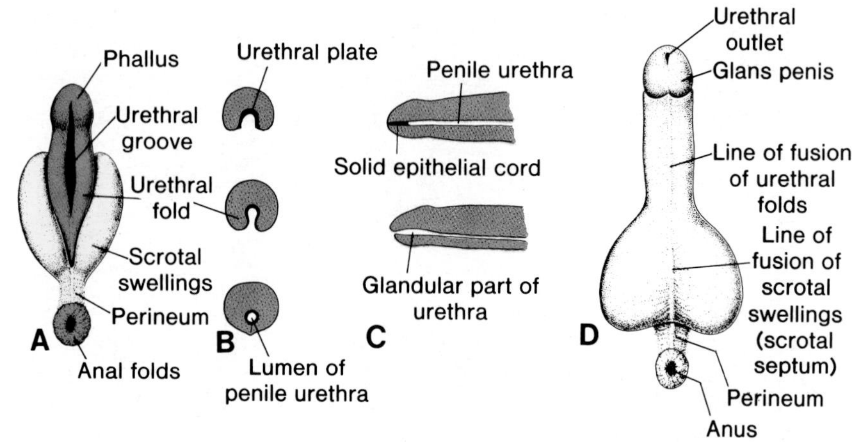

Figure 15-27. *A*, Development of the external genitalia in the male at 10 weeks. Note the deep urethral groove flanked by the urethral folds. *B*, Transverse sections through phallus during the formation of the penile urethra. The urogenital groove is bridged over by the two urethral folds. *C*, Development of the glandular portion of the penile urethra. *D*, In the newborn.

subdivided into the **urethral folds** anteriorly, and the **anal folds** posteriorly (fig. 15-26*B*).

In the meantime another pair of elevations, the **genital swellings**, become visible on each side of the urethral folds. In the male these swellings later form the **scrotal swellings** (fig. 15-27*A*), and in the female the **labia majora** (fig. 15-29*B*). By the end of the sixth week, however, it is impossible to distinguish between the two sexes (fig. 15-26*B*).

External Genitalia in the Male. Development in the male is under influence of androgens secreted by the fetal testes and is characterized by rapid elongation of the genital tubercle which now is called the **phallus** (fig. 15-27*A*). During this elongation the phallus pulls the urethral folds forward so that they form the lateral walls of the **urethral groove.** This groove extends along the caudal aspect of the elongated phallus, but does not reach the most distal part, known as the glans. The epithelial lining of the groove is of endodermal origin and forms the **urethral plate** (fig. 15-27*B*).

At the end of the third month the two urethral folds close over the urethral plate, thus forming the **penile urethra** (fig. 15-27*B*).[18] This canal does not extend to the tip of the phallus. The most distal portion of the urethra is formed during the fourth month when ectodermal cells from the tip of the glans penetrate inward and form a short epithelial cord. This cord later obtains a lumen, thus forming the **definitive external urethral meatus** (fig. 15-27*C*).

The genital swellings known in the male as the scrotal swellings are initially located in the inguinal region. With further development they move caudally, and each swelling then makes up half of the scrotum. The two are separated from each other by the **scrotal septum** (fig. 15-27*D*).

Hypospadias. When fusion of the urethral folds is incomplete, abnormal openings of the urethra may be found along the inferior aspect of the penis. Most frequently the abnormal orifices are near the glans, along the shaft, or near the base of the penis (fig. 15-28*A*). In rare cases the urethral meatus may be located along the scrotal raphe. When fusion of the urethral folds fails entirely, a wide sagittal slit is found along the entire length of the penis and the scrotum. The two scrotal swellings then closely resemble the labia majora.

Epispadias. In this abnormality the urethral meatus is found on the dorsum of the penis. Instead of having developed at the cranial margin of the cloacal membrane, the genital tubercle seems to have formed in the region of the urorectal septum. Hence a portion of the cloacal membrane is then found cranial to the genital tubercle, and when this membrane ruptures the outlet of the urogenital sinus comes to lie on the cranial aspect of the penis (fig. 15-28*B*).[18] Occasionally, the two sides of the genital tubercle do not fuse at all, and this results in a so-called **divided penis.**

Ectopia or Exstrophy of the Bladder. Ectopia of the bladder, seen frequently in combination with epispadias, is a condition in which the posterior wall of the bladder is exposed to the outside (fig. 15-28*B*). Under normal

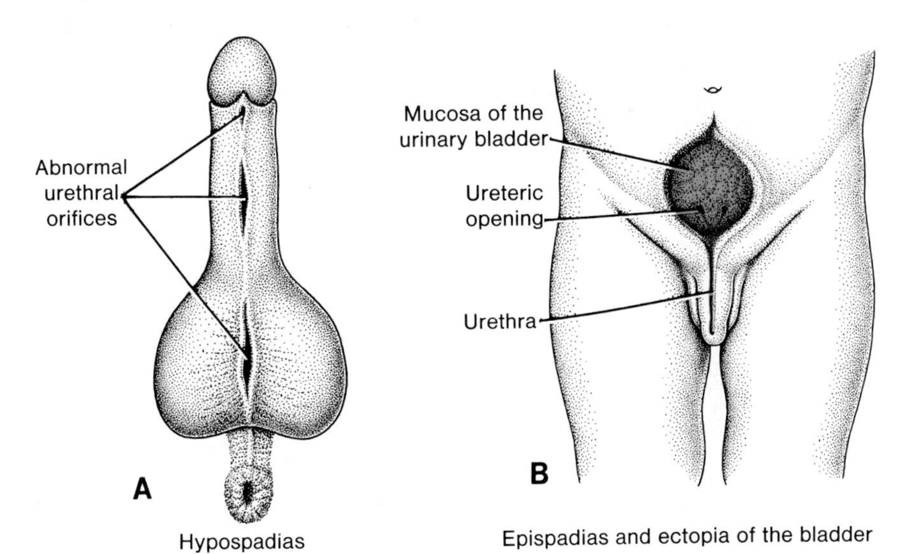

Figure 15-28. *A*, Hypospadias. The drawing shows the various locations of abnormal urethral orifices. *B*, Drawing of epispadias combined with exstrophy of the bladder. The mucosa of the bladder is exposed to the surface and the orifices of the ureters can readily be seen.

conditions the abdominal wall in front of the bladder is formed by primitive streak mesoderm, which migrates around the cloacal membrane.[19, 20] When this migration does not occur, the rupture of the cloacal membrane may extend in the cranial direction further than normal, thus establishing an **ectopia of the bladder.** The posterior aspect of the bladder is then exposed to the surface of the body, and the ureters and urethra can be readily seen (fig. 15-28*B*).

External Genitalia in the Female. The external genitalia of the female change under influence of the estrogens produced by the placenta and the mother (see table 15-2). The genital tubercle elongates only slightly and forms the **clitoris** (fig. 15-29*A*, *B*); the urethral folds do not fuse as in the male, but develop into the **labia minora**. The genital swellings enlarge greatly and form the **labia majora**. The urogenital groove is open to the surface and forms the **vestibule** (fig. 15-29*B*).

SEX ABNORMALITIES AS REFLECTED IN DUCT SYSTEM AND EXTERNAL GENITALIA

Ovarian Hypoplasia. Ovarian hypoplasia is found in patients with 44 autosomes and one X-chromosome[21, 22] (Turner's syndrome; see Chapter 8). The primordial germ cells are present and migrate toward the undifferentiated gonad, but few if any true follicles develop. Many of the germ cells degenerate and six months after birth no germ cells are visible in the gonad.[23, 24] Since the Y-chromosome is absent, the placental and maternal estrogens will stimulate the paramesonephric duct system and external genitalia as in the normal female (see table 15-2). As the gonad does not produce any hormones after birth, differentiation of the paramesonephric ducts and external genitalia ceases following birth and the sex characteristics remain infantile (fig. 15-30).

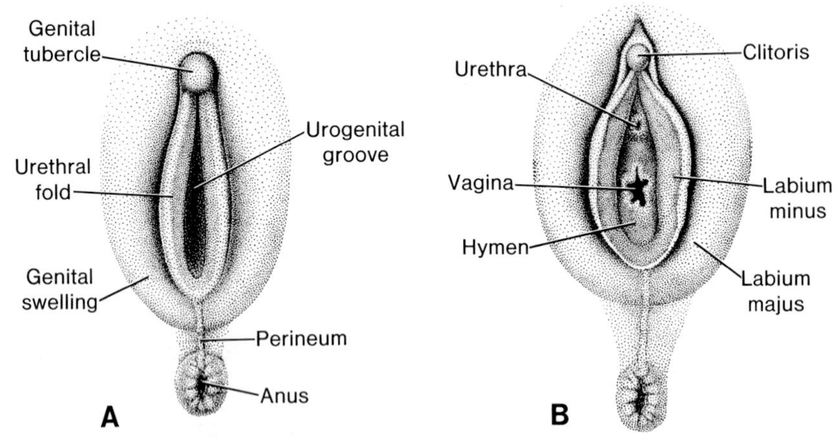

Figure 15-29. Development of the external genitalia in the female at five months (*A*), and in the newborn (*B*).

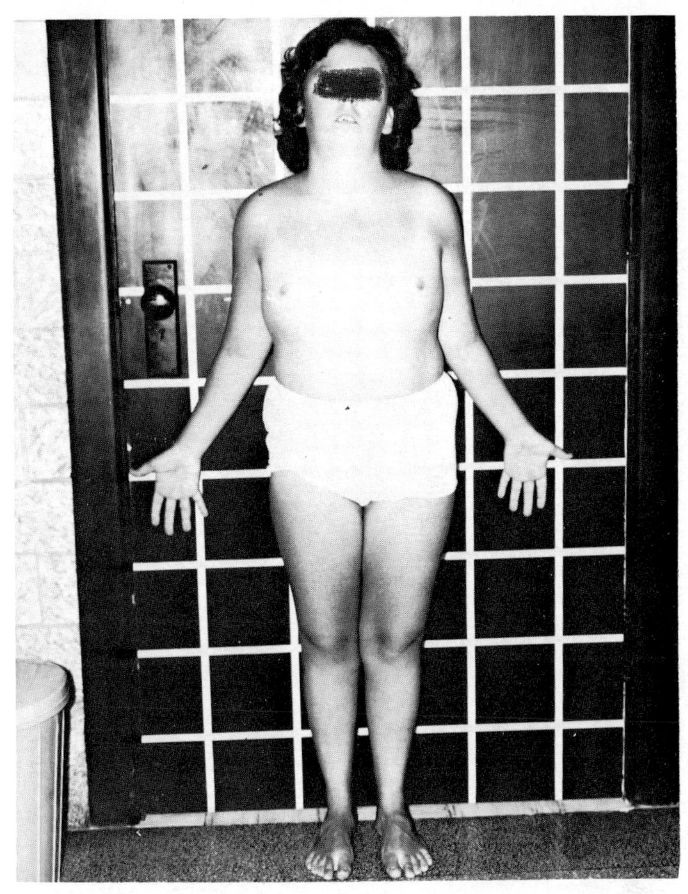

Figure 15-30. *A*, Patient with Turner's syndrome, which is characterized by an XO sex chromosome complement. Note the absence of sexual maturation. Other typical features are the webbed neck, broad chest with widely spaced nipples, and the short stature.

Pure Gonadal Dysgenesis. Patients with this syndrome have no abnormality of the chromosomes in either number or structure and the chromosomal complement may be 44 + 2X, or 44 + XY.[25, 26] In these patients the primordial germ cells **do not form** or **do not migrate** into the gonadal area and neither ovary nor testis develops. Without the presence of inducer substance and androgens, the paramesonephric system and the external genitalia are stimulated by the maternal and placental estrogens. Differentiation of the duct system and external genitalia ceases after birth. None of the other symptoms characteristic for Turner's syndrome are present in these patients, since the chromosomes are normal.

Testicular Feminization Syndrome. Patients with this syndrome have a 44

+ XY chromosome complement, but have the external appearance of normal females (fig. 15-31).[27] The tissues of the external genitalia are unresponsive to the androgens produced by the testes and they develop and differentiate as in the normal female under influence of the estrogens. Since these patients

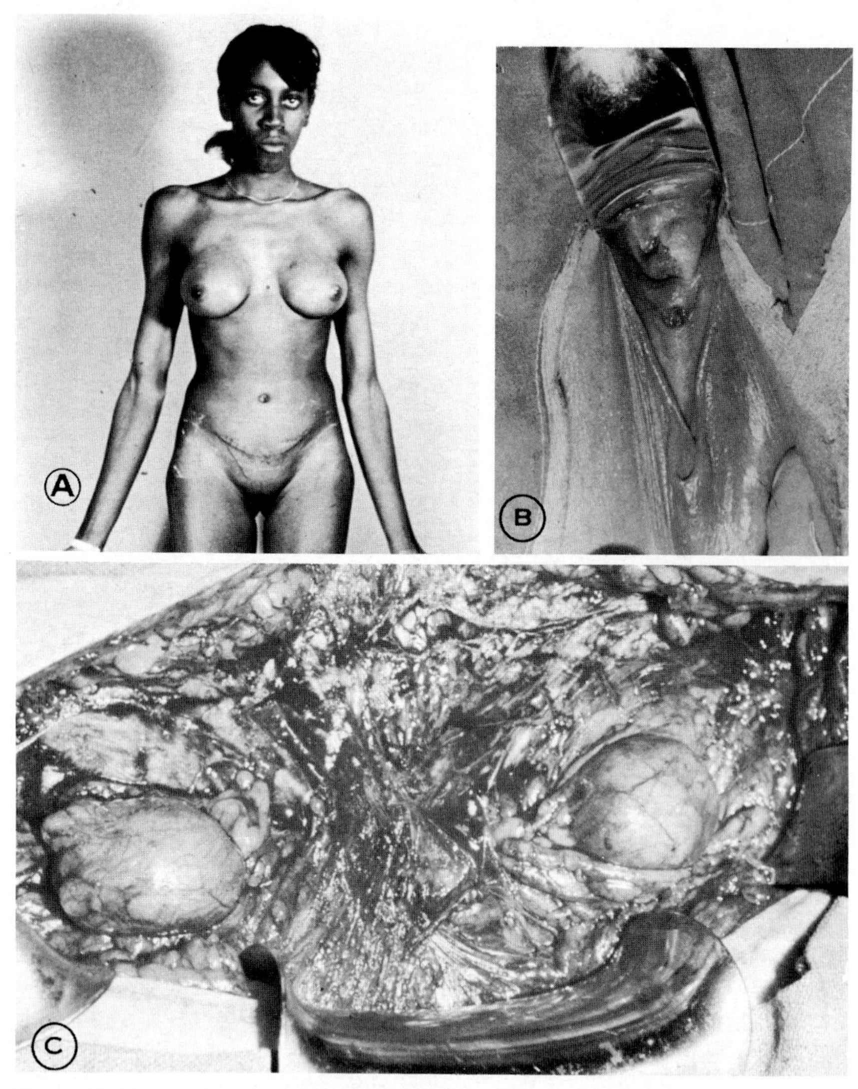

Figure 15-31. *A*, Patient with testicular feminization syndrome, which is characterized by a 44 + XY chromosome complement. *B*, External genitalia. A small vaginal groove is visible under the underdeveloped penis. Uterus and uterine tubes are absent. *C*, The testes were in the inguinal canal. (Courtesy Dr. J. Kitchin, Department of Obstetrics and Gynecology, University of Virginia.)

have testes and the inducer substance is present, the paramesonephric system is suppressed and the uterine tubes and uterus are absent; the vagina is short and ends blindly. The testes are frequently found in the inguinal or labial regions, but spermatogenesis does not occur. They have an increased risk of tumor formation.

Pseudohermaphroditism and Intersexuality. A **true hermaphrodite** is an individual in whom the gonads and external genitalia of both sexes are present. Although patients with both active testicular and ovarian tissue exist, true hermaphrodites with male and female genitalia are rarely observed.

In **pseudohermaphrodites** the genotypic sex is masked by a phenotypic appearance that closely resembles the other sex. When the pseudohermaphrodite has a testis, the patient is called a **male pseudohermaphrodite**; when an ovary is present the patient is called a **female pseudohermaphrodite**. Patients with ovarian hypoplasia, gonadal dysgenesis, and Klinefelter's syndrome are not considered hermaphrodites. Their external genitalia usually are not ambiguous.

Female Pseudohermaphrodites. The most common cause of female pseudohermaphroditism is the **adrenogenital syndrome**. The patients have a 44 + XX chromosome complement, have chromatin-positive nuclei, and have ovaries, but the **excessive production of androgens by the adrenals** causes the external genitalia to develop in a male direction. The masculinization may vary from enlargement of the clitoris to almost male genitalia (fig. 15-32).[28] Frequently there is clitoris hypertrophy, partial fusion of the labia majora giving the appearance of a scrotum and a small persistent urogenital sinus. The syndrome is caused by an abnormal steroid metabolism in the fetal adrenal cortex.

Although progestins administered during pregnancy may cause similar abnormalities as in the adrenogenital syndrome, the great majority of female pseudohermaphrodites fall in the first category.[29, 30]

Male Pseudohermaphrodites. These patients have a 44 + XY chromosome complement and their cells are usually chromatin-negative. It is thought that androgenic hormones are produced in insufficient amounts or that they are formed after the tissue sensitivity of the sexual structures has passed. The internal and external sex characteristics may vary considerably, depending on the degree of phallus development and the presence of paramesonephric representatives.

DESCENT OF THE TESTIS

Toward the end of the second month the testis and the mesonephros are attached to the posterior abdominal wall by the **urogenital mesentery** (fig. 15-3*A*). With degeneration of the mesonephros, the attachment band serves mainly as a mesentery for the gonad (fig. 15-22*B*). In caudal direction it becomes ligamentous and is known as the **caudal genital ligament** (fig. 15-

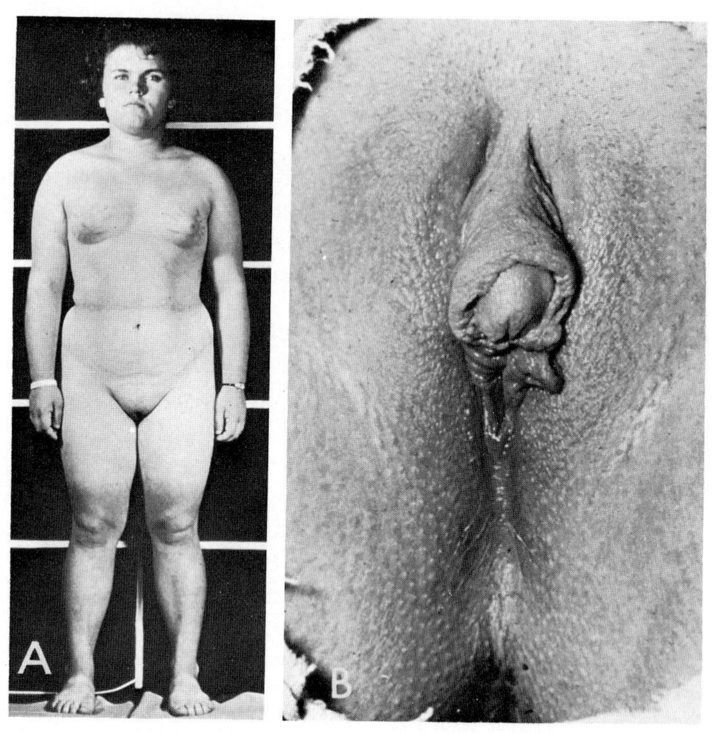

Figure 15-32. *A*, Patient with female pseudohermaphroditism caused by congenital virilizing adrenal hyperplasia (adrenogenital syndrome). *B*, External genitalia showing fusion of the labia majora and enlargement of clitoris. (Courtesy Dr. J. Kitchin, Department of Obstetrics and Gynecology, University of Virginia.)

33*A*). In the inguinal region the caudal genital ligament is continuous with a band of mesenchyme which in turn continues into a mesenchymal condensation in the genital (scrotal) swelling. Together the three components are referred to as the **gubernaculum testis** (fig. 15-33*C*).

As a result of the rapid growth of the body and the failure of the gubernaculum to elongate correspondingly, the testis descends below its level of origin. By the beginning of the third month it lies close to the inguinal region (fig. 15-33*B*). Hence the **descent of the testis** is not an active migration, but is a relative shift in position with regard to the body wall. The blood supply from the aorta is retained and the testicular vessels descend from the original lumbar level to the inguinal region.

Independent from the descent of the testis the peritoneum of the coelomic cavity forms an evagination on each side of the midline into the ventral abdominal wall. The evagination follows the course of the gubernaculum testis into the scrotal swellings (fig. 15-33*B*) and is known as the **vaginal process**. Hence, the vaginal process, accompanied by the muscular and

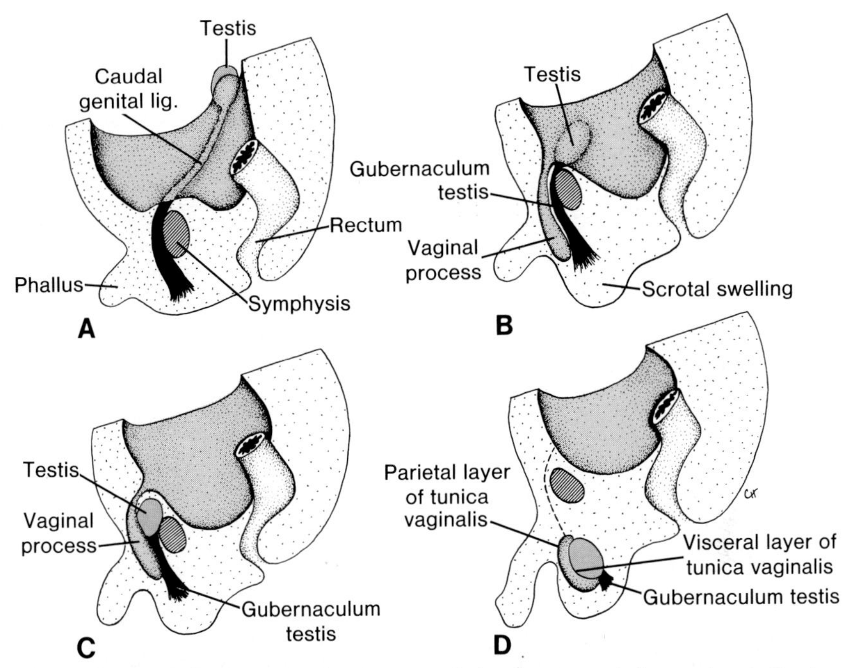

Figure 15-33. Schematic representation of the descent of the testis. *A*, During the second month; *B*, middle of the third month; note that the coelomic cavity evaginates into the scrotal swelling, where it forms the vaginal process (tunica vaginalis); *C*, seventh month; *D*, shortly after birth.

fascial layers of the body wall, evaginates into the scrotal swelling, thus forming the **inguinal canal** (fig. 15-34). The **gubernaculum testis remains ventral** to and **outside** the vaginal process at all times.

The testis descends through the inguinal ring and over the rim of the pubic bone into the scrotal swelling at the time of birth. The testis is then covered by a reflected fold of the vaginal process (fig. 15-33*D*). The peritoneal layer covering the testis is known as the **visceral layer of the tunica vaginalis**; the remainder of the peritoneal sac forms the **parietal layer of the tunica vaginalis** (fig. 15-33*D*). The narrow canal, connecting the lumen of the vaginal process with the peritoneal cavity, is obliterated at birth or shortly thereafter.

The final descent of the testis is accompanied by a shortening of the gubernaculum, but whether or not this shortening causes the descent of the testis is still controversial.[31] Undoubtedly, the descent of the testis is also controlled by hormones such as gonadotrophins and androgens.

Congenital Inguinal Hernia. The connection between the coelomic cavity and the vaginal process in the scrotal sac normally closes in the first year after birth (fig. 15-33*D*). If this passageway remains open intestinal loops may descend into the scrotum, thus causing a **congenital inguinal hernia** (fig. 15-34*B*). Sometimes the obliteration of this passageway is irregular, leaving

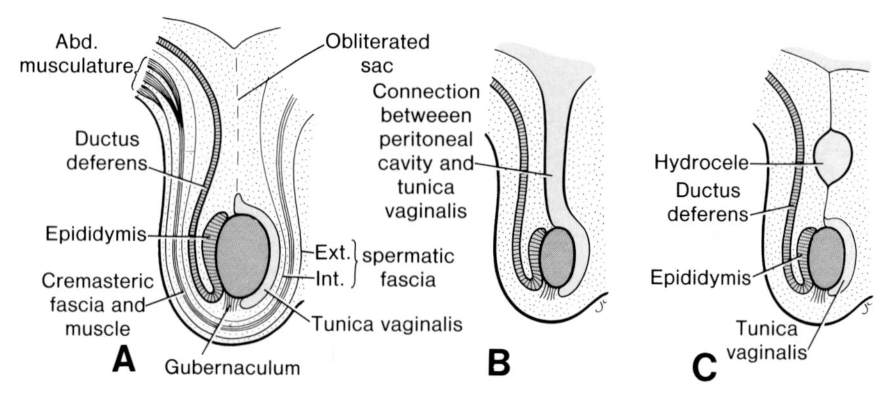

Figure 15-34. *A*, Diagrammatic drawing of the testis, epididymis, ductus deferens, and the various layers of the abdominal wall which surround the testis in the scrotum. *B*, Vaginal process in open communication with the peritoneal cavity. In such a case, portions of the intestinal loops often descend toward the scrotum, thus causing an inguinal hernia. *C*, Hydrocele.

small cysts along its course. Later these cysts may secrete excess fluid, resulting in the formation of a **hydrocele** (fig. 15-34*C*).

Cryptorchism or Undescended Testis. At about the time of birth, but with wide individual variation, the testes arrive in the scrotum. In certain cases one or both testes may remain in the pelvic cavity or somewhere in the inguinal canal until puberty and then descend or remain indefinitely in the abnormal position.[33] This condition is known as **cryptorchism**, and seems to be due to abnormal androgen production. An undescended testis is unable to produce mature spermatozoa, most likely because of the high temperature in the abdominal cavity.

DESCENT OF THE OVARY

In the female the descent of the gonad is considerably less than in the male and the ovary is finally located just below the rim of the true pelvis. The cranial genital ligament is thought to form the **suspensory ligament of the ovary**, whereas the caudal genital ligament forms the **ligament of the ovary proper** and the **round ligament of the uterus** (fig. 15-21). The latter extends into the labia majora.

SUMMARY

The urinary and genital systems both develop from mesodermal tissue and are, particularly in the male, closely related. Because of their different functions they are usually treated separately.

The **urinary system** develops from three successive systems: (1) the **pronephros** forms in the cervical region and is vestigial; (2) the

mesonephros forms in the thoracic and lumbar regions, is large, and is characterized by excretory units—nephrons—and its own collecting duct, the mesonephric or Wolffian duct. In the human its function is doubtful and most of the system disappears; (3) the **metanephros** or **permanent kidney** develops from two sources. It forms its own excretory tubules or nephrons as the other systems but its collecting system originates from the **ureteric bud**, an outgrowth of the mesonephric duct. This bud gives rise to the ureter, the renal pelvis, the calyces, and the entire collecting system (fig. 15-5). Connection between the collecting and excretory tubule systems is essential for normal development (fig. 15-6) and failure to connect may cause congenital cystic diseases and renal agenesis. Similarly, early division of the ureteric bud may lead to bifid or supernumerary kidneys with ectopic ureters (fig. 15-7). Abnormal positions of the kidney, such as pelvic and horseshoe kidney, are also well known (fig. 15-9).

The **genital system** consists of (1) the gonads or **primitive sex glands**, (2) the **genital ducts**, and (3) the **external genitalia**. All three components go through an indifferent stage in which they may develop in male or female direction. Although the genetic sex of the embryo is established at fertilization, the influence of the chromosome complement of the primordial germ cells, which appear in the third week of development, pushes the indifferent gonad in a male or female direction. The ***Y-chromosome is testis determining*** and causes (1) the development of the medullary (testis) cords, (2) the formation of the tunica albuginea, and (3) the failure of the cortical (ovarian) cords to develop. The double XX-chromosome stimulates the formation of the ovary with its (1) typical cortical cords, (2) the disappearance of the medullary (testis) cords, and (3) the failure of the tunica albuginea to develop (table 15-1). When the primordial germ cells fail to reach the indifferent gonad, the gonad remains indifferent or is absent. The indifferent duct system and external genitalia develop under influence of hormones. The **inducer substance** produced by the testis stimulates the development of the mesonephric ducts (vas deferens—epididymis) and suppresses the paramesonephric ducts (female duct system). The **androgens**, also produced by the testis, stimulate the external genitalia, penis, scrotum, and prostate (table 15-2). The maternal and placental estrogens formed in the presence of 2X chromosomes stimulate the paramesonephric female system such as (1) the uterine tube, (2) the uterus, and (3) the upper third of the vagina, as well as the external genitalia such as clitoris, labia, and lower two-thirds of the vagina (table 15-2). Errors in the production of or sensitivity to testes hormones lead to a predominance of the female characteristics under influence of the maternal and placental estrogens (figs. 15-30 and 15-31). Disorders leading to the production of excessive amounts of androgens (androgenital syndrome) produce in females enlargement of the clitoris and partial fusion of the labia majora (female pseudohermaphrodites; fig. 15-32).

REFERENCES

1. Leeson, T. S. The fine structure of the mesonephros of the 17-day rabbit embryo. *Exp. Cell Res., 12:* 670, 1957.
2. Grobstein, C. Some transmission characteristics of the tubule inducing influence on mouse metanephrogenic mesenchyme. *Exp. Cell Res., 13:* 575, 1957.
3. Saxen, L. Embryonic induction. *Clin. Obstet. Gynecol., 18:* 149, 1975.
4. Osathanondh, V., and Potter, E. L. Pathogenesis of polycystic kidneys. *Arch. Pathol., 77:* 459, 1964.
5. Baxter, T. J. Cysts arising in the renal tubules. *Arch. Dis. Child., 40:* 464, 1965.
6. Davidson, W. M., and Ross, G. I. M. Bilateral absence of kidneys and related congenital anomalies. J. Pathol., 68: 456, 1954.
7. Johnston, J. H. Problems in the diagnosis and management of ectopic ureters and ureteroceles. In *Problems in Paediatric Urology*, edited by J. H. Johnston and R. J. Scholtmeyer. Excerpta Medica, Amsterdam, 1972.
8. Gyllensten, L. Contributions to embryology of the urinary bladder; development of definitive relations between openings of the Wolffian ducts and ureters. *Acta Anat., 7:* 305, 1949.
9. Gray, S. W., and Skandalakis, J. E. *Embryology for Surgeons: The Embryological Basis for the Treatment of Congenital Defects.* W. B. Saunders, Philadelphia, 1972.
10. Gillman, J. The development of the gonads in man, with a consideration of the role of fetal endocrines and histogenesis of ovarian tumors. *Contrib. Embryol., 32:* 81, 1948.
11. Mintz, B. Embryological phases of mammalian gametogenesis. *J. Cell. Comp. Physiol., 56:* 31, 1960.
12. Pinkerton, J. H. M., McKay, D. G., Adams, C., and Hertig, A. T. Development of the human ovary—a study using histochemical techniques. *Obstet. Gynecol., 18:* 152, 1961.
13. Clermont, Y., and Huckins, C. Microscopic anatomy of the sex glands and seminiferous tubules in growing adult male albino rats. *Am. J. Anat., 108:* 79, 1961.
14. Mancini, R. E., Narbaitz, R., and Lavieri, J. C. Origin and development of the germinative epithelium and Sertoli cells in the human testis: cytological, cytochemical and quantitative study. *Anat. Rec., 136:* 477, 1960.
15. Jost, A. Development of sexual characteristics. *Science, 6:* 67, 1970.
16. O'Rahilly, R. The development of the vagina in the human. In *Morphogenesis and Malformation of the Genital Systems*, edited by R. J. Blandau and D. Bergsma, pp. 123–136. Original article series. Alan R. Liss, New York, 1977.
17. Cunha, G. R. The dual origin of vaginal epithelium. *Am. J. Anat., 143:* 387, 1975.
18. Glenister, T. W. A correlation of the normal and abnormal development of the penile urethra and of the infra-umbilical abdominal wall. *Br. J. Urol., 30:* 117, 1958.
19. Marshall, V. F., and Muecke, E. C. Variations in exstrophy of the bladder. *J. Urol., 88:* 766, 1962.
20. Muecke, E. C. The role of the cloacal membrane in exstrophy: the first successful experimental study. *J. Urol., 92:* 659, 1964.
21. Moore, K. L. Sex determination, sexual differentiation and intersex development. *Can. Med. Assoc. J., 97:* 292, 1967.
22. Ford, C. E., Jones, K. W., Polani, P. E. de Almeida, J. C., and Briggs, J. H. A sex chromosome anomaly in a case of gonadal dysgenesis (Turner's syndrome). *Lancet, 1:* 711, 1959.
23. Hamerton, J. L. *Human Cytogenetics.* Vol. II. *Clinical Cytogenetics.* Academic Press, New York, 1971.

24. Carr, D. H., Haggar, R. A., and Hart, A. G. Germ cells in the ovaries of XO female infants. *Am. J. Clin. Pathol., 49:* 521, 1968.
25. Morishima, A., and Grumbach, M. M. The interrelationship of sex chromosome constitution and phenotype in the syndrome of gonadal dysgenesis and its variants. *Ann. N.Y. Acad. Sci., 155:* 695, 1968.
26. Boczkowski, K. The syndrome of pure gonadal dysgenesis. In *Medical Gynaecology and Sociology.* Pergamon Press, Oxford, England, 1968.
27. Morris, J. M., and Mahesh, V. B. Further observations on the syndrome "Testicular feminization." *Am. J. Obstet. Gynecol., 87:* 731, 1963.
28. Schlegel, R. J., and Gardner, L. I. Ambiguous and abnormal genitalia in infants: differential diagnosis and clinical management. In *Endocrine and Genetic Diseases of Childhood,* edited by L. I. Gardner. W. B. Saunders, Philadelphia, 1975.
29. Wilkins, L. *The Diagnosis and Treatment of Endocrine Disorders in Childhood and Adolescence.* Charles C Thomas, Springfield, Ill., 1957.
30. Hayles, A. B., and Nolan, R. B. Masculinization of female fetus, possibly related to administration of progesterone during pregnancy: report of two cases. *Proc. Staff Meetings Mayo Clin., 33:* 300, 1958.
31. Backhouse, K. M., and Butler, H. The gubernaculum testis of the pig. *J. Anat., 94:* 107, 1960.
32. Brunet, J., Mowbray, R. R., and Bishop, P. M. F. Management of the undescended testis. *Br. Med. J., 1:* 1367, 1958.
33. Scorer, C. G. The descent of the testis. *Arch. Dis. Child., 39:* 605, 1964.

chapter 16

Head and Neck

The most typical feature in the development of the head and neck is formed by the **branchial** or **pharyngeal arches**. These arches appear in the fourth and fifth week of development and contribute greatly to the characteristic external appearance of the embryo (fig. 16-1). Initially, they consist of bars of mesenchymal tissue separated by deep clefts, known as **branchial** or **pharyngeal clefts** (figs. 16-1*C* and 16-4). Simultaneously with the development of the arches and clefts, a number of outpocketings, the **pharyngeal pouches**, appear along the lateral walls of the pharyngeal gut, the most cranial part of the foregut (figs. 16-2 and 16-4). The pouches gradually penetrate the surrounding mesenchyme, but do not establish an open communication with the external clefts (fig. 16-8). Hence, although the development of pharyngeal arches, clefts, and pouches resembles the formation of gills in fishes and amphibia, in the human embryo real gills—branchia—are never formed. Therefore, the term **pharyngeal** arches, clefts, and pouches has been adopted for the human embryo.

The pharyngeal arches not only contribute to the formation of the neck, but also play an important role in the formation of the head. At the end of the fourth week the center of the face is formed by the stomodeum, surrounded by the first pair of pharyngeal arches (fig. 16-3). When the

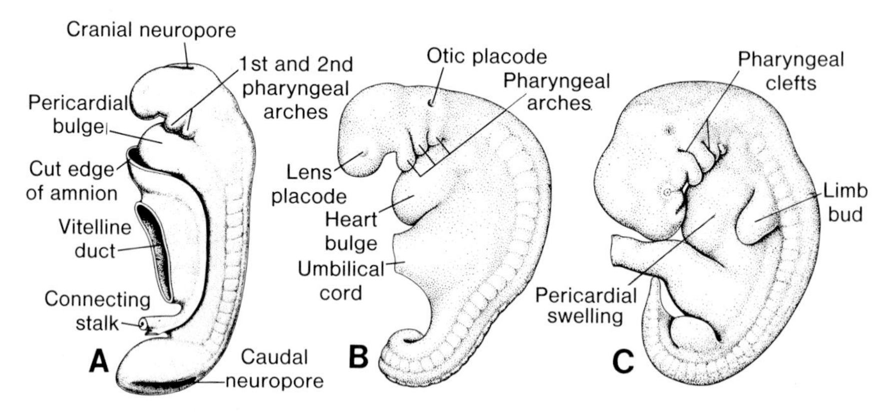

Figure 16-1. Series of human embryos to show the development of the pharyngeal arches. *A*, Approximately 25 days. *B*, 28 days. *C*, Five weeks.

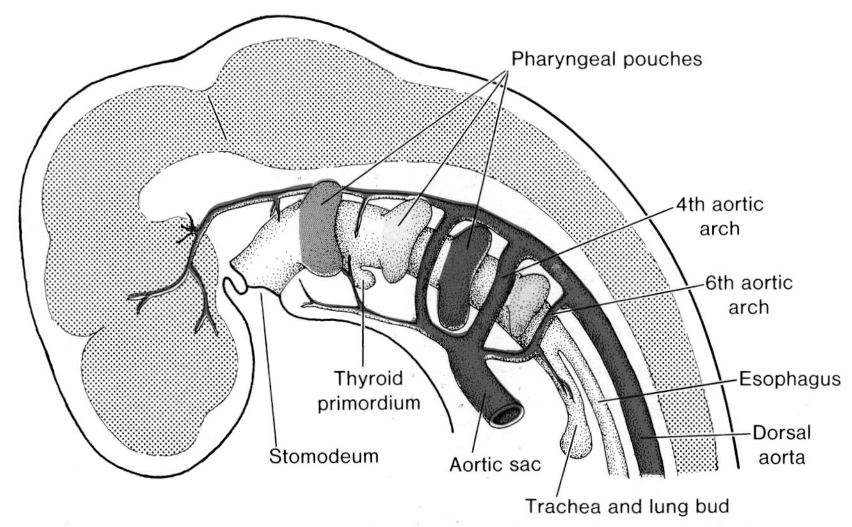

Figure 16-2. Drawing of the pharyngeal pouches as outpocketings of the foregut. Note also the primordium of the thyroid gland and the aortic arches.

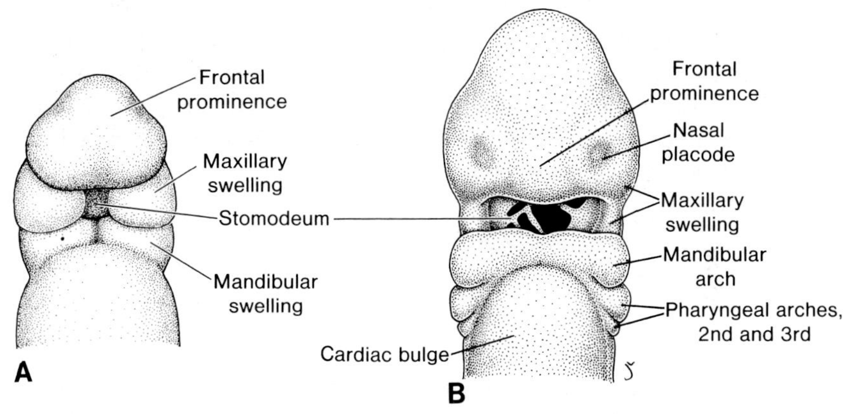

Figure 16-3. *A*, Frontal view of an embryo of approximately 24 days. The stomodeum, temporarily closed by the buccopharyngeal membrane, is surrounded by five mesenchymal swellings. *B*, Frontal view of a slightly older embryo, showing the rupture of the buccopharyngeal membrane.

embryo is 4½ weeks old, five mesenchymal swellings can be recognized: (1 and 2) the **mandibular swellings** (first pharyngeal arch) can be distinguished caudally to the stomodeum; (3 and 4) the **maxillary swellings** (dorsal portion of the first pharyngeal arch) laterally to the stomodeum; and (5) the **frontal prominence,** a slightly rounded elevation, cranially to the stomodeum. The development of the face is later complemented by the formation of the **nasal swellings** (fig. 16-7).

Since the formation of the face and neck is greatly influenced by the development of the pharyngeal arches and their derivatives, each of the arches will be discussed separately in the following paragraphs.

Pharyngeal Arches

Each pharyngeal arch consists of a core of mesodermal tissue, covered on the outside by surface ectoderm and on the inside by epithelium of endodermal origin (fig. 16-4). In addition to local mesenchyme, the core of the arches receives substantial numbers of crest cells, which migrate into the arches to contribute to the **skeletal components** of the face. The original mesoderm of the arches gives rise to the musculature of the face and neck. Each pharyngeal arch is thus characterized by its own **muscular components** (fig. 10-3). The muscular components of each arch carry their own nerve, and wherever the muscular cells may migrate, they carry their cranial **nerve component** with them (fig. 16-5). In addition, each arch has its own **arterial component** (fig. 16-2).

FIRST PHARYNGEAL ARCH

The **cartilage** of the **first pharyngeal arch** consists of a dorsal portion, known as the **maxillary process**, extending forward beneath the region of the eye, and a ventral portion, the **mandibular process** or **Meckel's cartilage** (fig. 16-6*A*). During further development both the maxillary process and Meckel's cartilage retrogress and disappear, except for two small portions at their dorsal ends which persist and form the **incus** and **malleus**, respectively

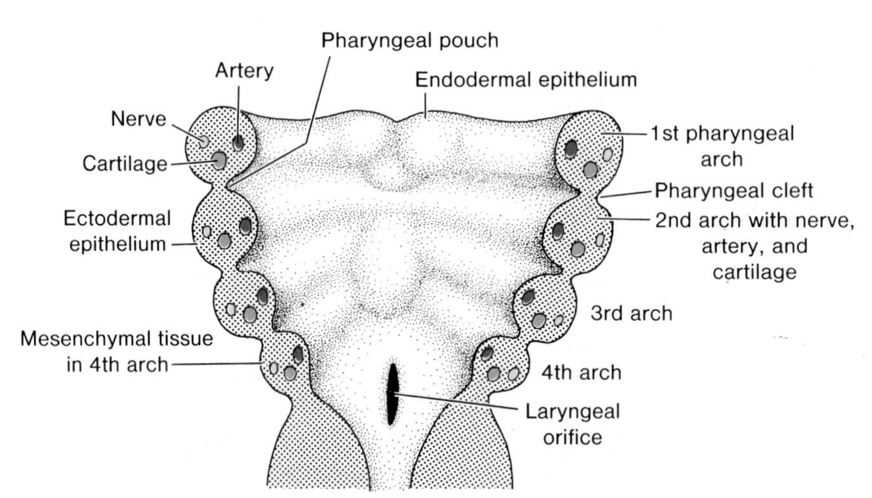

Figure 16-4. Schematic drawing of the pharyngeal arches. Each arch contains a cartilaginous component, a nerve, an artery, and a muscular component (see fig. 10-3).

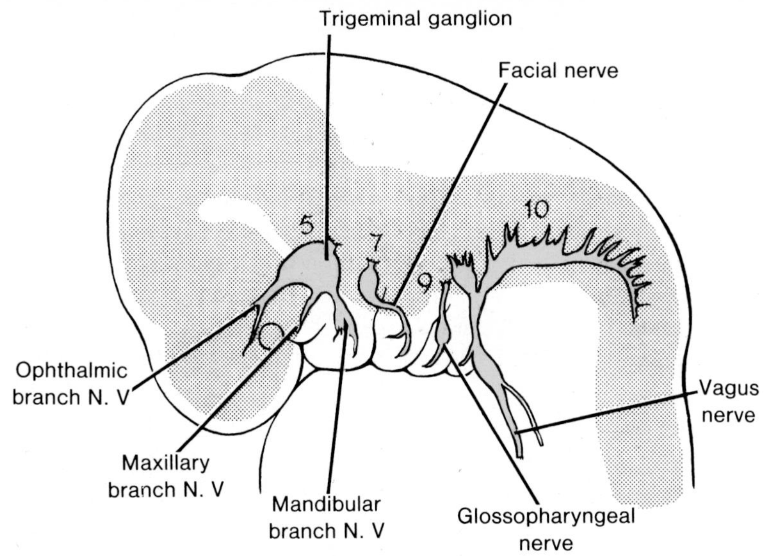

Figure 16-5. Each pharyngeal arch carries its own cranial nerve. The trigeminal nerve, supplying the first branchial arch, has three branches: the ophthalmic, maxillary, and mandibular branches. The nerve of the second arch is the facial nerve; that of the third the glossopharyngeal nerve. The musculature of the fourth arch is supplied by the superior laryngeal branch of the vagus nerve, and that of the sixth arch by the recurrent branch of the vagus nerve.

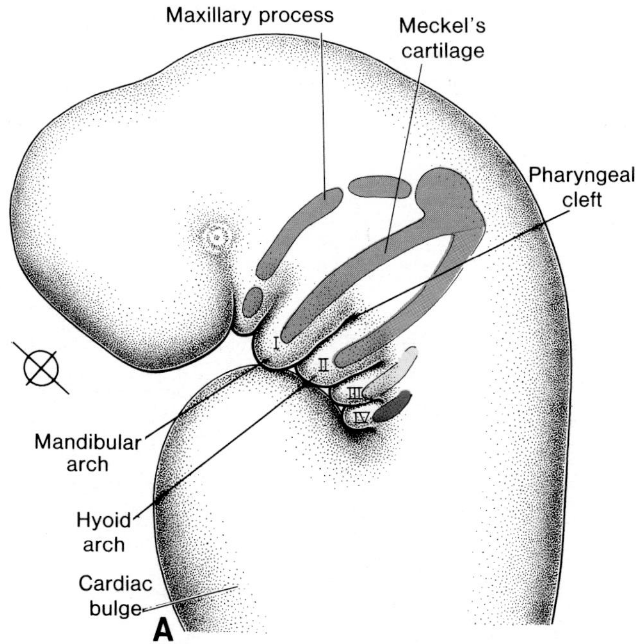

Figure 16-6*A* Lateral view of the head and neck region of a four-week embryo to demonstrate the cartilages of the pharyngeal arches participating in the formation of the bones of the face and neck. (Figure 16-6*B* on p. 272.)

(figs. 16-6*B* and 16-7).[1] The mesenchyme of the maxillary process subsequently gives rise to the **premaxilla, maxilla, zygomatic bone** and **part of the temporal bone** through membranous ossification (fig. 16-6*B*). The **mandible** is similarly formed by membranous ossification of the mesenchymal tissue surrounding Meckel's cartilage. Only a small portion of Meckel's cartilage undergoes a fibrous transformation. Hence, the maxillary and mandibular processes contribute greatly to the formation of the facial skeleton through membranous ossification. In addition, the first arch contributes to the formation of the bones of the middle ear (see "Ear," Chapter 17).

The **musculature** of the first pharyngeal arch is formed by the **muscles of mastication** (temporal, masseter, and pterygoids), the **anterior belly of the digastric**, the **mylohyoid**, the **tensor tympani**, and **tensor palatini**.

The muscles of the different arches do not always attach to the bony or cartilaginous components of their own arch, but sometimes migrate into surrounding regions. The origin of these muscles, however, can always be traced, since their nerve supply comes from the arch of origin.

The **nerve** supply to the muscles of the first arch is provided only by the mandibular branch of the trigeminal nerve (fig. 16-5). As mesenchyme from the first arch also contributes to the dermis of the face, the sensory supply of

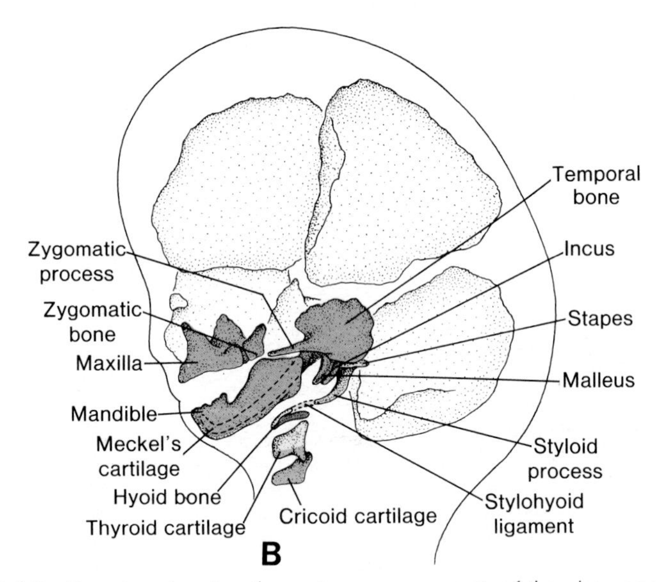

Figure 16-6*B* Drawing showing the various components of the pharyngeal arches at a later stage of development. Some of the components ossify, while others disappear or become ligamentous. The maxillary process and Meckel's cartilage are replaced by the definitive maxilla and mandible, which both develop by membranous ossification.

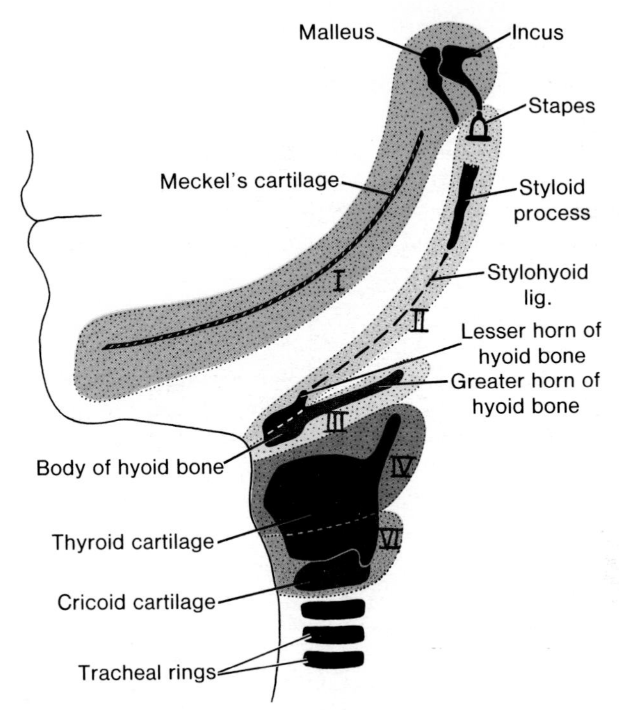

Figure 16-7. Drawing showing the definitive structures formed by the cartilaginous components of the various pharyngeal arches.

the skin of the face is provided by the ophthalmic, maxillary, and mandibular branches.

SECOND PHARYNGEAL ARCH

The cartilage of the **second** or **hyoid arch** (**Reichert's cartilage**) (fig. 16-6*B*) gives rise to the **stapes,** the **styloid process of the temporal bone,** the **stylohyoid ligament,** and ventrally to the **lesser horn** and the **upper part of the body of the hyoid bone** (fig. 16-7).

The muscles of the hyoid arch are the **stapedius**, the **stylohyoid**, the **posterior belly of the digastric**, the **auricular**, and the **muscles of facial expression**. The **facial nerve**, the nerve of the second arch, supplies all these muscles.

THIRD PHARYNGEAL ARCH

The **cartilage** of this arch produces the **lower part of the body** and the **greater horn of the hyoid bone** (fig. 16-7). The **musculature** is limited to the **stylopharyngeal muscle** and possibly the upper pharyngeal constrictors. They are innervated by the **glossopharyngeal nerve**, the nerve of the third arch (fig. 16-5).

FOURTH AND SIXTH PHARYNGEAL ARCHES

The **cartilaginous components** of these arches fuse to form the **thyroid, cricoid, arytenoid, corniculate, and cuneiform cartilages** of the larynx (fig. 16-7).

The **muscles** of the fourth arch (the **cricothyroid, the levator palatini**, and the **constrictors of the pharynx**) are innervated by the **superior laryngeal branch of the vagus**, the nerve of the fourth arch. The intrinsic muscles of the larynx, however, are supplied by the **recurrent laryngeal branch of the vagus**, the nerve of the sixth arch.

Pharyngeal Pouches

The human embryo has five pairs of pharyngeal pouches (fig. 16-8). The last one of these is atypical and often considered as part of the fourth. Since the **epithelial endodermal lining** of the pouches gives rise to a number of important organs, the fate of each pouch is discussed separately.

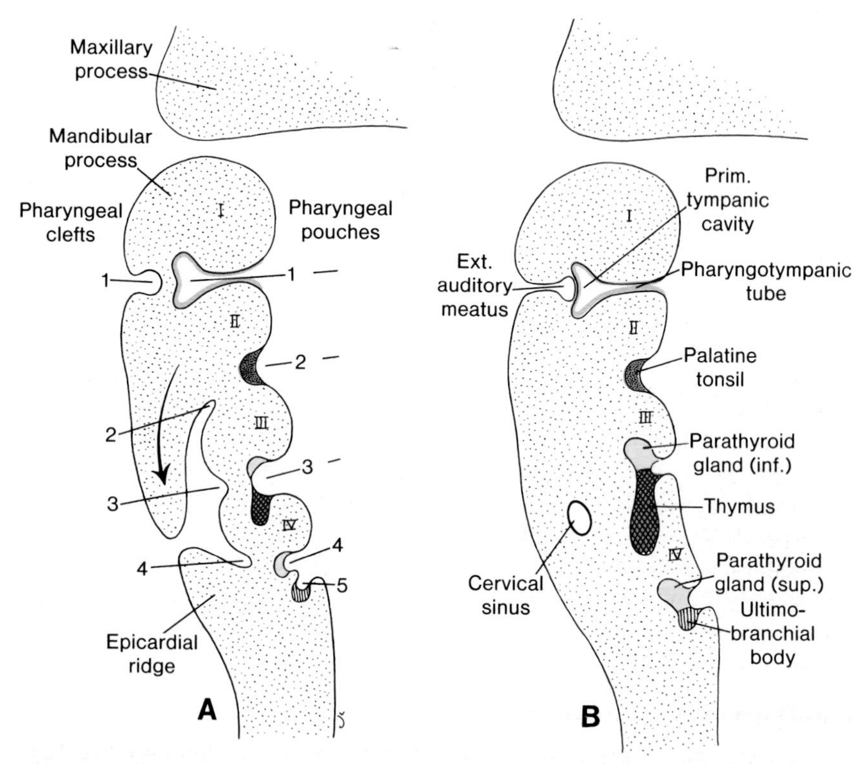

Figure 16-8. *A*, Schematic representation of the development of the pharyngeal clefts and pouches. Note that the second arch grows over the third and fourth arches, thereby burying the second, third, and fourth pharyngeal clefts. *B*, The remnants of the second, third, and fourth pharyngeal clefts form the cervical sinus. Note the structures formed by the various pharyngeal pouches.

FIRST PHARYNGEAL POUCH

The first pharyngeal pouch forms a stalk-like diverticulum, the **tubotympanic recess**, which comes in contact with the epithelial lining of the first pharyngeal cleft, the future **external auditory meatus** (fig. 16-8*A*,*B*). The distal portion of the outpocketing widens into a sac-like structure, the **primitive tympanic** or **middle ear cavity**, whereas the proximal part remains narrow, forming the **pharyngotympanic tube**. The lining of the tympanic cavity later aids in the formation of the **tympanic membrane** or **eardrum** (see "Ear," Chapter 17).

SECOND PHARYNGEAL POUCH

The epithelial lining of this pouch proliferates and forms buds which penetrate into the surrounding mesenchyme. The buds are secondarily invaded by mesodermal tissue, thus forming the primordium of the **palatine tonsil** (fig. 16-8*A*, *B*). During the third to fifth months the tonsil is gradually infiltrated by lymphatic tissue. Part of the pouch remains and is found in the adult as the **tonsillar fossa**.

THIRD PHARYNGEAL POUCH

The third and fourth pouches are characterized at their distal extremity by a so-called dorsal and ventral wing (fig. 16-2). In the fifth week the epithelium of the dorsal wing differentiates into **parathyroid tissue**, while that of the ventral part forms the **thymus** (fig. 16-8*A*, *B*). Both gland primordia lose their connection with the pharyngeal wall,[2] and the thymus then migrates in a caudal and medial direction, pulling the parathyroid with it (fig. 16-9). While the main portion of the thymus moves rapidly to its final position in the thorax, where it fuses with its counterpart from the opposite side, its tail portion becomes thin and eventually breaks up into small fragments. These parts sometimes persist, either embedded in the thyroid gland or as isolated thymic nests.[3, 4]

Growth and development of the thymus continue after birth until puberty. In the young child the gland occupies considerable space in the thorax and lies behind the sternum and anterior to the pericardium and great vessels. In older persons the gland is difficult to recognize since it is atrophied and replaced by fatty tissue.

The parathyroid tissue of the third pouch finally comes to rest on the dorsal surface of the thyroid gland and, in the adult, forms the **inferior parathyroid gland** (fig. 16-9).

FOURTH PHARYNGEAL POUCH

The epithelium of the dorsal wing of this pouch forms the **superior parathyroid gland**. Although the fate of the ventral portion of the pouch is uncertain, it is believed that it gives rise to a small amount of thymus tissue which, soon after its formation, disappears.[2]

When the parathyroid gland loses its contact with the wall of the pharynx,

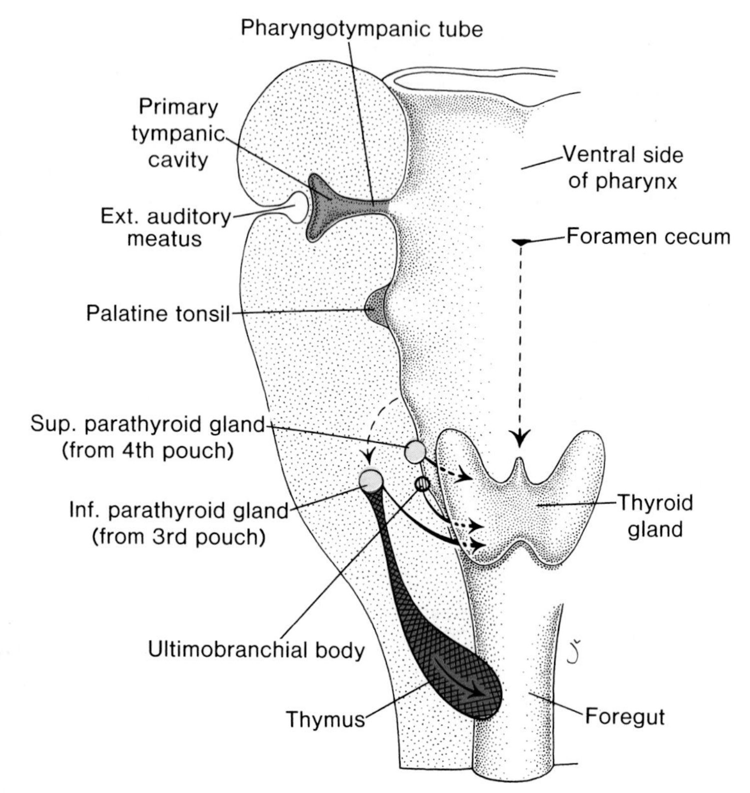

Figure 16-9. Schematic representation of the migration of the thymus, parathyroid glands, and ultimobranchial body. The thyroid gland originates at the level of the foramen cecum and descends to the level of the first tracheal rings.

it attaches itself to the caudally migrating thyroid and finally is found on the dorsal surface of this gland as the **superior parathyroid gland** (fig. 16-9).

FIFTH PHARYNGEAL POUCH

This is the last of the pharyngeal pouches to develop and it is usually considered to be a part of the fourth pouch. It gives rise to the **ultimobranchial body**, which is later incorporated in the thyroid gland.[3-5] In the adult, the cells of the ultimobranchial body give rise to the **parafollicular** or **C cells** of the thyroid gland. These cells secrete **calcitonin**, a hormone involved in the regulation of the calcium level in the blood.

Pharyngeal Clefts

The five-week embryo is characterized by the presence of four pharyngeal clefts (fig. 16-4) of which only one contributes to the definitive structure of the embryo. The dorsal part of the first cleft penetrates the underlying mesoderm and gives rise to the **external auditory meatus** (figs. 16-8 and 16-9). The epithelial lining at the bottom of this meatus participates in the formation of the **eardrum** (see "Ear," Chapter 17).

Active proliferation of the mesodermal tissue in the second arch causes it to overlap the third and fourth arches. Finally, it fuses with the so-called **epicardial ridge** in the lower part of the neck (fig. 16-8*A*, *B*) and the second, third, and fourth clefts lose contact with the outside (fig. 16-8*B*). Temporarily, the clefts form a cavity lined with ectodermal epithelium, the **cervical sinus**, but with further development this sinus usually disappears entirely.

LATERAL CYSTS OF THE NECK (BRANCHIAL CYSTS)

When the second pharyngeal arch fails to grow caudally over the third and fourth arches, the remnants of the second, third, and fourth clefts remain in contact with the surface by a narrow canal, known as the **branchial fistula** (fig. 16-10*A*). Such a fistula, found on the lateral aspect of the neck directly anterior to the sternocleidomastoid muscle, usually provides drainage for a **lateral cervical cyst** (fig. 16-10*B*). These cysts are remnants of the cervical sinus and are most often located just below the angle of the jaw (fig. 16-11*A*).[6] They may, however, be found anywhere along the anterior border of the sternocleidomastoid muscle. Frequently, a lateral cervical cyst is not visible at birth, but becomes evident as the result of enlargement in life.

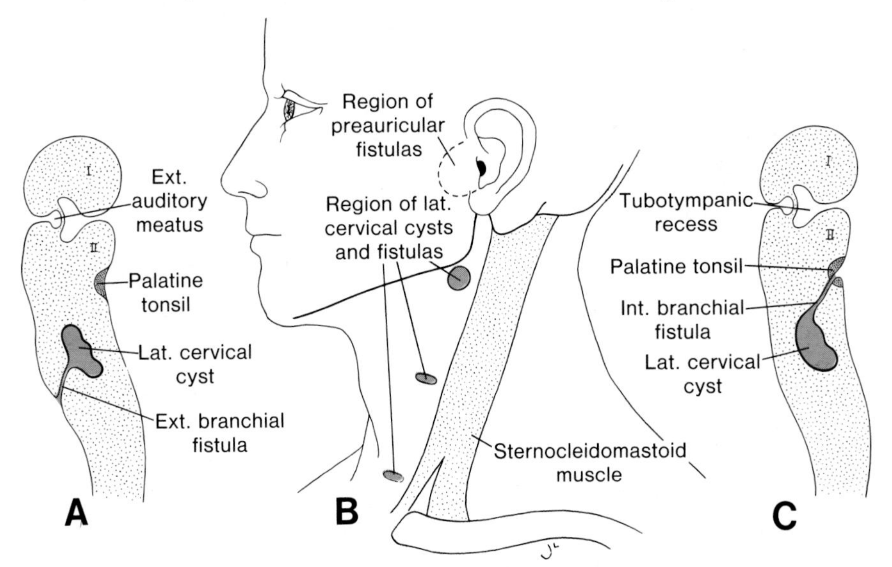

Figure 16-10. *A*, Schematic drawing of a lateral cervical cyst opening at the lateral side of the neck by way of a fistula. *B*, Localization of lateral cervical cysts and fistulas in front of the sternocleidomastoid muscle. Note also the region of localization of the preauricular fistulas. *C*, A lateral cervical cyst opening into the pharynx at the level of the palatine tonsil.

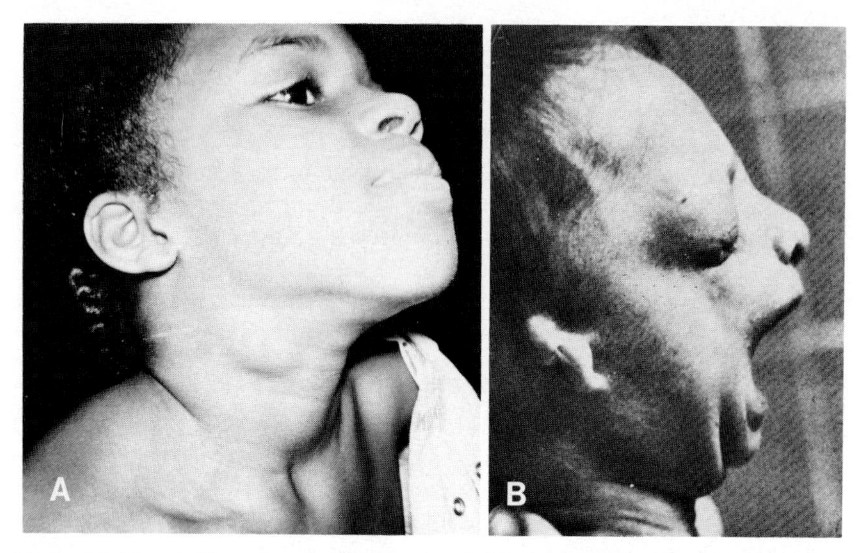

Figure 16-11. *A*, Photograph of patient with lateral cervical cyst. These cysts are always located on the lateral side of the neck in front of the sternocleidomastoid muscle. Frequently they are found under the angle of the mandible and do not enlarge until tne 20th year. (Courtesy Dr. A. Shaw, Department of Surgery, University of Virginia.) *B*, Patient with first arch syndrome. Note the hypoplasia of the mandible, large mouth, defect in cheek, preauricular appendix, and the abnormal external ear. (Courtesy Dr. J. Warkany. From *Congenital Malformations*, by J. Warkany. Copyright 1971, by Year Book Medical Publishers, Inc., Chicago. Used by permission.)

A rare anomaly is the **internal branchial fistula**. The cervical sinus is then connected to the lumen of the pharynx by a small canal, which usually opens in the tonsillar region (fig. 16-10*C*). Such a fistula indicates a rupture of the membrane between the second pharyngeal cleft and pouch at some time during development.

Sometimes a branchial fistula is confused with a **preauricular fistula** or **pit**. These fistulas, found in front of the ear, have probably no relationship to the branchial fistulas (fig. 16-10*B*). It is thought that they result from an incomplete disappearance of one of the sulci between the auricular hillocks (see "Ear," Chapter 17).[7]

FIRST ARCH SYNDROME

This syndrome consists of a number of malformations resulting from the disappearance or abnormal development of various components of the first pharyngeal arch (fig. 16-11*B*).[8] In the Treacher-Collins syndrome (**mandibulofacial dysostosis**) the following abnormalities are found: abnormal external ear, anomalies of the middle and inner ear, hypoplasia of the malar bone and mandible, and defects of the lower eye lid. Since most of these tissues receive contributions from the cephalic neural crest, it is believed that this abnormality is caused by insufficient migration of the first arch neural crest cells. In the **Pierre Robin syndrome**, another first arch syndrome, the abnormalities are more restricted: (1) hypoplasia of the mandible, (2) cleft palate, and (3) defects of ear and eye.[9]

Tongue

The tongue appears in embryos of approximately four weeks in the form of two **lateral lingual swellings** and one medial swelling, the **tuberculum impar** (fig. 16-12*A*). These three swellings originate from the first pharyngeal arch. A second median swelling, the **copula** or **hypobranchial eminence**, is formed by mesoderm of the second, third, and part of the fourth arch. Finally, a third median swelling, formed by the posterior part of the fourth arch, marks the development of the **epiglottis**. Immediately behind this swelling is the **laryngeal orifice**, which is flanked by the **arytenoid swellings** (fig. 16-12*A*).

As a result of growth of the lateral lingual swellings, they overgrow the tuberculum impar and merge with each other, thus forming the anterior two-thirds or the **body of the tongue** (fig. 16-12*B*). Since the mucosa covering the body of the tongue originates from the first pharyngeal arch, it is innervated by the **mandibular branch of the trigeminal nerve**. The anterior two-thirds or body of the tongue is separated from the posterior third by a V-shaped groove, the **terminal sulcus** (fig. 16-12*B*).

The posterior part or **root of the tongue** originates from the second, third,

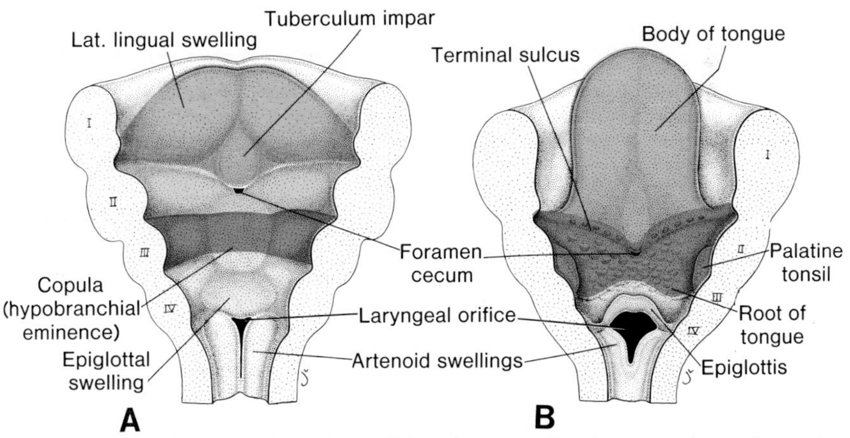

Figure 16-12. The ventral portions of the pharyngeal arches seen from above, to show development of the tongue. The cut pharyngeal arches are indicated by numbers I to IV, respectively. *A*, At five weeks (approximately 6 mm). *B*, At five months. Note the foramen cecum, the site of origin of the thyroid primordium, and the terminal sulcus, which forms the dividing line between the first and second pharyngeal arches.

and part of the fourth pharyngeal arch. Since in the adult the sensory innervation of this part of the tongue is supplied by the **glossopharyngeal nerve**, it seems likely that tissue of the third arch has overgrown that of the second.

The extreme posterior part of the tongue as well as the epiglottis are innervated by the **superior laryngeal nerve**, indicating their development from the fourth arch.

Some of the tongue muscles probably differentiate in situ, but most are derived from myoblasts originating in the **occipital somites**. This is supported by the fact that the tongue musculature is innervated by the **hypoglossal nerve**.

The **innervation** of the tongue is easy to understand. The sensory innervation of the anterior two-thirds must be supplied by the trigeminal nerve, the nerve of the first arch; that of the posterior third is supplied by the glossopharyngeal and vagus nerves, the nerves of the third and fourth arch, respectively. Since tissue of the second arch is overgrown by the third arch, the facial nerve, the nerve of the second arch, does not participate in the sensory innervation of the tongue.

TONGUE-TIE

This abnormality, also known as **ankyloglossia**, indicates that the tongue is not freed from the floor of the mouth.[9] Normally extensive cell degeneration occurs and the frenulum is the only tissue persisting, tying the tongue to the floor of the mouth. In the most common form of tongue-tie the frenulum extends to the tip of the tongue.

Thyroid Gland

The thyroid gland appears as an epithelial proliferation in the floor of the pharynx between the tuberculum impar and the copula, at a point later

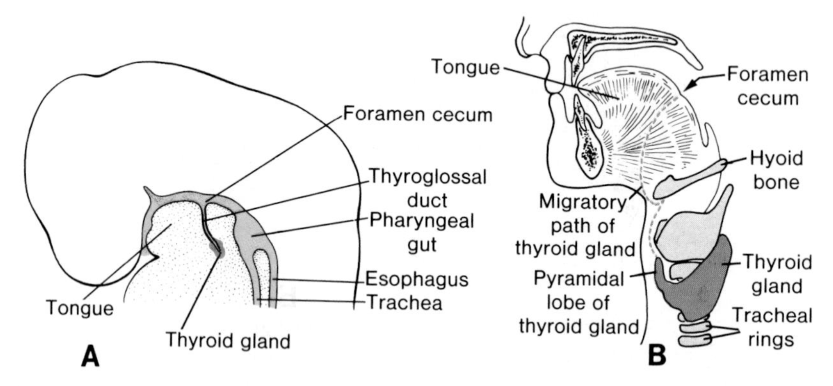

Figure 16-13. *A*, The thyroid primordium arises as an epithelial diverticulum in the midline of the pharynx immediately caudal of the tuberculum impar. *B*, Position of thyroid gland in the adult. Broken line indicates path of migration.

indicated by the **foramen cecum** (figs. 16-12 and 16-13*A*). Subsequently, the thyroid descends in front of the pharyngeal gut as a bilobed diverticulum (fig. 16-13). During this migration the gland remains connected to the tongue by a narrow canal, the **thyroglossal duct**. This duct later becomes solid and finally disappears.

With further development, the thyroid gland descends in front of the hyoid bone and the laryngeal cartilages. It reaches its final position in front of the trachea in the seventh week (fig. 16-13*B*). By then it has acquired a small median isthmus and two lateral lobes.

The thyroid begins to function at approximately the end of the third month, at which time the first follicles containing colloid become visible.

THYROGLOSSAL CYST AND FISTULA

A thyroglossal cyst may be found at any point along the migratory path followed by the thyroid gland, but is always located close to or in the midline of the neck. As indicated by its name, it is a cystic remnant of the thryoglossal duct. Although approximately 50 per cent of these cysts are located close to or behind the hyoid bone (fig. 16-14), they may also be found at the base of the tongue, or close to the thyroid cartilage.[10] Sometimes a thyroglossal cyst is connected to the outside by a fistulous canal, the **thyroglossal fistula**. Such a fistula usually arises secondarily after rupture of a cyst, but may be present at birth (fig. 16-15).

Aberrant thyroid tissue may be found anywhere along the path of the

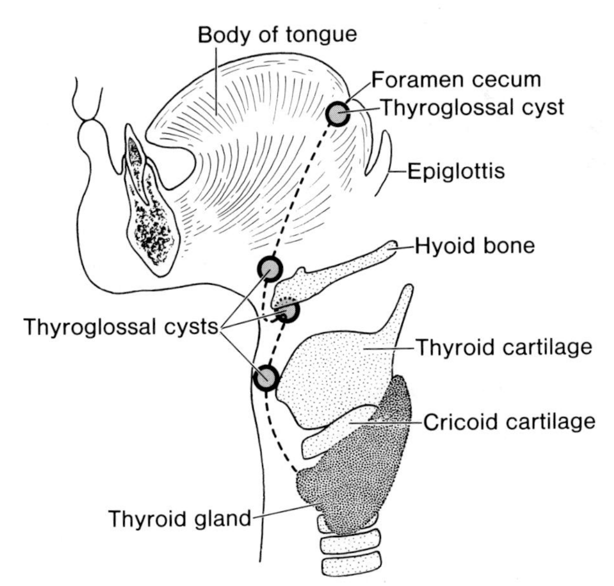

Figure 16-14. Schematic drawing indicating the localization of the thyroglossal cysts. These cysts, most frequently found in the thyroid region, are always located close to the midline.

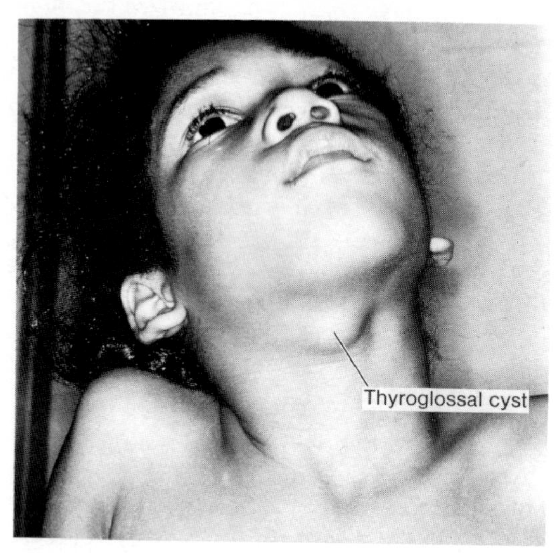

Figure 16-15. Photograph of patient with thyroglossal cyst. These cysts are remnants of the thyroglossal duct and may be located at any place along the migration path of the thyroid gland. They are frequently located behind the arch of the hyoid bone. An important diagnostic characteristic is the midline location. (Courtesy Dr. A. Shaw, Department of Surgery, University of Virginia.)

descent of the thyroid gland. It is most commonly found in the base of the tongue, just behind the foramen cecum, and is subject to the same diseases as the gland itself.[11]

Facial Swellings and Upper Lip

At the end of the fourth week the facial swellings are formed mainly by the first pair of pharyngeal arches. The maxillary swellings can be distinguished laterally to the stomodeum and the mandibular swellings caudally (fig. 16-16). The frontal prominence, formed by proliferation of mesenchyme ventral to the brain vesicles, constitutes the upper border of the stomodeum. On each side of the frontal prominence and just above the stomodeum is a local thickening of the surface ectoderm, the **nasal placode** (fig. 16-16*B*).

During the fifth week, two fast-growing ridges, the **lateral** and **medial nasal swellings**, surround the nasal placode, which then forms the floor of a depression, the **nasal pit** (fig. 16-17).

During the following two weeks the maxillary swellings continue to increase in size. Simultaneously they grow in medial direction, thereby compressing the medial nasal swellings toward the midline. Subsequently, the cleft between the medial nasal swelling and the maxillary swelling is overbridged and the two fuse (fig. 16-18). Hence, the upper lip is formed by the two medial nasal swellings and the two maxillary swellings. The lateral

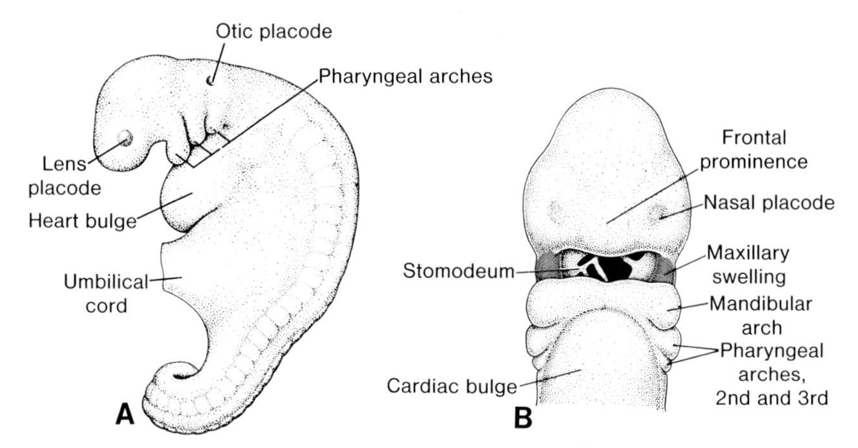

Figure 16-16. *A*, Lateral view of an embryo at the end of the fourth week showing the position of the pharyngeal arches. *B*, Frontal view of a 4½-week embryo. Note the location of the mandibular and maxillary swellings. The nasal placodes are visible on either side of the frontal prominence.

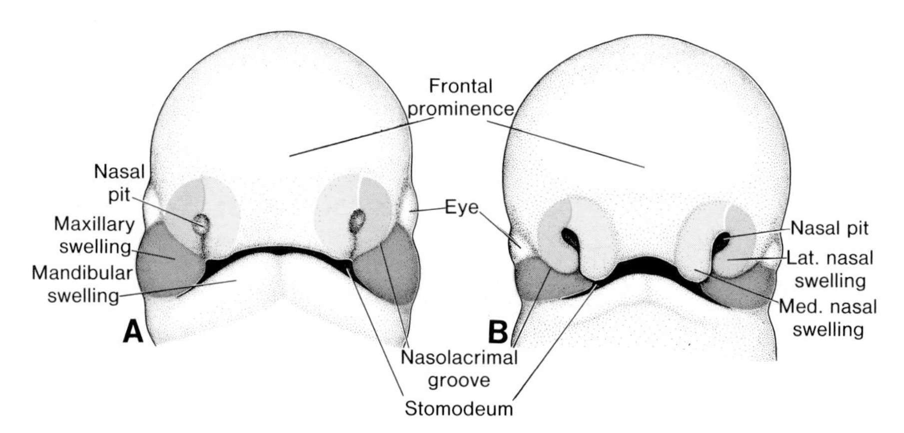

Figure 16-17. Frontal aspect of the face. *A*, Five-week embryo. *B*, Six-week embryo. The nasal swellings are gradually separated from the maxillary swelling by deep furrows.

nasal swellings do not participate in the formation of the upper lip, but form the alae of the nose.

Initially it was believed that the maxillary swellings merged over a short distance with the mandibular swellings, thus forming the cheeks. Careful examination of the interrelationship of the various components of the oral cavity, however, showed that the width of the mouth is not determined by fusion of the maxillary and mandibular swellings and that the cheeks develop by positional changes of the tongue, the floor of the mouth, and broadening of the mandible.[12]

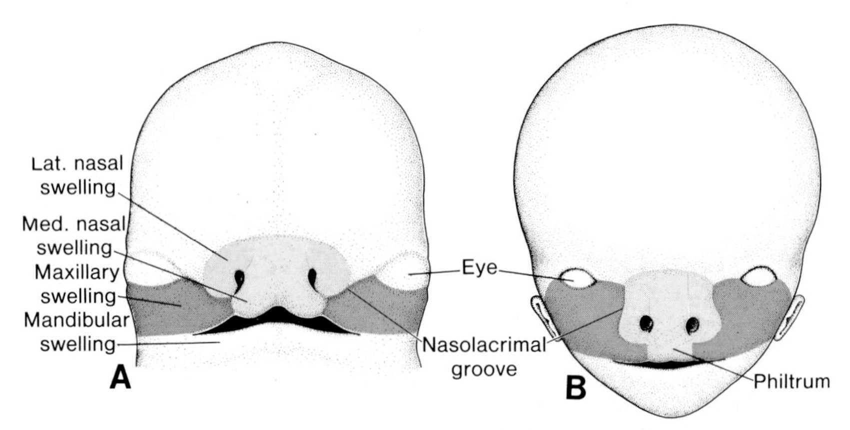

Figure 16-18. Frontal aspect of the face. *A*, Seven-week embryo. The maxillary swellings have fused with the medial nasal swellings. *B*, Ten-week embryo.

The manner in which the maxillary swellings unite with the lateral nasal swellings is slightly more complicated. Initially these structures are separated by a deep furrow, the **nasolacrimal groove** (fig. 16-17). The ectoderm in the floor of this groove forms a solid epithelial cord, which detaches from the overlying ectoderm. After canalization the cord forms the **nasolacrimal duct**: its upper end widens to form the **lacrimal sac**. Following detachment of the cord the maxillary and lateral nasal swellings fuse with each other. The nasolacrimal duct then runs from the medial corner of the eye to the inferior meatus of the nasal cavity.

Intermaxillary Segment

As a result of the medial growth of the maxillary swellings, the two medial nasal swellings merge not only at the surface but also at the deeper level. The structures formed by the two merged swellings are together known as the **intermaxillary segment.**[13] It is comprised of: (1) a **labial component**, which forms the philtrum of the upper lip; (2) an **upper jaw component**, which carries the four incisor teeth; and (3) a **palatal component**, which forms the triangular **primary palate** (fig. 16-19). A small portion of the external middle part of the nose is probably also derived from the intermaxillary segment. Cranially the intermaxillary segment is continuous with the rostral portion of the nasal septum, which is formed by the frontal prominence.

Secondary Palate

While the primary palate is derived from the intermaxillary segment (fig. 16-19), the main part of the definitive palate is formed by two shelf-like outgrowths from the maxillary swellings. These outgrowths, the **palatine**

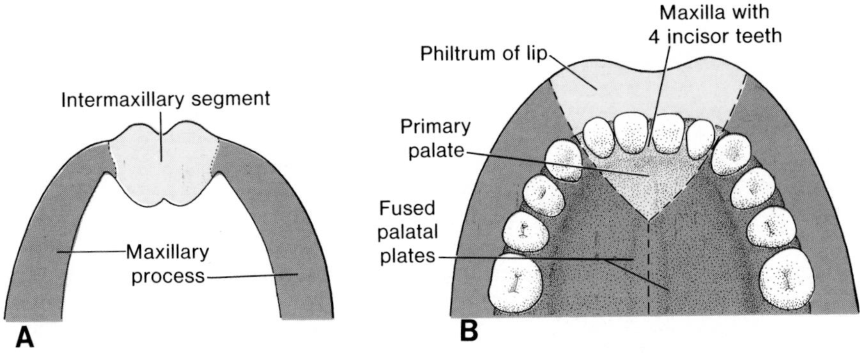

Figure 16-19. *A*, Schematic drawing of the intermaxillary segment and maxillary processes. *B*, The intermaxillary segment gives rise to the philtrum of the upper lip; the median part of the maxillary bone and its four incisor teeth; and the triangular primary palate.

shelves, appear in the sixth week of development and are directed obliquely downward on either side of the tongue (fig. 16-20*A*). In the seventh week, however, the palatine shelves ascend to attain a horizontal position above the tongue and fuse with each other, thus forming the **secondary palate** (figs. 16-21 and 16-22).

Anteriorly, the shelves fuse with the triangular primary palate, and the **incisive foramen** may be considered the midline landmark between the primary and secondary palates (fig. 16-22B). At the same time as the palatine shelves fuse, the nasal septum grows down and joins with the cephalic aspect of the newly formed palate (fig. 16-22*A*).

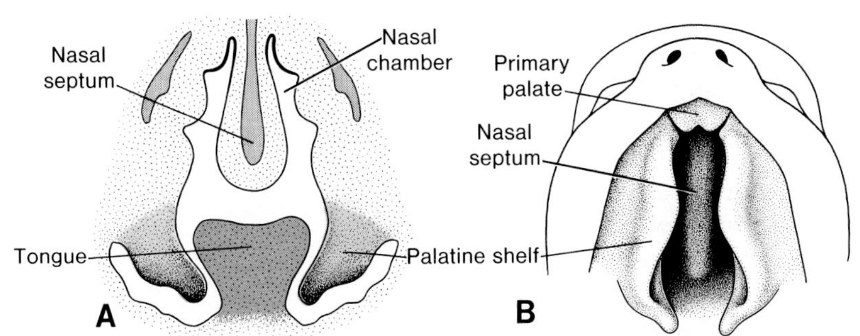

Figure 16-20. *A*, Frontal section through the head of a 6½-week-old embryo. The palatine shelves are located in the vertical position on each side of the tongue. *B*, Ventral view of the palatine shelves after removal of the lower jaw and the tongue. Note the clefts between the primary triangular palate and the palatine shelves, which are still in a vertical position.

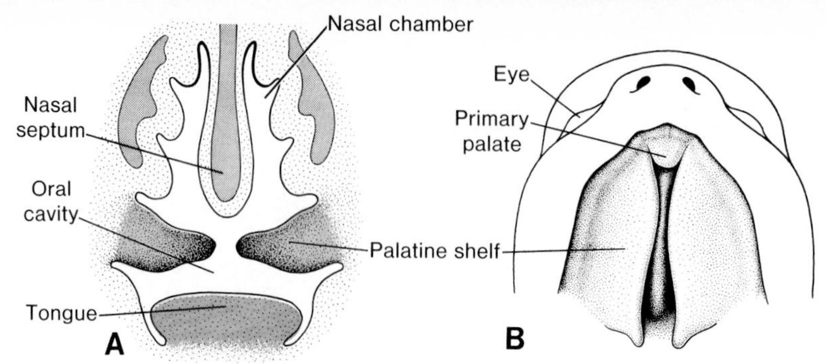

Figure 16-21. *A*, Frontal section through the head of a 7½ week embryo. The tongue has moved downward and the palatine shelves have reached a horizontal position. *B*, Ventral view of the palatine shelves after removal of the lower jaw and tongue. The shelves are in a horizontal position. Note the nasal septum.

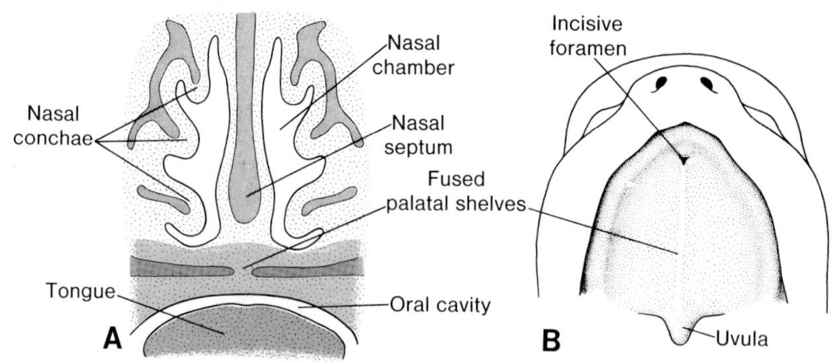

Figure 16-22. *A*, Frontal section through the head of a 10-week embryo. The two palatine shelves have fused with each other and with the nasal septum. *B*, Ventral view of the palate.

CLEFT LIP AND PALATE

The **incisive foramen** is considered the dividing landmark between the anterior and posterior cleft deformities. Those anterior to the incisive foramen include the **lateral cleft lip**, the **cleft upper jaw**, and the **cleft** between the **primary and secondary palates** (figs. 16-23*B–D* and 16-24 *A*,*B*). Those which lie posterior to the incisive foramen are caused by failure of the palatine shelves to fuse and include the **cleft (secondary) palate** and the **cleft uvula** (figs. 16-23*E* and 16-24*C*). The third category is formed by a combination of clefts lying anterior as well as posterior to the incisive foramen (fig. 16-23*F*). Since the palatine shelves fuse approximately one week after completion of the upper lip, and since the closing mechanisms of the lip and secondary palate differ greatly, the anterior and posterior cleft must be considered as separate entities.

The anterior clefts may vary in severity from barely visible defects in the vermilion of the lip to clefts extending into the nose (fig. 16-24*A*).

In more severe cases, the cleft extends to a deeper level, thereby forming a cleft of the upper jaw. The maxilla is then split between the lateral incisor

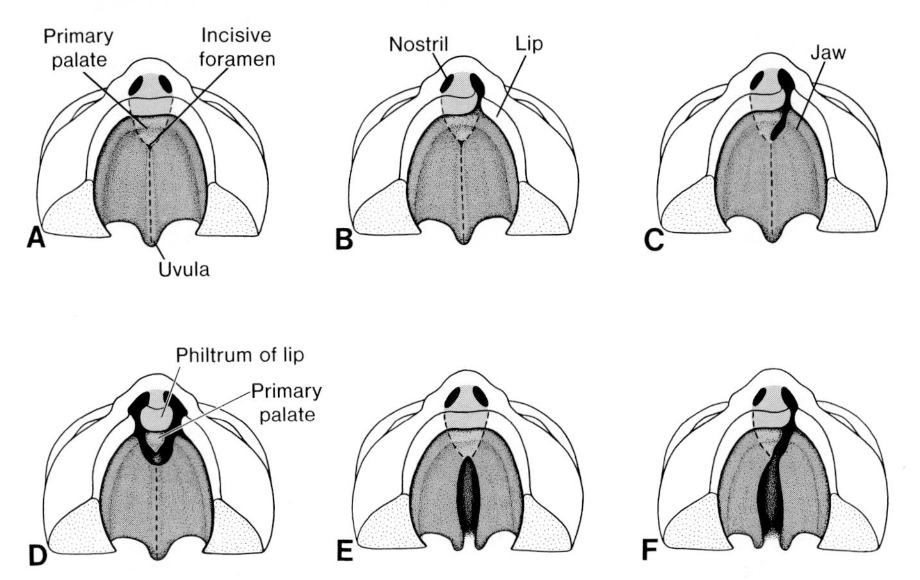

Figure 16-23. Ventral view of the palate, gum, lip and nose. *A*, Normal; *B*, unilateral cleft lip extending into the nose; *C*, unilateral cleft involving lip and jaw, and extending to incisive foramen; *D*, bilateral cleft involving lip and jaw; *E*, isolated cleft palate; *F*, cleft palate combined with unilateral anterior cleft.

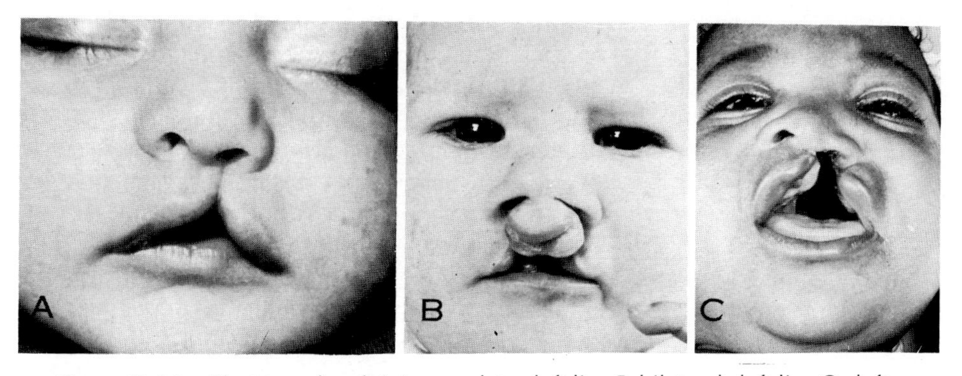

Figure 16-24. Photographs of *A*, incomplete cleft lip; *B*, bilateral cleft lip; *C*, cleft lip, cleft jaw, and cleft palate. (Courtesy Dr. M. Edgerton, Department of Plastic Surgery, University of Virginia.)

and the canine tooth. Frequently, such a cleft extends to the incisive foramen (fig. 16-23*C*,*D*).

The posterior clefts likewise may vary in severity from clefts involving the entire secondary palate to clefts of the uvula only.

MEDIAN CLEFT LIP

The median cleft lip, a rare abnormality, is caused by the incomplete merging of the two medial nasal swellings in the midline. This anomaly is usually accompanied by a deep groove between the right and left sides of the nose (fig. 16-25*D*).

OBLIQUE FACIAL CLEFT

Failure of the maxillary swelling to merge with its corresponding lateral nasal swelling results in an oblique facial cleft. When this occurs the nasolacrimal duct is usually exposed to the surface (fig. 16-25*A*).

HEREDITARY FACTORS

It is now generally accepted that the main etiological factor of the cleft lip and the cleft palate is genetic in nature, although in some cases a mixed genetic and environmental causation has been suggested.[14, 15, 16] The cleft lip (approximately 1:1000 births) is seen more in males than females; its incidence is slightly higher with increasing maternal age and varies in different population groups.[17]

If the parents are normal and have had one child with a cleft lip, the chance that the next baby will have the same defect is 4 per cent.[18, 19] If two siblings are affected the risk for the next child increases to 9 per cent. If, however, one of the parents has a cleft lip, and they have one child with the same defect, the probability that the next baby will be affected rises to 17 per cent.

The frequency of the cleft palate is much lower than that of the cleft lip (1:2500 births);[20] it is seen more in females than in males and is not related to maternal age. If the parents are normal and have one child with a cleft palate, the probability of the next child being affected is about 2 per cent. If, however, there is a similarly affected relative or a parent and child with a cleft palate, the probability increases to 7 and 15 per cent, respectively.[21]

In a recent study it was shown that in the female the palatal shelves fuse approximately one week later than in the male.[22] This may explain why isolated cleft palate is seen more frequently in the female than in the male.

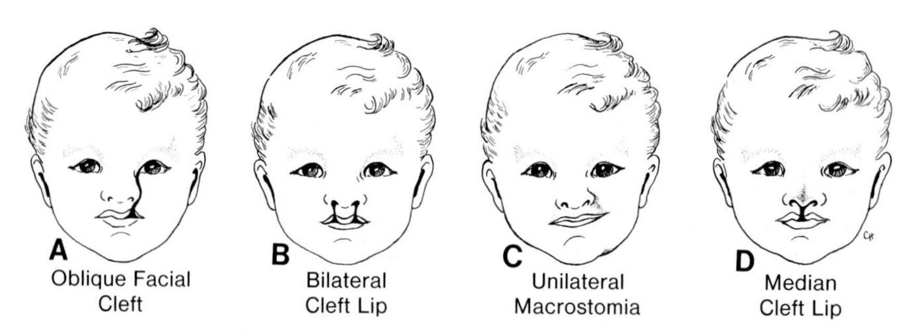

Figure 16-25. *A*, Oblique facial cleft; *B*, bilateral cleft lip; *C*, unilateral macrostomia; *D*, median cleft lip with partially cleft nose.

ENVIRONMENTAL FACTORS

Recently it has become evident that anticonvulsant drugs given during pregnancy increase the risk of cleft palate.[21,22] In a retrospective study on 427 pregnancies in 186 epileptic women the frequency of heart malformations, facial clefts, and microcephaly was twice as high as expected. In a similar study in Holland facial clefts were 29 times more frequent than in a control population. It is presently accepted that phenobarbital and diphenylhydantoin, when given to epileptic women during pregnancy, increase the incidence of cleft lip and palate two or three times when compared with a control population (see also Chapter 8).

EXPERIMENTAL DATA

In experimental work, it has been possible to cause cleft palates in the offspring of the rat and mouse by a variety of teratogenic agents.[23–28] From these studies it has become evident that congenital clefts of the palate can be produced in several different ways, each influenced by multiple genetic and environmental factors.[23] In addition, these studies have been valuable in analyzing the mechanisms of palate closure, and the possible role of the tongue, the growth of the mandible and the head, and the inherent movements of the palatine shelves in palate closure.[29–33]

Johnston showed that the neural crest cells, which migrate above and below the optic cup into the area of the face, greatly contribute to the formation of the facial swellings.[34,35] Extirpation of neural crest on one side of the forebrain frequently resulted in clefts of the primary palate on the same side. Hence, it may be that interference with the migration of neural crest cells at a stage long before the formation of the facial swellings may result in a cleft lip at a much later stage of development.[36]

Nasal Cavities

During the sixth week the nasal pits deepen considerably, partly because of the growth of the surrounding nasal swellings and partly because of their penetration into the underlying mesenchyme (fig. 16-26*A*). At first, the **oronasal membrane** separates the pits from the primitive oral cavity, but after its rupture the primitive nasal chambers open into the oral cavity by way of the newly formed foramina, the **primitive choanae** (figs. 16-20*B* and 16-26*C*). These choanae are located on each side of the midline and immediately behind the primary palate. Later, with the formation of the secondary palate and further development of the primitive nasal chambers (fig. 16-26*D*), the **definitive choanae** are located at the junction of the nasal cavity and the pharynx.[37]

The **paranasal airsinuses** develop as diverticula of the lateral nasal wall and extend into the maxilla, ethmoid, frontal, and sphenoid bones. They

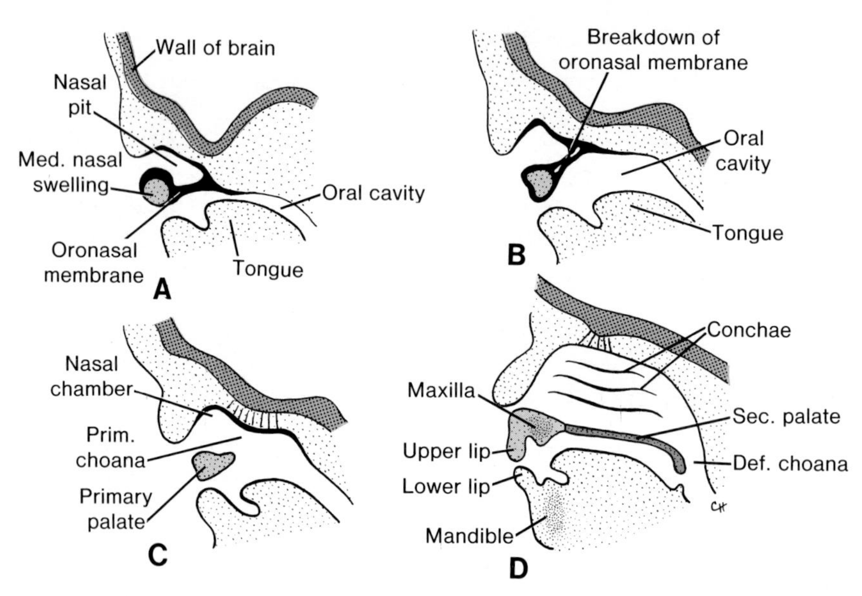

Figure 16-26. *A*, Drawing of a sagittal section through the nasal pit and the lower rim of the medial nasal swelling of a six week embryo. The primitive nasal cavity is separated from the oral cavity by the oronasal membrane. *B*, Similar section as in *A*, showing the oronasal membrane breaking down, *C*, In a seven-week embryo the primitive nasal cavity is in open connection with the oral cavity. *D*, Sagittal section through the face of a nine-week embryo, showing the intermaxillary segment, comprised of a labial component, a maxillary component, and the primary palate.

reach their maximum size during puberty and thus contribute greatly to the definitive shape of the face.

Teeth

The shape of the face is not only determined by the expansion of the paranasal sinuses, but also by the growth of the mandible and maxilla to accommodate the teeth. Approximately by the sixth week of development the basal layer of the epithelial lining of the oral cavity forms a C-shaped structure, the **dental lamina**, along the length of the upper and lower jaw. This lamina subsequently gives rise to a number of outbuddings (fig. 16-27*A*), 10 in each jaw, which form the primordia of the ectodermal components of the teeth. Soon the deep surface of the buds invaginates, resulting in the so-called **cap stage of tooth development** (fig. 16-27*B*). Such a cap consists of an outer layer, the **outer dental epithelium**, an inner layer, the **inner dental epithelium**, and a central core of loosely woven tissue, the **stellate reticulum**. The mesenchyme located in the indentation forms the **dental papilla** (fig. 16-27*B*).

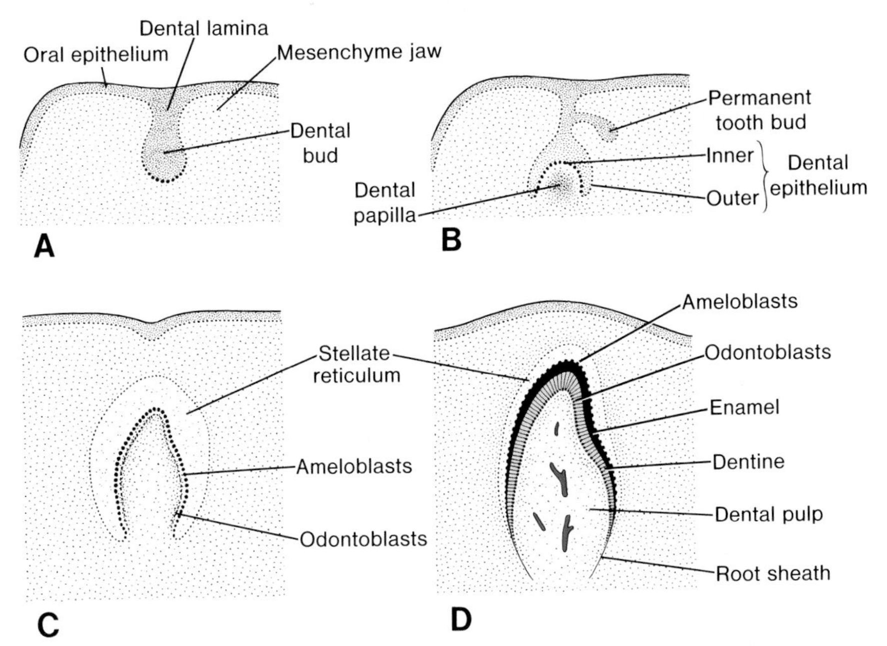

Figure 16-27. Schematic representation of the formation of the tooth at successive stages of development. *A*, At eight weeks; *B*, 10 weeks; *C*, three months; *D*, six months.

As the dental cap grows and the indentation deepens, the tooth takes on the appearance of a bell (**bell stage**) (fig. 16-27*C*). The mesenchyme cells of the papilla adjacent to the inner dental layer then differentiate into **odontoblasts**, which later produce the **dentine**. With the thickening of the dentine layer, the odontoblasts retreat into the dental papilla, thereby leaving a thin cytoplasmic process (**dental process**) behind in the dentine (fig. 16-27*D*). The odontoblast layer persists throughout the life of the tooth and continuously provides predentine, which is subsequently transformed into **dentine**. The remaining cells of the dental papilla form the **pulp** of the tooth (fig. 16-27*D*).

In the meantime the epithelial cells of the outer dental epithelium differentiate into the **ameloblasts** (enamel formers). These cells produce long enamel prisms which are deposited over the dentine (fig. 16-27*D*). The contact layer between the enamel and dentine layers is known as the **enamel dentine junction**.[38]

The enamel is first laid down at the apex of the tooth and from here spreads gradually toward the neck. When the enamel thickens, the ameloblasts retreat into the stellate reticulum. Here they regress, temporarily leaving a thin membrane (the **dental cuticle**) on the surface of the enamel. After eruption of the tooth this membrane gradually sloughs off.

The formation of the root of the tooth begins when the dental epithelial

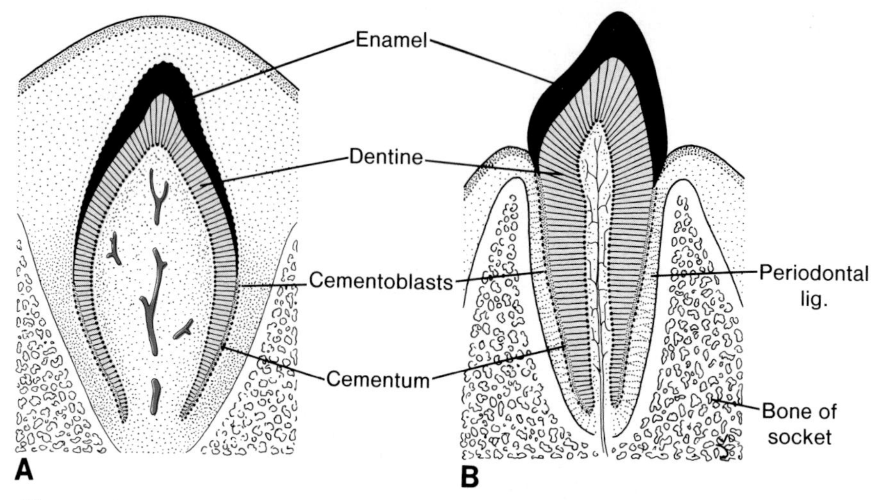

Figure 16-28. *A*, The tooth just before birth; *B*, after eruption.

layers penetrate into the underlying mesenchyme and form the **epithelial root sheath** (fig. 16-27*D*).[9] The cells of the dental papilla lay down a layer of dentine continuous with that of the crown (fig. 16-28*A*,*B*). As more and more dentine is deposited the pulp chamber narrows and finally forms a canal containing the blood vessels and nerves of the tooth.

The mesenchymal cells located on the outside of the tooth and in contact with the dentine of the root differentiate into **cementoblasts** (fig. 16-28*A*). These cells produce a thin layer of specialized bone, the **cementum**. Outside the cement layer, the mesenchyme gives rise to the **periodontal ligament** (fig. 16-28*A*,*B*), which holds the tooth firmly in position and simultaneously functions as a shock absorber.

With further lengthening of the root, the crown is gradually pushed through the overlying tissue layers into the oral cavity (fig. 16-28*B*). The eruption of the **deciduous** or **milk teeth** occurs 6 to 24 months after birth.

The buds for the permanent teeth are located on the lingual aspect of the milk teeth and are formed during the third month of development. These buds remain dormant until approximately the sixth year of postnatal life (fig. 16-29). Then they begin to grow, thereby pushing against the underside of the corresponding milk teeth and aiding in their shedding. As a permanent tooth grows, the root of the overlying deciduous tooth is resorbed by osteoclasts.

Occasionally the two lower central incisors have already erupted at birth. They are then usually abnormally formed and have little enamel and no roots. Although abnormalities of the teeth are mainly hereditary in nature, environmental factors such as rubella, syphilis, and irradiation have been described as causes for tooth anomalies.[10, 11]

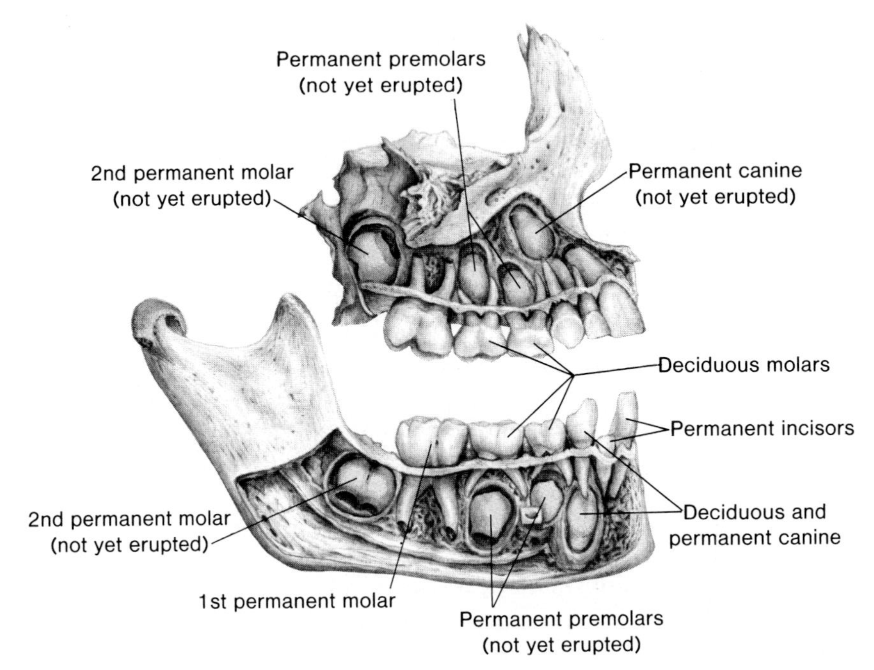

Figure 16-29. Replacement of deciduous teeth by permanent teeth in child of eight to nine years. (From *Atlas of Medical Anatomy* by Langman and Woerdeman. W. B. Saunders, Philadelphia, 1978.)

SUMMARY

The pharyngeal arches, consisting of bars of mesenchymal tissue and separated from each other by pharyngeal pouches and clefts, initially give the head and neck their typical appearance (fig. 16-1). Postnatally, the appearance of the teeth and paranasal sinuses provides the face with its own personal characteristics.

Each arch contains its own artery (fig. 16-2), nerve (fig. 16-5), muscle element (fig. 10-3), and cartilage bar or skeletal element (figs. 16-6 and 16-7). The endoderm of the pharyngeal pouches gives rise to a number of endocrine glands and part of the middle ear. In subsequent order the pouches give rise to (1) middle ear cavity and pharyngotympanic tube (fig. 16-8); (2) the palatine tonsil; (3) the parathyroid glands; (4) the thymus; and (5) the ultimobranchial bodies (fig. 16-9). The pharyngeal clefts give rise to only one structure: the external auditory meatus. The thyroid gland originates from an epithelial proliferation in the floor of the tongue and descends to its level in front of the tracheal rings in the course of development.

The complicated development of the face through the pharyngeal arches leads to many craniofacial abnormalities (fig. 16-11), while, in addition, the development of abnormal lateral and midline cysts (figs. 16-10, 16-14, and 16-15) such as lateral branchial cysts, fistulas, thyroglossal ducts, sinuses, and cysts are well-known congenital defects.

Particularly the pharyngeal swellings of the facial region are of great importance since they determine, through fusion and special growth, the size of the mandible, upper lip, palate, and nose. Though hypoplasia of the mandible is well-known, the formation of the upper lip by fusion of two maxillary swellings and two lateral nasal swellings (figs. 16-17 and 16-18) is of utmost importance since the cleft lip is a serious morphological and psychological handicap (figs. 16-23 and 16-24). In the deeper region of the face, fusion of the maxillary swellings gives rise to the hard and soft palate (figs. 16-20 to 16-22) and the derivatives of the **intermaxillary segment** such as (1) the philtrum, (2) the upper jaw component, which carries the four incisor teeth, and (3) the palatal component, which forms the triangular primary palate. As is understandable, a series of cleft deformities may result from partial or incomplete fusion of these mesenchymal tissues, which may be caused by hereditary as well as environmental (diphylhydantoin) factors.

The final adult form of the face is greatly contributed to by the development of paranasal sinuses, nasal conchae, and the teeth. The teeth develop from an ectodermal and mesodermal component. The **enamel**, the white layer on the surface, is made by the ameloblasts (figs. 16-27 and 16-28). It lies on a thick layer of **dentine** produced by the **odontoblasts**, a mesenchymal derivative. The cementum is formed by the **cementoblasts**, another mesenchymal derivative found in the root of the tooth. Although the first teeth (**deciduous** or **milk teeth**) appear 6 to 24 months after birth, the **definitive** or **permanent teeth**, which appear postnatally, are formed mainly during the third month of development (fig. 16-29).

REFERENCES

1. Hanson, J. R., Anson, B. J., and Best, T. H. The early embryology of the auditory vesicles in man. *Bull. Northwest Univ. Med. Sch., 33:* 350, 1959.
2. Moore, M. A. S., and Owen, J. J. T. Experimental studies on the development of the thymus. *J. Exp. Med., 126:* 715, 1967.
3. Samuel, F. T. Ultrastructure of differentiating cells during thymus histogenesis. A light and

electron microscopic study of epithelial and lymphoid cell differentiation during thymus histogenesis in C57 black mice. *Z. Zellforsch.*, *83:* 8, 1967.
4. Anast, C. S. Calcitonin. In *Endocrine and Genetic Diseases of Childhood and Adolescence*, ed. 2, edited by L. T. Gardner. W. B. Saunders, Philadelphia, 1975.
5. Moseley, J. M., Mathews, E. W., Breed, R. H., Galante, E., Tse, A., and MacIntyre, I. The ultimobranchial origin of calcitonin. *Lancet*, *1:* 108, 1968.
6. Albers, G. D. Branchial anomalies. *J.A.M.A.*, *183:* 399, 1963.
7. Martins, A. G. Lateral cervical sinus and preauricular sinuses. *Br. Med. J.*, *5:* 255, 1961.
8. McKenzie, J. The first arch syndrome. *Dev. Med. Child Neurol.*, *8:* 55, 1966.
9. MacCollum, D. W., and Witkop, C. J. Lateral and oblique facial clefts and Robin syndrome. In *Handbook of Congenital Malformations*, edited by A. Rubin. W. B. Saunders, Philadelphia, 1967.
10. Marshall, S. F., and Becker, W. F. Thyroglossal cysts and sinuses. *Ann. Surg.*, *129:* 642, 1949.
11. Shepard, T. H. Development of the thyroid gland. In *Endocrine and Genetic Diseases of Childhood and Adolescence*, ed. 2, edited by L. T. Gardner. W. B. Saunders, Philadelphia, 1975.
12. Zenker, W. Ueber die Bedeutung der Kautasche für die Entwicklung der menschlichen Wange. Z. für Anat. u. Entwicklungsgesch., *124:* 289, 1964.
13. Patten, B. M. The normal development of the facial region. In *Congenital Anomalies of the Face and Associated Structures*, edited by S. Pruzansky, p. 11. Charles C. Thomas, Springfield, Ill., 1961.
14. Stark, R. B. The pathogenesis of harelip and cleft palate. *Plast. Reconstr. Surg.*, *13:* 20, 1954.
15. Carter, C. D. Incidence and aetiology. In *Congenital Abnormalities in Infancey*, edited by A. P. Norman. Blackwell Scientific Publications, Oxford, England, 1971.
16. Fogh-Anderson, P. Inheritance patterns for cleft lip and cleft palate. In *Congenital Anomalies of the Face and Associated Structures*, edited by S. Pruzansky, p. 123. Charles C. Thomas, Springfield, Ill., 1961.
17. Neel, J. R. A study of major congenital defects in Japanese infants. *Am. J. Hum. Genet.*, *10:* 398, 1958.
18. Curtis, E. J., Fraser, F. C., and Warburton, D. Congenital cleft lip and palate. *Am. J. Dis. Child.*, *102:* 853, 1961.
19. Fraser, F. C. Genetics and congenital malformations. In *Progress in Medical Genetics*, edited by A. G. Steinberg, p. 38. Grune & Stratton, New York, 1961.
20. Gorlin, R. J., Cervenka, J., and Pruzansky, S. Facial clefting and its syndromes. *Birth Defects*, *8:* 3, 1971.
21. Fogh-Andersen, P. Genetic and non genetic factors in the etiology of facial clefts. *Scand. J. Plast. Reconstr. Surg.*, *1:* 22, 1967.
22. Burdi, A. R. Sexual differences in closure of the human palatal shelves. *Cleft Palate J.*, *6:* 1, 1969.
23. Fraser, F. C. Workshop on embryology of cleft lip and cleft palate (report of symposium). *Teratology*, *1:* 353, 1968.
24. Wilson, J. G. Present status of drugs as teratogens in man. *Teratology*, *7:* 3, 1973.
25. Preisler, O. Is prolonged cortisone treatment in pregnancy damaging to the infant? *Zentralbl. Gynaekol.*, *18:* 675, 1960.
26. Bongiovanni, A. M., and McPadden, A. J. Steroids during pregnancy and possible fetal consequences. *Fertil. Steril.*, *11:* 181, 1960.
27. Speidel, B. D., and Meadow, S. R. Maternal epilepsy and abnormalities of the fetus and newborn. *Lancet*, *2:* 839, 1972.
28. Seip, M. The effects of antiepileptic drugs in pregnancy on the fetus and new born infant. *Ann. Clin. Res.*, *5:* 205, 1973.
29. Humphrey, T. The relation between human fetal mouth opening reflexes and closure of the palate. *Am. J. Anat.*, *125:* 317, 1969.
30. Walker, B. E. Correlation of embryonic movement with palate closure in mice. *Teratology*, *2:* 191, 1969.
31. Harris, J. W. S. Experimental studies on closure and cleft formation in the secondary palate. *Sci. Basis Med. Annu. Rev.*, pp. 356–369, 1967.

32. Long, S. Y., Larsson, K. S., and Lohmander, S. Cell proliferation in the cranial base of A/J mice with 6-AN-induced cleft palate. *Teratology, 8:* 127, 1973.
33. Pratt, R. M., and King, C. T. G. Inhibition of collagen cross-linking associated with aminoproprionitrile-induced cleft palate in the rat. *Dev. Biol., 27:* 322, 1972.
34. Johnston, M. C. A radioautographic study of the migration and fate of cranial neural crest cells in the chick embryo. *Anat. Rec., 156:* 143, 1966.
35. Johnston, M. C., and Listgarten, M. A. Observations on the migration, interaction, and early differentiation of orofacial tissues. In *Developmental Aspects of Oral Biology*, edited by H. Slavkin and L. A. Bavetta. Academic Press, New York, 1972.
36. Johnston, M. C. and Sulik, K. K. Development of face and oral cavity. In *Orban's Oral Histology and Embryology*, ed. 9, edited by S. N. Bhaskar. C. V. Mosby Co., St. Louis, 1979.
37. Remmick, H. *Embryology of the Face and Oral Cavity.* Fairleigh Dickinson University Press, Rutherford, N.J., 1970.
38. Sperber, G. H. In *Craniofacial Embryology*, ed. 2, edited by G. H. Sperber. Year Book Medical Publishers, Chicago, 1976.

chapter 17

Ear

In the adult the ear forms one anatomical unit serving both hearing and equilibrium. In the embryo, however, it develops from three distinctly different parts: (1) the **external ear**, which serves as the sound-collecting organ; (2) the **middle ear**, which functions as a sound conductor from the external to the internal ear; and (3) the **internal ear**, which converts the sound waves into nerve impulses and registers changes in equilibrium.

Internal Ear

OTIC VESICLE

The first indication of the developing ear can be found in embryos of approximately 22 days as a thickening of the surface ectoderm on each side of the rhombencephalon (figs. 17-1*A*, *B*).[1] These thickenings, the **otic placodes**, invaginate rapidly and form the **otic** or **auditory vesicles** (otocysts) (fig. 17-2*A*, *B*, and *C*). During later development each vesicle divides into (1) a ventral component which gives rise to the **saccule** and the **cochlear duct**, and (2) a dorsal component which forms the **utricle, semicircular canals,** and **endolymphatic duct** (figs. 17-3 to 17-6). The epithelial structures so formed are known collectively as the **membranous labyrinth**.

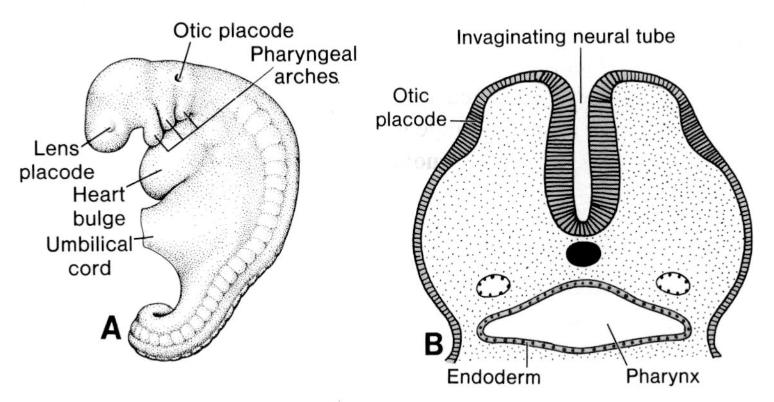

Figure 17-1. *A*, Drawing of the left side of a 28-day-old embryo showing the otic vesicle. *B*, Schematic transverse section through the region of the rhombencephalon to show the otic placode in a 22-day-old embryo.

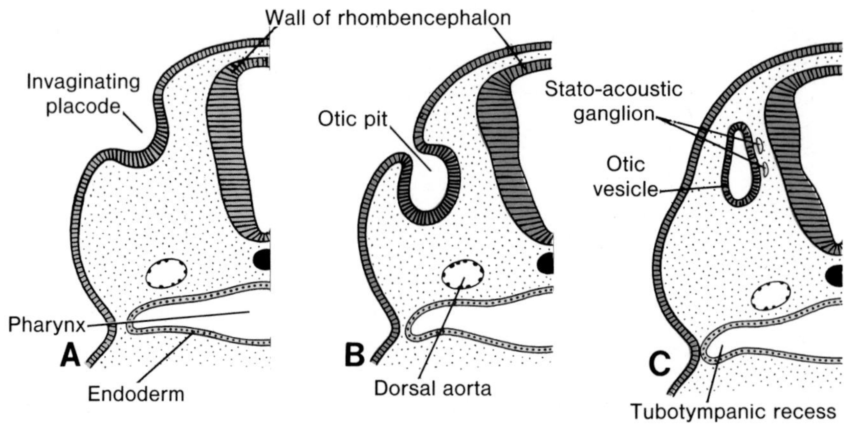

Figure 17-2. Schematic transverse sections through the region of the rhombencephalon showing the formation of the otic vesicles. *A*, 24 days; *B*, 27 days; *C*, 4½ weeks. Note the appearance of the stato-acoustic ganglion.

SACCULE, COCHLEA, AND ORGAN OF CORTI

In the sixth week of development the saccule forms a tubular-shaped outpocketing at its lower pole (fig 17-3*C*). This outgrowth, the **cochlear duct**, penetrates the surrounding mesenchyme in spiral fashion until, at the end of the eighth week, it has completed 2½ turns (fig. 17-3*D, E*). Its connection with the remaining portion of the saccule is then confined to a narrow pathway, the **ductus reuniens** (fig. 17-3*E*).

The mesenchyme surrounding the cochlear duct soon differentiates into cartilage (fig. 17-4*A*). In the 10th week this cartilaginous shell undergoes vacuolization and two perilymphatic spaces, the **scala vestibuli** and **scala tympani**, are formed (figs. 17-4*B, C*). The cochlear duct is then separated from the scala vestibuli by the **vestibular membrane**, and from the scala tympani by the **basilar membrane** (fig. 17-4*C*). The lateral wall of the cochlear duct remains attached to the surrounding cartilage by the **spiral ligament**, whereas its median angle is connected to, and partly supported by, a long cartilaginous process, the **modiolus**, the future axis of the bony cochlea (fig. 17-4*B*).

The epithelial cells of the cochlear duct are initially all alike (fig. 17-4*A*). With further development, however, they form two ridges: the **inner ridge** (the future **spiral limbus**) and the **outer ridge** (fig. 17-4*B*). The latter forms one row of inner and three or four rows of outer **hair cells**, the sensory cells of the auditory system (fig. 17-5). They are covered by the **tectorial membrane**, a fibrillar gelatinous substance, which is carried by the spiral limbus and rests with its tip on the hair cells (fig 17-5). The sensory cells and the covering tectorial membrane together are known as the **organ of Corti**. The impulses received by this organ are transmitted to the spiral ganglion and

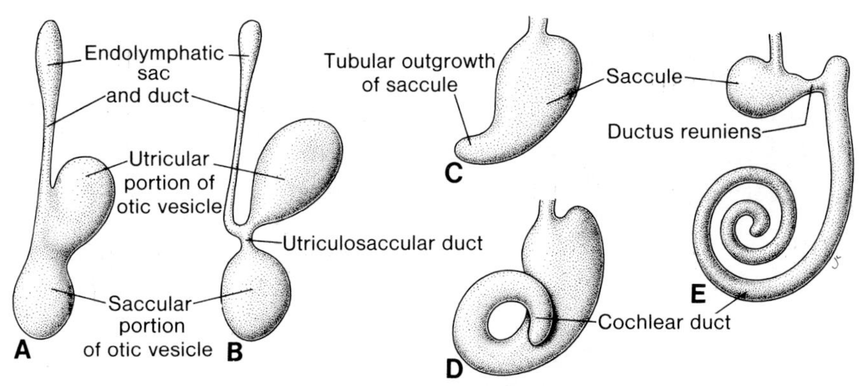

Figure 17-3. *A* and *B*, Development of the otocyst showing a dorsal utricular portion with the endolymphatic duct, and a ventral saccular portion. *C*, *D*, and *E*, The cochlear duct at six, seven, and eight weeks, respectively. Note the formation of the ductus reuniens and the utriculosaccular duct.

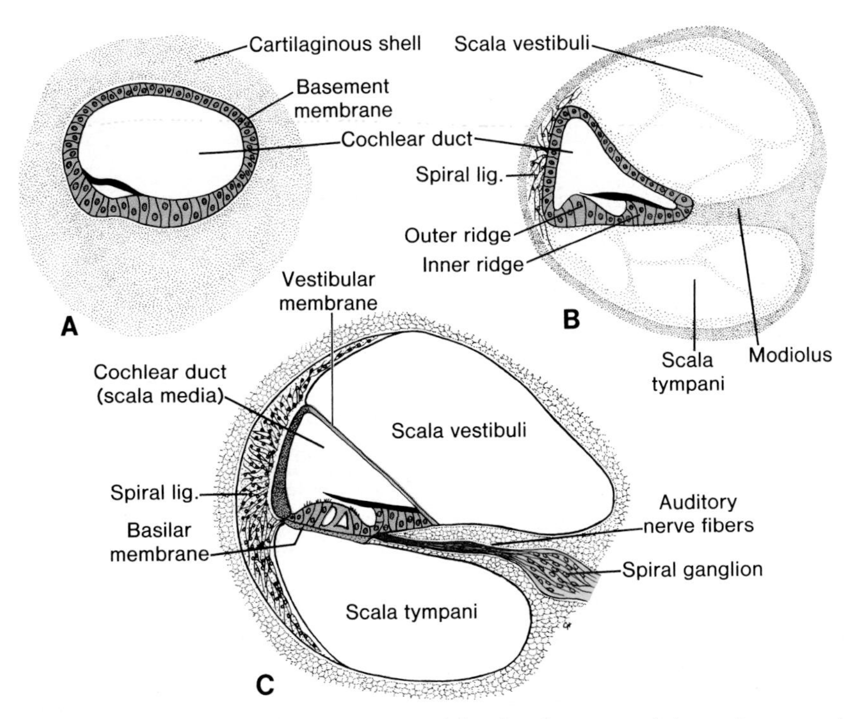

Figure 17-4. Schematic representation of the development of the scala tympani and scala vestibuli. *A*, The cochlear duct is surrounded by a cartilaginous shell. *B*, During the 10th week, large vacuoles appear in the cartilaginous shell. *C*, The cochlear duct (scala media) is separated from the scala tympani and the scala vestibuli by the basilar and vestibular membranes, respectively. Note the auditory nerve fibers and the spiral (cochlear) ganglion.

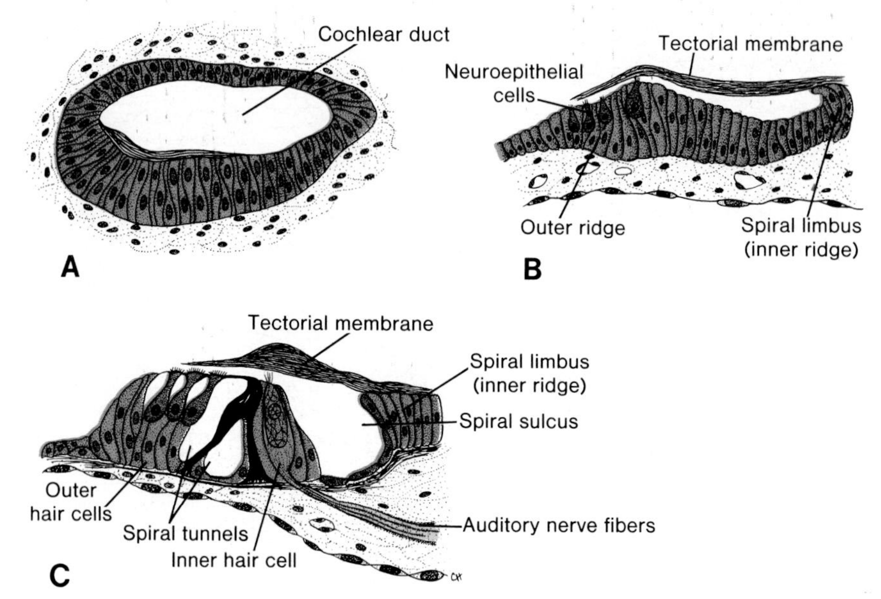

Figure 17-5. Development of the organ of Corti. *A*, At 10 weeks; *B*, approximately five months; *C*, full term. Note the appearance of the spiral tunnels in the organ of Corti.

then to the nervous system by the auditory fibers of the eighth cranial nerve (figs. 17-4 and 17-5).

UTRICLE AND SEMICIRCULAR CANALS

During the sixth week of development the semicircular canals appear as flattened outpocketings of the utricular part of the otic vesicle (fig. 17-6*A*, *B*). The central portions of the walls of these outpocketings eventually become apposed to each other (fig. 17-6*C*, *D*) and disappear, thus giving rise to the three semicircular canals (fig. 17-6*E*, *F*). While one end of each canal dilates to form the **crus ampullare**, the other does not widen and is known as the **crus nonampullare** (fig. 17-6*E*). However, since two of the latter type fuse, only five crura enter the utricle—three with an ampulla and two without.

The cells in the ampullae form a crest, the **crista ampullaris**, containing the sensory cells for the maintenance of equilibrium. Similar sensory areas develop in the walls of utricle and saccule, where they are known as **maculae acousticae**. Impulses generated in the sensory cells of the cristae and maculae as a result of a change in position of the body are carried to the brain by the **vestibular fibers of the eighth cranial nerve**.

During formation of the otic vesicle a small group of cells breaks away from its wall and forms the **stato-acoustic ganglion** (fig. 17-2*C*).[2, 3] Other cells of this ganglion probably are derived from the neural crest. The ganglion subsequently splits into cochlear and vestibular portions which supply the sensory cells of the organ of Corti and those of the saccule, utricle, and semicircular canals, respectively (fig. 17-2).

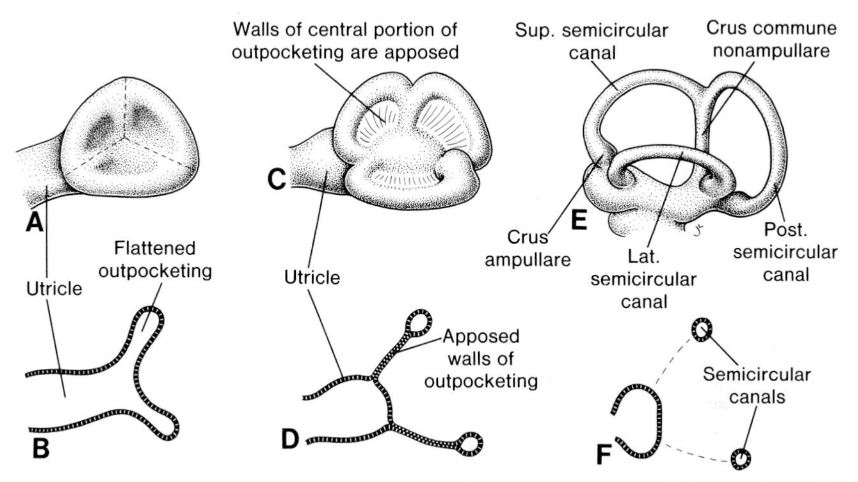

Figure 17-6. Schematic representation of the development of the semicircular canals. *A*, At five weeks; *C*, six weeks; *E*, eight weeks. *B*, *D*, and *F* show diagrammatically the apposition, fusion, and disappearance of the central portions of the walls of the semicircular outpocketings. Note the ampullae in the semicircular canals.

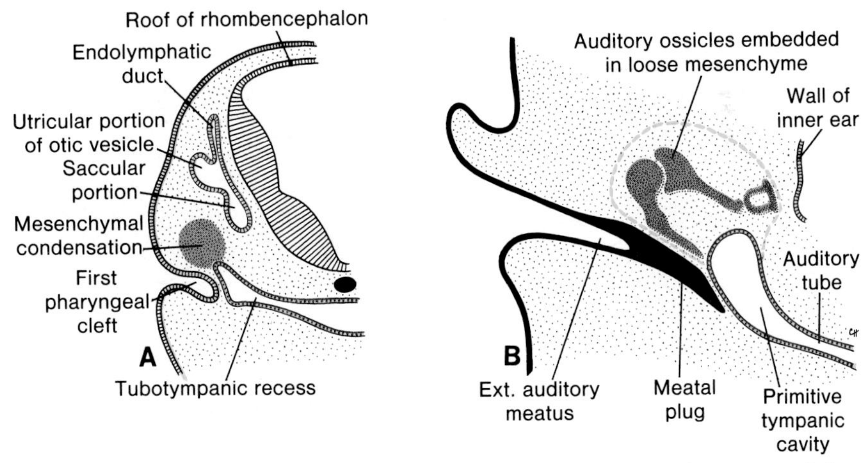

Figure 17-7. *A*, Transverse section of a seven-week embryo in the region of the rhombencephalon, showing the tubotympanic recess, the first pharyngeal cleft, and the mesenchymal condensation, foreshadowing the development of the ossicles. *B*, Schematic representation of the middle ear, showing the cartilaginous precursors of the auditory ossicles. Thin yellow line in mesenchyme indicates future expansion of the primitive tympanic cavity. Note the meatal plug extending from the primitive auditory meatus to the tympanic cavity.

Middle Ear

TYMPANIC CAVITY AND EUSTACHIAN TUBE

The tympanic cavity is of **endodermal** origin. It is derived from the first pharyngeal pouch (figs. 17-2 and 17-7). This pouch grows rapidly in lateral direction, and temporarily comes in contact with the floor of the first pharyngeal cleft. The distal part of the pouch, the **tubotympanic recess**, widens and gives rise to the **primitive tympanic cavity**, while the proximal part remains narrow and forms the **auditory** or **Eustachian tube** (fig. 17-7*B*).[4] The latter is the channel through which the tympanic cavity communicates with the nasopharynx.

OSSICLES

The malleus and incus are derived from the cartilage of the first pharyngeal arch, and the stapes from that of the second arch (fig. 17-8*A*).[5, 6] Although the ossicles appear during the first half of fetal life they remain embedded in mesenchyme until the eighth month (fig. 17-7*B*), when the surrounding tissue dissolves (fig. 17-8*B*). The **endodermal** epithelial lining of the primitive tympanic cavity then gradually extends along the wall of the newly developing space. The tympanic cavity is now at least twice as large as before.

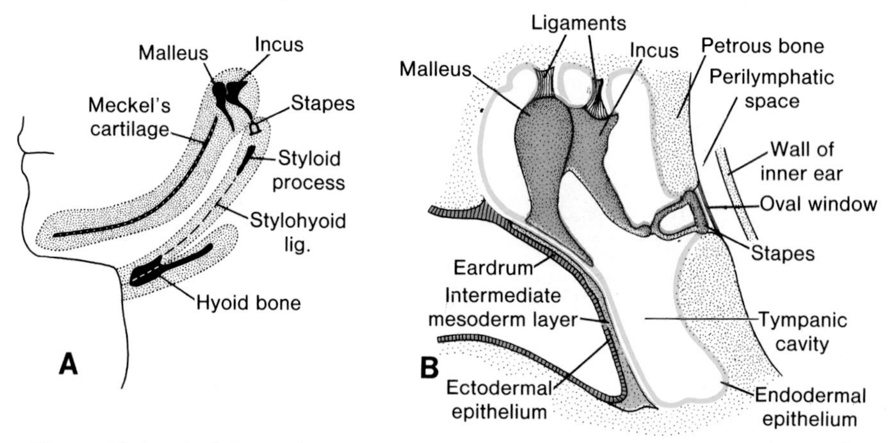

Figure 17-8. *A*, Schematic representation of the derivatives of the first three branchial arches. Note the malleus and incus at the dorsal tip of the first arch and the stapes at that of the second arch. *B*, Schematic representation of the middle ear, showing the handle of the malleus in contact with the eardrum. The stapes will establish contact with the membrane in the oval window. The wall of the tympanic cavity is lined with epithelium of endodermal origin.

When the ossicles are entirely free of surrounding mesenchyme the endodermal epithelium connects them in a mesentery-like fashion to the wall of the cavity (fig. 17-8*B*). The supporting ligaments of the ossicles later develop within these mesenteries.

Since the malleus is derived from the first pharyngeal arch, its muscle, the **tensor tympani**, is innervated by the mandibular branch of the trigeminal nerve. Similarly, the **stapedius muscle**, which is attached to the stapes, is innervated by the facial nerve.

During late fetal life the tympanic cavity expands dorsally by vacuolization of the surrounding tissue to form the **tympanic antrum**. After birth the bone of the developing mastoid process is also invaded by epithelium of the tympanic cavity and epithelial lined air sacs are formed (**pneumatization**). Later most of the mastoid air sacs come in contact with the antrum and tympanic cavity. Expansion of inflammations of the middle ear into the antrum and mastoid air cells is a rather common complication of middle ear infections.[7]

External Ear

EXTERNAL AUDITORY MEATUS

The external auditory meatus develops from the dorsal portion of the first pharyngeal cleft (fig. 17-7*A*). At the beginning of the third month the epithelial cells at the bottom of the meatus proliferate, thereby forming a solid epithelial plate, the **meatal plug** (fig. 17-7*B*). In the seventh month this plug dissolves and the epithelial lining of the floor of the meatus then participates in the formation of the definitive eardrum. Occasionally, the meatal plug persists until birth, resulting in congenital deafness.

EARDRUM OR TYMPANIC MEMBRANE

The eardrum is made up of (1) the ectodermal epithelial lining at the bottom of the auditory meatus; (2) the endodermal epithelial lining of the tympanic cavity; and (3) an intermediate layer of connective tissue (fig. 17-8*B*), which forms the **fibrous stratum**. The major part of the eardrum is firmly attached to the handle of the malleus (fig. 17-8*B*), while the remaining portion forms the separation between the external auditory meatus and the tympanic cavity.

AURICLE

The auricle develops from six mesenchymal proliferations located at the dorsal ends of the first and second pharyngeal arches and surrounding the first pharyngeal cleft (fig. 17-9*A*). These swellings, three on each side of the external meatus, later fuse and gradually form the definitive auricle. As the

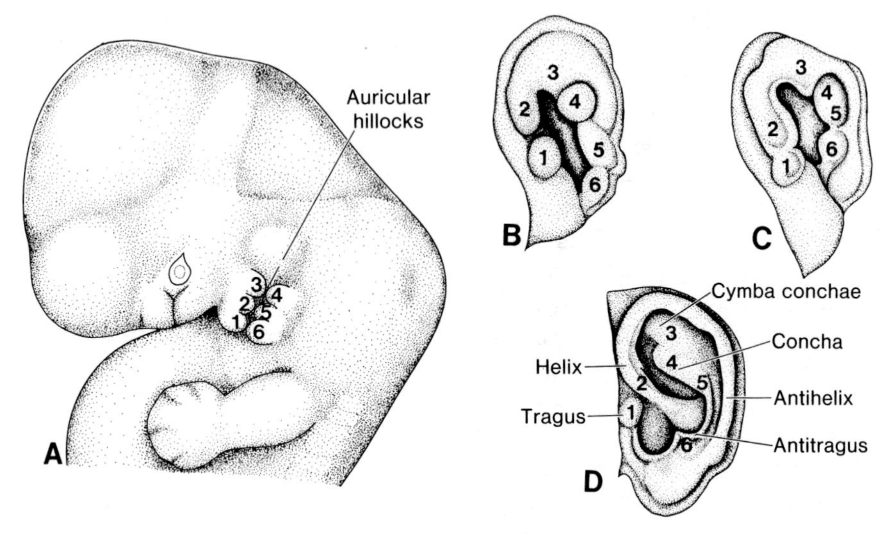

Figure 17-9. *A*, Lateral view of the head of an embryo showing the six auricular hillocks surrounding the dorsal end of the first pharyngeal cleft. *B*, *C*, and *D* show the fusion and progressive development of the hillocks into the adult auricle.

fusion of the auricular hillocks is rather complicated, developmental abnormalities of the auricle are not uncommon. Initially the external ears are located in the lower neck region, but with development of the mandible they ascend to the side of the head at the level of the eyes.

CONGENITAL DEAFNESS

Congenital deafness, usually associated with deaf-mutism, may be caused by abnormal development of the membranous and bony labyrinths, as well as by malformations of the auditory ossicles and eardrum. In the most extreme cases the tympanic cavity and external meatus are absent.

Although congenital deafness was believed to be mainly hereditary,[8, 9] it is now known that environmental factors may likewise interfere with normal development of the internal and middle ear. Rubella virus, affecting the embryo in the seventh to eighth week, will cause severe damage to the organ of Corti.[10] Although it has been suggested that poliomyelitis, erythroblastosis fetalis, diabetes, hypothyroidism, and toxoplasmosis may likewise cause congenital deafness, the observations await more evidence.[11-14]

Although minor variations of the auricle are clinically not important, major changes are sometimes associated with serious internal abnormalities such as kidney malformations.[15] Ear abnormalities may also be part of the first arch syndrome or other chromosomally determined syndromes (see fig. 16-11B).

SUMMARY

The ear consists of three parts which are of different origin but function as one unit. The **internal ear** originates from the **otic vesicle**, which in the fourth week of development splits off from the **surface ectoderm**. This otic vesicle divides in a ventral component which gives rise to the **saccule** and **cochlear** duct, and a dorsal component which gives rise to the **utricle**, **semicircular canals**, and **endolymphatic duct** (fig. 17-3 to 17-6). The epithelial structures so formed are known collectively as the **membranous labyrinth**. With exception of the cochlear duct, from which will develop the **organ of Corti**, all structures derived from the membranous labyrinth serve the equilibrium.

The **middle ear** consisting of the **tympanic cavity** and **auditory tube**, is lined with epithelium of endodermal origin and is derived from the first pharyngeal pouch. The auditory tube maintains contact between the tympanic cavity and nasopharynx. The **ossicles**, used to transfer sound fibrations from the tympanic membrane to the oval window, are derived from the first (malleus and incus) and second (stapes) pharyngeal arches.

The **external auditory meatus** develops from the first pharyngeal cleft and at its bottom is lined by the tympanic membrane. The eardrum consists of (1) an ectodermal epithelial lining; (2) an intermediate layer of mesenchyme; and (3) an endodermal lining from the first pharyngeal pouch.

The **auricle** develops from six mesenchymal hillocks (fig. 17-9).

REFERENCES

1. O'Rahilly, R. The early development of the otic vesicle in staged human embryos. *J. Embryol. Exp. Morphol., 11:* 741, 1963.
2. Politzer, G. Die Entstehung des Ganglion Acusticum beim Menschen. *Acta Anat.* (*Basel*), *26:* 1, 1956.
3. Batten, E. H. The origin of the acoustic ganglion in the sheep. *J. Embryol. Exp. Morphol., 6:* 597, 1958.
4. Kanagasuntheram, R. A note on the development of the tubotympanic recess in the human embryo. *J. Anat., 101:* 731,1967.
5. Anson, B. J., and Bast, T. H. Development of the stapes of the human ear. *Q. Bull. Northw. Univ. Med. Sch., 33:* 44,1959.
6. Anson, B. J., Hanson, J. S., and Richany, S. F. Early embryology of the auditory ossicles and associated structures in relation to certain anomalies observed clinically. *Ann. Otol., 69:* 427, 1960.
7. Proctor, B. The development of the middle ear spaces and their surgical significance. *J. Laryngol. Otol., 78:* 631, 1964.

8. Stevenson, A. C., and Cheeseman, E. A. Hereditary deaf-mutism with particular reference to Northern Ireland. *Ann. Hum. Genet., 20:* 177, 1956.
9. Konigsmark, B. W., and Gorlin, R. J. *Genetic and Metabolic Deafness.* W. B. Saunders, Philadelphia, 1976.
10. Gray, J. E. Rubella in pregnancy; fetal pathology in the internal ear. *Ann. Otol., 68:* 170, 1959.
11. Keleman, J. Acute poliomyelitis of the mother with aural lesions of the premature infant. *Arch. Otolaryngol., 62:* 602, 1955.
12. Keleman, J. Erythroblastosis fetalis. Pathologic report on the hearing organs of newborn infant. *Arch. Otolaryngol., 63:* 392, 1956.
13. Jorgensen, M. B. The influence of maternal diabetes on the inner ear of the fetus. *Acta Otolaryngol. (Stockh.), 53:* 49, 1961.
14. Thould, A. K., and Scowen, E. F. The syndrome of congenital deafness and goiter. *J. Endocrinol., 30:* 69, 1964.
15. Smith, D. W. *Recognizable Patterns of Human Malformation: Genetic, Embryologic, and Clinical Aspects.* W. B. Saunders, Philadelphia, 1970.

chapter 18

Eye

Optic Cup and Lens Vesicle

The developing eye appears in the 22-day embryo as a pair of shallow grooves on each side of the invaginating forebrain (fig 18-1*A*). With closure of the neural tube, these grooves form outpocketings of the forebrain, the **optic vesicles**. These vesicles subsequently come in contact with the surface ectoderm and induce changes in the ectoderm necessary for lens formation (fig 18-1*B*).[1] Shortly thereafter, the optic vesicle begins to invaginate and forms the double-walled **optic cup** (figs. 18-1*C* and 18-2*A*). The inner and outer layers of this cup are initially separated by a lumen, the **intraretinal space** (fig. 18-2*B*), but soon this lumen disappears and the two layers are then apposed to each other (fig. 18-6). The invagination is not restricted to the central portion of the cup but also involves a part of the inferior surface (fig. 18-2*A*). Here is formed the **choroid fissure**. Formation of this fissure allows the **hyaloid artery** to reach the inner chamber of the eye (figs. 18-3 and 18-7). During the seventh week the lips of the choroid fissure fuse, and the mouth of the optic cup then becomes a round opening, the future **pupil**.[2]

While these events occur, the cells of the surface ectoderm, initially in contact with the optic vesicle, begin to elongate and form the **lens placode**

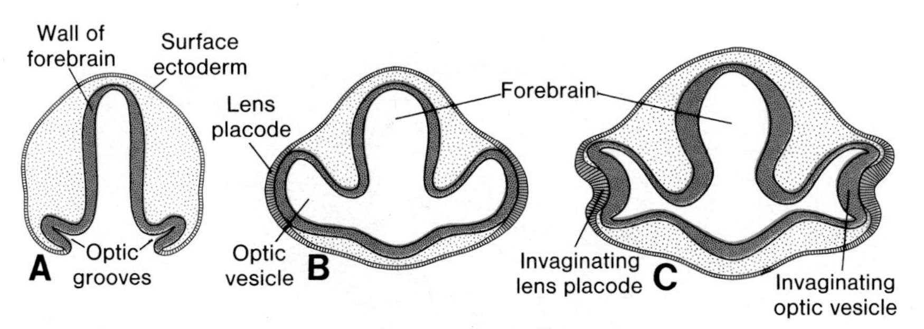

Figure 18-1. *A*, Transverse section through the forebrain of a 22-day embryo (approximately 14 somites), showing the optic grooves. *B*, Transverse section through the forebrain of a four-week embryo, showing the optic vesicles in contact with the surface ectoderm. Note the slight thickening of the ectoderm (lens placode). *C*, Transverse section through the forebrain of a 5-mm embryo, showing the invagination of the optic vesicle and the lens placode.

(fig. 18-1*B*, *C*). This placode subsequently invaginates and develops into the **lens vesicle**. During the fifth week the lens vesicle loses contact with the surface ectoderm and is then located in the mouth of the optic cup (figs. 18-2*C* and 18-3).[3]

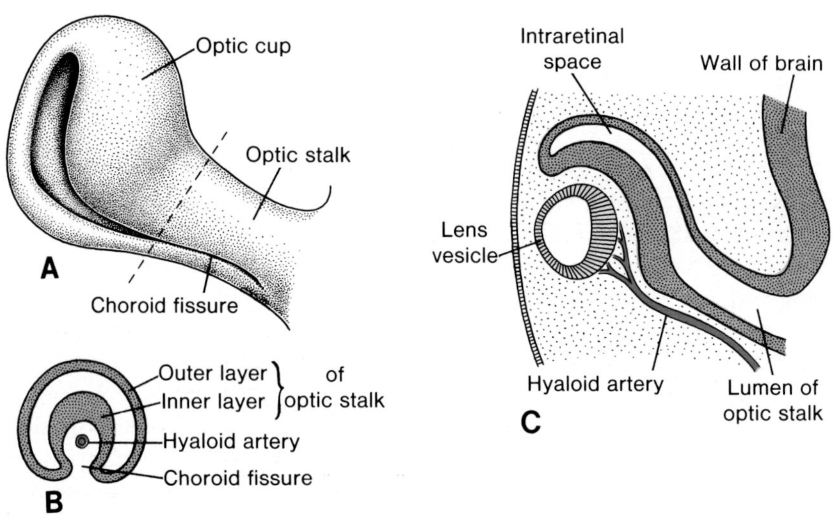

Figure 18-2. *A*, Ventrolateral view of the optic cup and optic stalk of a six-week embryo. The choroid fissure located on the undersurface of the optic stalk gradually tapers off. *B*, Transverse section through the optic stalk indicated as in *A*, showing the hyaloid artery in the choroid fissure. *C*, Section through the lens vesicle, the optic cup, and optic stalk at the plane of the choroid fissure (after Mann[2]).

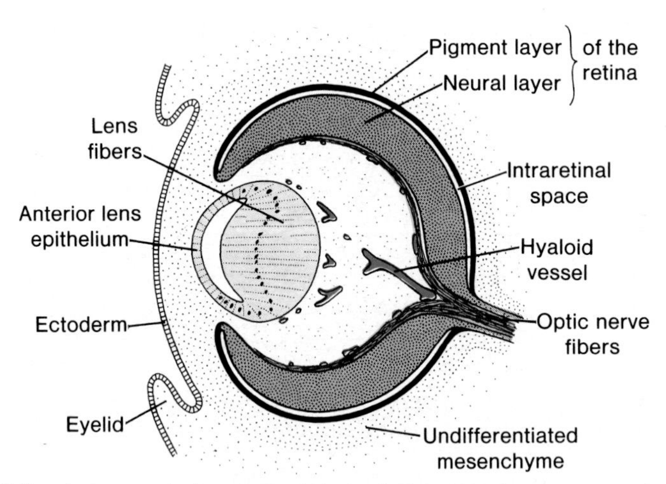

Figure 18-3. Anteroposterior section through the eye of a seven-week embryo. The eye primordium is completely embedded in mesenchyme. The fibers of the neural retina converge toward the optic nerve (modified after Mann[2]).

Retina, Iris, and Ciliary Body

The outer layer of the optic cup is characterized by the appearance of small pigment granules and is known as the **pigment layer of the retina** (figs. 18-3 and 18-6).

The development of the inner layer of the optic cup is more complicated. In the posterior four-fifths, known as the **pars optica retinae**, the cells bordering the intraretinal space differentiate into the light-receptive elements, the **rods** and **cones** (fig. 18-4). Adjacent to the photoreceptive layer is the mantle layer, which, as in the brain, gives rise to the neurons and supporting cells. In the adult the **outer nuclear layer**, the **inner nuclear layer**, and the **ganglion cell layer** can be distinguished (fig. 16-4). On the surface is found a fibrous layer, which contains the axons of the nerve cells of the deeper layers. The nerve fibers in this zone converge toward the optic stalk,

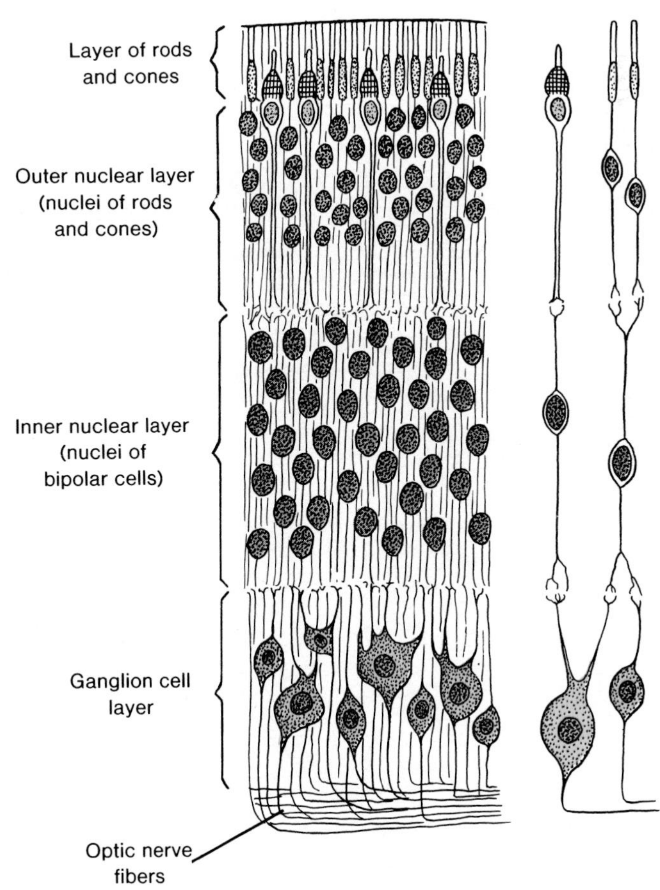

Figure 18-4. Schematic representation of the various layers of the pars optica retinae in a fetus of approximately 25 weeks (modified after Mann[2]).

which develops gradually into the optic nerve (figs. 18-3 and 18-6). Hence, light impulses pass through most layers of the retina before they reach the rods and cones.

The anterior one-fifth of the inner layer, known as the **pars caeca retinae,** does not change much and remains one cell layer thick. It is later divided into the **pars iridica retinae**, which forms the inner layer of the iris, and the **pars ciliaris retinae**, which participates in the formation of the **ciliary body** (fig. 18-5).

Meanwhile, the region between the optic cup and the overlying surface epithelium is filled with loose mesenchyme (figs. 18-3 and 18-6). In this tissue appear the **sphincter** and **dilator pupillae** muscles (fig. 18-5). In the human embryo these muscles develop from the underlying ectoderm of the optic cup in a similar manner as has been shown for lower animals.[4] In the adult, the iris is formed by the pigment-containing external and the unpigmented internal layer of the optic cup as well as by a layer of richly vascularized connective tissue, which contains the pupillary muscles (fig. 18-5).

The **pars ciliaris retinae** is easily recognized by its marked folding (figs. 18-5*B* and 18-6). Externally it is covered by a layer of mesenchyme which forms the **ciliary muscle**; on the inside, it is connected to the lens by a network of elastic fibers, the **suspensory ligament** or zonula (fig. 18-6).[5] Contraction of the ciliary muscle changes the tension in the ligament and controls the curvature of the lens.

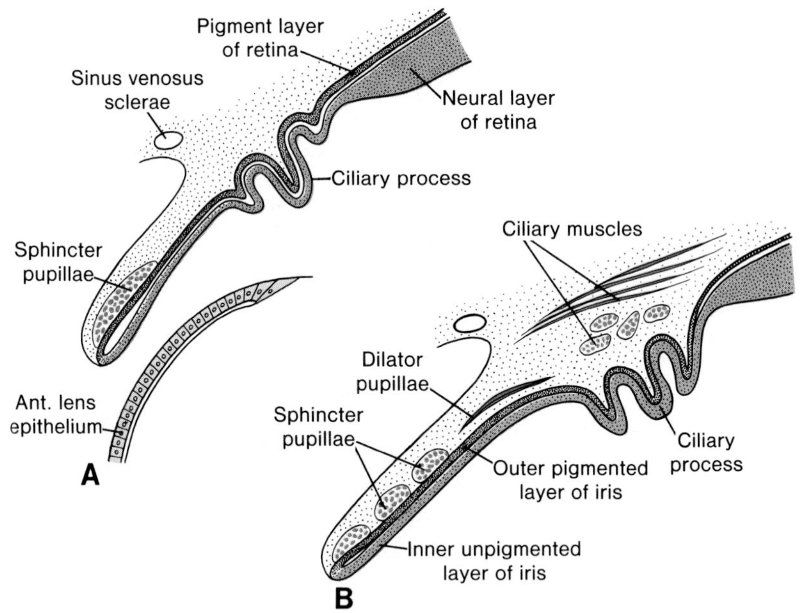

Figure 18-5. Development of the iris and the ciliary body. The rim of the optic cup is covered by mesenchyme, in which the sphincter and dilator pupillae develop from the underlying ectoderm.

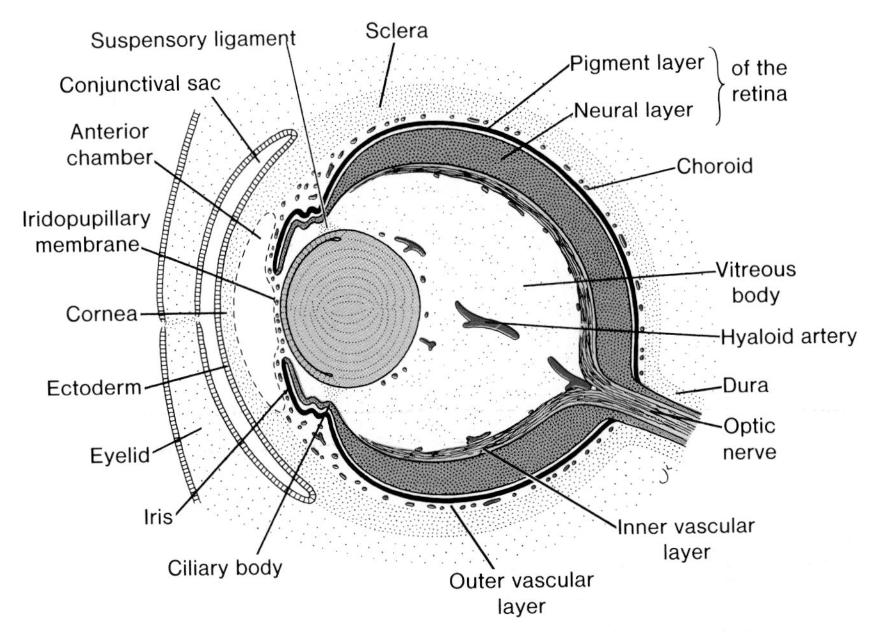

Figure 18-6. Anteroposterior section through the eye of a 15-week fetus. Note the anterior chamber, the iridopupillary membrane, the inner and outer vascular layers, the choroid, and the sclera.

Lens

Shortly after the formation of the lens vesicle (fig. 18-2*C*), the cells of the posterior wall begin to elongate in an anterior direction and form long fibers which gradually fill the lumen of the vesicle (fig. 18-3). By the end of the seventh week these **primary lens fibers** reach the anterior wall of the lens vesicle (fig. 18-6). Growth of the lens, however, is not finished at this stage, but new (secondary) lens fibers are continuously added to the central core.

Choroid, Sclera, and Cornea

At the end of the fifth week the eye primordium is completely surrounded by loose mesenchyme (fig. 18-3). This tissue soon differentiates into an inner layer comparable to the pia mater of the brain and an outer layer comparable to the dura mater. While the inner layer later forms a highly vascularized pigmented layer, known as the **choroid**, the outer layer develops into the sclera and is continuous with the dura mater around the optic nerve (fig. 18-6).

The differentiation of the mesenchymal layers overlying the anterior aspect of the eye is different. Through vacuolization is formed a space, known as the **anterior chamber**, which splits the mesenchyme into an inner layer in front of the lens and iris, the **iridopupillary membrane**, and an outer layer continuous with the sclera, the **substantia propria** of the **cornea** (fig. 18-6). The anterior chamber itself is lined by flattened mesenchymal cells.

Hence, the cornea from outside in is formed by: (1) an epithelial layer derived from the surface ectoderm; (2) the **substantia propria** or **stroma**, which is continuous with the sclera; and (3) an epithelial layer, which borders the anterior chamber. The iridopupillary membrane in front of the lens disappears completely, thus providing a communication between the anterior and posterior eye chambers. Sometimes resorption of the **iridopupillary membrane** is not complete and connective tissue fibers are suspended in front of the pupil (fig. 18-8*B*).

Vitreous body

Mesenchyme not only surrounds the eye primordium from the outside, but also invades the inside of the optic cup by way of the choroid fissure. Here, it forms the hyaloid vessels, which, during intra-uterine life, supply the lens and form the vascular layer located on the inner surface of the retina (fig. 18-6). In addition, it forms a delicate network of fibers between the lens and the retina. The interstitial spaces of this network are later filled with a transparent gelatinous substance, thus forming the **vitreous body** (fig. 18-6). The hyaloid vessels obliterate and disappear during fetal life.

Optic Nerve

The optic cup is connected to the brain by the optic stalk, which, on its ventral surface, has a groove, the **choroid fissure** (fig. 18-2). In this groove are found the hyaloid vessels. The nerve fibers of the retina returning to the brain are found among the cells of the inner wall of the stalk (fig. 18-7*A*). During the seventh week the choroid fissure closes and a narrow tunnel is formed inside the optic stalk (fig. 18-7*B*). As a result of the continuously increasing number of nerve fibers the inner wall of the stalk increases in size and the inside and outside walls of the stalk fuse (fig. 18-7*C*). The cells of the inner layer provide a network of neuroglia cells which support the optic nerve fibers.

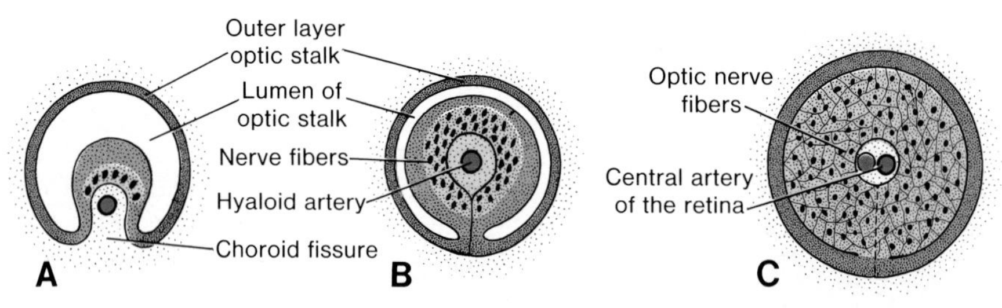

Figure 18-7. Diagrams showing the transformation of the optic stalk into the optic nerve. *A*, At sixth week (9 mm); *B*, seventh week (15 mm); *C*, ninth week. Note the central artery of the retina in the optic nerve (adapted from several sources).

The optic stalk is thus transformed into the **optic nerve**. In its center it contains the hyaloid artery, which is later called the **central artery of the retina**. On the outside the optic nerve is surrounded by a continuation of the choroid and sclera, which are known as the **pia-arachnoid** and **dura** layer of the nerve, respectively.

Congenital Malformations

COLOBOMA IRIDIS

Under normal conditions the choroid fissure closes during the seventh week of development (fig. 18-7). When this fails to occur, a cleft persists. Although such a cleft is usually located in the iris only and is known as **coloboma iridis** (fig. 18-8*A*), it may extend into the ciliary body, the retina, the choroid, and the optic nerve.[6] This malformation is frequently seen in combination with other eye abnormalities.

CONGENITAL CATARACT

This is a condition in which the lens has become opaque during intrauterine life. Although this anomaly is usually genetically determined, in 1941, Gregg[8] observed that children of mothers who suffered from German measles between the fourth and seventh weeks of pregnancy often showed cataract. If, however, the mother was infected after the seventh week of pregnancy, then the lens escaped damage, but the child was often deaf as a result of imperfect differentiation of the cochlea.[9] This seems to indicate that the most actively differentiating parts of the embryo are the most sensitive ones. Indeed, the lens goes through one of its most active stages of development during the sixth week of development, when the primary lens fibers fill the lumen of the lens vesicle.

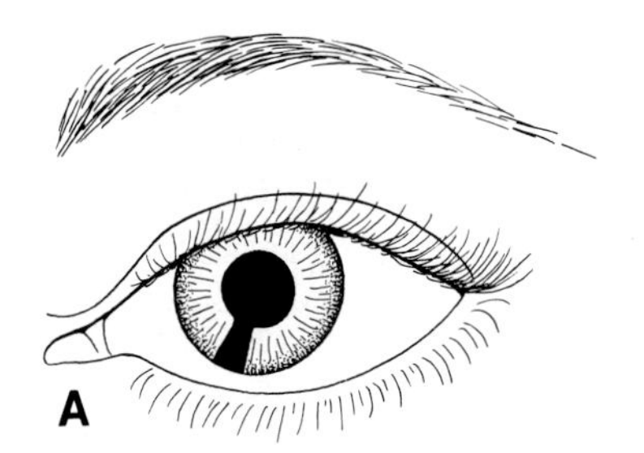

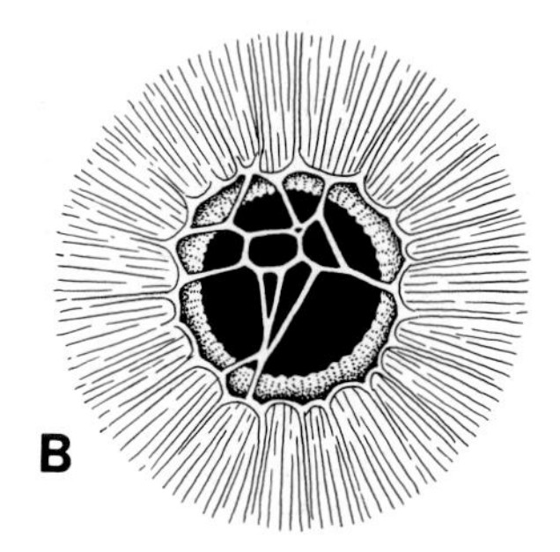

Figure 18-8. *A*, Coloboma iridis. *B*, Partially persistent iridopupillary membrane.

MICROPHTHALMIA

The over-all size of the eye is too small, and the eyeball may be reduced to two-thirds of its normal volume. Usually it is associated with other ocular abnormalities. Microphthalmia frequently results from intra-uterine infections such as cytomegalovirus or toxoplasmosis (See Chapter 8).

ANOPHTHALMIA

Sometimes the eye is grossly absent and it is impossible to detect any trace of the eyeball except by histological means. It is usually accompanied by other serious craniocerebral abnormalities.

SUMMARY

The eyes begin to develop as a pair of **optic vesicles** on each side of the forebrain at the end of the fourth week of development. The optic vesicles, outpocketings of the brain, contact the **surface ectoderm** and induce changes necessary for **lens formation**. When the optic vesicle begins to invaginate to form the **pigment** and **neural layer of the retina**, the **lens placode invaginates to form the lens vesicle**. Through a groove at the inferior aspect of the optic vesicle, the **choroid fissure**, the hyaloid artery (later the central artery of the retina) enters the eye (figs. 18-2 and 18-3), while the nerve fibers of the eye will move through this groove to reach the optic areas of the brain. The cornea is formed from outside in by (1) a layer of surface ectoderm, (2) the stroma which is continuous with the sclera; and (3) an epithelial layer bordering the anterior chamber (fig. 18-6).

There are many congenital ocular abnormalities, particularly congenital cataract of the lens. Similarly, in some cases the choroid fissure does not close, resulting in coloboma iridis or retinae (fig. 18-8).

REFERENCES

1. Langman, J. The first appearance of specific antigens during induction of the lens. *J. Embryol. Exp. Morphol.*, *7:* 193, 1959.
2. Mann, I. C. *The Development of the Human Eye*, ed 3. British Medical Association, Grune & Stratton, N.Y., 1974.
3. O'Rahilly, R. The early development of the eye in staged embryos. *Contrib. Embryol.*, *38:* 1, 1966.
4. Tamura, T., and Smelser, J. K. Development of the sphincter and dilator muscles of the iris. *Arch. Ophthalmol.*, *89:* 332, 1973.
5. Gloor, B. P. Zur Entwicklung des Glaskörpers und der Zonula. VI. Autoradiographische Untersuchungen zur Entwicklung der Zonula der Maus mit ^{3}H-markierten Aminosäuren und ^{3}H-Glucose. *Albrecht von Graefes Arch. klin. exp. Ophthalmol.*, *189;* 105, 1974.
6. Mann, I. C. *Developmental Abnormalities of the Eye.* J. B. Lippincott Co., Philadelphia, 1957.
7. Francois, J. *Heredity in Ophthalmology.* C. V. Mosby Co., St. Louis, 1961.
8. Gregg, N. M. Congenital cataract following German measles in the mother. *Trans. Ophthalmol. Soc. Aust.*, *3:* 35, 1941.
9. Töndury, L., and Smith, D. W. Fetal rubella pathology. *Z. Pediatr.*, *68:* 867, 1966.

chapter 19

Integumentary System

Skin

The skin has a twofold origin: (1) a superficial layer, the **epidermis**, which develops from the surface ectoderm; and (2) a deep layer, the **dermis**, developing from the underlying mesoderm.

EPIDERMIS

Initially, the embryo is covered by a single layer of ectodermal cells (fig. 19-1*A*). In the beginning of the second month this epithelium divides and a layer of flattened cells, the **periderm** or **epitrichium**, is laid down on the surface (fig. 19-1*B*). With further proliferation of the cells in the basal layer, a third, so-called intermediate zone is formed (fig. 19-1*C*). Finally, at the end of the fourth month the epidermis acquires its definitive arrangement and four layers can be distinguished (fig. 19-1*D*): (1) The basal layer, responsible for the production of new cells and known as the **germinative layer**. This layer later forms ridges and hollows, which are reflected on the surface of the skin in the fingerprint.[1] (2) A thick **spinous layer** consisting of large polyhedral cells connected by fine tonofibrils. (3) The **granular layer**, the cells of which contain small keratohyaline granules. (4) The **horny layer**, forming the tough, scale-like surface of the epidermis. It is made up of closely packed dead cells, loaded with keratin. The cells of the periderm are usually cast off during the second part of intra-uterine life, and can be found in the amniotic fluid.

During the first three months of development the epidermis is invaded by cells of neural crest origin.[2] These cells synthesize **melanin** pigment, which can be transferred to other cells of the epidermis by way of dendritic processes. They are known as **melanocytes**, and after birth cause pigmentation of the skin (fig. 19-1*D*).[3, 4]

The epidermal ridges, which produce typical patterns on the surface of the finger tips, the palms of the hand, and the soles of the foot, are genetically determined. They form the basis for many studies in medical genetics and criminal investigations—**dermatoglyphics**.[5] In children with chromosomal abnormalities the epidermal pattern on the hand and fingers is used as a diagnostic tool.

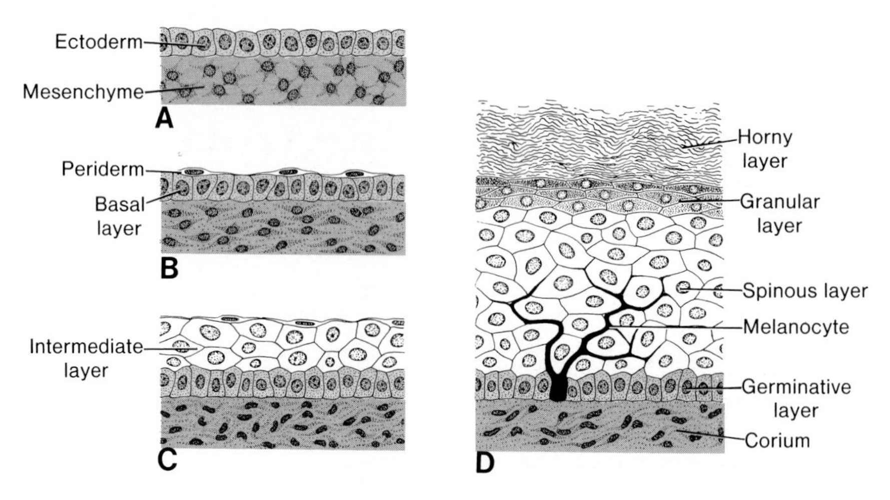

Figure 19-1. Schematic drawing showing the formation of the skin at various stages of development. *A*, At five weeks; *B*, seven weeks; *C*, four months; *D*, at birth.

DERMIS

The dermis is derived from the somatic mesoderm. During the third and fourth months this tissue, the **corium** (fig. 19-1*D*), forms many irregular papillary structures, the **dermal papillae**, which project upward into the epidermis. These papillae usually contain a small capillary or a sensory nerve end organ. The deeper layer of the dermis, the **subcorium**, contains large amounts of fatty tissue.

At birth, the skin is covered by a whitish paste, the **vernix caseosa**, formed by secretion of the sebaceous glands and degenerated epidermal cells and hairs. It protects the skin against the macerating action of amniotic fluid.

The skin of the newborn may show varying degrees of keratinization. Sometimes the superficial layers show excessive cornification, giving the skin a scale-like appearance. Such a condition is known as **ichthyosis**.

Hair

The hairs appear as solid epidermal proliferations penetrating the underlying dermis (fig. 19-2*A*). At their terminal ends the hair buds invaginate. The invaginations, the **hair papillae**, are rapidly filled with mesoderm in which vessels and nerve endings develop (fig. 19-2*B*, *C*).[6]

Soon, the cells in the center of the hair buds become spindle-shaped and keratinized, forming the **hair shaft**, while the peripheral cells become cuboidal, giving rise to the **epithelial hair sheath** (fig. 19-2*B* and *C*).

The **dermal root sheath** is formed by the surrounding mesenchyme. A small smooth muscle, also derived from mesenchyme, is usually attached to

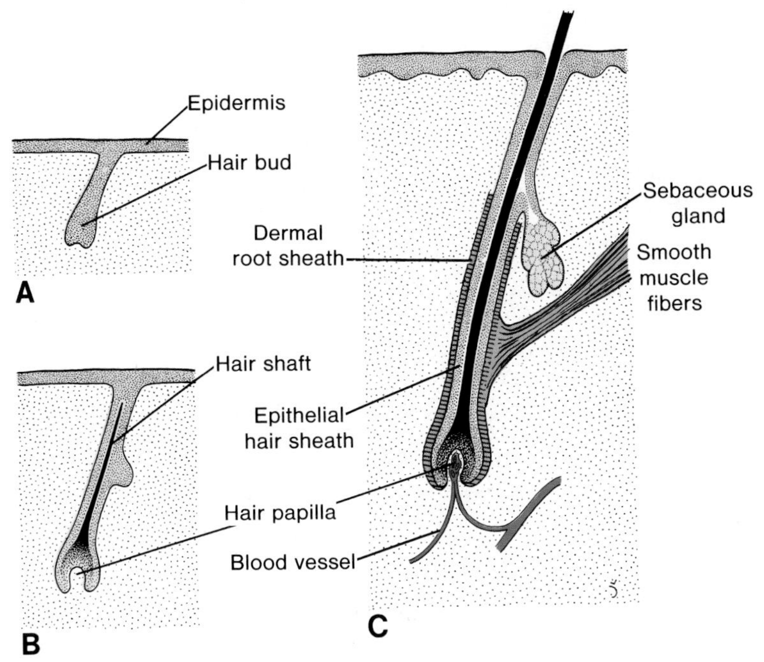

Figure 19-2. Schematic representation of the development of a hair and a sebaceous gland. *A*, At four months; *B*, six months; *C*, newborn.

the dermal root sheath. The muscle is known as **arrector pili muscle**. Continuous proliferation of the epithelial cells at the base of the shaft pushes the hair upward, and by the end of the third month the first hairs appear on the surface in the region of the eyebrow and upper lip. These hairs, **lanugo hairs**, are shed at about the time of birth and are later replaced by coarser hairs arising from new hair follicles.

The epithelial wall of the hair follicle usually shows a small outbudding penetrating the surrounding mesoderm (fig. 19-2*C*). The cells in the center of these outbuddings, the **sebaceous glands**, degenerate, thereby forming a fat-like substance which is secreted into the hair follicle, from where it reaches the skin.

Excessive hairiness, known as **hypertrichosis**, caused by the increased formation of hair follicles, may be localized in certain areas of the body (dorsal midline region) or may be general over the whole body. **Atrichia**, the congenital absence of hair, is usually associated with abnormalities of other ectodermal derivatives such as teeth and nails.

Mammary Glands

The first indication of the mammary glands is found in the form of a band-like thickening of epidermis, the **mammary line** or **ridge**. In a seven-

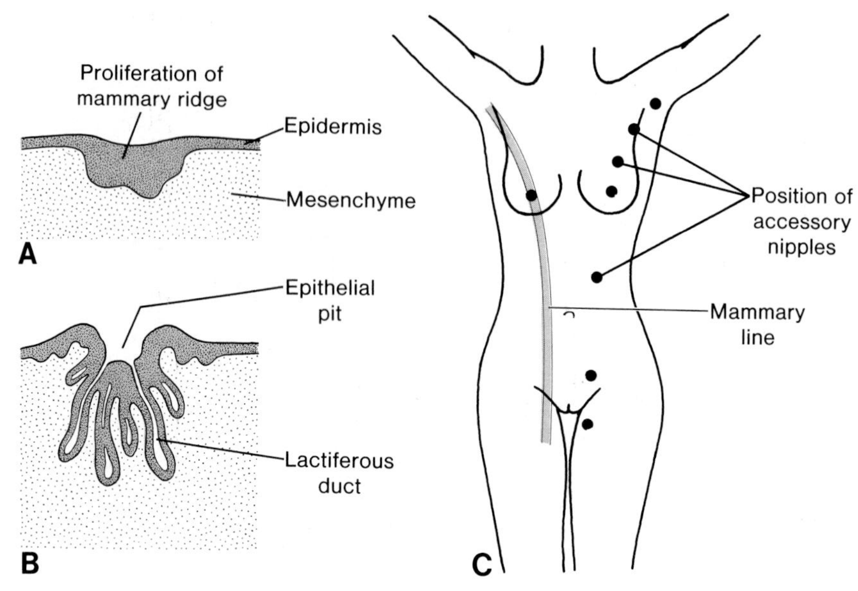

Figure 19-3. *A, B,* Sections through the developing mammary gland at the third and eighth months, respectively. *C,* Diagram showing the positions of accessory nipples (blue line indicates position of mammary line).

week embryo, this line extends on each side of the body from the base of the forelimb to the region of the hindlimb (fig. 19-3*C*). Although the major part of the mammary line disappears shortly after its formation, a small portion in the thoracic region persists and penetrates the underlying mesenchyme (fig. 19-3*A*). Here it forms 16 to 24 sprouts, which in turn give rise to small, solid outbuddings. By the end of prenatal life the epithelial sprouts are canalized, forming the **lactiferous ducts**, while the outbuddings form the small ducts and alveoli of the gland. The **lactiferous ducts** at first open into a small epithelial pit (fig. 19-3*B*). Shortly after birth this pit is transformed into the nipple by proliferation of the underlying mesenchyme.

Normally only a small part of the mammary line persists in the midthoracic region. Occasionally other fragments persist which then give rise to accessory nipples. Such a condition is known as **polythelia**. Accessory nipples may develop anywhere along the original mammary line, but are most frequently seen in the axillary region (fig. 19-3*C*). Sometimes an abnormally located remnant of the mammary line may develop into a complete mammary gland, an abnormality known as **polymastia**.

Occasionally the lactiferous ducts open into the original epithelial pit, which has failed to evert as a nipple. Such a condition, known as **inverted nipple**, is usually of congenital origin, but may be caused by retraction of the nipple as a result of the presence of a fast-growing tumor in the gland. (For development of the teeth, another componemt of the integumentary system, see chapter 16).

SUMMARY

The skin and its appendices such as hairs, nails, and glands are derived from surface ectoderm. The melanocytes which give the skin its color are derived from neural crest cells, which migrate into the epidermis. The production of new cells occurs in the **germinative** layer; after moving to the surface the cells are sloughed off in the horny layer (fig. 19-1). The dermis, the deep layer of the skin, is of mesodermal origin.

Hairs develop from downgrowth of epidermal cells in the underlying dermis. By about 20 weeks the fetus is covered by downy hair, **lanugo hairs**, which are shed at the time of birth. The **sebaceous glands**, **sweat glands**, and **mammary glands** all develop from epidermal downgrowths. Supernumerary nipples (polythelia) and breast (polymastia) are relatively common (see fig. 19-3).

REFERENCES

1. Hale, A. R. Morphogenesis of volar skin in the human fetus. *Am. J. Anat.*, *91:* 147, 1952.
2. Rawles, M. E. Origin of melanophores and their role in development of color patterns in vertebrates. *Physiol. Rev.*, *28:* 383, 1948.
3. Boyd, J. D. The embryology and comparative anatomy of the melanocyte. In *Progress in the Biological Sciences in Relation to Dermatology*, edited by A. Rook. Cambridge University Press, London, 1960.
4. Billingham, R. E., and Silvers, W. K. Melanocytes of mammals. *Q. Rev. Biol.*, *35:* 1, 1960.
5. Uchida, T. A., and Soltan, H. C. Dermatoglyphics in medical genetics. In *Endocrine and Genetic Diseases of Childhood*, edited by L. T. Gardner. W. B. Saunders, Philadelphia, 1969.
6. Pinkus, H. The embryology of hair. In *Biology of Hairgrowth*, edited by W. Montague and R. A. Ellis. Academic Press, New York, 1958.

chapter 20

Central Nervous System

The central nervous system appears at the beginning of the third week as a slipper-shaped plate of thickened ectoderm, the **neural plate**. This plate is located in the mid-dorsal region in front of the **primitive pit**. Its lateral edges soon become elevated to form the **neural folds** (fig. 20-1).

With further development the neural folds become more elevated, approach each other in the midline, and finally fuse, thus forming the **neural tube** (figs. 20-2 and 20-3). The fusion begins in the cervical region and proceeds in a somewhat irregular fashion in cephalic and caudal directions (fig. 20-3*A*).[1, 2] At the cranial and caudal ends of the embryo, however, fusion is delayed, and the **cranial** and **caudal neuropores** temporarily form open connections between the lumen of the neural tube and the amniotic cavity (fig. 20-3*B*). Closure of the cranial neuropore occurs at the 18- to 20-

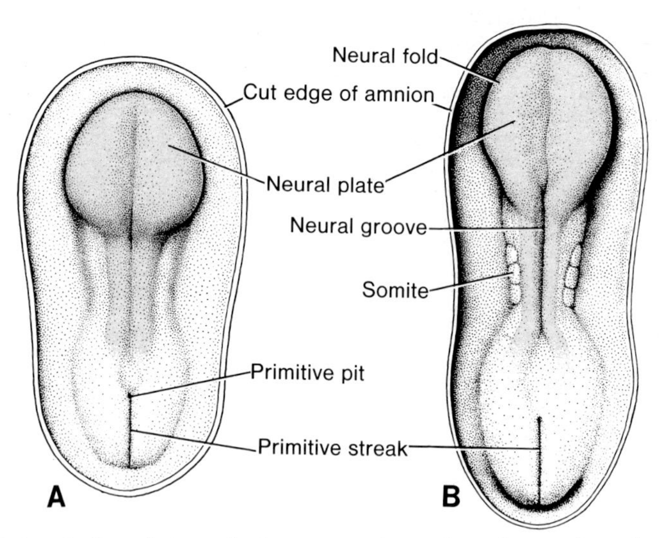

Figure 20-1. *A*, Dorsal view of a late presomite embryo (approximately 18 days) (modified after Davis). The amnion has been removed. The neural plate is clearly visible. *B*, Dorsal view at approximately 20 days (modified after Ingalls). Note the somites and the neural groove and neural folds.

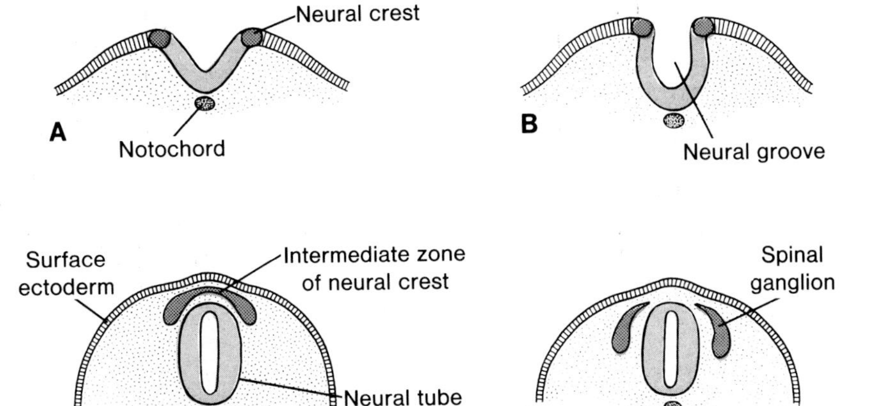

Figure 20-2. Transverse sections through successively older embryos, showing the formation of the neural groove, neural tube, and neural crest. The cells of the neural crest, initially forming an intermediate zone between the neural tube and surface ectoderm (*C*), develop into the spinal and cranial sensory ganglia (*D*).

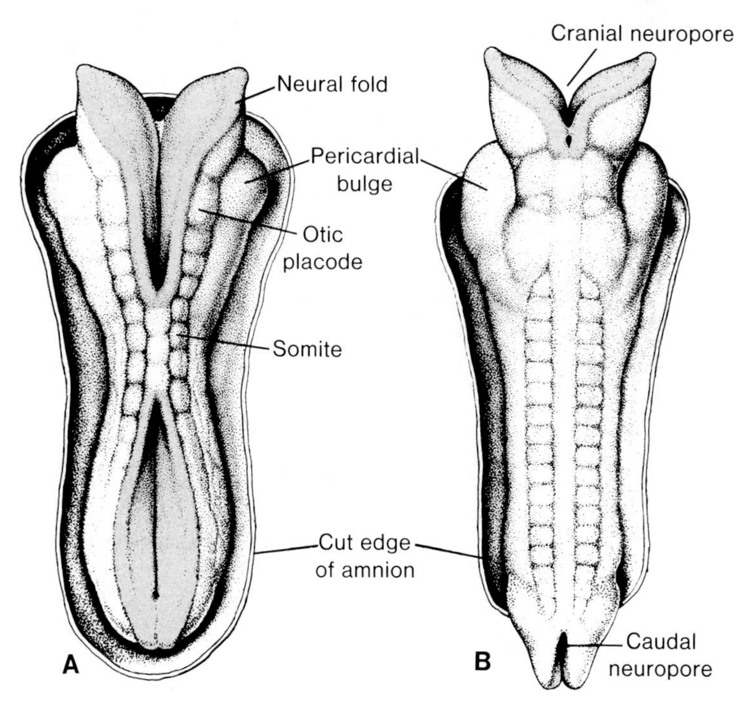

Figure 20-3. *A*, Dorsal view of a human embryo at approximately day 22 (modified after Payne). Seven distinct somites are visible in each side of the neural tube. *B*, Dorsal view of a human embryo at approximately day 23 (modified after Corner). The nervous system is in connection with the amniotic cavity through the cranial and caudal neuropores.

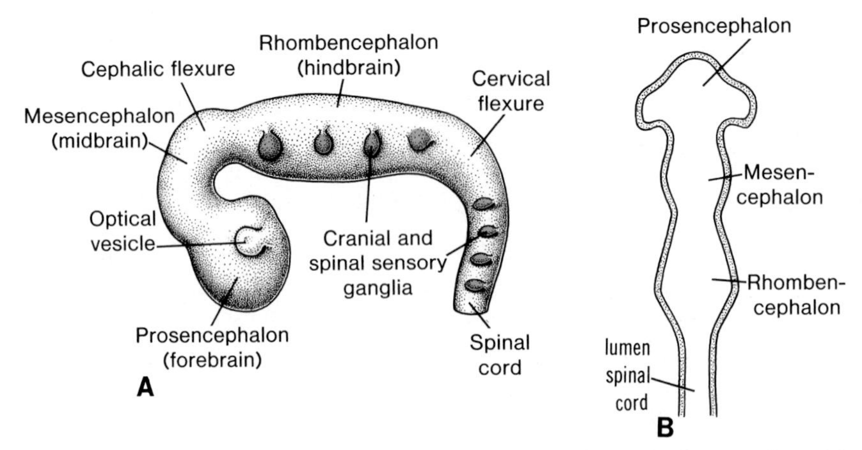

Figure 20-4. *A*, Lateral view of the brain vesicles and part of the spinal cord in a four-week embryo (modified after Hochstetter). Note the sensory ganglia formed by the neural crest on each side of the rhombencephalon and spinal cord. *B*, Diagram to show the lumina of the three brain vesicles and spinal cord.

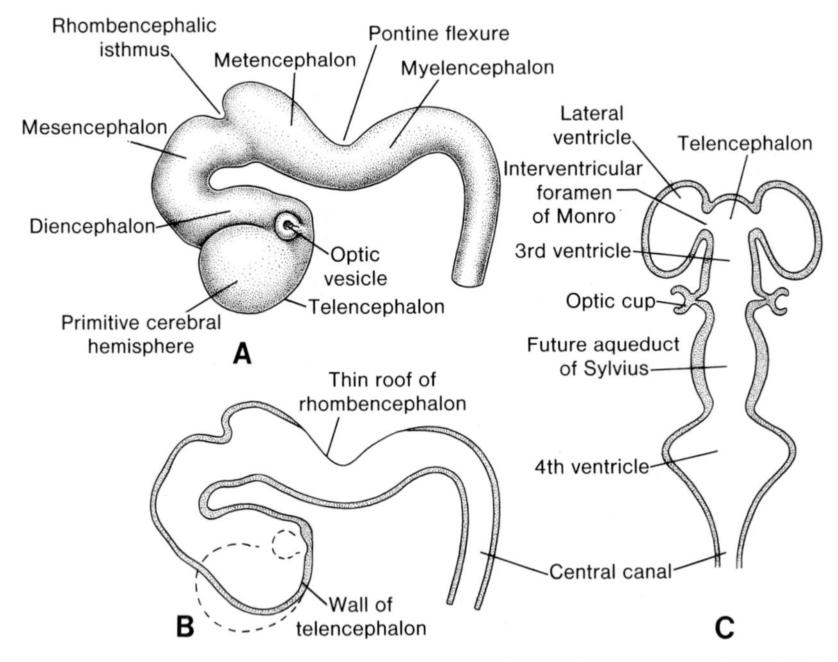

Figure 20-5. *A*, Lateral view of the brain vesicles of the human embryo in the beginning of the sixth week (modified after Hochstetter). *B*, Midline section through the brain vesicles and spinal cord of an embryo of the same age as shown in *A*. Note the thin roof of the rhombencephalon. *C*, Diagram to show the lumina of the spinal cord and brain vesicles.

somite stage (25th day); closure of the caudal neuropore about two days later.

The cephalic end shows three dilatations, the **primary brain vesicles**: (1) the **prosencephalon** or **forebrain**; (2) the **mesencephalon** or **midbrain**; and (3) the **rhombencephalon** or **hindbrain** (fig. 20–4). Simultaneously it forms two flexures: the **cervical flexure** at the junction of the hindbrain and the spinal cord; and the **cephalic flexure** located in the midbrain region (fig. 20–4*A*).

When the embryo is five weeks old the prosencephalon consists of two parts: (1) the **telencephalon** or **endbrain**, formed by a midportion and two lateral outpocketings, the **primitive cerebral hemispheres,** and (2) the **diencephalon**, characterized by the outgrowth of the optic vesicles (fig. 20–5*A*). The mesencephalon is separated from the rhombencephalon by a deep furrow, the **rhombencephalic isthmus.**

The rhombencephalon also consists of two parts: (1) the **metencephalon,** which later forms the **pons** and **cerebellum**; and (2) the **myelencephalon.** The boundary between these two portions is marked by a flexure, known as the **pontine flexure** (fig. 20–5*A*).

The lumen of the spinal cord, the **central canal**, is continuous with that of the brain vesicles. The cavity of the rhombencephalon is known as the **fourth ventricle**, that of the diencephalon as the **third ventricle**, and those of the cerebral hemispheres as the **lateral ventricles** (fig. 20–5*C*). The third and fourth ventricles are connected to each other through the lumen of the mesencephalon. This lumen becomes very narrow and is then known as the **aqueduct of Sylvius.** The lateral ventricles communicate with the third ventricle through the **interventricular foramina of Monro** (fig. 20–5*C*).

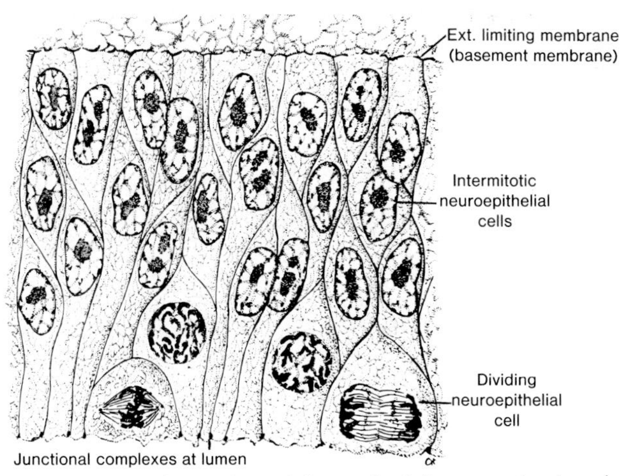

Figure 20-6. Drawing of a section of the wall of the recently closed neural tube. The neuroepithelial cells, which form a pseudostratified epithelium extending over the full width of the wall, can be recognized. Note the dividing cells at the lumen of the tube.

Spinal Cord

NEUROEPITHELIAL, MANTLE, AND MARGINAL LAYERS

The wall of a recently closed neural tube consists of the **neuroepithelial cells.**[3] These cells extend over the entire thickness of the wall and form a thick pseudostratified epithelium (fig. 20–6). They are connected to each other by junctional complexes at the lumen.[4] During the neural groove stage and immediately after closure of the tube, they divide rapidly, resulting in the production of more and more neuroepithelial cells. Collectively they are referred to as the **neuroepithelial layer** or **neuroepithelium.**

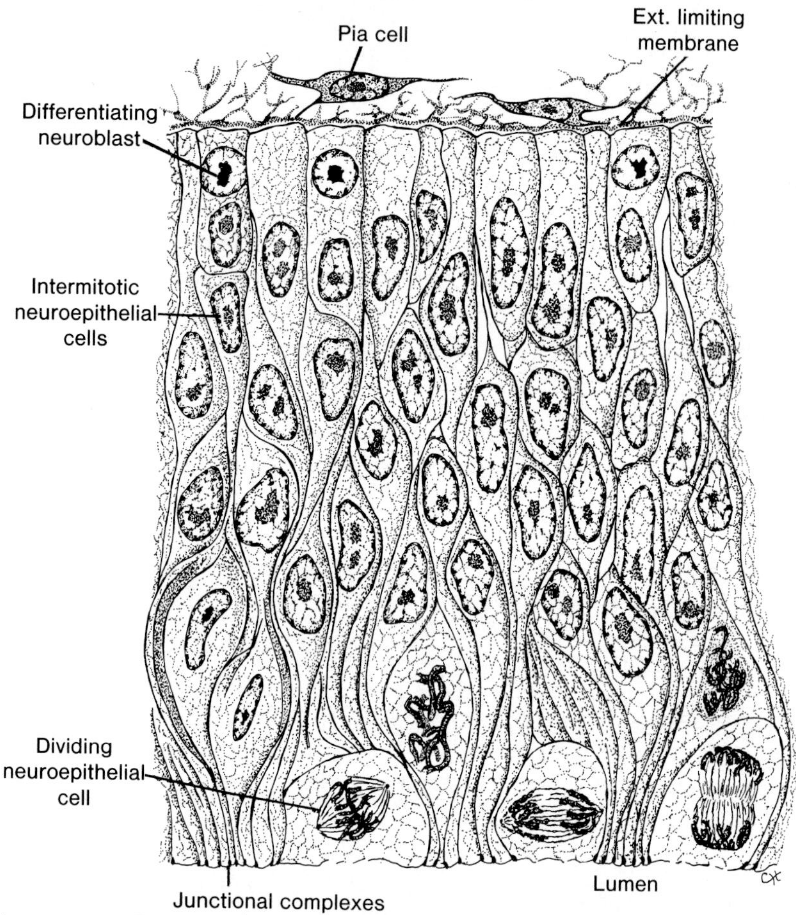

Figure 20-7. Drawing of a section of the neural tube at a slightly more advanced stage than represented in figure 20-6. The major portion of the wall consists of neuroepithelial cells. On the periphery, however, immediately adjacent to the external limiting membrane are found differentiating neuroblasts. These cells, which are produced by the neuroepithelial cells in ever-increasing numbers, will form the mantle zone.

Once the neural tube is closed, the neuroepithelial cells begin to give rise to another cell type, which is characterized by a large round nucleus with pale nucleoplasm and a dark-staining nucleolus. These cells are the primitive nerve cells or **neuroblasts** (fig. 20–7).[6, 5] They form a zone around the neuroepithelial layer known as the **mantle layer** (fig. 20–8). The mantle zone later forms the **gray matter of the spinal cord.**

The outermost layer of the spinal cord contains the nerve fibers emerging from the neuroblasts in the mantle layer and is known as the **marginal layer.** As a result of the myelination of the nerve fibers, this layer obtains a white appearance and is therefore referred to as the **white matter of the spinal cord** (fig. 20–8).

BASAL, ALAR, ROOF, AND FLOOR PLATES

As a result of the continuous addition of neuroblasts to the mantle layer, each side of the neural tube shows a ventral and dorsal thickening. The ventral thickenings, the **basal plates**, contain the ventral motor horn cells and **form the motor areas** of the spinal cord; the dorsal thickenings, the **alar plates, form the sensory areas** (fig. 20–8*A*). A longitudinal groove, the **sulcus limitans**, marks the boundary between the two. The dorsal and ventral midline portions of the neural tube, known as the **roof** and **floor plates**, respectively, do not contain neuroblasts and serve primarily as pathways for nerve fibers crossing from one side to the other.

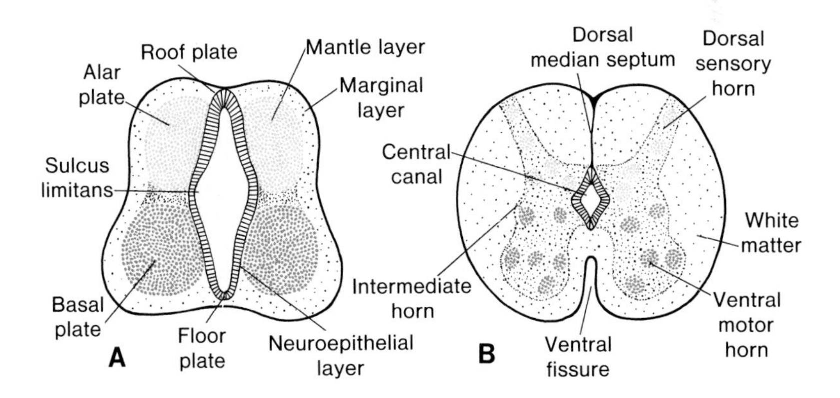

Figure 20-8. Diagrams to show two successive stages in the development of the spinal cord. Note the formation of the ventral motor and dorsal sensory horns, and the intermediate column.

In addition to the ventral motor horn and the dorsal sensory horn, a group of neurons accumulates between the two areas and causes the formation of a small **intermediate horn** (fig. 20–8*B*). This horn contains mainly neurons of the autonomic nervous system.

HISTOLOGICAL DIFFERENTIATION

Nerve Cells. The **neuroblasts** or primitive nerve cells arise exclusively by division of the neuroepithelial cells. Initially they have a central process extending to the lumen—**transient dendrite**[5, 7]—but when they migrate into the mantle zone this process disappears and the neuroblasts are temporarily round, **apolar neuroblasts** (fig. 20–9*A*). With further differentiation two new cytoplasmic processes appear on opposite sides of the cell body, thus forming the **bipolar neuroblast** (fig. 20–9*B*). The process at one end of the cell elongates rapidly to form the **primitive axon**, while the process at the other end shows a number of cytoplasmic arborizations, the **primitive dendrites** (fig. 20–9*C*). The cell is then known as a **multipolar neuroblast** and with further development becomes the adult nerve cell or **neuron.** Once the neuroblasts are formed, they lose their ability to divide. The axons of the neurons in the basal plate break through the marginal zone and become visible on the ventral aspect of the cord. They are known collectively as the **ventral motor root of the spinal nerve**, and conduct motor impulses from the spinal cord to the muscles (fig. 20–10).

The axons of the neurons in the dorsal sensory horn (alar plate) behave differently from those in the ventral horn. They penetrate into the marginal layer of the cord, where they either ascend or descend to a higher or lower level to form **association neurons.**

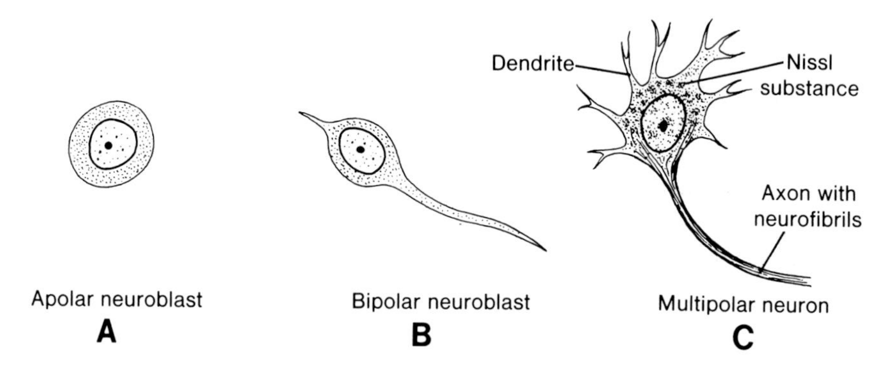

Figure 20-9. Various stages of the development of a neuroblast. A neuron is structural and functional unit, consisting of the cell body and all its processes.

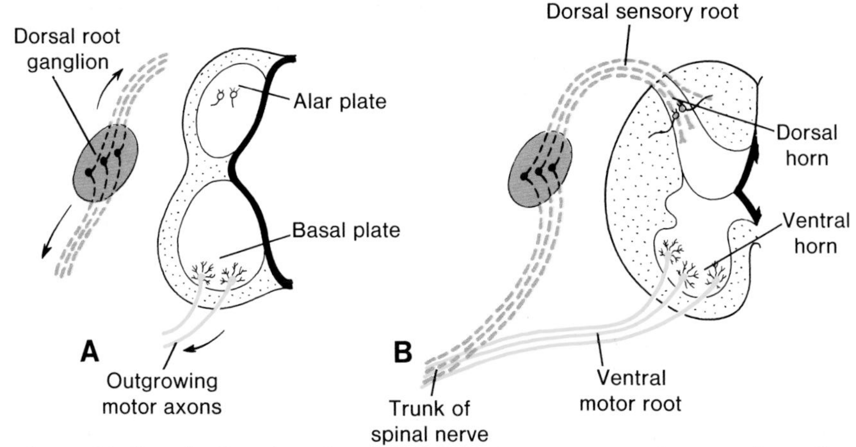

Figure 20-10. *A*, Drawing showing the motor axons growing out from the neurons in the basal plate, and the centrally and peripherally growing fibers of the nerve cells in the dorsal root ganglion. *B*, The nerve fibers of the ventral motor and dorsal sensory roots join to form the trunk of the spinal nerve.

Glia Cells. The majority of the primitive supporting cells, the **gliablasts**, are formed by the neuroepithelial cells after the production of neuroblasts has ceased.[8] From the neuroepithelial layer the gliablasts migrate to the mantle and marginal layer. In the mantle layer they differentiate into the **protoplasmic** and **fibrillar astrocytes** (fig. 20–11).[9]

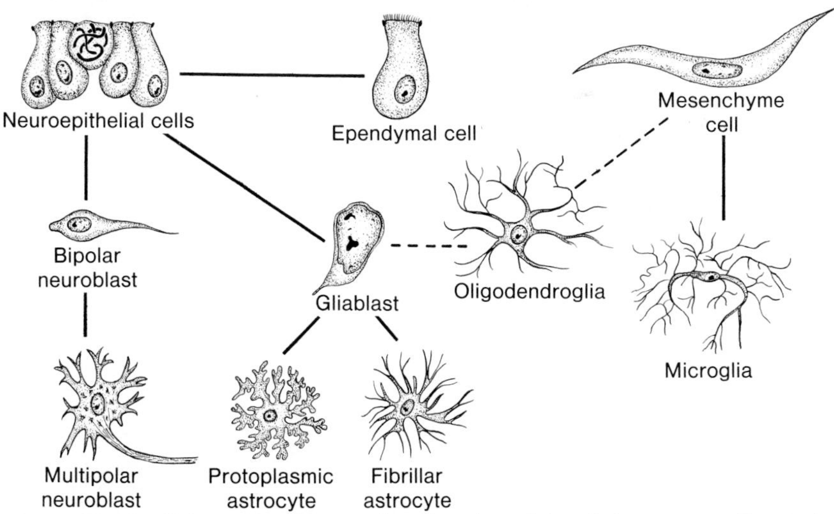

Figure 20-11. Schematic diagram showing the origin of the nerve cell and the various types of glia cells. The neuroblasts, the fibrillar and protoplasmic astrocytes, and the ependymal cells originate from the neuroepithelial cells. The microglia develops from mesenchyme cells. The origin of the oligodendroglia remains in doubt.

Another type of supporting cell possibly derived from the gliablasts, is the **oligodendroglia cell.**[10] This cell, which is mainly found in the marginal layer, forms the myelin sheaths around the ascending and descending axons in the marginal layer.

In the second half of development a third type of supporting cell, the **microglia cell**, appears in the central nervous system. This cell type is believed to originate from the mesoderm surrounding the neural tube (fig. 20–11).[9]

When the neuroepithelial cells cease to produce neuroblasts and gliablasts, they finally differentiate into the ependymal cells.

Neural Crest Cells. During the invagination of the neural plate, a group of cells appears along each edge of the neural groove (figs. 20–2 and 20–4*A*). These cells, ectodermal in origin, and known as the **neural crest cells**, temporarily form an intermediate zone between the tube and the surface ectoderm (fig. 20–2*C*). This zone extends from the prosencephalon to the levels of the caudal somites, and on each side migrates to the dorsolateral aspect of the neural tube.[11] Here the cells give rise to the **sensory ganglia** or **dorsal root ganglia** of the spinal and cranial nerves (5th, 7th, 9th, and 10th cranial nerves) (figs. 20–2 and 20–4).

During further development the neuroblasts of the sensory ganglia form two processes (fig. 20–10*A*). The centrally growing processes penetrate the dorsal portion of the neural tube. In the spinal cord they either end in the dorsal horn or ascend through the marginal layer to one of the higher brain centers. These processes are known collectively as the **dorsal sensory root of the spinal nerve** (fig. 20–10*B*). The peripherally growing processes join the fibers of the ventral motor root and thus participate in the formation of the trunk of the spinal nerve. Eventually these processes terminate in the sensory receptor organs. Hence, the neuroblasts of the sensory ganglia give rise to the **dorsal root neurons.**

In addition to forming the sensory ganglia, the cells of the neural crest differentiate into sympathetic neuroblasts (see fig. 20–32), Schwann cells, pigment cells, odontoblasts, meninges, and cartilage cells of the branchial arches.[12, 13] Extirpation of neural crest cells of the trigeminal region resulted in facial abnormalities, among which were clefts of the primary palate[14] (see Chapter 16). In recent years great attention has been given to the neural crest cells and their possible importance in formation of the face.[15]

Myelination. Myelination of the peripheral nerves is accomplished by the **neurilemma cells** or **cells of Schwann.** These cells originate from the neural crest, migrate peripherally and wrap themselves around the axons, thus

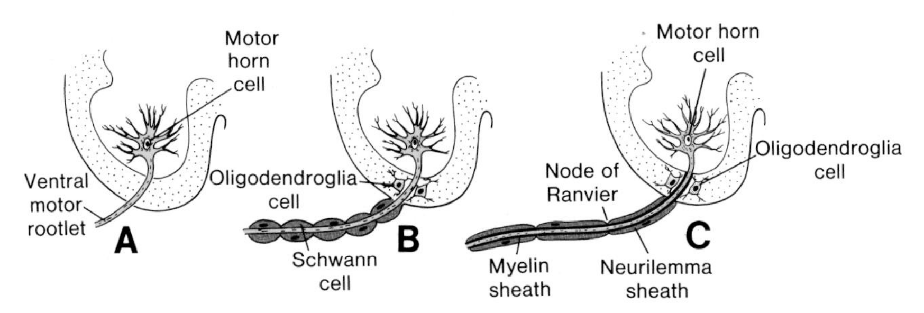

Figure 20–12. *A*, Drawing of motor horn cell with naked rootlet. *B*, In the spinal cord oligodendroglia cells approach the ventral rootlet; outside the spinal cord the Schwann cells begin to surround the rootlet. *C*, In the spinal cord the myelin sheath is formed by the oligodendroglia cells; outside the spinal cord by the Schwann cells.

forming the **neurilemma sheath** (fig. 20–12*B*). In this manner axons, varying in number from 1 to 20, may be enwrapped by one neurilemma cell.

Beginning at the fourth month of fetal life, the nerve fibers gradually obtain a whitish appearance as a result of the deposition of **myelin** between the axon and the neurilemma. This substance is formed by repeated coiling of the membrane around the axon.[12, 16] Hence, both the neurilemma and the myelin sheath of the peripheral nerve fibers are formed by the cells of Schwann (fig. 20–12*C*).

The myelin sheath surrounding the nerve fibers in the spinal cord is of completely different origin, since it is formed by the oligodendroglia cells (fig. 20–12*B, C*). Although myelination of the nerve fibers in the spinal cord begins in approximately the fourth month of intrauterine life, some of the motor fibers descending from the higher brain centers to the spinal cord do not become myelinated until the first year of postnatal life. It seems that the tracts in the nervous system become myelinated at about the time they start to function.[9]

POSITIONAL CHANGES OF THE CORD

In the third month of development, the spinal cord extends the entire length of the embryo and the spinal nerves pass through the intervertebral foramina at their level of origin (fig. 20–13*A*). With increasing age, however, the vertebral column and the dura lengthen more rapidly than the neural tube and the terminal end of the spinal cord gradually shifts to a higher level. At birth this end is located at the level of the third lumbar vertebra (fig. 20–13*C*). As a result of this disproportionate growth, the spinal nerves run obliquely from their segment of origin in the spinal cord to the corresponding level of the vertebral column. The dura remains attached to the vertebral column at the coccygeal level.

In the adult, the spinal cord terminates at the level of L2. Below this point the central nervous system is represented only by the **filum terminale internum**, which marks the tract of regression of the spinal cord. The nerve fibers below the terminal end of the cord are known collectively as the **cauda equina.** When cerebrospinal fluid is tapped during a **lumbar puncture,** the needle is inserted at the lower lumbar level, thus avoiding the lower end of the cord.

ABNORMALITIES OF THE CORD

By the end of the fourth week the central nervous system forms a closed tubular structure detached from the overlying ectoderm (fig. 20-2). Occasionally, however, the neural groove fails to close, either because of faulty induction by the underlying notochord,[17] or because of the action of environmental teratogenic factors on the neuroepithelial cells.[18] The neural tissue then remains exposed to the surface. Such a defect may extend the total length of the embryo or may be restricted to a small area only (**complete** or **partial rachischisis**). If localized in the region of the spinal cord, the abnormality is commonly referred to as **spina bifida**, whereas failure of closure in the cephalic region is known as **anencephalus.**

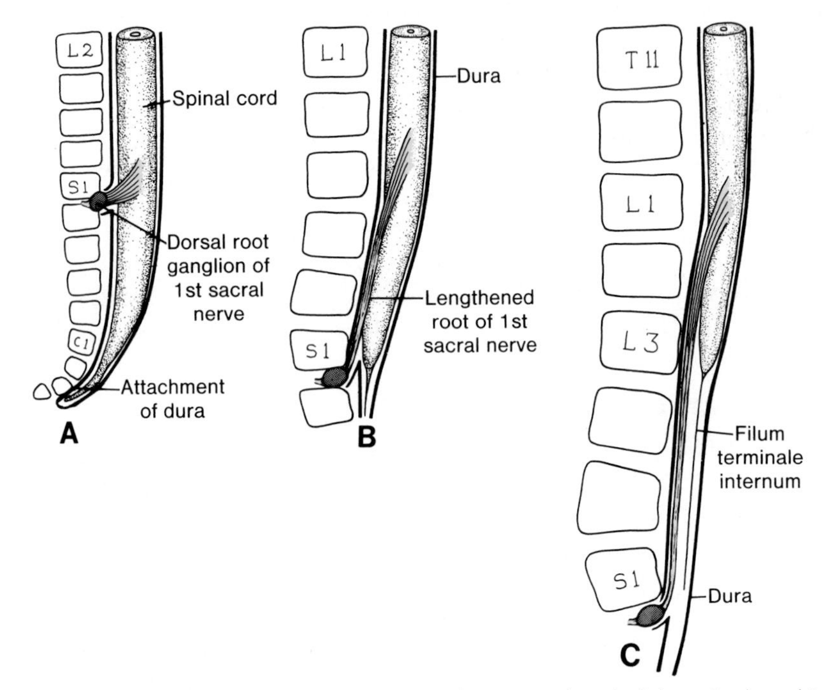

Figure 20-13. Schematic drawing showing the terminal end of the spinal cord in relation to that of the vertebral columnn at various stages of development. *A*, Approximately at third month; *B*, end of fifth month; *C*, in the newborn (modified after Streeter).

SPINA BIFIDA

The term spina bifida is used to cover a wide range of defects. Literally translated, it indicates a bifid spine and in its most simple form is seen as a failure of the dorsal portions of the vertebrae to fuse with one another. Such an abnormality, usually localized in the sacrolumbar region, is mostly covered by skin and is not noticeable on the surface except for the presence of a small tuft of hair over the affected area **(spina bifida occulta)** (fig. 20-14*A*). In such a case the spinal cord and nerves are usually normal and neurological symptoms are absent.

If more than one or two vertebrae are involved, the meninges of the spinal cord bulge through the opening and a sac covered with skin is visible on the surface **(meningocele)** (figs. 20-14*B* and 20-15*A*). Sometimes this sac is so large that it contains not only the meninges but also the spinal cord and its nerves. The abnormality is then known as **meningomyelocele**, and is usually covered by a thin, easily torn membrane (figs. 20-14*C* and 20-15*B*). Neurological symptoms are usually present.

Another type of spina bifida results from failure of the neural groove to close, and the nervous tissue is then widely exposed to the surface **(myelocele** or **rachischisis)** (fig. 20-14*D, E*). Occasionally the neural tissue shows considerable overgrowth; usually, however, the excess tissue becomes necrotic shortly before or after birth.

Myelomeningoceles are usually associated with a caudal displacement of the medulla (oblongata) and of a portion of the cerebellum into the spinal

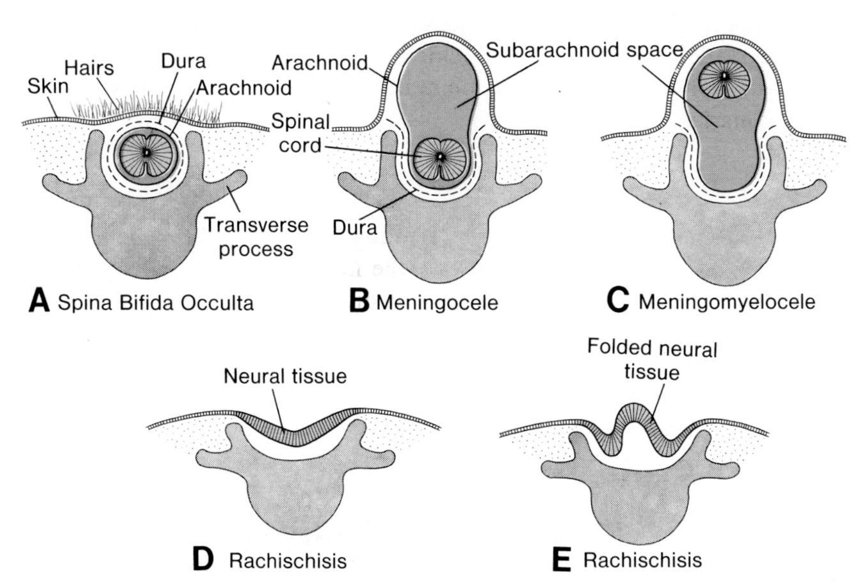

Figure 20-14. Schematic drawings to show the various types of spina bifida.

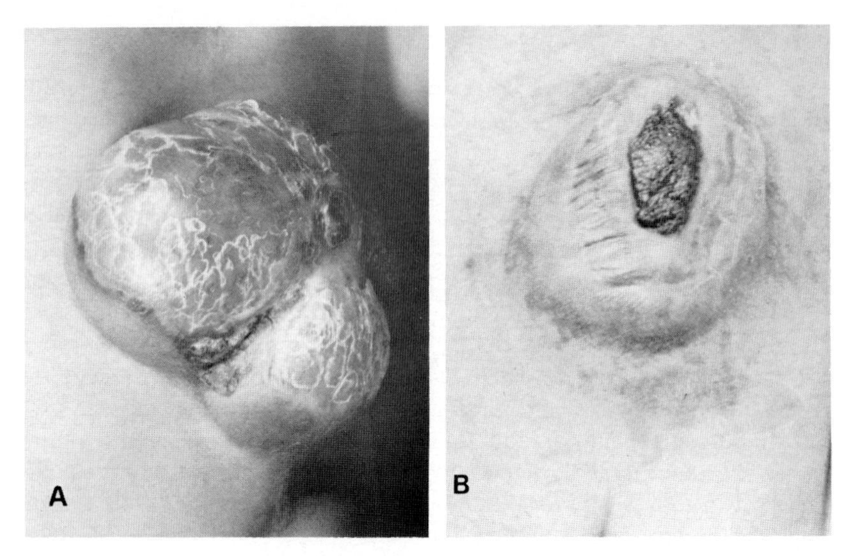

Figure 20-15. *A*, Photograph of patient with large meningocele covered by skin. This is a rather common abnormality. *B*, Photograph of large meningomyelocele covered by membranes. (Courtesy Dr. J. Warkany. From *Congenital Malformations*, by Warkany, J. Copyright 1971, by Year Book Medical Publishers, Inc., Chicago. Used by permission.)

canal. Frequently the upper cervical nerve roots descend from their intervertebral foramina toward the spinal cord, which at its caudal end is fixed to the sacral region. Since the foramen magnum is obstructed either by the medulla or the cerebellum, the myelomeningocele is frequently combined with hydrocephaly. The combination of these abnormalities is known as the **Arnold-Chiari malformation**.[19]

Brain

Distinct **basal** and **alar plates**, representing the motor and sensory areas, respectively, are found on each side of the midline in the majority of the brain vesicles. Even the **sulcus limitans**, the dividing line between motor and sensory areas, is present in the rhombencephalon and mesencephalon (figs. 20-17 and 20-22).

A brief description of the fundamental components of each of the five brain vesicles is given in the following account.

RHOMBENCEPHALON

The rhombencephalon consists of the **myelencephalon**, the most caudal of the brain vesicles, and the **metencephalon**, which extends from the pontine flexure to the rhombencephalic isthmus (figs. 20-16 and 20-5).

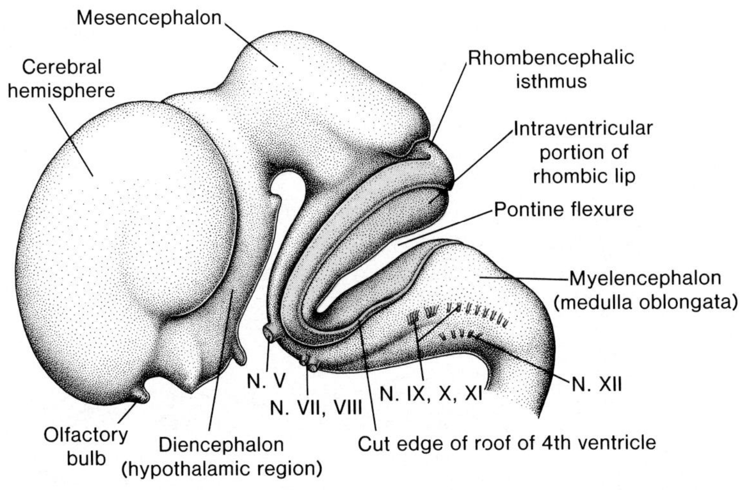

Figure 20-16. Lateral view of the brain vesicles in an eight-week embryo (crown-rump length approximately 27 mm) (after Hochstetter). The roof plate of the rhombencephalon has been removed to show the intraventricular portion of the rhombic lip. Note the origin of the cranial nerves.

MYELENCEPHALON

This brain vesicle gives rise to the **medulla oblongata** and differs from the spinal cord in that its lateral walls rotate around an imaginary longitudinal axis in the floor plate, a movement comparable to the opening of a book (fig. 20-17*B, C*). The alar and basal plates separated by the sulcus limitans can be clearly distinguished. The basal plate, similar to that of the spinal cord, contains the motor nuclei. These nuclei are divided into three groups: (1) a medial **somatic efferent** group; (2) an intermediate **special visceral efferent** group; and (3) a lateral **general visceral efferent group** (fig. 20-17*C*).

The first group contains the motor neurons, which form the **cephalic continuation of the anterior horn cells.** Since this somatic efferent group continues rostrally into the mesencephalon, it is referred to as the **somatic efferent motor column**. In the myelencephalon it represents the neurons of the **hypoglossal nerve**, which supply the tongue musculature. In the metencephalon and mesencephalon the column represents the neurons of the **abducens** (fig. 20-18) and **trochlear and oculomotor nerves** (fig. 20–22) respectively. These nerves supply the eye musculature, thought to be derived from the preotic myotomes.

The **special visceral efferent** group extends into the metencephalon, thus forming the **special visceral efferent motor column**. Its motor neurons supply the **striated muscles** of the branchial arches. In the myelencephalon the column is presented by the neurons of the **accessory, vagus,** and **glossopharyngeal nerves**.

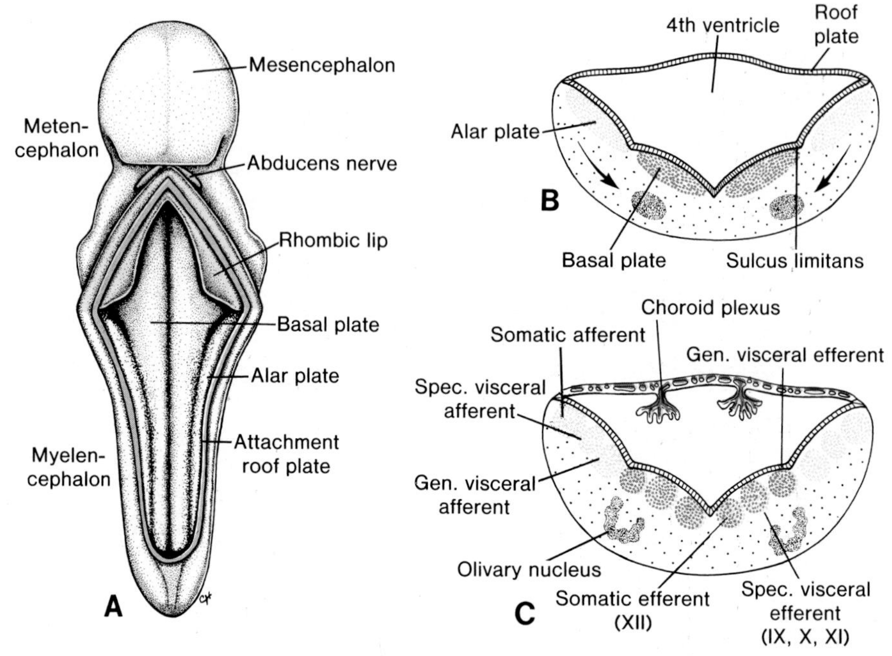

Figure 20-17. *A*, Dorsal view of the floor of the fourth ventricle in a 10-mm embryo after removal of the roof plate. Note the alar and basal plates in the myelencephalon. The rhombic lip is visible in the metencephalon. *B*, *C*, Diagrams showing the position and differentiation of the basal and alar plates of the myelencephalon at different stages of development. Note the formation of the nuclear groups in the basal and alar plates. The arrows indicate the path followed by the cells of the alar plate to the olivary nuclear complex. The choroid plexus produces the cerebrospinal fluid.

The **general visceral efferent** group contains the motor neurons, which supply the **involuntary musculature** of the respiratory tract, intestinal tract, and heart.

The alar plate contains three groups of **sensory relay nuclei** (fig. 20-17*C*). The most lateral of these, the **somatic afferent** (sensory) group, receives impulses from the ear and the surface of the head by way of the **statoacoustic** and **trigeminal nerves**. The intermediate or special visceral sensory group receives impulses from the taste buds of the tongue and from the palate, oropharynx, and epiglottis. The medial or **general visceral afferent** group receives interoceptive information from the gastrointestinal tract and the heart.

The roof plate of the myelencephalon consists of a single layer of ependymal cells covered by vascular mesenchyme, the **pia mater**. The two combined are known as the **tela choroidea**. Owing to the active proliferation of the vascular mesenchyme, a number of sac-like invaginations project into

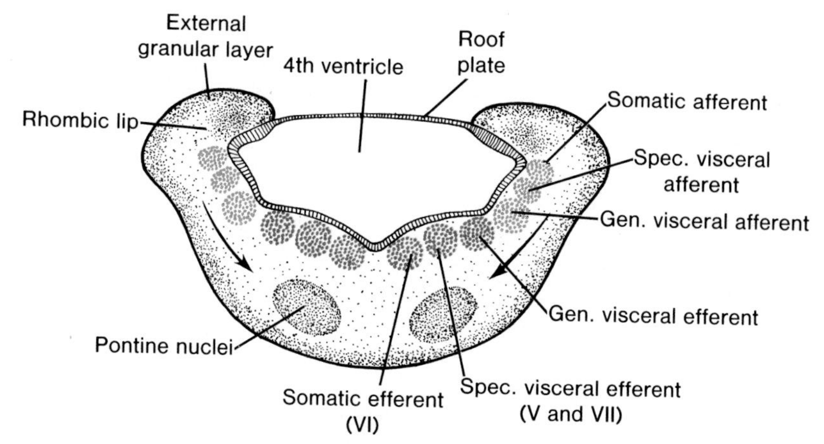

Figure 20-18. Schematic drawing of a transverse section through the caudal part of the metencephalon. Note the differentiation of the various motor and sensory nuclear areas in the basal and alar plates, respectively. Note the position of the rhombic lips, which project partly into the lumen of the fourth ventricle and partly above the attachment of the roof plate.

the underlying ventricular cavity (fig. 20-17*C*). These tuft-like invaginations form the **choroid plexus** which produces the cerebrospinal fluid of the central nervous system.

METENCEPHALON

The metencephalon, similar to the myelencephalon, is characterized by basal and alar plates (fig. 20-18). Two new components, however, are formed: (1) the **cerebellum**, which functions as a coordination center for posture and movement (fig. 20-20); and (2) the **pons**, which serves as the pathway for nerve fibers between the spinal cord and the cerebral and cerebellar cortices.

Each basal plate of the metencephalon (fig. 20-18) contains three groups of motor neurons: (1) the medial **somatic efferent** group, which gives rise to the nucleus of the **abducens nerve**; (2) the **special visceral** motor group containing the nuclei of the **trigeminal** and **facial nerves**, which innervate the musculature of the first and second branchial arches; and (3) the **general visceral** motor group, whose axons supply the submandibular and sublingual glands.

The marginal layer of the basal plates of the metencephalon expands considerably as it serves as a bridge for the nerve fibers connecting the cerebral cortex and the cerebellar cortex with the spinal cord. Hence, this portion of the metencephalon is known as the **pons.** In addition to the nerve fibers, the pons contains the **pontine nuclei**, which originate in the alar plates of the metencephalon and myelencephalon (see arrow, fig. 20-18).

The alar plates of the metencephalon contain three groups of sensory nuclei: (1) a lateral **somatic sensory** group, which contains neurons **of the trigeminal nerve**, and a small portion of the **vestibulocochlear complex**; (2) the **special visceral sensory** group; and (3) the **general visceral sensory** group (fig. 20-18).

CEREBELLUM

The dorsolateral parts of the alar plates bend medially and form the **rhombic lips** (fig. 20-17). In the caudal portion of the metencephalon, the rhombic lips are widely separated, but immediately below the mesencephalon they approach each other in the midline (fig. 20-19). As a result of a further deepening of the pontine flexure, the rhombic lips become compressed in a cephalo-caudal direction and form the **cerebellar plate** (fig. 20-19). In a 12-week embryo this plate shows a small midline portion, the **vermis**, and two lateral portions, the **hemispheres.** A transverse fissure soon separates the **nodule** from the vermis and the lateral **flocculus** from the hemispheres (fig. 20-19*B*). This **flocculonodular lobe** is phylogenetically the most primitive part of the cerebellum.

Initially the **cerebellar plate** consists of a neuroepithelial, a mantle, and a marginal layer (fig. 20-20*A*). During further development a number of cells formed by the neuroepithelium migrate to the surface of the cerebellum to form the **external granular layer.**[20] The cells of this layer retain their ability to divide and form a proliferative zone on the surface of the cerebellum (fig. 20-20*B*, *C*).

In the sixth month of development the external granular layer begins to release various cell types, which migrate toward the differentiating Purkinje

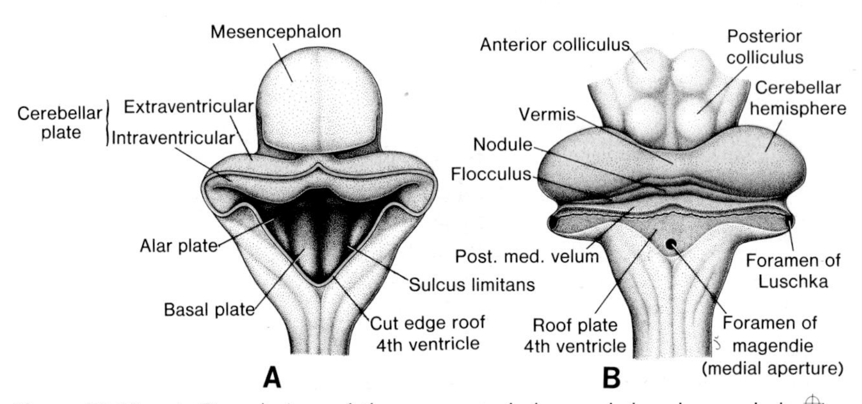

Figure 20-19. *A*, Dorsal view of the mesencephalon and rhombencephalon in an eight week embryo. The roof of the fourth ventricle has been removed, allowing a view of the floor of the fourth ventricle (modified after Hochstetter). *B*, Similar view in a four-month embryo. Note the choroidal fissure and the lateral and medial apertures in the roof of the fourth ventricle.

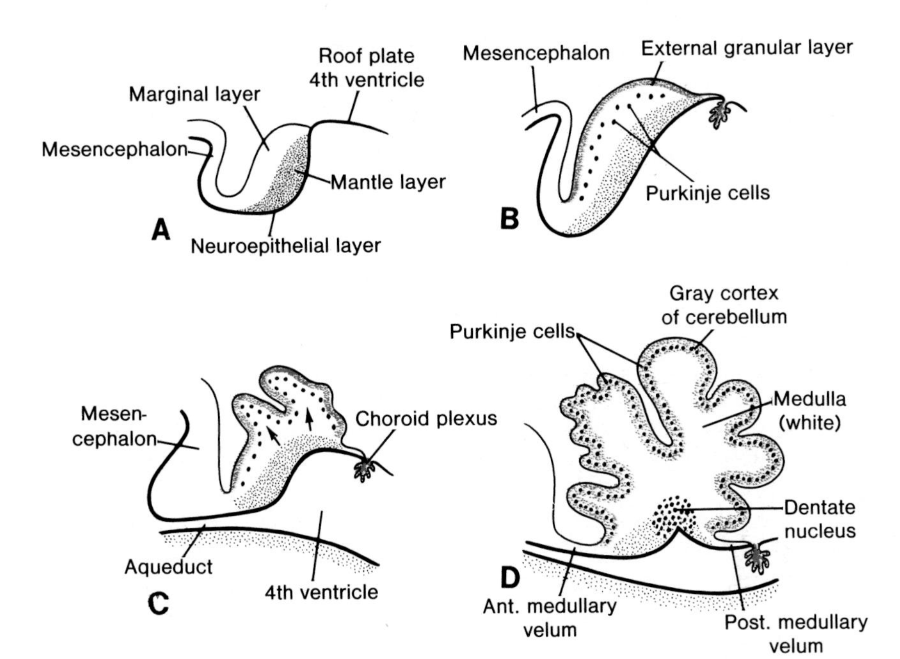

Figure 20-20. Sagittal sections through the roof of the metencephalon, showing the development of the cerebellum. *A*, At eight weeks (approximately 30 mm); *B*, 12 weeks (70 mm); *C*, 13 weeks; *D*, 15 weeks (modified after Keibel and Mall). Note the formation of the external granular layer on the surface of the cerebellar plate (*B* and *C*). During later stages the cells of the external granular layer migrate inwards to mingle with the Purkinje cells and thus form the definitive cortex of the cerebellum. The dentate nucleus is one of the deep cerebellar nuclei. Note the anterior and posteroir velum.

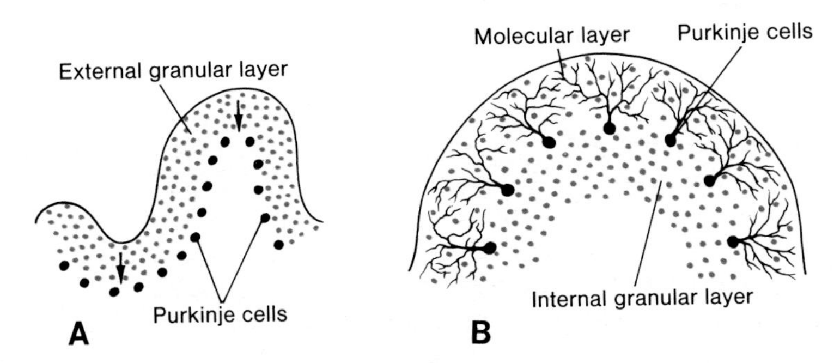

Figure 20-21. Stages in the development of the cerebellar cortex. *A*, The external granular layer located on the surface of the cerebellum forms a proliferative layer, from which granule, basket, and stellate cells arise. They migrate inwards from the surface as indicated by the arrows. *B*, Postnatal cerebellar cortex showing the differentiated Purkinje cells, the molecular layer on the surface, and the internal granular layer beneath the Purkinje cells.

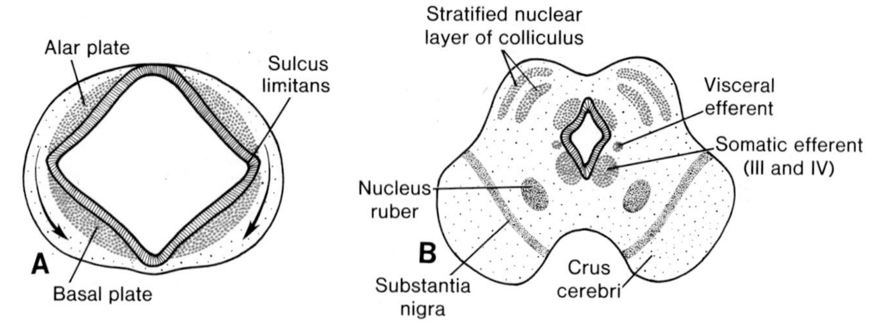

Figure 20-22. Diagram showing the position and differentiation of the basal and alar plates in the mesencephalon at various stages of development. The arrows in *A* indicate the path followed by the cells of the alar plate to form the nucleus ruber and substantia nigra. Note the various motor nuclei in the basal plate.

cells (fig. 20-21) and which give rise to the **granule cells**, **basket cells**, and **stellate cells.**[20] The cortex of the cerebellum, consisting of Purkinje cells, Golgi II neurons, and the neurons produced by the external granular layer, reaches its definitive size after birth (fig. 20-21*B*). The deep cerebellar nuclei, such as the **dentate nucleus**, reach their final position long before birth (fig. 20-20*D*).

In recent years young infants suffering from virus diseases have been treated with various chemicals which block the synthesis of DNA.[21, 22] Since DNA-synthesis occurs in the cells of the external granular layer to produce basket and stellate cell neurons, it is likely that the anti-viral therapy caused heavy damage to the production of cerebellar neurons.[23]

MESENCEPHALON

The mesencephalon is morphologically the most primitive of the brain vesicles (Fig. 20-22). Each basal plate contains two groups of motor nuclei: (1) a medial **somatic** motor group represented by the **oculomotor** and **trochlear nerves**, which innervate the eye musculature, and (2) a small **general visceral efferent** group, represented by the **nucleus of Edinger-Westphal**, which innervates the **sphincter pupillary muscle** (fig. 20-22*B*).

The marginal layer of each basal plate enlarges and forms the **crus cerebri.** These crura serve as pathways for the nerve fibers descending from the cerebral cortex to the lower centers in the pons and spinal cord.

The alar plates of the mesencephalon initially appear as two longitudinal elevations separated by a shallow midline depression (fig. 20-19). With further development, a transverse groove divides each elevation into an **anterior** (superior) and a **posterior** (inferior) **colliculus** (fig. 20-19*B*). The posterior colliculi serve as synaptic relay stations for the auditory reflexes; the anterior colliculi function as correlation and reflex centers for the visual impulses.

The colliculi are formed by waves of neuroblasts produced by the neuro-epithelial cells and migrating into the overlying marginal zone. Here they become arranged in stratified layers (fig. 20-22*B*).

DIENCEPHALON

Roof Plate and Epiphysis. This part of the brain develops from the median portion of the prosencephalon (figs. 20–5 and 20–16), and is thought to consist of a roof plate and two alar plates, but to lack floor and basal plates.

The roof plate of the diencephalon consists of a single layer of ependymal cells covered by vascular mesenchyme. The two combined give rise to the **choroid plexus** of the third ventricle (fig. 20-28). The most caudal part of the roof plate develops into the **pineal body** or **epiphysis.** This body initially appears as an epithelial thickening in the midline, but by the seventh week begins to evaginate (figs. 20-23 and 20-24). Eventually it becomes a solid organ located on the roof of the mesencephalon (fig. 20-28). Though many theories have been proposed as to its function, no satisfactory answer has yet been found.[24, 25] In the adult, calcium is frequently deposited in the epiphysis and it then serves as a landmark on an x-ray of the skull.

Alar Plate, Thalamus, and Hypothalamus. The alar plates form the lateral walls of the diencephalon. A groove, the **hypothalamic sulcus**, divides the plate into a dorsal and ventral region and **thalamus** and **hypothalamus**, respectively (figs. 20-23, and 20-24).

As a result of a high proliferative activity, the thalamus gradually projects into the lumen of the diencephalon. Frequently, this expansion is so great that the thalamic regions from the right and left sides fuse in the midline, thereby forming the **massa intermedia** or **interthalamic connexus** (see neuro-anatomy textbooks).

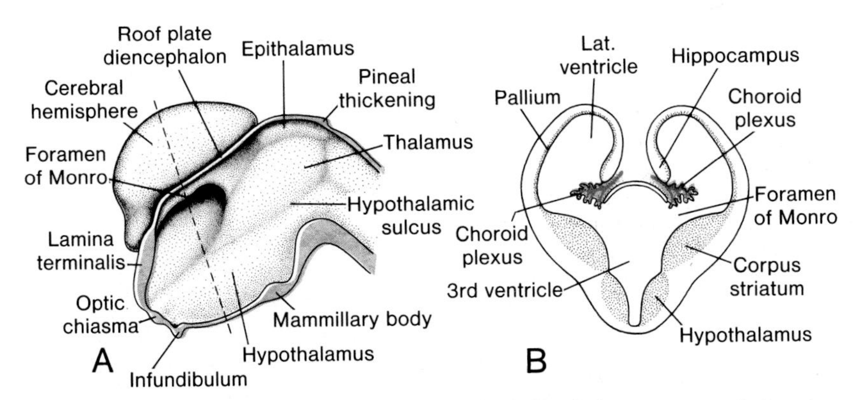

Figure 20-23. *A*, Medial surface of the right half of the prosencephalon in a seven-week embryo. *B*, Schematic transverse section through the prosencephalon at the level of the broken line in *A*. The corpus striatum bulges out in the floor of the lateral ventricle and the foramen of Monro.

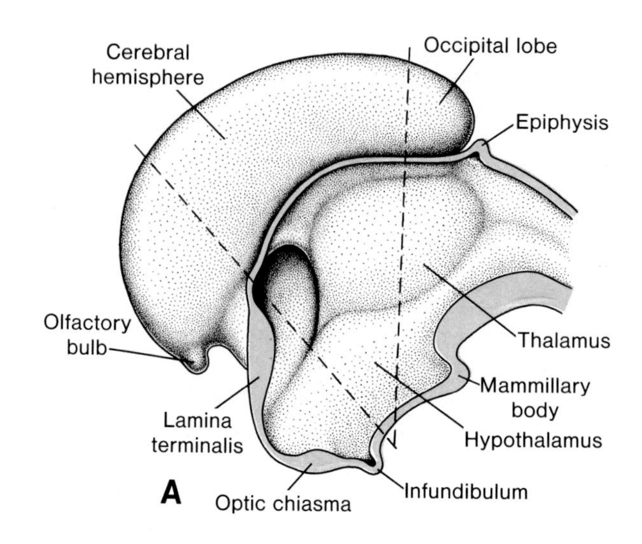

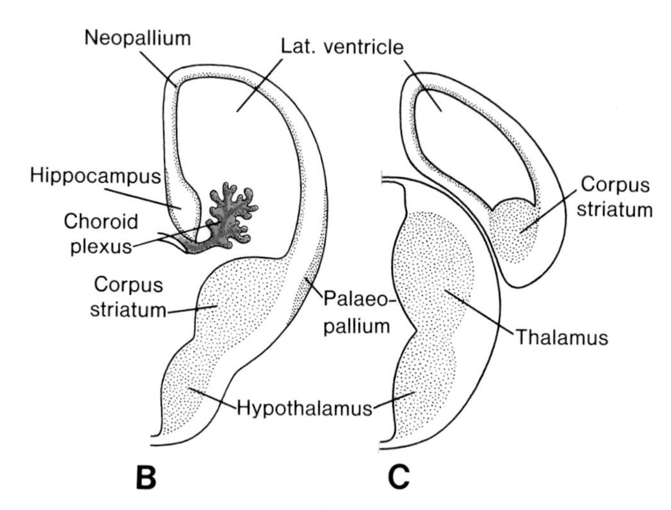

Figure 20-24. *A*, Medial surface of the right half of the telencephalon and diencephalon in an eight-week embryo. *B* and *C*, Schematic transverse section through the right half of the telencephalon and diencephalon at the level of the broken lines as indicated in *A*.

The hypothalamus, forming the lower portion of the alar plate, differentiates into a number of nuclear areas, which serve as regulation centers of the visceral functions such as sleep, digestion, body temperature, and emotional behavior. One of these groups, the **mammillary body**, forms a distinct protuberance on the ventral surface of the hypothalamus on each side of the midline (figs. 20-23*A* and 20-24*A*).

Hypophysis or Pituitary Gland. The hypophysis or pituitary gland develops from two completely different parts: (1) an ectodermal outpocketing of the stomodeum immediately in front of the buccopharyngeal membrane, known as **Rathke's pouch**; and (2) a downward extension of the diencephalon, the **infundibulum** (fig. 20-25*A*).

When the embryo is approximately three weeks old, Rathke's pouch appears as an evagination of the stomodeum, and subsequently grows dorsally toward the infundibulum. By the end of the second month it loses its connection with the oral cavity and is then in close contact with the infundibulum. Occasionally a small portion of the pouch persists in the wall of the pharynx (**pharyngeal hypophysis**).

During further development the cells in the anterior wall of Rathke's pouch increase rapidly in number and form the **anterior lobe of the hypophysis** or **adenohypophysis** (fig. 20-25*B*). A small extension of this lobe, the **pars tuberalis**, grows along the stalk of the infundibulum and eventually surrounds it (fig. 20-25*C*). The posterior wall of Rathke's pouch develops into the **pars intermedia**, which in man seems to have little significance.

The infundibulum gives rise to the **stalk** and the **pars nervosa** or **posterior lobe of the hypophysis** (neurohypophysis) (fig. 20-25*C*). It is composed of neuroglia cells. In addition, it contains a number of nerve fibers which come from the hypothalamic area.

A not uncommon remnant of Rathke's pouch is the **craniopharyngioma** or **Rathke's pouch tumor.** These tumors are found intracranially and their symptoms are greatly similar to those caused by tumors in the anterior lobe of the hypophysis. The onset of symptoms is usually before the age of 15.

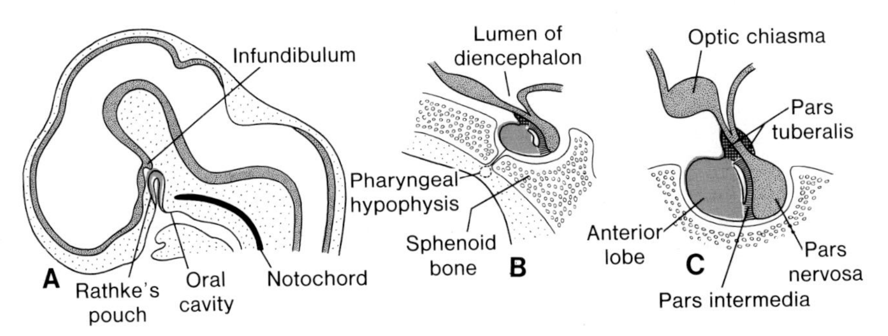

Figure 20-25. *A*, Sagittal section through the cephalic part of a six-week embryo, showing Rathke's pouch as a dorsal outpocketing of the oral cavity, and the infundibulum as a thickening in the floor of the diencephalon. *B* and *C*, Sagittal sections through the developing hypophysis in the 11th and 16th weeks of development, respectively. Note the formation of the pars tuberalis, encircling the stalk of the pars nervosa. Sometimes a remnant of Rathke's pouch is found in the pharynx. It is known as pharyngeal hypophysis.

TELENCEPHALON

The telencephalon, the most rostral of the brain vesicles, consists of two lateral outpocketings, the **cerebral hemispheres**, and a median portion, the **lamina terminalis** (figs. 20-4, 20-5, 20-23, and 20-24). The cavities of the hemispheres, the **lateral ventricles**, communicate widely with the lumen of the diencephalon through the **interventricular foramina of Monro** (fig. 20-23).

Cerebral Hemispheres. The cerebral hemispheres arise at the beginning of the fifth week of development as bilateral evaginations of the lateral wall of the prosencephalon (fig. 20-23). By the middle of the second month the basal part of the hemispheres (that is, the part which initially formed the forward extension of the thalamus) (fig. 20-23*A*) begins to increase in size. As a result this area bulges into the lumen of the lateral ventricle as well as into the floor of the foramen of Monro (figs. 20-23*B* and 20-24*A*, *B*). In transverse sections, the rapidly growing region has a striated appearance and therefore is known as the **corpus striatum** (fig. 20-24*B*).

In the region where the wall of the hemisphere is attached to the roof of the diencephalon, it fails to develop neuroblasts and remains very thin (fig. 20-23*B*). Here the hemisphere wall consists of a single layer of ependymal cells covered by vascular mesenchyme, and together they form the **choroid plexus.** The choroid plexus should have formed the roof of the hemisphere, but as a result of the disproportionate growth of the various parts of the hemisphere it protrudes into the lateral ventricle along a line known as the **choroidal fissure** (figs. 20-24 and 20-26*B*). Immediately above the choroidal fissure the wall of the hemisphere is thickened, thus forming the **hippocampus** (figs. 20-23*B* and 20-24*B*). This structure, which mainly has an olfactory function, gradually bulges out into the lateral ventricle.

With further expansion the hemispheres gradually cover the lateral aspect of the diencephalon, mesencephalon, and the cephalic portion of the metencephalon (figs. 20-23 to 20-28). The corpus striatum (fig. 20-23*B*), being a part of the wall of the hemisphere, likewise expands posteriorly and is divided into two parts: (1) a dorsomedial portion, the **caudate nucleus**; and (2) a ventrolateral portion, the **lentiform nucleus** (fig. 20-26*B*). This division is accomplished by axons passing to and from the cortex of the hemisphere and breaking through the nuclear mass of the corpus striatum. The fiber bundle so formed is known as the **internal capsule** (fig. 20-26*B*).[27]

At the same time, the medial wall of the hemisphere and the lateral wall of the diencephalon fuse, and the caudate nucleus and thalamus come in close contact (fig. 20-26*B*).[28]

Continuous growth of the cerebral hemispheres in anterior, dorsal, and inferior directions results in the formation of the frontal, temporal, and occipital lobes. As the region overlying the corpus striatum lags in growth,

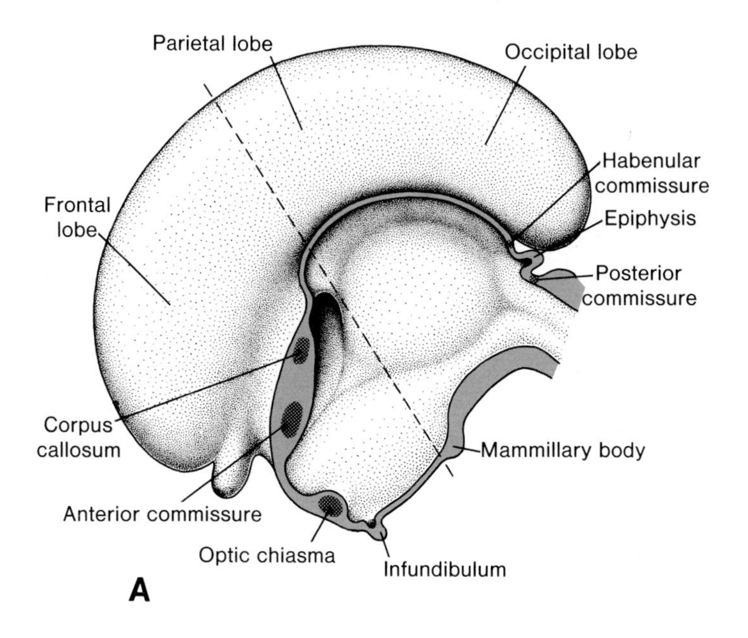

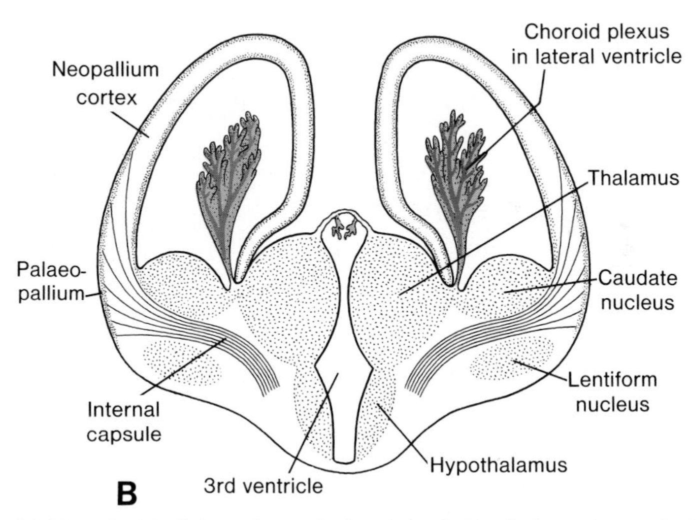

Figure 20-26. *A*, Medial surface of the right half of the telencephalon and diencephalon in a 10-week embryo. *B*, Schematic transverse section through the hemisphere and diencephalon at the level of the broken line as indicated in *A*.

however, the area between the frontal and temporal lobes becomes depressed and is known as the **insula** (fig. 20-27*A*). This region is later overgrown by the adjacent lobes and at the time of birth is almost completely covered. During the final part of fetal life the surface of the cerebral hemispheres

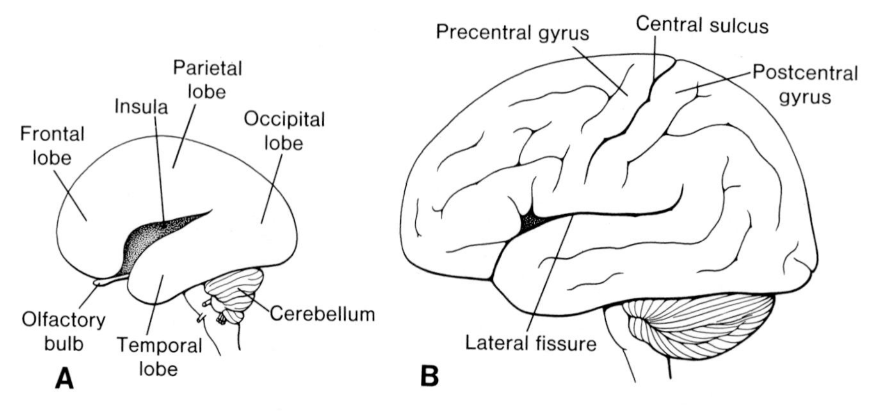

Figure 20-27. Schematic drawing to show the development of gyri and sulci on the lateral surface of the cerebral hemisphere. *A*, At seven months; *B*, nine months.

grows so rapidly that a great many convolutions (**gyri**) separated by fissures and sulci appear on its surface (fig. 20-27*B*).

Cortex Development. The cerebral cortex develops from the pallium (fig. 20-23), which may be divided into two regions: (1) the **palaeo-** or **archipallium,** an area located immediately lateral to the corpus striatum (fig. 20-24*B*); and (2) the **neopallium,** between the hippocampus and the palaeopallium (figs. 20-24*B* and 20-26*B*).

In the neopallium the neuroepithelial cells begin to release wave after wave of neuroblasts. These neuroblasts migrate to a subpial position and then differentiate into fully mature neurons. When the next wave of neuroblasts arrives they migrate through the earlier formed layers of cells until they reach the subpial position. Hence, the early formed neuroblasts obtain a deep position in the cortex, while those formed at later times obtain a more superficial position.[29, 30]

At birth the cortex has a stratified appearance due to the specific differentiation of the cells: the motor cortex contains a large number of **pyramidal cells**, and the sensory areas are characterized by **granular cells.**[31]

Commissures. In the adult the right and left halves of the hemispheres are connected by a number of fiber bundles, the **commissures**, which cross the midline. The most important of these fiber bundles make use of the **lamina terminalis** (figs. 20-23, 20-26 and 20-28). The first of the crossing bundles to appear is the **anterior commissure.** It consists of fibers connecting the olfactory bulb and related brain areas of one hemisphere to those of the opposite side (fig. 20-26 and 20-28).

The second commissure to appear is the **hippocampal** or **fornix commissure.** Its fibers arise in the hippocampus and converge on the lamina terminalis close to the roof plate of the diencephalon. From here the fibers

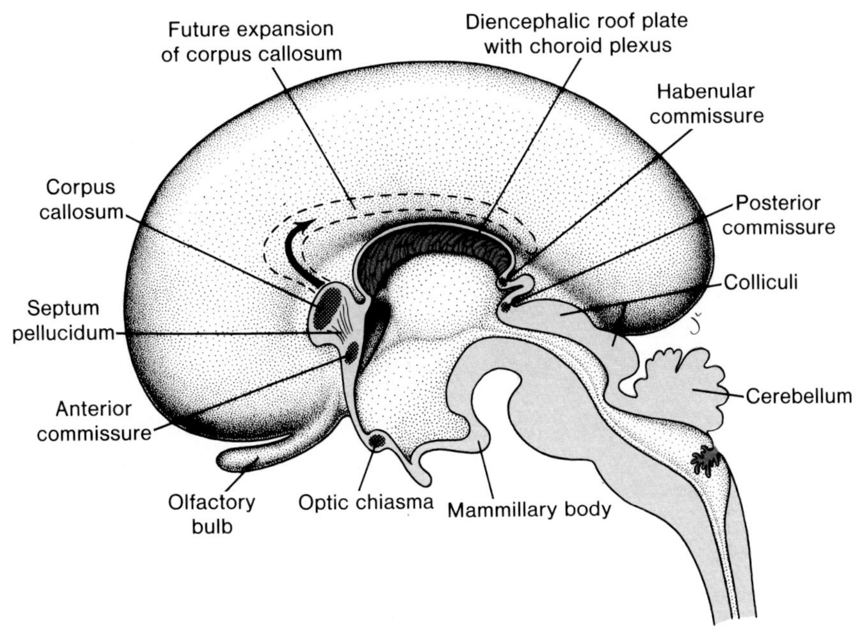

Figure 20-28. View of the medial surface of the right half of the brain in a four-month embryo, showing the various commissures. The broken line indicates the future expansion of the corpus callosum. The hippocampal commissure is not indicated.

continue, forming an arching system immediately outside the choroid fissure, to the mammillary body and the hypothalamus.

The most important commissure is the **corpus callosum.** It appears by the 10th week of development and connects the non-olfactory areas of the right and left cerebral cortex. Initially it forms a small bundle in the lamina terminalis. However, as a result of the continuous expansion of the neopallium, it rapidly extends first anteriorly and then posteriorly, thereby arching over the thin roof of the diencephalon (fig. 20-28).

In addition to the three commissures developing in the lamina terminalis, three more appear. Two of these, the **posterior** and **habenular commissures**, are found just below and rostral to the stalk of the pineal gland. The third, the **optic chiasma**, appears in the rostral wall of the diencephalon and contains fibers from the medial halves of the retinae (fig. 20-28).

Congenital Malformations

MENINGOCELE, MENINGOENCEPHALOCELE, MENINGOHYDROENCEPHALOCELE

The primary cause of all these malformations is an ossification defect in the bones of the skull. The most frequently affected bone is the squamous part of the occipital bone, which may be partially or totally lacking. The thus formed foramen is often confluent with the foramen magnum. If the opening of the occipital bone is small, only the meninges bulge through it (**meningocele**), but if the defect is large, part of the brain and even part of the ventricle may penetrate through the opening into the meningeal sac (figs. 20-29 and 20-30). The latter two malformations are known as **meningoencephalocele** and **meningohydroencephalocele**, respectively.

ANENCEPHALUS

Anencephalus is characterized by failure of the cephalic part of the neural tube to close and at birth the brain is represented by a mass of degenerated tissue exposed to the surface. The defect is almost always continuous with an open cord in the cervical region. The vault of the skull is absent, giving the head a characteristic appearance: the eyes bulge forward, the neck is absent, and the surfaces of the face and chest form a continuous plane (fig. 20-31). Since the fetus lacks the control mechanism for swallowing, the last two months of pregnancy are characterized by **hydramnios**. On an x-ray of the fetus, the abnormality can easily be recognized, since the vault of the skull is absent. Anencephalus is a common abnormality (1:1000) and is seen four times more frequently in females than in males.[32] Similarly, it is seen four times more frequently in whites than in blacks.

HYDROCEPHALUS

Hydrocephalus is characterized by an abnormal accumulation of cerebrospinal fluid within the ventricular system or, as in the case of external hydrocephalus, between the brain and dura mater. In the majority of cases, hydrocephalus in the newborn is thought to be due to an obstruction of the aqueduct of Sylvius. This prevents the cerebrospinal fluid of the lateral and third ventricles from passing into the fourth ventricle and form there into

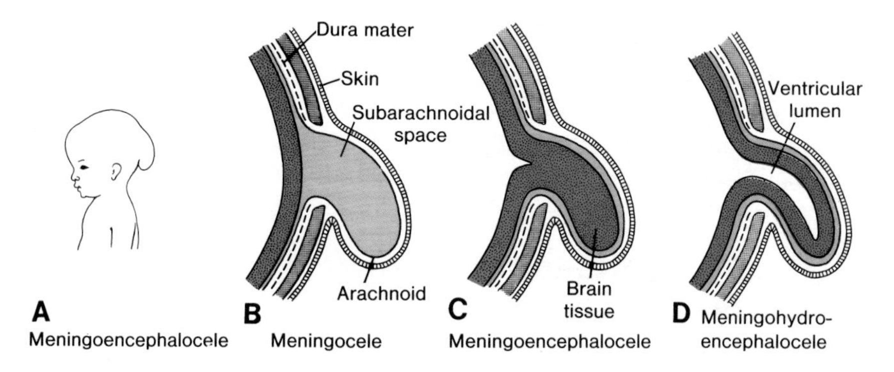

Figure 20-29. Schematic drawings to show the various types of brain herniation due to abnormal ossification of the skull.

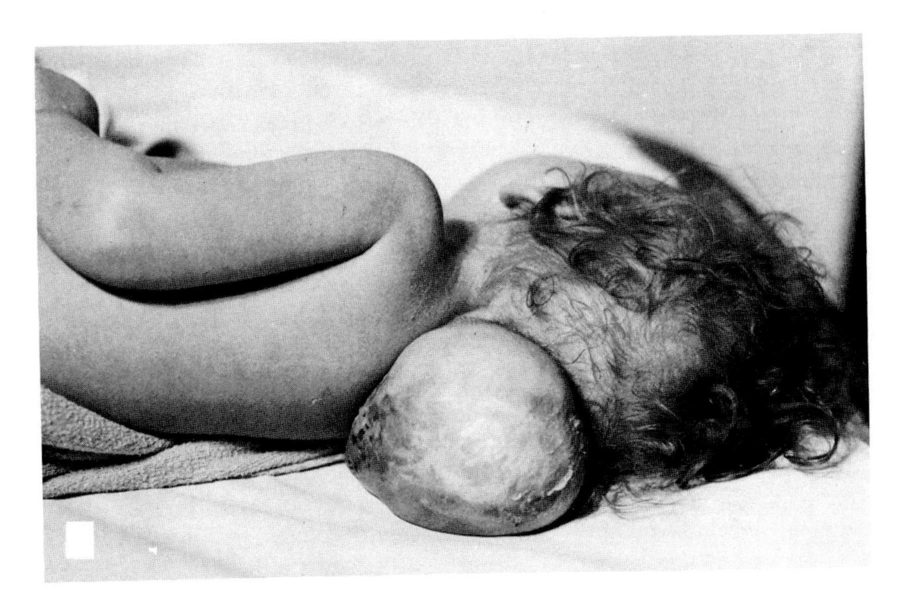

Figure 20-30. Photograph of patient with meningohydroencephalocele. (Courtesy Dr. J. Warkany. From *Congenital Malformations*, by Warkany, J. Copyright 1971, by Year Book Medical Publishers, Inc., Chicago. Used by permission.)

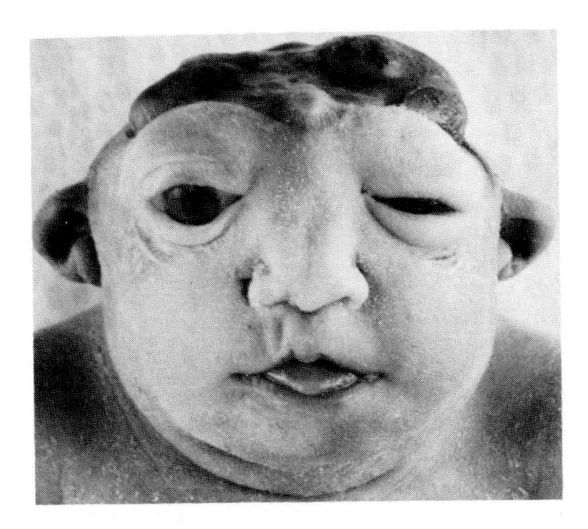

Figure 20-31. Photograph of anencephalic child. Ventral view. This abnormality is frequently seen. Usually the child dies a few days after birth. (Courtesy Dr. J. Warkany. From *Congenital Malformations*, by Warkany, J. Copyright 1971, by Year Book Medical Publishers, Inc., Chicago. Used by permission.)

the subarachnoidal space.[33] The malformation is often accompanied by a widening of the sutures of the skull, while the bones themselves gradually become thin. In extreme cases the skull may be three times the normal size.

It is obvious that the above-mentioned abnormalities are only the most serious ones and usually incompatible with life. A great many other defects of the central nervous system, however, may occur without much external manifestation. For example, the corpus callosum may be partially or completely absent without much functional disturbance. Likewise, partial or complete absence of the cerebellum may show only a little disturbance of coordination. On the other hand, cases of severe imbecility or idiocy may show hardly any morphological brain abnormalities.

ENVIRONMENTAL AND CHROMOSOMAL FACTORS

It is now well established that abnormalities of the central nervous system can occur as the result of fetal infection by the **toxoplasma** organism. The affected child may then suffer from cerebral calcification, mental retardation, hydrocephalus, or microcephalus.[34] Likewise, radiation during the early stages of development may account for microcephalus.[35, 36] Some authors have suggested that maternal infection with Asian influenza in the early stages of pregnancy may occasionally result in anencephalus, but that the risk is low;[37] others, however, have reported no significant increase in the malformation rate following maternal influenza.[38]

The hypothesis has been presented that the seasonal peak incidence during successive years and geographic distribution of potato blight had concordance with the incidence of anencephaly and spina bifida.[39] The original epidemiological data have been sharply criticized and the correlation between potato blight and anencephaly is no longer accepted.

The discovery that congenital malformations may be caused by chromosomal abnormalities has had important implications with regard to mental retardation. That a human defect could be caused by a chromosomal imbalance was first demonstrated by the discovery that children with mongolism had 47 chromosomes, chromosome number 21 being represented three times instead of twice (trisomy).[40] Likewise, 47 chromosomes were found in patients with Klinefelter's syndrome, the extra chromosome being an X. This syndrome accounts for about 1 per cent of mentally defective children.[41] In addition to Klinefelter's syndrome, it has been discovered that many sex chromosome abnormalities are associated with mental retardation.[42] Likewise, many of the congenital metabolic defects such as phenylketonuria cause or are accompanied by mental retardation.

Although little is known about the etiology of malformations of the central nervous system in man, over the past years many environmental teratogens have been detected as causing malformations of the central nervous system in the offspring of experimental animals. Anencephalus, hydrocephalus, spina bifida, cranium bifidum, exencephalus, meningocele, microcephalus, and cerebellar defects have been produced by a variety of teratogenic factors (vitamin A, riboflavin, folic acid, pantothenic acid, vitamin E, and nicotinamide deficiencies; maternal fasting; hypervitaminosis A; trypan blue; hypoxia; x-irradiation; and antibody treatment).[18, 43–48]

Although the great majority of malformations of the central nervous system have been produced by treating pregnant animals during early stages of gestation, it was recently found that excess vitamin A when given to mice during the later stages of pregnancy may also cause abnormalities of the cerebral cortex.[49] These abnormalities were only detectable with the microscope and concerned degeneration or abnormal differentiation of the neuroblasts in certain areas of the cortex. Recently it has been possible to produce in mice abnormal behavioral patterns by treating the animals with certain drugs on specific days late in gestation.[50, 51]

Autonomic Nervous System

Functionally, the autonomic nervous system can be divided into two parts: a **sympathetic** portion which is localized in the thoracolumbar region, and a **parasympathetic** portion, found in the cephalic and sacral regions.

SYMPATHETIC NERVOUS SYSTEM

In the fifth week of development, cells originating in the **neural crest** of the thoracic region migrate on each side of the spinal cord toward the region immediately behind the dorsal aorta (fig. 20-32). Here they form a bilateral

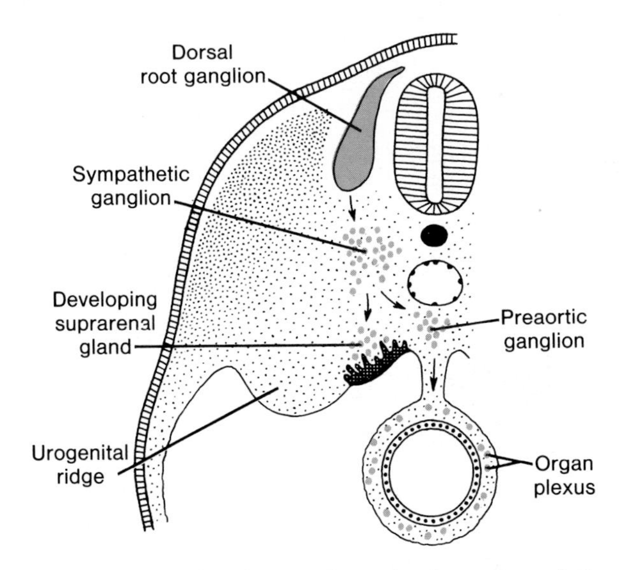

Figure 20-32. Schematic drawing to show the formation of the sympathetic ganglia. A portion of the sympathetic neuroblasts migrates toward the proliferating mesothelium to form the medulla of the suprarenal gland.

chain of segmentally arranged sympathetic ganglia interconnected by longitudinal nerve fibers. Together they form the **sympathetic chains**, located on each side of the vertebral column. From their position in the thorax, the neuroblasts migrate toward the cervical and lumbosacral regions, thus extending the sympathetic chains to their full length. Although initially the ganglia are arranged segmentally, this arrangement is later obscured, particularly in the cervical region, by the fusion of ganglia (see the gross anatomy books).

Some of the sympathetic neuroblasts migrate in front of the aorta to form the **preaortic ganglia**, such as the **celiac** and **mesenteric ganglia**. Still other sympathetic cells migrate to the heart, lungs, and gastrointestinal tract, where they give rise to the **sympathetic organ plexuses**. (fig. 20-32).

Once the sympathetic chains have been established, nerve fibers originating in the viscero-efferent column (intermediate horn) of the thoracolumbar

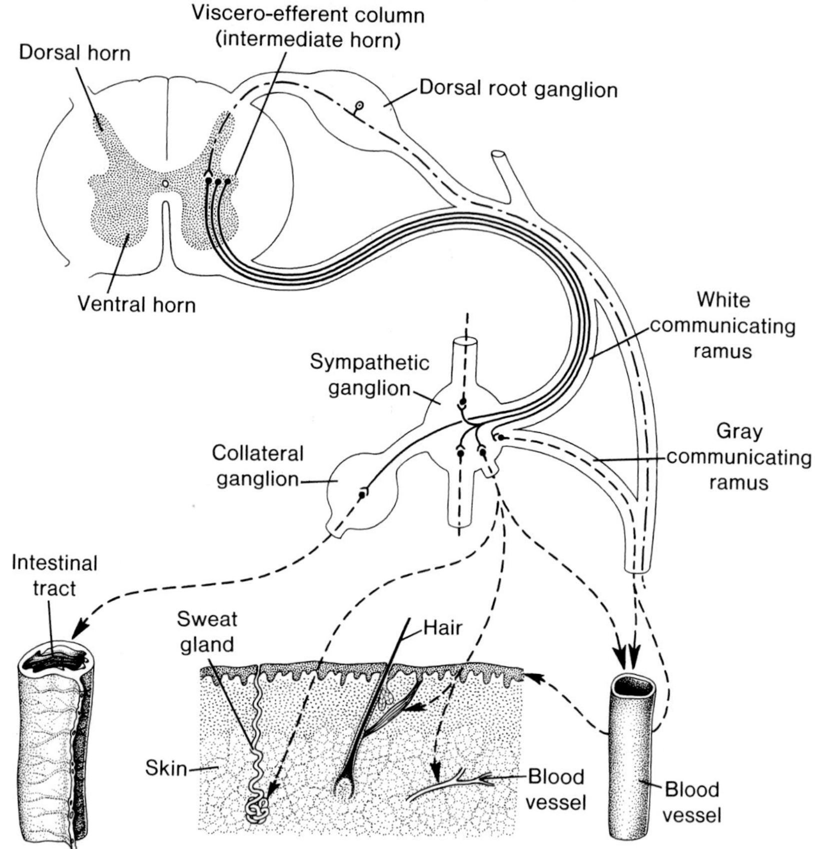

Figure 20-33. Schematic drawing to show the relationship of the preganglionic and postganglionic nerve fibers of the sympathetic nervous system to the spinal nerves. Note the origin of the preganglionic fibers in the viscero-efferent motor column of the spinal cord.

segments penetrate the ganglia of the chain and form synapses around the developing neuroblasts (fig. 20-33). Some of these nerve fibers either extend to higher or lower levels in the sympathetic chains or to the **preaortic** or **collateral ganglia** before synapsing (fig. 20-33). They are known as **preganglionic fibers**, have a myelin sheath, and stimulate the sympathetic ganglion cells into action. Passing from the spinal nerves to the sympathetic ganglia, they form the so-called **white communicating rami**. Since the viscero-efferent column extends only from the first thoracic to the second lumbar segment of the spinal cord, the white rami are found only between these levels.

The axons of the sympathetic ganglion cells are called **postganglionic fibers** and have no myelin sheath. They pass either to other levels of the sympathetic chain or extend to the heart, lungs, and intestinal tract (broken lines in fig. 20-33). Other fibers known as the **gray communicating rami** pass from the sympathetic chain to the spinal nerves and from there to the peripheral blood vessels, hair and sweat glands. The gray communicating rami are found at all levels of the spinal cord.

Suprarenal Gland. The suprarenal gland develops from two components: (1) a mesodermal portion which forms the **cortex**, and (2) an ectodermal portion which forms the **medulla**.

During the fifth week of development, mesothelial cells located between the root of the mesentery and the developing gonad begin to proliferate and penetrate the underlying mesenchyme (fig. 20-32). Here they differentiate into large acidophilic organs which form the **fetal** or **primitive cortex** of the suprarenal gland (fig. 20-34*A*). Shortly afterward a second wave of cells from the mesothelium penetrates the mesenchyme and surrounds the original acidophilic cell mass. These cells, smaller than those of the first wave, later form the **definitive cortex** of the gland (fig. 20-34 *A*, *B*). After birth the fetal

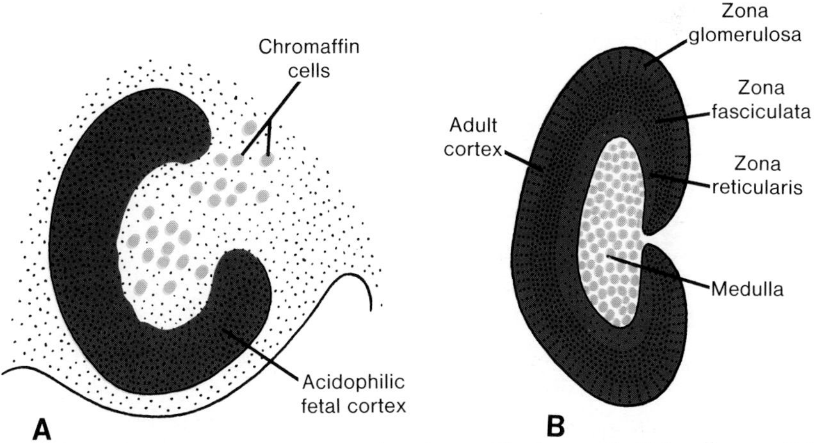

Figure 20-34. *A*, Drawing showing the chromaffin (sympathetic) cells penetrating the fetal cortex of the suprarenal gland. *B*, At a later stage of development, the definitive cortex surrounds the medulla almost completely.

cortex regresses rapidly, except for its outermost layer, which differentiates into the reticular zone. The adult structure of the cortex is not achieved until the approach of puberty.

While the fetal cortex is being formed, cells originating in the sympathetic system invade its medial aspect, where they become arranged in cords and clusters. These cells give rise to the medulla of the suprarenal gland. They stain yellow-brown with chrome salts and hence are called **chromaffin cells** (fig. 20-34). During embryonic life the chromaffin cells are widely scattered throughout the embryo, but in the adult the only persisting group is found in the medulla of the adrenal glands.

PARASYMPATHETIC NERVOUS SYSTEM

The origin of the parasympathetic ganglia found along the oculomotor, facial, glossopharyngeal, and vagus nerves is rather controversial. Some believe that the cells migrate out of the central nervous system along the preganglionic fibers of the above-mentioned nerves, whereas others feel that they arise from neuroblasts originating in the sensory ganglia of the fifth, seventh, and ninth nerves.

The postganglionic fibers of the parasympathetic ganglia pass to the branchial arches and to the cardiac, pulmonary, and intestinal plexuses. The action of these fibers is thought to be antagonistic to those of the sympathetic nervous system.

SUMMARY

The central nervous system is of **ectodermal origin** and appears as the **neural plate** around the middle of the third week (fig. 20-1). After the edges of plate become folded, these **neural folds** approach each other in the midline to fuse into the **neural tube** (figs. 20-2 and 20-3). The cranial end closes approximately at day 25 and the caudal end at day 27. The CNS then forms a tubular structure with a broad cephalic portion, the **brain**, and a long caudal portion, the **spinal cord**. Failure of the neural tube to close results in frequently seen defects such as **spina bifida** (figs. 20-14 and 20-15) and **anencephaly** (figs. 20-29 to 20-31).

The **spinal cord** forms the caudal end of the CNS and is characterized by the **basal plate** containing the **motor neurons**; the **alar plate** for the **sensory neurons**; and a **floor** and a **roof plate** as connecting plates between the two sides (fig. 20-8). These basic features can be recognized through most of the brain vesicles.

The brain forms the cranial part of the CNS and consists originally of three brain vesicles: (1) the rhombencephalon, (2) the mesencephalon, and (3) the prosencephalon or forebrain.

The **rhombencephalon** is underdivided in (a) the myelencephalon which gives rise to the **medulla oblongata**, which has a basal plate for somatic and visceral motor neurons and an alar plate for somatic and visceral sensory neurons (fig. 20-17); (b) the **metencephalon** with its typical basal (motor) and alar sensory plates (fig. 20-18). This brain vesicle is, in addition, characterized by the formation of the **cerebellum** (fig. 20-19), a coordination center for posture and movement, and the **pons**, the pathway for nerve fibers between the spinal cord and cerebral and cerebellar cortices (fig. 20-18).

The **mesencephalon** or **midbrain** is the most primitive brain vesicle and resembles most the spinal cord with its **basal** motor and alar sensory plates. Its alar plates form the anterior and posterior colliculi as relay stations for auditory and visual reflex centers. (fig. 20-22).

The **diencephalon**, the posterior portion of the forebrain, consists of a thin roof plate and a thick alar plate in which the **thalamus** and **hypothalamus** develop (figs. 20-23 and 20-24). It participates in the formation of the **pituitary gland**, which develops from Rathke's pouch and the infundibulum (fig. 20-25). While Rathke's pouch forms the adenohypophysis, the intermediate lobe, and pars tuberalis, the diencephalon forms the posterior lobe, which contains neuroglia cells and receives nerve fibers from the hypothalamus. The **telencephalon**, the most rostral of the brain vesicles, consists of two lateral outpocketings, the **cerebral hemispheres**, and a median portion, the **lamina terminalis** (fig. 20-26). The lamina terminalis is mainly used by the commissures as a connection pathway for fiber bundles between the right and left hemispheres (fig. 20-28). The cerebral hemispheres, originally two small outpocketings (figs. 20-23 and 20-24), expand gradually and cover the lateral aspect of the diencephalon, mesencephalon, and metencephalon (figs. 20-25 to 20-27). Eventually, nuclear regions of the telencephalon come in close contact with those of the diencephalon (fig. 20-26).

The ventricular system, containing the cerebrospinal fluid, extends from the lumen in the spinal cord to the fourth ventricle in the rhombencephalon, the narrow aqueduct in the mesencephalon, and subsequently to the third ventricle in the diencephalon. By way of the foramina of Monro the ventricular system extends into the lateral ventricles of the hemispheres. The cerebrospinal fluid is produced in the choroid plexus of the fourth, third, and lateral ventricles. Blockage of the cerebrospinal fluid in the ventricular system or subarachnoid space may lead to internal or external hydrocephalus.

REFERENCES

1. Geelen, J. A. G., and Langman, J. Closure of the neural tube in the cephalic region of the mouse embryo. *Anat. Rec., 189:* 625, 1977.
2. Waterman, R. E. Topographical changes along the neural fold associated with neurulation in the hamster and the mouse. *Am. J. Anat., 146:* 151, 1976.
3. Fujita, S. Kinetics of cellular proliferation. *Exp. Cell Res., 28:* 52, 1962.
4. Langman, J., Guerrant, R. L., and Freeman, B. G. Behavior of neuroepithelial cells during closure of the neural tube. *J. Comp. Neurol., 127:* 399, 1966.
5. Jacobson, S. In *An Introduction to the Neurosciences*, edited by B. A. Curtis, S. Jacobson, and E. M. Marcus. W. B. Saunders, Philadelphia, 1972.
6. Lyser, K. M. Early differentiation of motor neuroblasts in the chick embryo as studied by electron microscopy. *Dev. Biol., 17:* 117, 1968.
7. Fujita, H., and Fujita, S. Electron microscopic studies on neuroblast differentiation in the central nervous system of domestic fowl. *Z. Zellforsch. Mikrosk., 60:* 463, 1963.
8. Langman, J. Histogenesis of the central nervous system. In *The Structure and Function of Nervous Tissue*, edited by G. H. Bourne. Academic Press, New York, 1968.
9. Jacobson, M. *Developmental Neurobiology*, ed. 2, edited by M. Jacobson. Plenum Press, New York, 1978.
10. Kruger, L., and Maxwell, D. S. Electron microscopy of oligodendrocytes in normal rat cerebrum. *Am. J. Anat., 118:* 411, 1966.
11. Nichols, D. H. Neural crest formation in the head of the mouse as observed using a new histological technique. *J. Embryol. Exp. Morphol.* (in press).
12. Bunge, P. R. Structure and function of neuroglia: Some recent observations. In *Neurosciences: Second Study Program*, edited by F. O. Schmidt. Rockefeller University Press, New York, 1970.
13. Weston, J. A. The migration and differentiation of neural crest cells. In *Advances in Morphogenesis*, edited by M. Abercrombie, J. Brachet, and T. J. King. Academic Press, New York, 1970.
14. Johnston, M. C., and Listgarten, M. A. Observations on the migration, interaction, and early differentiation of orofacial tissues. In *Developmental Aspects of Oral Biology*, edited by H. C. Slavkin and L. A. Baretta. Academic Press, New York, 1972.
15. Noden, D. M. The control of avian cephalic neural crest cytodifferentiation. I. Skeletal and connective tissues. *Dev. Biol., 67:* 296, 1978.
16. Metuzalis, J. Ultrastructure of myelinated nerve fibers and nodes of Ranvier in the central nervous system of the frog. In *Proceedings of the European Regional Conference on Electron Microscopy, Delft*, The Netherlands, Vol. 2, p. 799, 1960.
17. Peters, P. W. J., Dormans, J. A. M. A., and Geelen, J. A. G. Light microscopic and ultrastructural observations in advanced stages of induced exencephaly and spina bifida. *Teratology, 19:* 183, 1979.
18. Geelen, J. A. G. The teratogenic effects of hypervitaminosis A on the formation of the neural tube. Doctoral thesis: University of Nimwegen. Krips Repro Meppel. 1980.
19. Warkany, J. *Congenital Malformations. Notes and Comments.* Year Book Medical Publishers, Chicago, 1971.

20. Rakic, P., and Sidman, R. L. Histogenesis of cortical layers in human cerebellum, particularly the lamina dissecans. *J. Comp. Neurol., 139:* 473, 1970.
21. Conchie, A. F., Barton, B. W., and Tobin, J. O. H. Congenital cytomegalovirus infection treated with iodoxuridine. *Br. Med. J., 4:* 162, 1968.
22. Feigin, R. D., Shakelford, P. G., DeVivo, D. C., and Haymond, M. W. Floxuridine treatment of congenital cytomegalic inclusion disease. *Pediatrics, 47:* 318, 1971.
23. Langman, J., Shimada, M., and Rodier, P. Floxuridine and its influence on postnatal cerebellar development. *Pediatr. Res., 6:* 758, 1972.
24. Gladstone, R. J., and Wakelay, C. P. G. *The Pineal Organ.* Baillière, Tindall & Cox, Ltd., London, 1940.
25. Kitay, J. L., and Altschule, M. D. *The Pineal Gland.* Harvard University Press, Cambridge, Mass., 1954.
26. Boyd, J. D. Observations on the human pharyngeal hypophysis. *J. Endocrinol., 14:* 66, 1956.
27. Hewitt, W. The development of the human internal capsule and lentiform nucleus. *J. Anat., 95:* 191, 1961.
28. Sharp, J. A. The junctional region of the cerebral hemisphere and the third ventricle in mammalian embryos. *J. Anat., 93:* 159, 1959.
29. Angevine, J. B., and Sidman, R. L. Autoradiographic study of cell migration during histogenesis of cerebral cortex in the mouse. *Nature, 192:* 766, 1961.
30. Shimada, M., and Langman, J. Cell proliferation, migration and differentiation in the cerebral cortex of the golden hamster. *J. Comp. Neurol., 139:* 227, 1970.
31. Conel, J. L. *Postnatal Development of the Human Cerebral Cortex.* Harvard University Press, Cambridge, Mass., 1959.
32. Laurence, K. M., and Weeks, R. Abnormalities in the central nervous system. In *Congenital Abnormalities in Infancy,* edited by A. P. Norman. Blackwell Scientific Publications, Oxford, 1971.
33. Lemire, R. J., Loeser, J. D., Leech, R. W., and Alvord, E. C. *Normal and Abnormal Development of the Human Nervous System.* Harper and Row, Hagerstown, Md., 1975.
34. Feldman, H. A. Toxoplasmosis. *Pediatrics, 22:* 559, 1958.
35. Plummer, G. Anomalies occurring in children exposed *in utero* to the atomic bomb in Hiroshima. *Pediatrics, 10:* 687, 1952.
36. Yamazaki, J. N., Wright, S. W., and Wright, P. M. Outcome of pregnancy in women exposed to the atomic bomb in Nagasaki. Am. *J. Dis. Child., 87:* 448, 1954.
37. Doll, R., and Hill, A. B. Asian influenza in pregnancy and congenital defects. *Br. J. Prev. Soc. Med., 14:* 167, 1960.
38. Wilson, M. G., Heins, H. L., Imagawa, D. T., and Adams, J. M. Teratogenic effects of Asian influenza. *J.A.M.A., 171:* 638, 1959.
39. Renwick, J. H. Prevention of anencephaly and spina bifida in man. *Teratology, 8:* 321, 1973.
40. Jacobs, P. A., Bailie, A. G., Brown, W. M. C., and Strong, J. A. The somatic chromosomes in mongolism. *Lancet, 1:* 710, 1959.
41. MacLean, N., et al. A survey of sex chromosome abnormalities among 4514 mental defectives. *Lancet, 1:* 293, 1962.
42. Harnden, D. G., and Jacobs, P. A. Cytogenetics of abnormal sexual development in man. *Br. Med. Bull., 17: 206, 1961.*
43. Miller, J. R. Clinical and experimental studies on the etiology of skull, vertebra, rib and palate malformations. Doctoral Thesis, McGill University, 1959.
44. Russell, L. B. X-ray induced developmental abnormalities in the mouse and their use in the analysis of embryological patterns. II. Abnormalities of the vertebral column and thorax. *J. Exp. Zool., 131:* 329, 1956.
45. Hicks, S. P., et al. Migrating cells in the developing nervous system studied by their radiosensitivity and tritiated thymidine uptake. *Brookhaven Symp. Biol., 14:* 246, 1961.
46. Pinsky, H., and Fraser, F. C. Congenital malformations following a two-hour inactivation of nicotinamide by its analogue, 6-aminonicotinamide in pregnant mice. *Br. Med. J., 2:* 195, 1960.
47. Runner, M. N., and Dagg, C. P. Metabolic mechanisms of teratogenic agents during morphogenesis (symposium on normal and abnormal differentiation and development). *Nat. Cancer Inst. Monogr., 2:* 41, 1960.

48. Brent, R. L., Averich, E., and Drapiewski, V. A. Production of congenital malformations using tissue antibodies. *Proc. Soc. Exp. Biol. Med., 106:* 523, 1961.
49. Langman, J., Cardell, E. L., Crowley, K. K. Cell degeneration and repair in the fetal mammalian CNS. In *Neural and Behavioral Teratology.* M.T.P. Press, Lancaster, England, 1980.
50. Rodier P. M., and Reynolds, S. S. Morphological correlates of behavioral abnormalities in experimental congenital brain damage. *Exp. Neurol., 57:* 81, 1977.
51. Rodier, P. M., Reynolds, S. S., and Roberts, W. N. Behavioral consequences of interference with CNS development in the early fetal period. *Teratology, 19:* 327, 1979.

ATLAS OF DEVELOPMENT — NORMAL AND ABNORMAL

Normal Limb Development

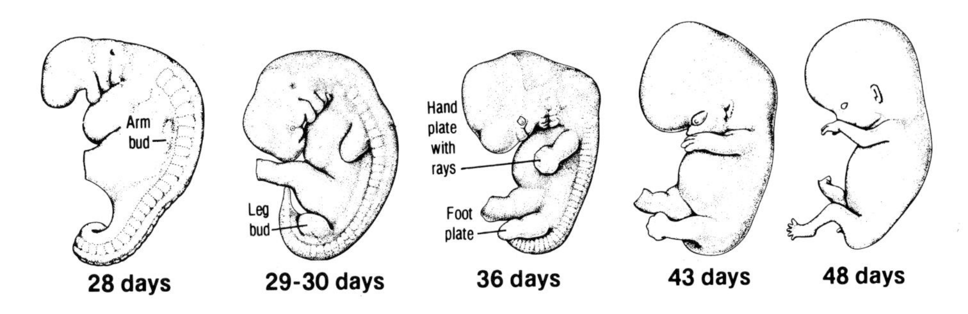

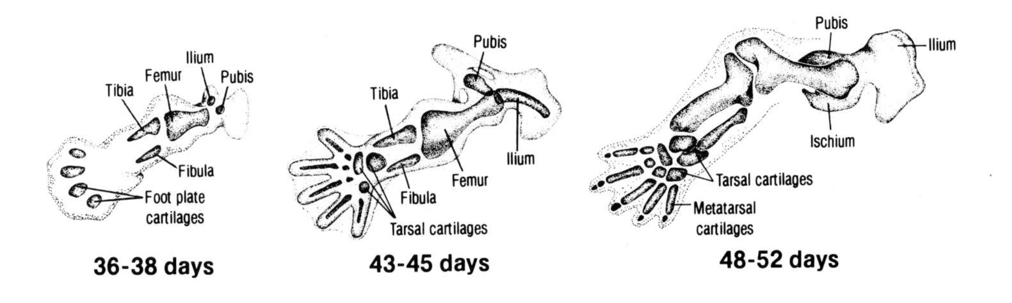

Abnormal limb development

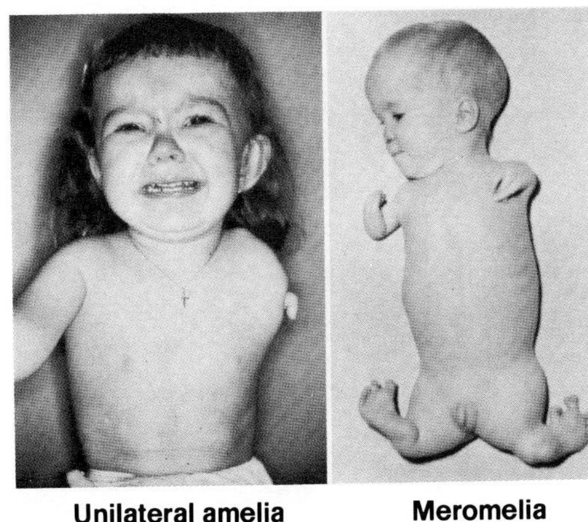

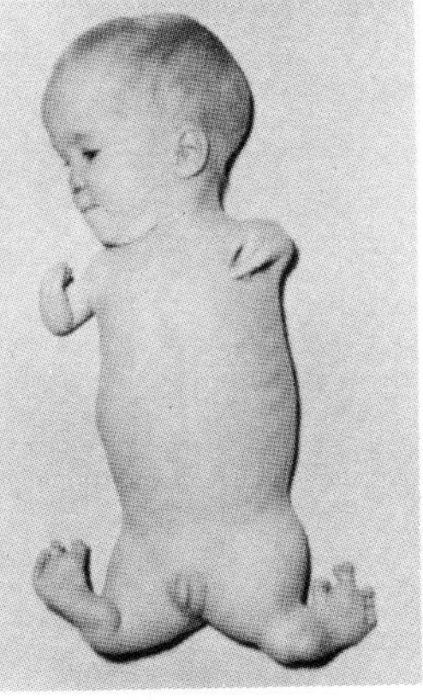

Unilateral amelia

Meromelia

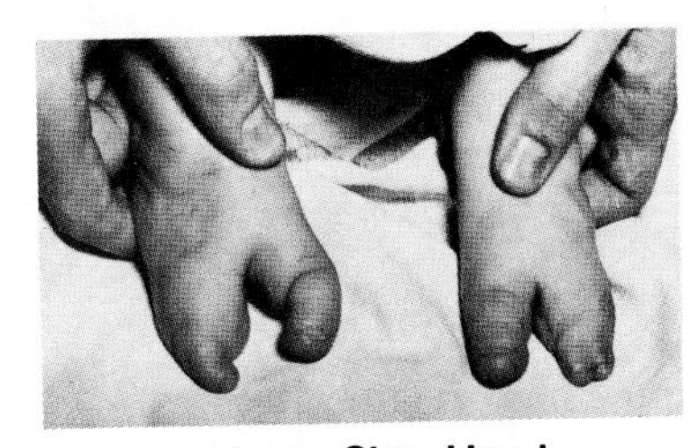

Lobster Claw Hand

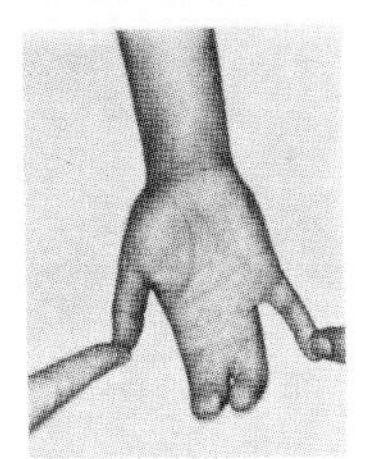

Syndactyly

Cardiovascular System Plate 1

Normal atrial septal development

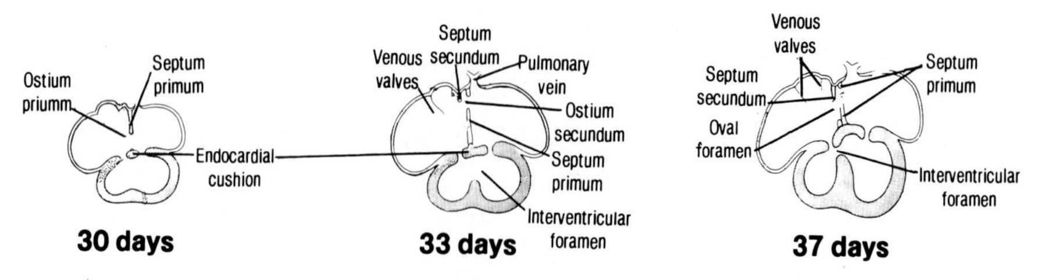

Abnormal atrial septal development

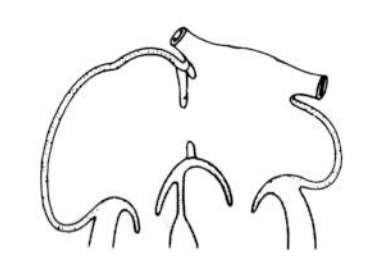

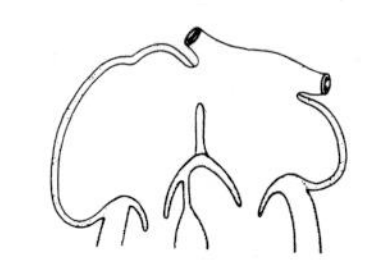

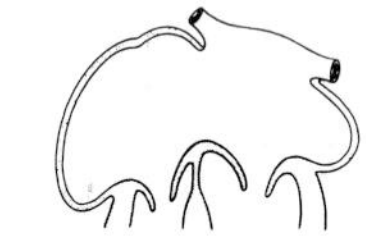

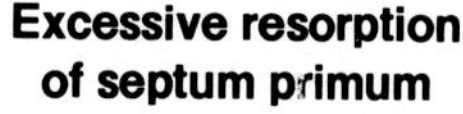

Excessive resorption of septum primum

Absence of septum secundum

Absence of septum primum and septum secundum

Normal division atrioventricular canal

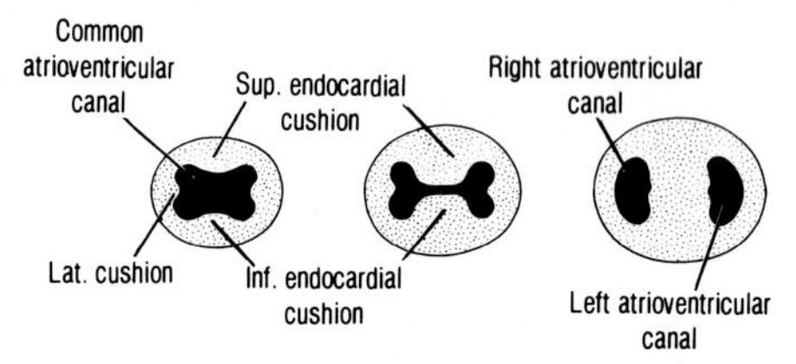

Beginning 5th week

End of 5th week

End of 6th week

Abnormal division atrioventricular canal

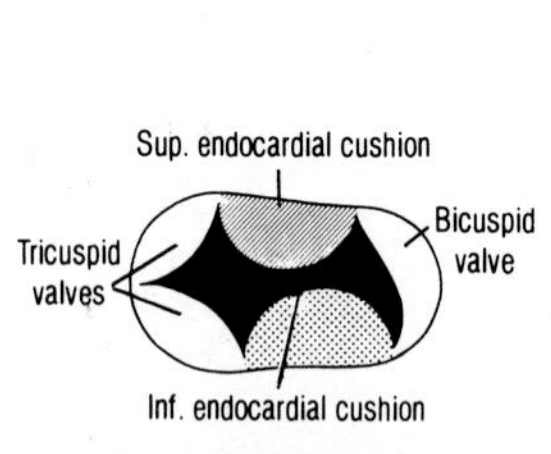

Persistent atrio-ventricular canal

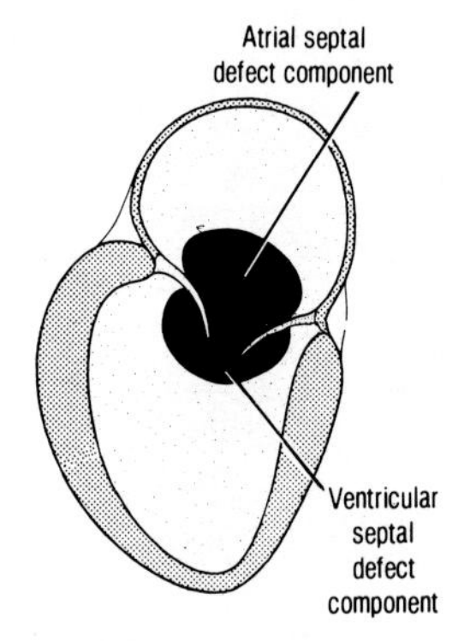

Persistent atrio-ventricular canal

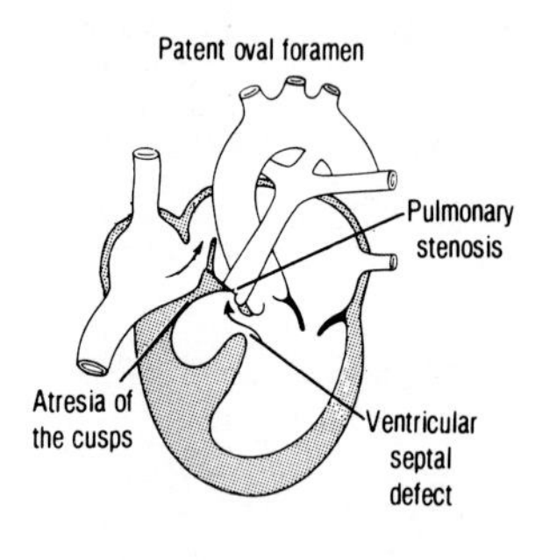

Tricuspid atresia

Cardiovascular System Plate 2

Normal septation of truncus and conus

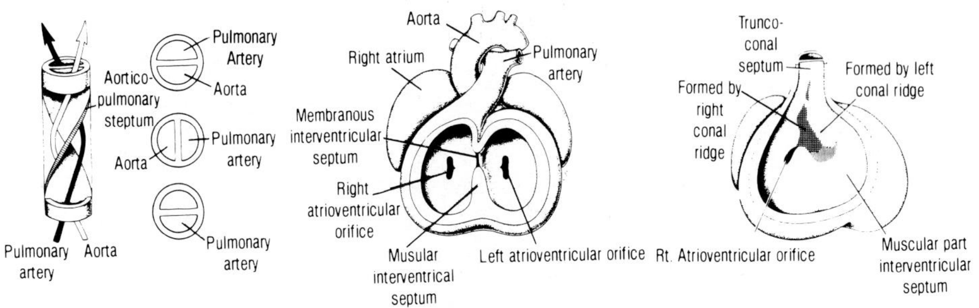

Events during 8th week

Abnormal partitioning of truncus and conus

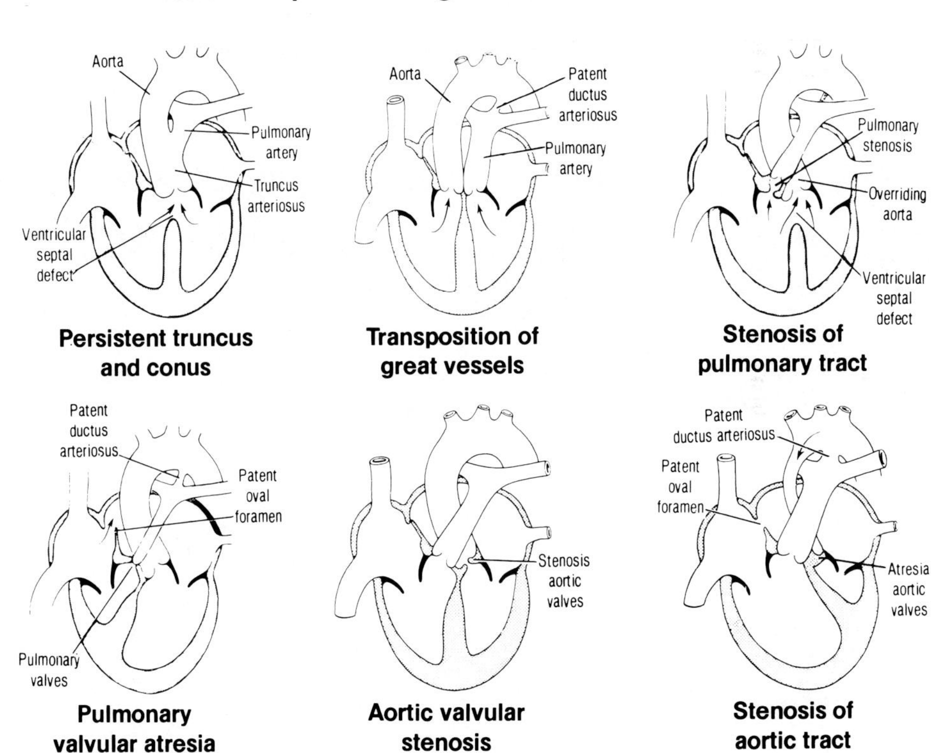

Cardiovascular System Plate 3

Normal development of aortic arches

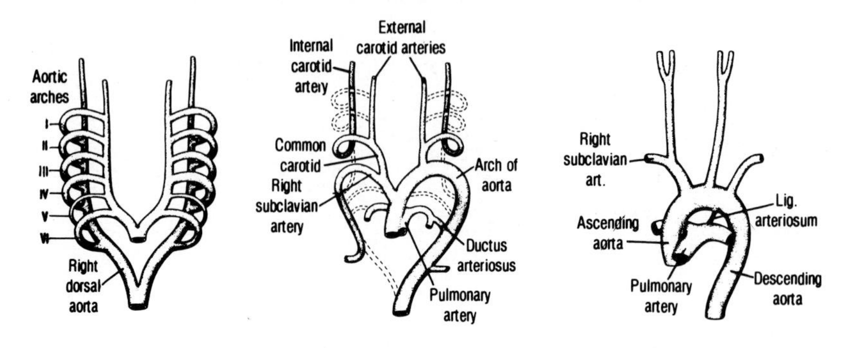

Abnormal development of aortic arches

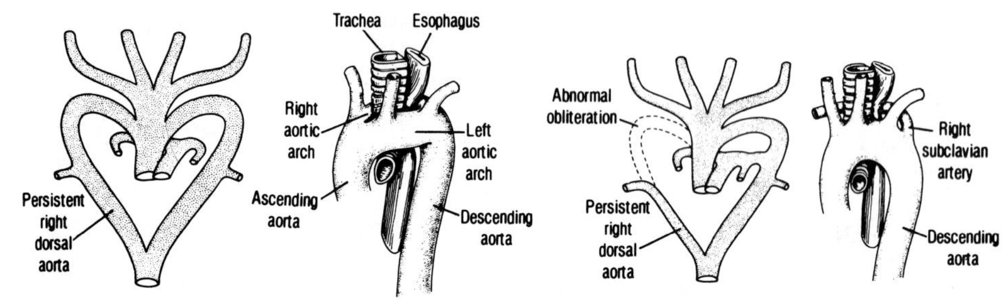

Double aortic arch compressing trachea and esophagus

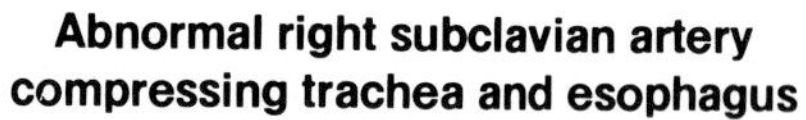

Abnormal right subclavian artery compressing trachea and esophagus

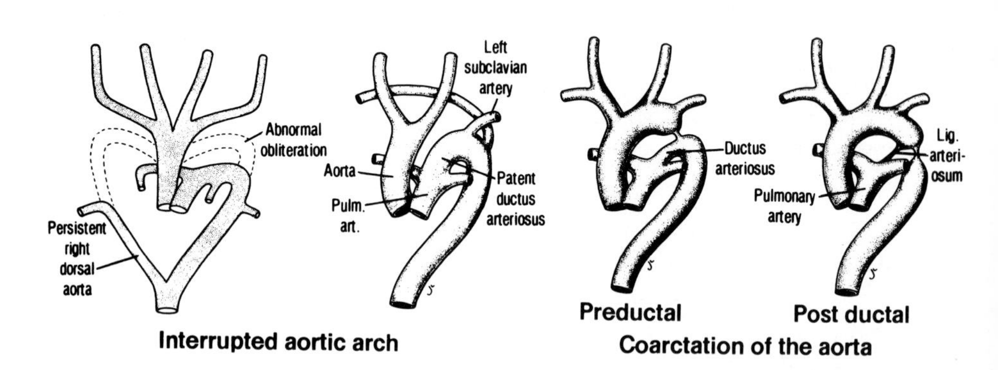

Interrupted aortic arch

Preductal **Post ductal**

Coarctation of the aorta

Digestive System Plate 1

Normal and abnormal division of foregut

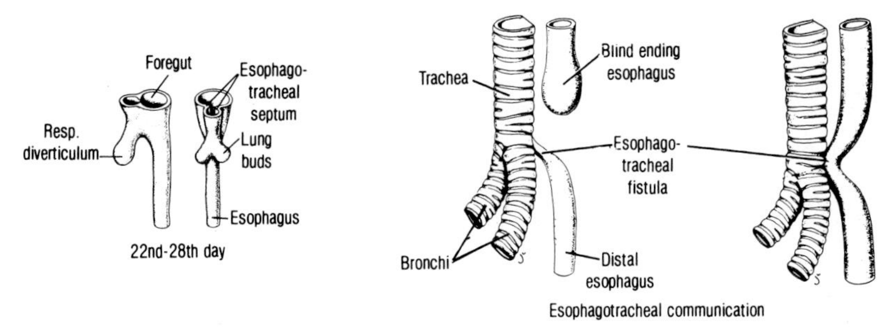

Normal rotation of the gut (6-12 weeks)

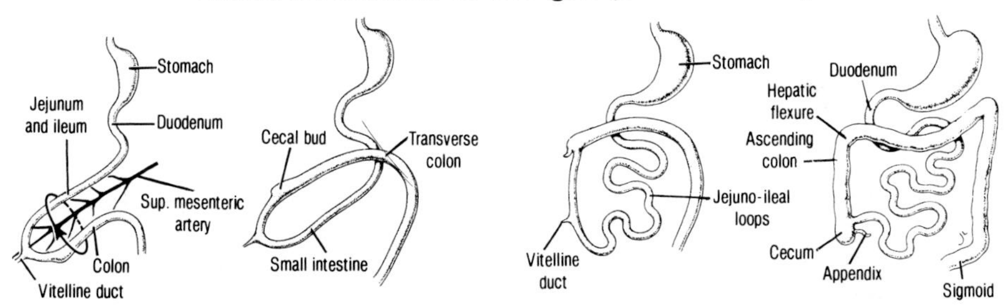

Abnormal rotation

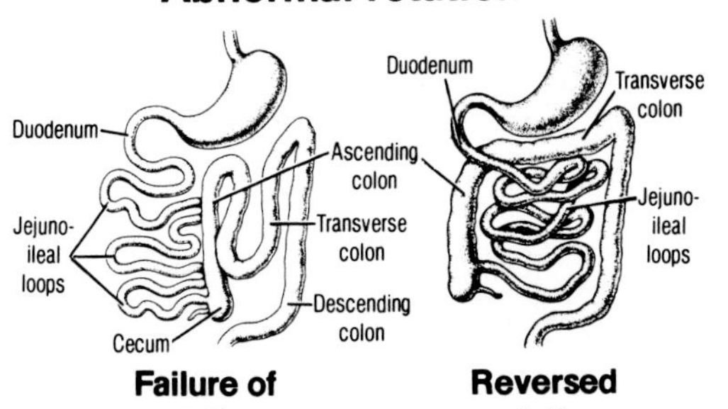

Failure of rotation

Reversed rotation

Remnants of vitelline duct

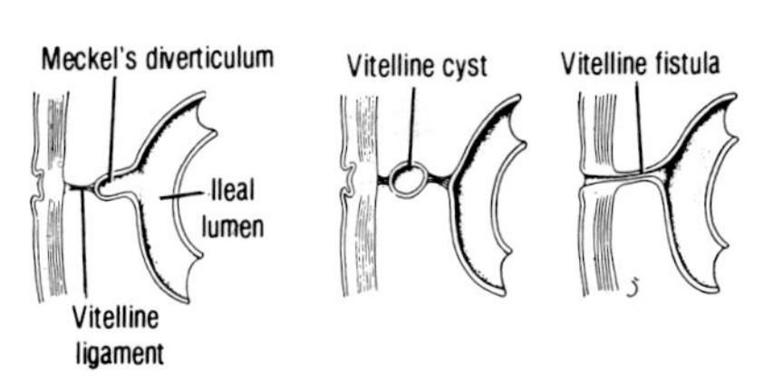

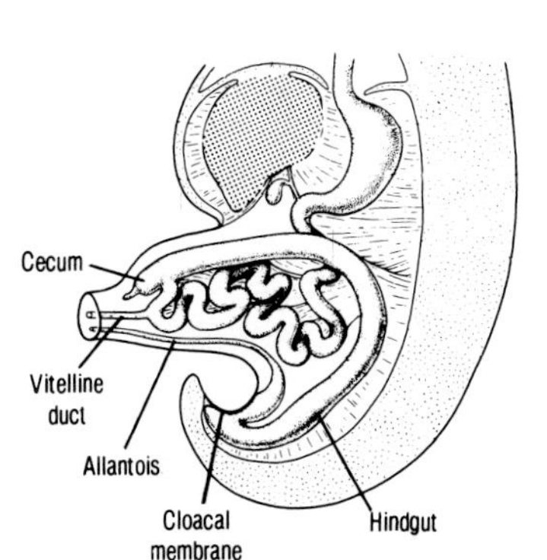

Physiological herniation (6-12 weeks)

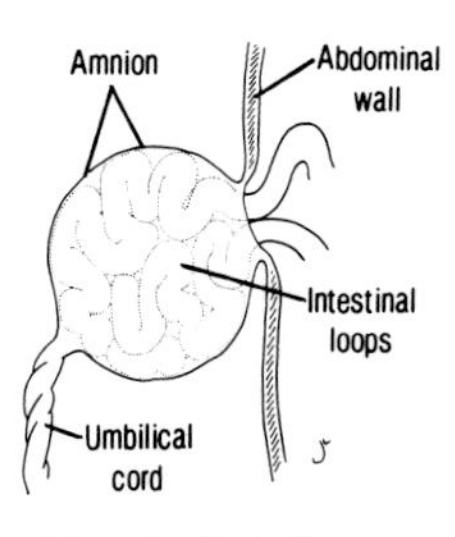

Omphalocele

Digestive System Plate 2

Normal septation of the hindgut

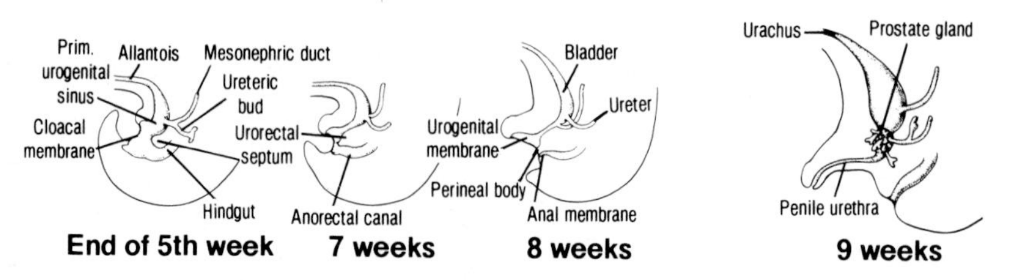

Abnormal septation of hindgut

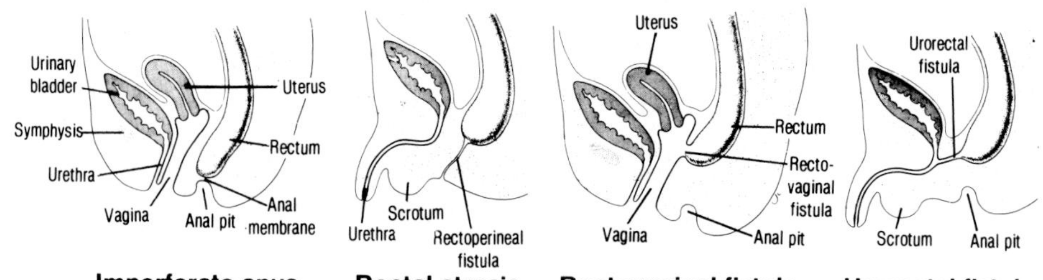

Normal and abnormal formation of diaphragm

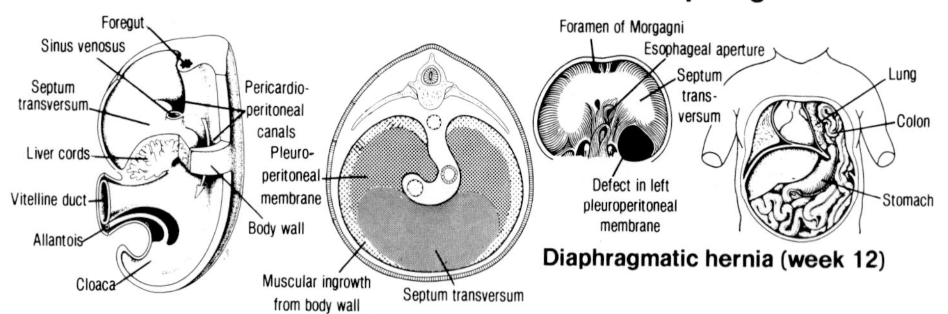

Normal and abnormal formation of the gut

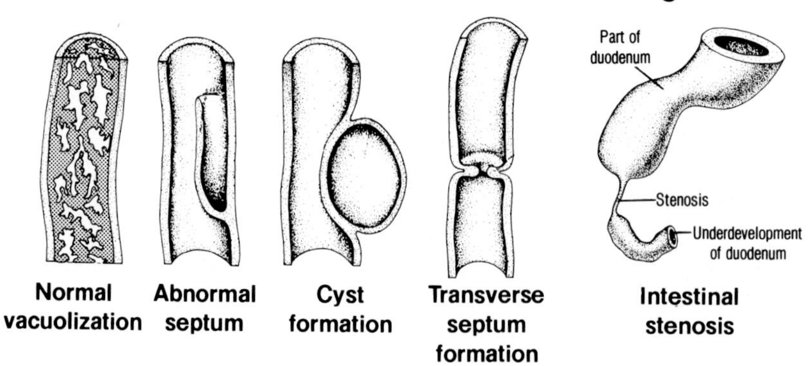

Normal ascent of kidney and development of bladder

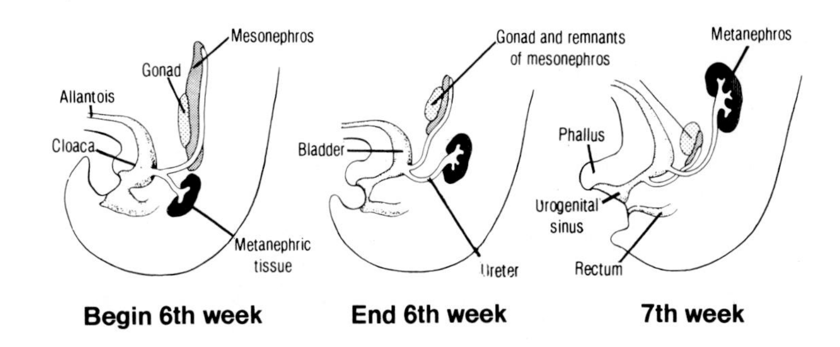

Abnormal position of kidney and ureter

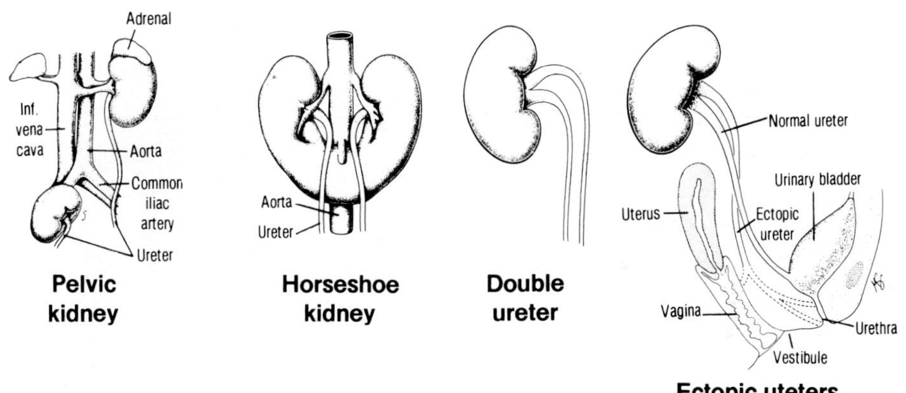

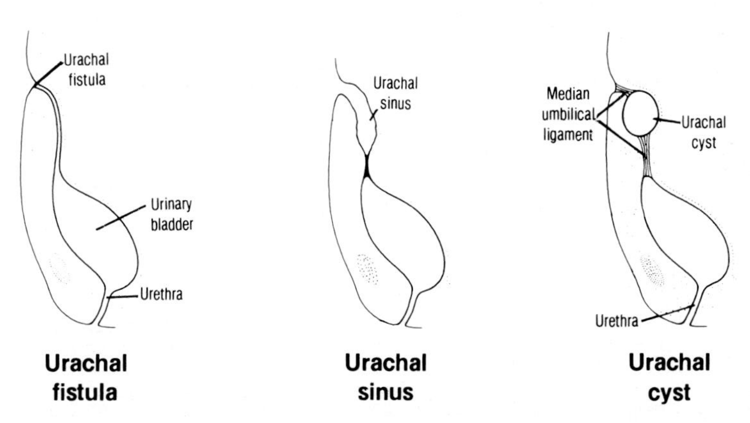

Urachal fistula **Urachal sinus** **Urachal cyst**

Genital System

Normal and abnormal development external genitalia

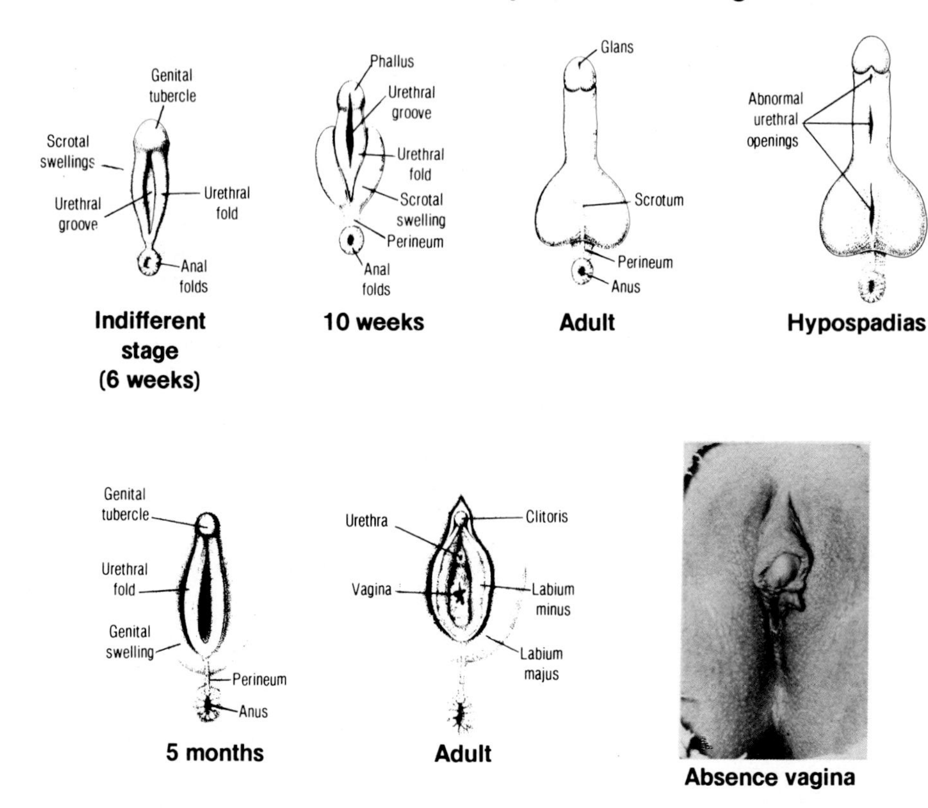

Indifferent stage (6 weeks) **10 weeks** **Adult** **Hypospadias**

5 months **Adult** **Absence vagina**

Normal and abnormal development of uterus and vagina

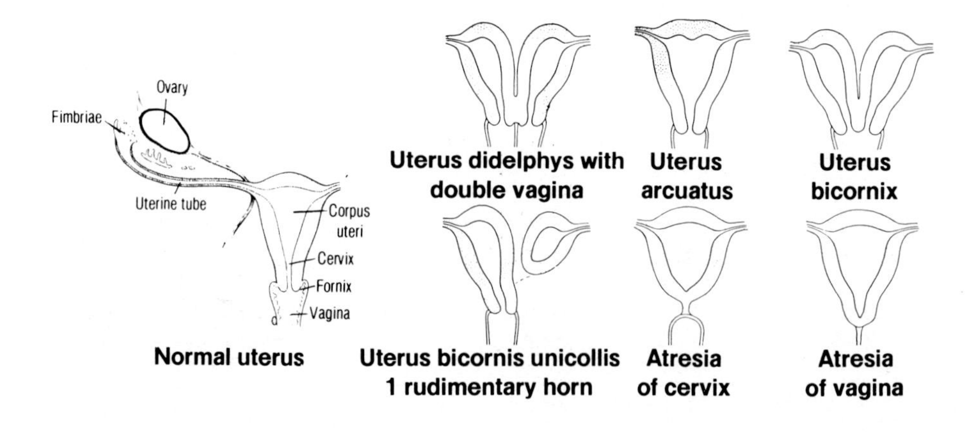

Normal uterus **Uterus didelphys with double vagina** **Uterus arcuatus** **Uterus bicornix** **Uterus bicornis unicollis 1 rudimentary horn** **Atresia of cervix** **Atresia of vagina**

Head and Neck Development

Normal eye development

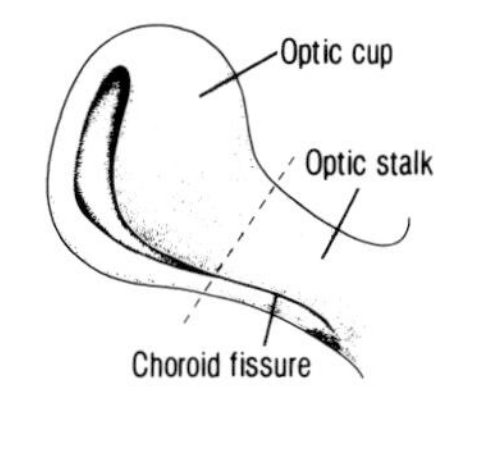

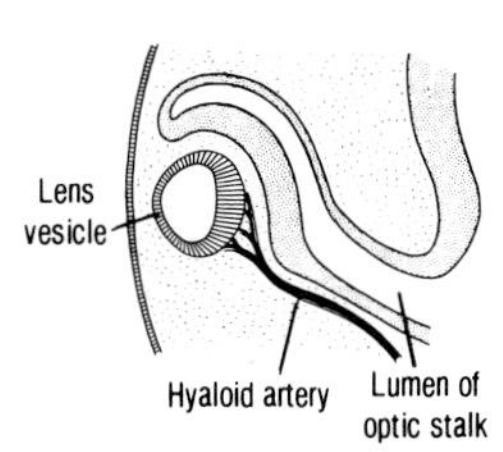

6 weeks

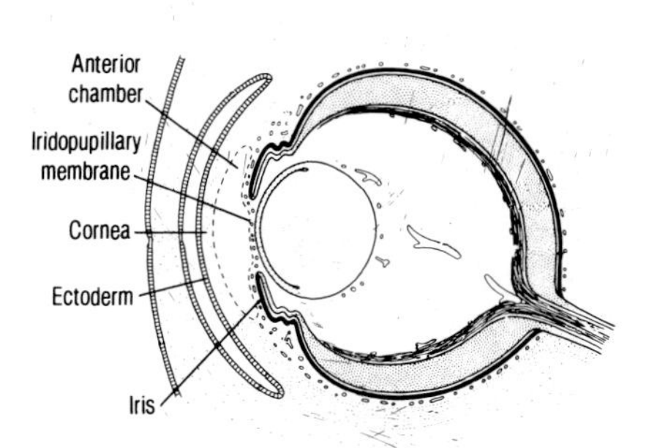

15 weeks

Abnormal eye

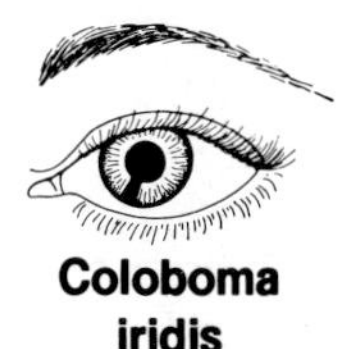

Coloboma iridis

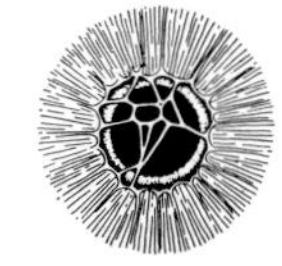

Persistent iridopupillary membrane

Development pharyngeal arches and pouches

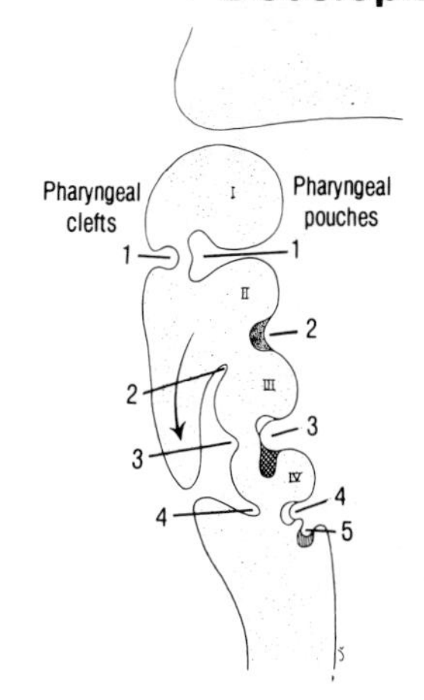

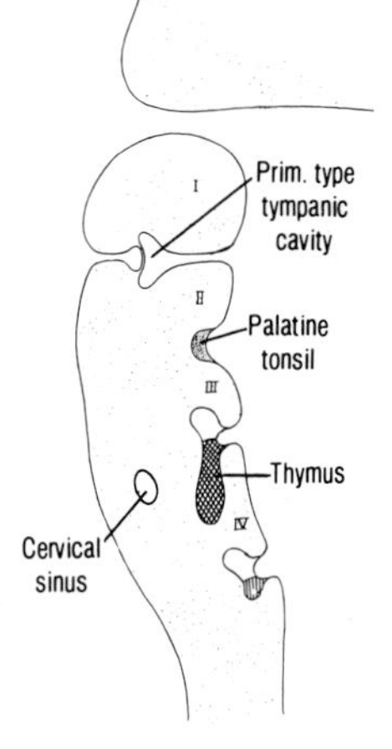

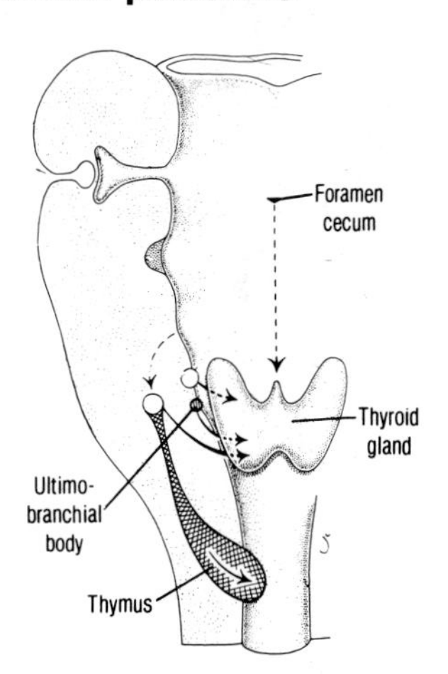

Location of cysts in neck

Region of periauricular fistulas

Region of lat. cervical cysts and fistulas

Sternocleidomastoid muscle

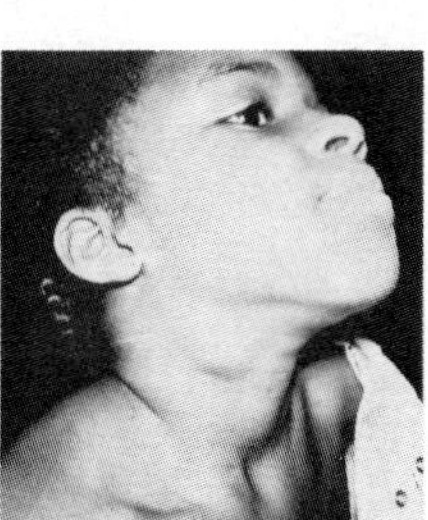

Location lat. cervical cysts

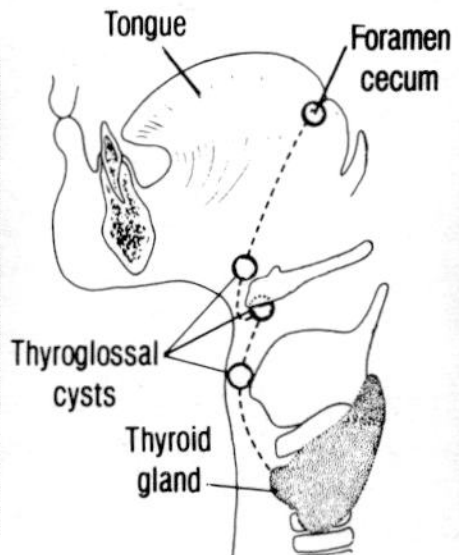

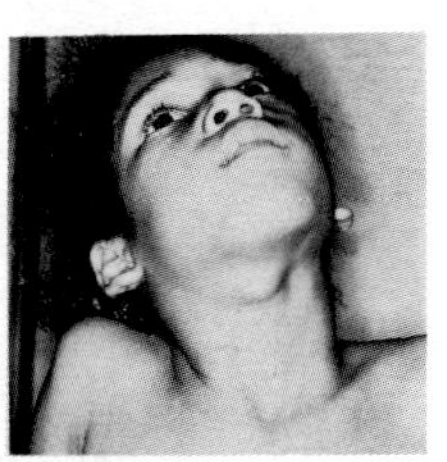

Location thyroglossal cysts

Normal Facial Development

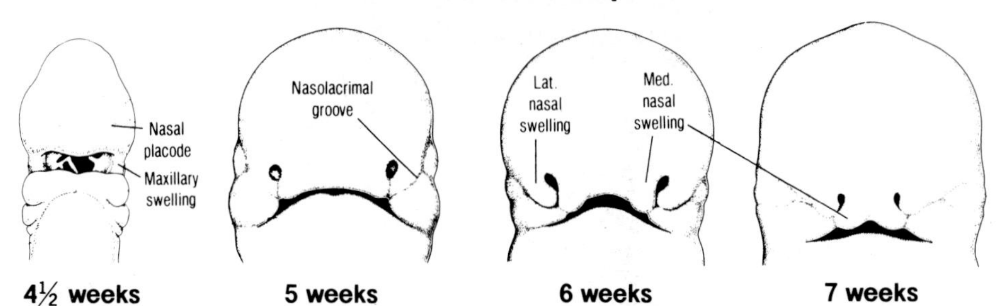

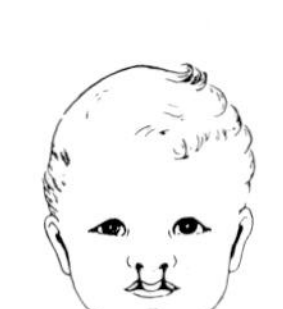

Bilateral cleft lip

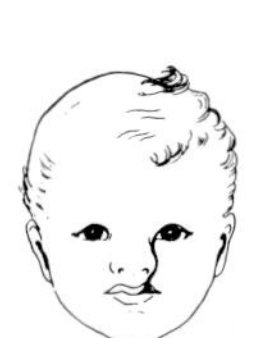

Oblique facial cleft

Median cleft lip

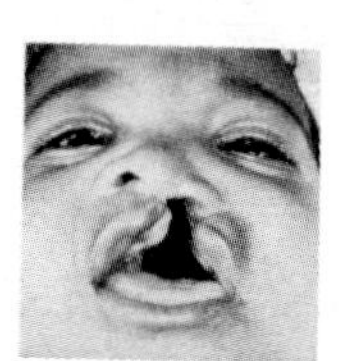

Cleft lip and cleft palate

Normal palate development

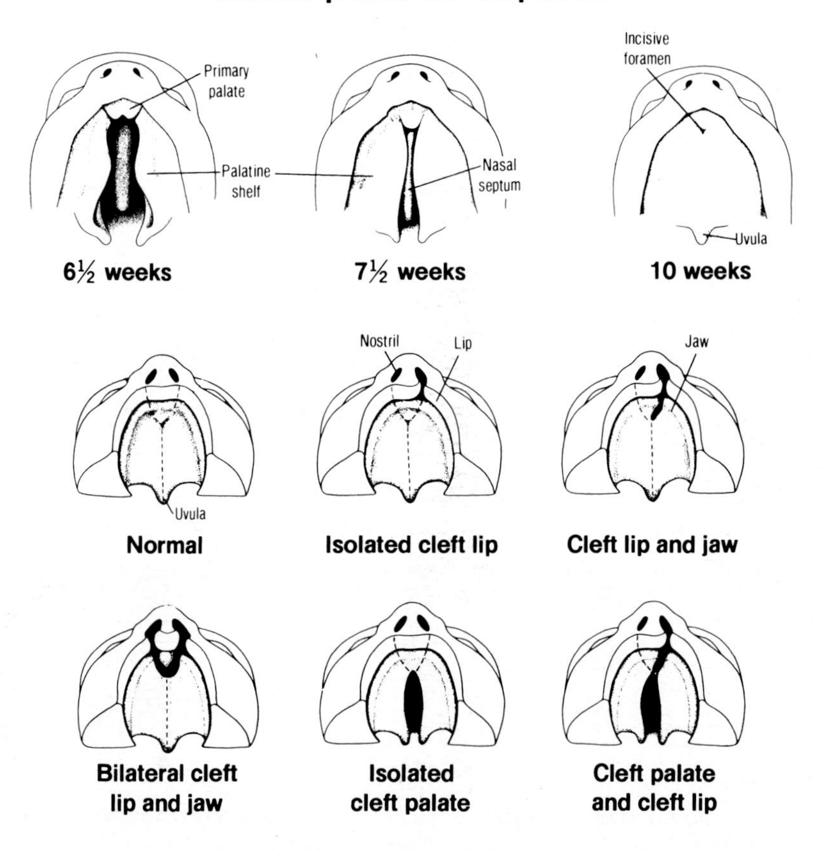

Normal Development Neural Tube

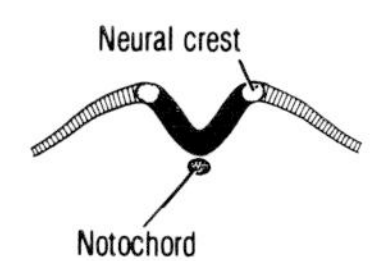

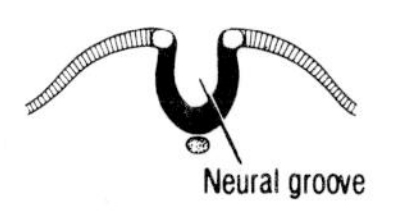

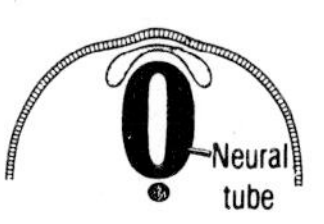

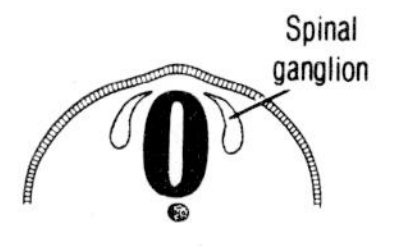

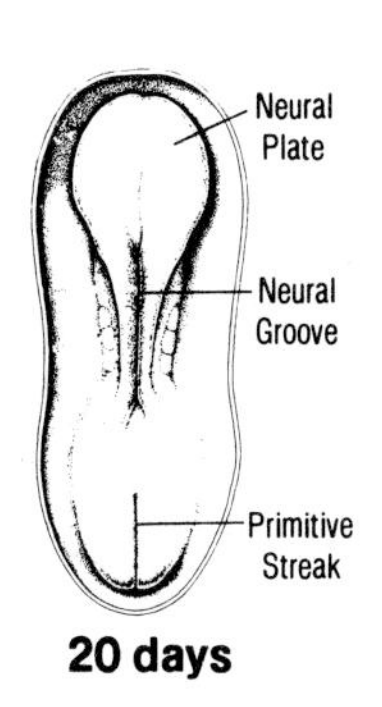

20 days

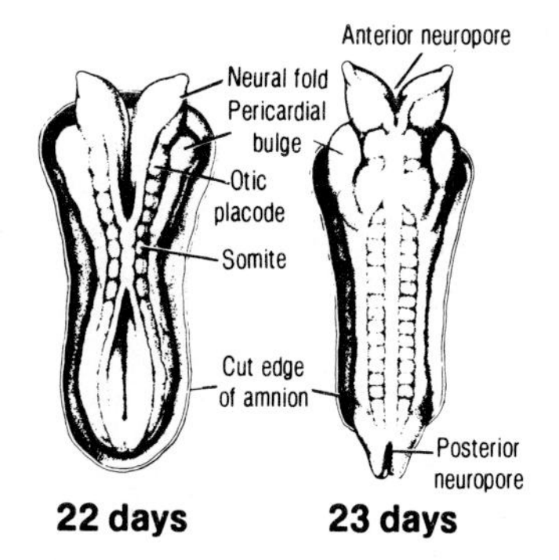

22 days **23 days**

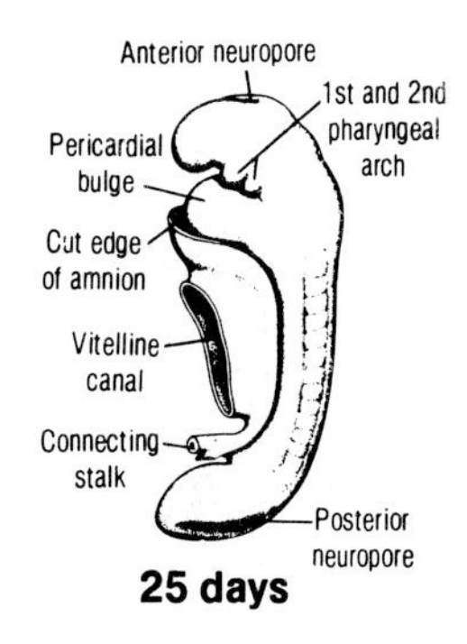

25 days

Abnormal Development of Neural Tube

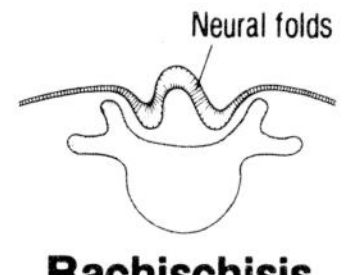

Rachischisis

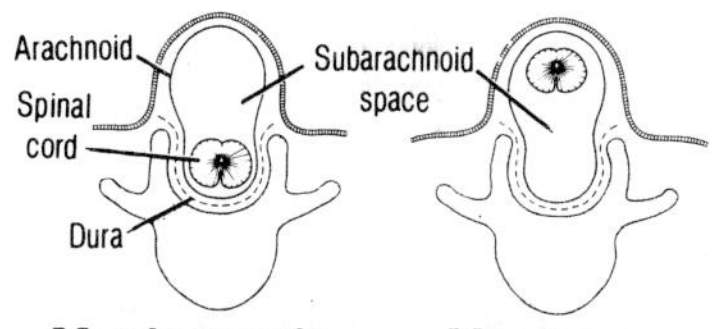

Meningocele **Meningo myelocele**

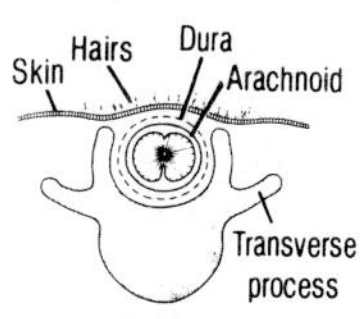

Spina bifida occulta

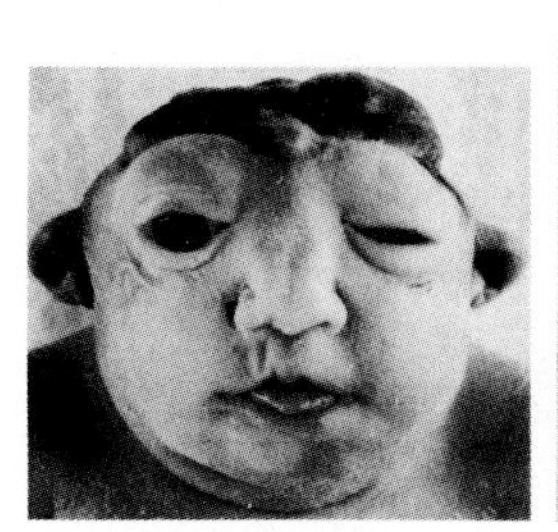

Anencephaly

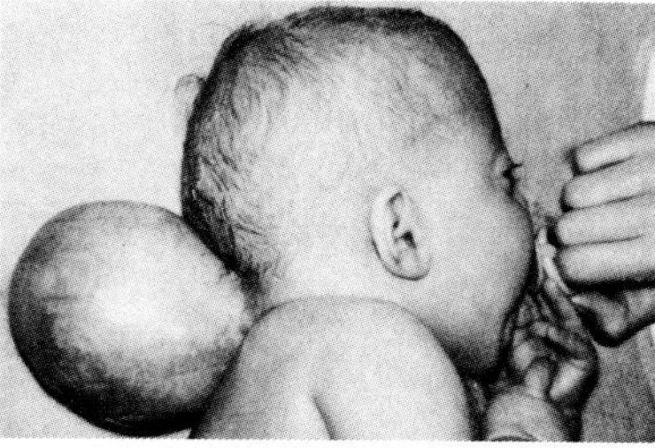

Meningocele

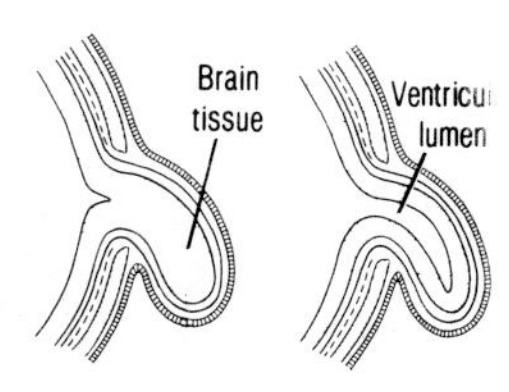

Meningoen-cephalocele **Meningo-hydroen-cephalocele**

Index

EMBRYONIC DEVELOPMENT IN DAYS

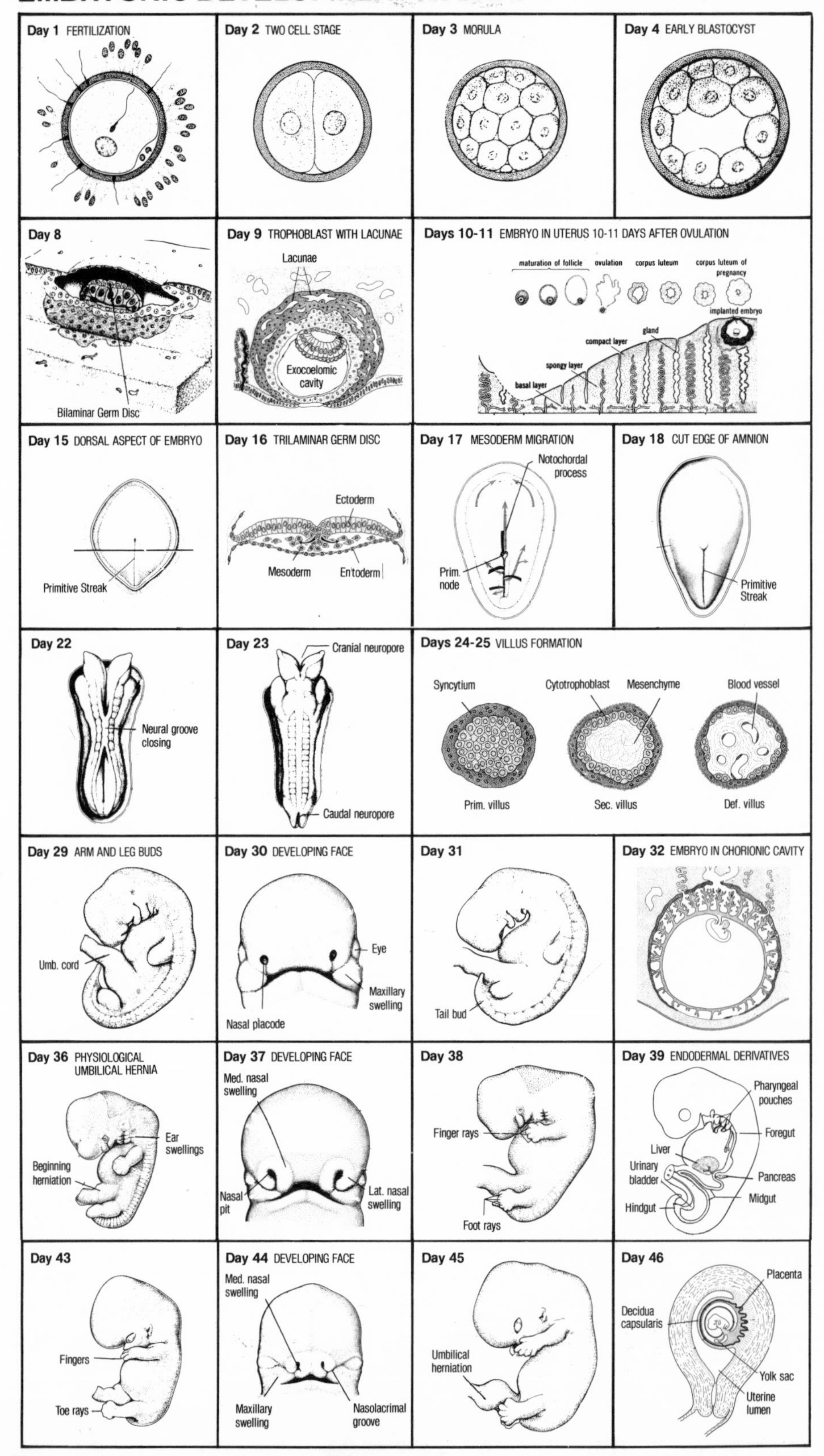